Ecology: *principles and applications*

Second Edition

J. L. Chapman, M.A., Ph.D.
Department of Earth Sciences
University of Cambridge

M. J. Reiss, M.A., Ph.D.
Homerton College
Cambridge

CAMBRIDGE
UNIVERSITY PRESS

PUBLISHED BY THE PRESS SYNDICATE OF THE UNIVERSITY OF CAMBRIDGE
The Pitt Building, Trumpington Street, Cambridge, United Kingdom

CAMBRIDGE UNIVERSITY PRESS
The Edinburgh Building, Cambridge, CB2 2RU, UK
40 West 20th Street, New York, NY 10011–4211, USA
10 Stamford Road, Oakleigh, VIC 3166, Australia
Ruiz de Alarcón 13, 28014 Madrid, Spain
Dock House, The Waterfront, Cape Town 8001, South Africa

http://www.cambridge.org

© Cambridge University Press 1992, 1999

This edition first published 1999
Reprinted 2000, 2001

Printed in the United Kingdom at the University Press, Cambridge

A catalogue record for this book is available from the British Library

ISBN 0 521 58802 2

Prepared for publication by Stenton Associates
Cover illustration by Linda Baker-Smith
Text illustrations by Chris Etheridge and Ian Evans

Acknowledgements 2.1 David T Grewcock / Frank Lane Picture Agency [99999-03510-056]; 2.2 Derek Bromhall / Oxford Scientific Films [85717]; 2.3 Mark Mattock / Planet Earth Pictures [02138751]; 2.4 Dr Jeremy Burgess / Science Photo Library [B610/024JBU21M]; 3.1 Biophoto Associates [DCH0043L]; 3.4 Prof J H Lawton / © Somerset Levels Project; 3.5 Biophoto Associates [TNT B/W 59]; 4.3 Forest Life Picture Library [1005778020]; 4.9 Heather Angel [HCA65a]; 4.10 Dr Alan Beaumont [8017]; 5.7 Rosemary Calvert / Planet Earth Pictures [02443872]; 6.3 Heather Angel [HTR8c]; 6.8 Stan Osolinski / Oxford Scientific Films [OSS/115815]; 6.11 David Phillips / Planet Earth Pictures [02698978]; 6.12 Biophoto Associates [TNT0610]; 7.3 Dr Norman Owen-Smith; 7.4 Dr Norman Owen-Smith; 7.5 Geoff du Feu / Planet Earth Pictures [01570064]; 7.11 Merlin D Tuttle, Bat Conservation International; 8.2 Stephen Dalton / Natural History Photographic Agency [IN/SDA06994B]; 8.4 David T Grewcock / Frank Lane Picture Agency [99999-03509-056]; 8.6 Gunter Ziesler / Bruce Coleman Collection [02323362]; 8.9 N M Collins / Oxford Scientific Films [73480]; 8.10 Dr Alan Beaumont [12173]; 8.11 Matthews and Purdy / Planet Earth Pictures [01415778]; 8.12 Dr C G Faulkes, Institute of Zoology; 9.1 NASA / Science Photo Library [S380/228USA02E]; 9.6 Dr Alan Beaumont [6577]; 9.8 Kunsthistorisches Museum, Vienna; 9.10 Heather Angel [HMAM251b]; 9.11 Jeremy Thomas / Biofotos; 10.1 Eric Soothill / Naturetrek Publications [58]; 10.15 P W / Photograph copyright English Nature [21,125]; 11.1 Robert Hessler / Planet Earth Pictures [01397664]; 13.8 Dr Jeremy Burgess / Science Photo Library [B238/02MJBU01G]; 13.10 Dr A Beckett [58]; 13.12 Terry Middleton / Oxford Scientific Films [138453]; 13.13 Eric & David Hosking / Frank Lane Picture Agency [99999-03516-075]; 13.14 Silvestris / Frank Lane Picture Agency [70135-00019-147]; 13.18 Vladimir Blinov / GSF Picture Library [VBC-66]; 14.2 John Lythgoe / Planet Earth Pictures [01220071]; 14.9 Mark Mattock / Planet Earth Pictures [01476017]; 14.11 Dr Morley Read / Science Photo Library [E640/127MRE00A]; 15.7 David George / Seaphot Limited: Planet Earth Pictures [02340461]; 15.8 Geoff Dore / Bruce Coleman Ltd [01799657]; 15.10 Gene Ahrens / Bruce Coleman Ltd; 15.11 Dr Alan Beaumont [7399]; 15.12 Dr B Booth, GSF Picture Library [KSG-2046]; 15.13 Martin Bond / Science Photo Library [G360/176MB004L]; 15.15 Prof John Coles / © Somerset Levels Project; 16.3 Dr R A Spicer; 16.4 Photograph copyright English Nature; 16.7 R P Lawrence / Frank Lane Picture Agency; 16.8 Gerald Cubitt / Bruce Coleman Ltd; 17.1 John Lythgoe / Planet Earth Pictures [01267418]; 17.3 Adam Jones / Planet Earth Pictures [02318709]; 17.4 Sims / Greenpeace [6/4/98]; 17.5 Tom McHugh, Photo Researchers Inc. / Oxford Scientific Films [5R8991B]; 17.6 Eric & David Hosking / Frank Lane Picture Agency [04302-00025-075]; 17.7 Mike Richards / RSPB Images; 17.8 L Lee Rue / Frank Lane Picture Agency [11272-00153-099]; 17.9 Tony Morrison / South American Pictures [XGA22]; 17.10 W Wisniewski / Frank Lane Picture Agency [10702-00015-170]; 17.11 Richard Packwood / Oxford Scientific Films [260635]; 17.12 Ronald Toms / Oxford Scientific Films [151449]; 17.14 Pete Oxford / Planet Earth Pictures [02216418]; 17.15a Dr B Booth / GSF Picture Library [EU36-17]; 17.15b Jim Clare / Oxford Scientific Films [PTF/0164104]; 17.17 Peter Clarke / Oxford Scientific Films [PC324]; 17.18 David Phillips / Planet Earth Pictures [01496190]; 17.20 Great Barrier Reef Marine Park Authority [60-49-4]; 17.21 Peter David / Planet Earth Pictures [01476009]; 17.22 Georgette Douwma / Planet Earth Pictures [01355430]; 17.23 NASA / Planet Earth Pictures [02640228]; 18.2 Jonathan Scott / Planet Earth Pictures [01087894]; 18.7 Len Robinson / Frank Lane Picture Agency [10620-00027-136]; 18.14 Dr Q C B Cronk; 18.16 Prof Olov Hedberg; 19.1 Biobest Trading; 19.4 Jen & Des Bartlett / Bruce Coleman Ltd; 19.6a Kathie Atkinson / Oxford Scientific Films [6219]; 19.6b Robert A Tyrell / Oxford Scientific Films [47881B]; 19.6c Mark Mattock / Planet Earth Pictures [01496247]; 19.6d Merlin D Tuttle, Bat Conservation International; 19.7 The Field Museum, Chicago, IL [Neg # 80031]; 19.8 Gerald Cubitt / Bruce Coleman Ltd [07008627]; 19.9 David Thompson / Oxford Scientific Films [38814]; 19.11 Alan P Dodd / Godfrey New Photographics Ltd; 19.12 Holt Studios International (Nigel Cattlin) [IC22037A]; 20.1 Steve Turner / Oxford Scientific Films [180915]; 20.2 Courtauld Gallery, London [B58/925]; 20.3 Dr B Booth / GSF Picture Library [RM1658]; 20.4a Mary Evans Picture Library [80024540]; 20.4b Mary Evans Picture Library [10016411/02]; 20.4c Natural History Museum, London [T01062/A]; 20.4d [Neg. No. 334128] Photo Logan, Courtesy Dept. Of Library Services, American Museum of Natural History; 20.5 Franz J Camenzind / Planet Earth Pictures [01004042]; 20.6 Saint Louis Zoological Park; 20.7 Gerard Lacz / Frank Lane Picture Agency [04230-00029-235]; 20.8 Mike Birkhead / Oxford Scientific Films [137957]; 20.11 G A MacLean / Oxford Scientific Films [OSF 32143]; 20.12 Heather Angel [HWYO/31/3]; 21.1 Richard Matthews / Planet Earth Pictures [01431412]; 21.2 F Hartmann / Frank Lane Picture Agency [99999-03514-061]; 21.3 John Peel; 21.4 Konrad Wothe / Oxford Scientific Films [281737]; 21.5 William M Smithey / Planet Earth Pictures [01712551]; 21.6, 21.7 Dr Bart R Johnson, Dept of Landscape Architecture, University of Oregon, Eugene; 21.8 Mike Read / Fauna & Flora International; 21.9 Mark Harrison; 21.12 Pilly Cowell / Environmental Images [1008461010]; 21.13 Joe B Blossom / Oxford Scientific Films [JBB 1040]; 21.14 National Trust Photographic Library / Joe Cornish [068003]; 21.15 Roland Mayr / Oxford Scientific Films [167295]; 21.16 Andreas Riedmiller / Das Fotoarchiv [E-08-110-18]. Every effort has been made to reach the copyright holders; the publishers would like to hear from anyone whose rights have been unwittingly infringed.

Contents

*To Doreen and John,
Ann and Herbert*

Preface

Ecology: principles and applications provides a clear and up-to-date introduction to ecology for students studying at first-year undergraduate level or at advanced level. We outline the principles of ecology and show how relevant applications follow from these principles. The chapters follow a sequence of ascending scale beginning with individual organisms and proceeding through communities and ecosystems to global considerations of biogeography, co-evolution and conservation. Human ecology and applied topics are considered throughout the text wherever appropriate and a wealth of examples are drawn from all five kingdoms and around the world. Each chapter contains a summary. Most chapters also include extra, somewhat peripheral, material, presented as boxed text. These boxes may contain a particular case history, an historical perspective, a worked example of a quantitative concept or the outline of a controversial theory. Sometimes they contain material that is more advanced than the main text.

Key terms are printed in bold throughout the text and are defined in a glossary at the back of the book. There is an extensive bibliography at the end of the book which allows particular aspects of ecology to be pursued in more detail.

The second edition shows a number of changes over the first. The number of diagrams and photographs has been increased and some sections are now in colour. New material has been added throughout the text and examples are now taken from a greater number of countries. The most extensive changes have been to the sections on nutrient cycling, pollution, endangered species, biodiversity and conservation. Indeed, conservation is now given two chapters rather than one, with greater attention being paid to its philosophy, underlying biology, politics and practice. Despite these additions we have striven to ensure that this remains an introductory text in ecology. We hope that the book, while thorough, is readable rather than daunting.

Our appreciation and thanks are due to many people for their help and encouragement. We are especially grateful to Stephen Tomkins whose encouragement was most welcome and much needed. Particular thanks go to Stephen and to Myra Black, both of whom provided valuable comments on the entire typescript, and to Francis Gilbert and Neil Ingram who, with Stephen and Myra, helped us when we were planning the book. Tony Seddon, Lucy Purkis and Lucy Harbron of Cambridge University Press proved to be helpful and understanding editors throughout the book's long gestation.

For the second edition we are especially grateful to Sarah Gardner and others who suggested improvements. Diane Abbott, Aiden Gill, Sue Kearsey and Tom Gamblin could not have been more helpful in the book's production.

ONE

Introduction

1.1 **What is ecology?**

Ecology is the study of **organisms** in relation to the surroundings in which they live. These surroundings are called the **environment** of the organism. This environment is made up of many different components, including other living organisms and their effects, and purely physical features such as the climate and soil type.

Ecologists, those who study ecology, are always aiming to understand how an organism fits into its environment. The environment is of supreme importance to an organism and its ability to exist in the environment where it lives will determine its success or failure as an individual. This means that much of this book is really about the study of environments from the point of view of various organisms (but see Chapter 9 which describes in detail what makes up and affects an environment).

There are several definitions of ecology. Many workers have produced their own description of this branch of biology. The word ecology was first used by a German called Ernst Haeckel in 1869. It comes from two Greek words *oikos* meaning home and *logos* meaning understanding. Haeckel described ecology as 'the domestic side of organic life' and 'the knowledge of the sum of the relations of organisms to the surrounding outer world, to organic and inorganic conditions of existence'. This 'surrounding outer world' is another way of saying the environment. In 1927 Charles Elton wrote that ecology is 'the study of animals and plants in relation to their habits and habitats'. Today an ecologist would probably substitute the word 'organisms' for 'animals and plants' because we now recognise other categories of organisms (fungi, protoctists and bacteria) which are not in the plant or animal kingdoms. Many of these are extremely important in ecology although they are seldom as well studied as the plants and animals. More recently Krebs (1985) has defined ecology as 'the scientific study of the interactions that determine the distribution and abundance of organisms'. You may take your pick as to which definition you prefer.

Ecology is like an enormous jigsaw puzzle. Each organism has requirements for life which interlock with those of the many other individuals in the area. Some of these individuals belong to the same species, but most are very different organisms with very different ways of living or interacting. Figure 1.1 is a diagrammatic representation of this interlocking jigsaw. It illustrates some of the ways in which a single individual fits in with others. In this case an animal is represented which catches other animals for food (it is a predator) and which in turn is hunted and may be killed by another species of predator. During the animal's lifetime it needs to find a mate of the same species to produce offspring. During its life it also competes with other animals (**competitors**) for food and will probably catch diseases.

The ability of the animal to avoid the predator, catch its prey, withstand disease and so on will

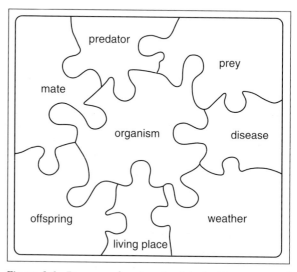

Figure 1.1 Diagram showing the interlocking nature of the features of the environment which influence an organism. In this case the organism is a predator, but to generalise the diagram, the word 'prey' could be replaced by 'food', or, for a plant, by 'light and nutrients'. For simplicity, this figure is two dimensional, but the ecological interactions of organisms and their environment are really multi-dimensional.

depend on the relationships it has with the organisms around it. Its life will also be affected by the weather, time of year and the quality of nesting or sleeping sites. In fact, this simplified example is already becoming complicated as more and more pieces of the puzzle are added. The study of these ecological relationships from the point of view of a single species, as is illustrated by Figure 1.1, is called **autecology** (see Chapter 3). If all the species living together are studied as a **community** (see Chapter 14) then this study is called **synecology**.

1.2 The nature of ecology

The only way to find out how any organism survives, reproduces and interacts with other organisms is to study it. This makes ecology a practical science. There are three main approaches to the study of ecology. The simplest method is to observe and record the organism in its natural environment. This is sometimes described as observation 'in the field' or **fieldwork**, although the term can be confusing as 'field' suggests open grasslands or the site of human cultivation. A second type of study is to carry out experiments in the field to find out how the organism reacts to certain changes in its surroundings. A third approach involves bringing organisms into a controlled environment in a laboratory, cage or greenhouse. This method is very useful as it is often easier to record information under controlled conditions. However, it must be remembered that the organisms may react differently because they have been removed from their natural home.

No single study can hope to discover everything there is to know about the relationships between an organism and its environment. These relationships are so varied that different kinds of investigation are needed to study them. Often both study in the natural environment and experiments in the laboratory are required to discover even part of the picture. Also, as the environment changes, so an organism may respond differently, with the result that an experiment under one set of conditions may well give different results to the same experiment carried out under different conditions.

So we have a picture of ecology as a subject full of complexity where an organism has many different responses and needs. Theoretically, therefore, there is an almost infinite amount to be discovered about the ecology of the world. Even after a century of ecological study we are just scratching the surface of possible knowledge. A large amount has been discovered over the years, but our knowledge is patchy; we know far more about northern hemisphere temperate woodland than we know about tropical rainforest, more about English rocky sea shores than the Australian barrier reef.

What makes ecology exciting, rather than an endless list of things to be learned about organisms, is that we are studying a living, working system. Because the system fits together so neatly it forms repeated patterns which can be recognised by the ecologist. Organisms with similar life styles often respond to their environment in similar ways. For example our predator in Figure 1.1 can only catch its prey in certain ways. If its prey becomes scarce it may starve, eat something else or migrate to where food is more plentiful. In other words it only has a certain number of options and its response to certain conditions may well be predictable. Understanding why organisms react to various conditions in one way rather than another takes us a long way towards an understanding of the principles of ecology.

These principles, with which this book is concerned, are only becoming understood because of the many studies of organisms both in the field and in the laboratory. Throughout this book you will find examples of how particular organisms relate to their environment given as evidence to support the principles being described. Because the relationship of organisms to their environment may be very subtle, it can often be difficult to unravel the situation to discover the principles involved. Yet finding out how organisms interact and applying these principles can be an absorbing and fascinating pursuit.

1.3 The study of ecology

An ecologist could start any study by asking the question: 'Why does this organism live or grow here and not there?'. In simplified terms this is the question ecological investigations often try to answer. Of course it is far too difficult to answer in one go and can be split up into many different questions. The first and most obvious question is: 'Where does the organism or species I am studying live?'. Usually this can be answered by careful observation in the field with the help of some sort of sampling method. Once we know this other questions may become obvious. For example an ecologist may ask:

How does the organism obtain its food?
Is a particular nutrient limiting its growth or numbers?
Is something else limiting its growth or numbers?
Does it reproduce in this site and if so how?
Is it absent from parts of the site due to some factor?
How and when do the young disperse?
What causes the death of the organisms?

There are numerous possible questions, some of which may be unique to a particular situation. Obviously, if an ecological investigation starts with specific questions, or objectives, it makes the task of studying the ecology of organisms much easier.

Box 1
Experimental design

As so much of our ecological knowledge comes from experiments it is worth thinking about how experiments are designed. The skill and care with which an experiment is planned and carried out will affect the accuracy of the information collected and the way in which the results are interpreted. Be critical when reading other people's experimental techniques and their interpretation of the results. They may have made mistakes. Even if they have not, you will be practising the important scientific skill of looking at and analysing information with an open mind. Watch out for poorly designed experiments – there are several about, even in the literature! If you want to use some findings, but only have a report of the work by a third party, try to get a copy of the original publication. A tentative suggestion by a careful worker is easily turned into an apparent certainty by someone else.

There are several things to consider when looking at reported experiments or when designing your own.

(1) Is there a 'control'?

A **control** is an experiment which is specially designed to show the effects of the actual experimental technique (other than the factor being investigated) on the organisms under consideration. A control group of organisms is treated in the same way as the experimental group, except for the factors being studied in the experiment. For example, in an experiment which involves digging up plants and planting them in another site to see how well they grow under different conditions, a control would involve digging up plants and replanting them in the original place. From this control, the effects of digging up the plant (root damage, soil disturbance, etc.) can be determined and this can be taken into account when looking at the plants which have been moved to the new site.

(2) Are there replicates in the experiment?

The confidence with which conclusions can be drawn from an experiment is increased by repeating it several times (having **replicate** experiments) and considering all the results together. The more times the experiment is repeated and the results averaged the more likely the results are to be reliable. Statistical analysis of the results is then possible. Obviously hundreds of repeats would be time consuming, possibly expensive and no doubt tedious! So many would not improve the results enough to be worthwhile. Some experiments are impossible to repeat because some unusual situation is being recorded or because the species being studied is rare.

(3) Are the right data being collected?

Sometimes the wrong data are collected from an experiment or wrong recordings made during observation of organisms in their natural surroundings. It may be very difficult to decide in advance what information is valuable to the study, so the best advice is to collect as much as possible and select what to analyse later. For example, it is no good applying a nutrient to the soil in which plants are growing, and then analysing only the leaves of the plant to see how the nutrient has been used. The plant may be concentrating the nutrient in its roots! 'Out of sight – out of mind' is a phrase we often use in everyday situations, but it can also apply to ecological investigations. Many experiments would be improved by taking into account 'invisible' factors such as root activity, the effects of microorganisms and so on. Unfortunately these factors are often very difficult to study.

(4) Are the data correctly interpreted?

Because data from ecological experiments are often quite variable and sometimes inconclusive, it is often quite difficult to interpret the results. Occasionally classic mistakes are made, as was the case for the worker who moved a plant into a different locality. When he went back to see how it was growing in its new environment he 'found' that it had turned into a different species! What had actually happened was that his original transplant had died and another plant, native to the area, had grown in the space left there. Of course bad experimental design was involved, the plant was not adequately labelled and so the wrong conclusions were reached. This is an obvious mistake and we might think the worker silly and that we would never make such an error. However, the interpretation of experimental results is seldom as simple as we might expect.

Summary

(1) Ecology is the study of how organisms live and how they interact with their environment.
(2) The environment includes other organisms and physical features.
(3) Autecology is the study of the ecology of a single species.
(4) Synecology is the study of the ecology of whole communities of organisms.
(5) Ecology is a practical science requiring observations and experiments to investigate organisms.
(6) There are underlying principles in ecology which predict how organisms will react in particular circumstances.
(7) Experimental design is extremely important and requires, wherever possible, controls, replicates, the accurate collection of data and careful interpretation of results.

TWO

The individual

2.1 Why look at individuals in ecology?

Although ecologists are often interested in the complex interactions between species, it is worth remembering that it is individual organisms that are the products of natural selection (Chapter 8). This chapter deals with the biology of individuals. Individuals are the fundamental units of populations, communities, ecosystems and biomes which are discussed in later chapters. In this chapter we will look at individuals from an ecological perspective. We will start with the essentials of how they obtain their energy and nutrients, and then consider how these are allocated to maintenance, growth and reproduction.

2.2 Autotrophs and heterotrophs

All organisms need energy to live and different organisms obtain this energy in different ways. There are many approaches to classifying the ways in which individuals obtain their food. A useful one is to divide organisms into **autotrophs** and **heterotrophs**. Autotrophs obtain only the simplest inorganic substances from their environment. Green plants are the most obvious autotrophs. These need only visible light, water, carbon dioxide and inorganic ions such as nitrate (NO_3^-) to survive, grow and reproduce. The process of **photosynthesis** enables most plants, the photosynthetic bacteria and some protoctists to synthesise all the complex organic molecules that they require from these simple building blocks. Because these organisms use light as their energy source, they are called **photoautotrophs**. **Chemoautotrophs** are autotrophs that obtain their energy not from sunlight but from certain specific chemical reactions involving only inorganic substrates. The ecological importance of chemoautotrophs and photoautotrophs is discussed in Chapter 11.

Animals, fungi, most bacteria, many protoctists and a few plants cannot synthesise their organic molecules from inorganic ones. They need to take in an external source of organic carbon. These organisms are heterotrophs.

2.2.1 Terms associated with heterotrophic nutrition

While autotrophs such as green plants and chemosynthetic bacteria obtain the few nutrients that they require from the environment around them, heterotrophs have first to **ingest** their food, if necessary **egesting** some of it too, and then to **digest** what they have ingested. Digestion is followed by **absorption**, and absorption by **assimilation**. Finally, some matter is **excreted**.

Ingestion

Ingestion is the process by which heterotrophs take in their food. The number of techniques used by heterotrophs to ingest their food is huge and is discussed in more detail in Section 2.2.2. For most animal species it is useful to distinguish two stages to ingestion. In the first, the food is captured; in the second, it is brought within the alimentary canal.

Egestion

In some animals not all the food that is ingested is eventually digested. For example, some time after an egg-eating snake has swallowed an egg it regurgitates the crushed shell. The eggshell is said to have been egested. Owls also egest pellets containing the fur, feathers and bones of the small animals they have eaten. In general, the parts of the food ingested by a heterotroph which have not been digested sufficiently to allow them to be absorbed into the tissues are disposed of by egestion. The faeces produced by animals also contain egested material. For example, the plant fibre in our diet, though important for the proper functioning of our large intestine, is not digested, but egested via the anus.

Digestion

Digestion is the process by which heterotrophs break down the food they have ingested into particles small enough to be absorbed. In most cases, this means that mastication and hydrolytic enzymes are used to break down food particles to their constituent monomers. Proteins are broken down to amino acids; carbohydrates to monosaccharides; fats to tiny fat globules or even to fatty acids, glycerol and other simple molecules. Salts and vitamins can be absorbed without needing to be digested.

Absorption

Absorption is the uptake from the gut of vitamins. salts and the products of digestion. In mammals these substances are absorbed by certain of the villi of the small intestine. The products of absorption are either respired or assimilated.

Assimilation

Assimilation occurs when the products of absorption are taken up by cells and synthesised into macro-molecules. These macromolecules may be stored or used for repair, growth and reproduction.

Excretion

Excretion is a characteristic of all organisms, not just heterotrophs. and occurs when the waste products of metabolism are expelled from an organism. Nitrogen is an important excretory product of heterotrophic organisms. When proteins are broken down, ammo-nia (NH_3) is produced. Ammonia is very toxic and needs to be diluted in a large volume of water. Freshwater fish, which obviously have water in abun-dance, are able to dilute and then excrete their ammonia directly. Other vertebrates convert the ammonia to compounds which can be concentrated so as to be less wasteful of water, such as trimethyl-amine oxide, urea ($CO(NH_2)_2$) or uric acid.

The excretory products of one organism are almost invariably utilised by other organisms (see Chapter 13). For instance, the nitrogenous waste products of heterotrophs are typically converted to nitrates by soil bacteria and taken up by plant roots. Oxygen is an excretory product of plants yet is vital for all aerobic organisms.

2.2.2 Ingestion by heterotrophs

The way an animal feeds profoundly influences many aspects of its ecology. Heterotrophs differ greatly in the sorts of food they ingest and in the ways they obtain their food. Three main types of heterotrophic nutrition are commonly recognised.

Holozoic nutrition

Holozoons feed on relatively large pieces of dead organic material. Most of the animals with which we are familiar can therefore be described as holozoic. **Carnivores**, such as the fox (*Vulpes vulpes*). feed on prey which they have caught (Figure 2.1). **Herbivores**. such as sheep, cattle and goats, feed on vegetation. Animals which feed on a mixture of plant

Figure 2.1 A five month old fox cub (*Vulpes vulpes*) feeding on a hen pheasant.

and animal food, such as the pig and most humans, are called **omnivores**.

Each of these terms may be subdivided. For exam-ple, herbivores include **granivores** (animals such as the world's most abundant bird, the redbilled quelea (*Quelea quelea*) which feed on grain or seeds), **frugi-vores** (which feed on fruit), **folivores** (which feed on the leaves of shrubs or trees), **grazers** (which feed on herbs and grasses) and **browsers** (which feed on the leaves, young shoots and fruit of shrubs and trees). Animals have anatomical, physiological and behav-ioural adaptations which are associated with their diets. Consider an animal such as a deer or antelope which grazes grass. Grasses contain very fine parti-cles of silica (SiO_2), the same hard substance of which sand is composed. This silica wears away teeth, and many grazers, if predation is infrequent. die of star-vation when their teeth literally wear away. Over the course of evolution grazers which can maintain the grinding surfaces of their teeth have been favoured by natural selection. The teeth of many grazers grow from their roots throughout their life. Elephants have responded to the problem of tooth abrasion in a remarkable manner. At any one point in their life they only have four grinding teeth in use, one in the upper and one in the lower jaw on each side. Each tooth, as it is worn down, is slowly replaced by another from behind. Should an elephant get to be more than about 60 years old, it will have used up all of its 24 teeth in this way.

Figure 2.2 One to two day old swifts (*Apus apus*) infected by lice.

Figure 2.3 Many-zoned polypore (*Coriolus versicolor*) on a dead tree trunk.

Parasitic nutrition

Parasites, unlike holozoons, feed off matter that is still alive. Because of this, parasites are usually much smaller than their **hosts** (Figure 2.2). This means that they generally have shorter lifespans than their hosts and are more numerous than them. As well as being smaller than their hosts, parasites harm their hosts. The host gains nothing from the relationship and is therefore constantly selected to avoid being parasitised.

Parasitism is an evolutionary trait which has evolved independently in a huge number of different taxa. Indeed, parasites may be found in all five kingdoms – bacteria, protoctists, fungi, plants and animals. Parasites can conveniently be divided into **endoparasites**, such as tapeworms, which live inside their hosts, and **ectoparasites**, such as fleas, which live outside them. Whether endoparasitic or ectoparasitic, most parasites obtain their food in liquid form. Parasites of plants may obtain their nourishment from the phloem, as aphids do. Animal parasites may live in the gut or tap into the lymph or blood system.

Saprotrophic nutrition

Saprotrophs feed on dead organic matter which they either absorb in solution or ingest as very small pieces. Many fungi, for instance, are saprotrophs, living off dead organisms (Figure 2.3). Most saprotrophs obtain their food by extracellular digestion – enzymes are secreted on to the food source and the soluble products are then absorbed. This is the technique used by bacteria, fungi and the house fly (*Musca domestica*) for instance. On the other hand, earthworms obtain their food in solid form including small bits of dead leaves and soil invertebrates. Animals which feed off dead organic matter, but ingest large pieces, may be classified as **scavengers**. Spotted hyaenas, for example, often scavenge, though they obtain a large amount of their energy by hunting and killing prey.

2.3   Metabolic rate

One of the most useful single pieces of information an ecologist can have about an individual is its **metabolic rate**. The metabolic rate of an organism is the amount of energy it needs per unit time. This indicates what the demands of that organism are on its environment. For instance, animals with large metabolic rates are generally found near the top of food webs and at quite low population densities. A convenient unit for measuring metabolic rate is the number of joules of energy an organism uses each day. For organisms that rely on aerobic respiration to supply their energy needs, oxygen consumption is directly proportional to energy requirements and metabolic rate may be measured in units of hourly oxygen consumption as this is fairly easy to measure.

Organisms require energy in order to replace their tissues and make new ones for growth and reproduction. Organisms which burn up a lot of energy for their size, such as birds, are said to have a high metabolic rate. Organisms such as snakes, which can survive on much less food, are said to have a low metabolic rate. Many measures of metabolic rate have been made, of which perhaps the best known is **basal metabolic rate**. The basal metabolic rate of an organism is the minimum amount of energy it needs to respire to maintain life when the body is at rest and just 'ticking over'. Three conditions need to be fulfilled for the metabolic rate of an organism to be its basal metabolic rate. First, the organism needs to be in a thermoneutral environment (so that no energy has to be used for thermoregulation). Second, the organism must be at rest (so that no energy is needed for locomotion or any other activities). Third, the organism has to be in a post-absorptive state (that is some time after food has been eaten) because the metabolic rate of an organism rises shortly after it eats.

Box 2
Difficulties in the classification of heterotrophic nutrition

This chapter is full of definitions of words which ecologists use to describe organisms and the processes which go on inside them. Humans like to have their world defined and classified into boxes. Such classification, and the naming that accompanies it, helps us to communicate with precision. We classify all living organisms into species which have Latin binomials that are traditionally printed in italics (e.g. *Homo sapiens* is the binomial we give ourselves) or underlined in non-printed text. The science of classifying organisms is called **taxonomy**. It could be said that taxonomists are members of the oldest profession, because in The Bible it was Adam who named all the animals (*Genesis* 2)!

Ecologists also try to categorise or classify many aspects of the environment and the species they are studying. Many such classifications are attempts to order the organisms using some of the principles of ecology which have been discovered. For instance, species can be aggregated into communities and these can be grouped into the major biomes of the world (Chapter 17). Such ecological categorisations can lead to problems because, although there are underlying principles which can be found in ecology, there are also many exceptions to general rules. These exceptions are sometimes very valuable in giving us further insights to the principles themselves, but they do cause problems to the precise mind. Ecology is a science of averages and possibilities, not an exact discipline like some branches of physics or chemistry.

Some examples of the problems that arise when organisms are forced into a system of precise definitions can be seen in this chapter. Organisms can be described as autotrophic or heterotrophic and heterotrophs can be divided into holozoons, saprotrophs and parasites. This suggests that all heterotrophs fall neatly into one of these three exclusive categories. The truth is more complex. For instance, some parasitic organisms, once their host is dead, feed saprotrophically on its body. Some holozoic feeders also take in soluble food at times during their lives.

The whole classification gets even more difficult when some plants are considered. Although most plants are true autotrophs, some are completely parasitic, lacking any chlorophyll. The broomrapes (*Orobanche* spp.) are such a group. Each broomrape exists as an underground system of roots. These are attached to the roots of photosynthetic plants from which the broomrape obtains sugars and other substances. Only the brown or purple flowering spike appears above ground to be pollinated by insects.

There are other species of plants which have such parasitic root connections, but also contain functioning chloroplasts. Such species are **hemiparasites**. This means that they obtain only some of their nutrients parasitically, the rest being made autotrophically. Eyebrights (*Euphrasia* spp.) fall into this intermediate category. It is quite probable that many more

Figure 2.4 Scanning electron micrograph of the leaf of a Cape sundew (*Drosera capensis*) showing a fly which it has caught on its sticky hairs.

plant species are hemiparasitic than we realise. Digging out complete root systems is a painstaking and unenviable job! However, once it is done, it quite often turns out that apparently normal autotrophic plants have root connections with other plants. Radioactive tracers have been used to see whether such root connections enable the movement of photosynthates from one individual to another.

Finally, what about plants like the sundew (*Drosera capensis*) shown in Figure 2.4? Sundews have modified leaves which bear sticky hairs. Should an unfortunate insect come into contact with these hairs it may be unable to break free. The leaf then curls up, holding the insect tight and enabling the plant to digest it. The sundews are often described as being carnivorous as are other plants which rely on animal matter, such as pitcher plants (*Nepenthes* spp.) which trap insects in their flask-like leaves, and the Venus fly-trap (*Dionaea muscipula*) which actively catches prey by snapping shut a modified leaf when an insect lands on it. In each of these cases the insect is digested outside the plant: extracellular enzymes are secreted by the plant and the soluble products are then absorbed. In some cases the insect is dead before digestion begins. Such nutrition is better described as saprotrophic rather than carnivorous. On the other hand, in those cases when the insect is alive at the start of digestion, the nutrition could be described as ectoparasitic – the parasitic plant attacking the live prey from the outside! It is probably best not to spend much time puzzling about how to classify heterotrophic nutrition. It is more fruitful to spend the time investigating it.

To an ecologist, measurements of *basal* metabolic rates are not especially useful. For one thing, the concept of a thermoneutral environment only applies to homoiotherms (endotherms). Anyway, ecologists are more interested in what an organism's energy expenditure is in real life. Nagy (1987) provides a useful catalogue of the various terms that have been used to refer to realistic estimates or measurements of the **daily energy expenditure** of organisms in the field. Daily energy expenditure includes the energy spent on locomotion, thermoregulation, growth and reproduction. It is usually about 1.5 to 3.0 times basal metabolic rate (Gessaman, 1973; King, 1974; Lucas *et al.*, 1993). In humans and other vertebrates, the maximal sustained energy consumption over 24 hours is about 7 times basal metabolic rate (Hammond & Diamond, 1997). In principle, both basal metabolic rate and daily energy expenditure can be measured for any organism. In practice, almost all the measurements have been made on animals at least several millimetres in length. Ecologists have tended to concentrate on larger species, partly because of the difficulties of measuring the metabolic rates of very small organisms.

2.4 Factors affecting metabolic rate

2.4.1 Size

Perhaps the most important factor affecting the metabolic rate of an individual is its size. Figure 2.5 shows the basal metabolic rates and daily energy expenditures for 47 species of birds plotted against their body mass (Bennett & Harvey, 1987). Overall,

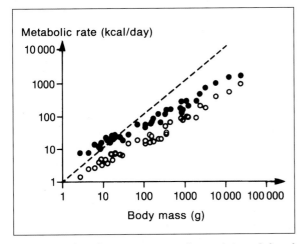

Figure 2.5 Daily energy expenditure (●) and basal metabolic rate (○) in kcal/day (one kcal is 4198J) on logarithmically scaled axes for 47 species of birds for which both values are available. Dotted line shows what the slope of the relationship would be if metabolic rate was directly proportional to body mass. (From Bennett & Harvey, 1987.)

larger birds need more energy each day to live. However, metabolic rates are not directly proportional to body mass. If they were, then the points on the graph would fall on a steeper line indicated by the thin dotted line. In fact the slope of the actual relationship is shallower than this. A bird ten times heavier than another one does not need ten times as much food each day. The relationship between metabolic rate and body mass can be expressed by what is called the **allometric equation**:

$$\text{Metabolic rate} = a(\text{body mass})^b$$

In this equation, b is referred to as the **exponent** that relates metabolic rate to body mass. If logarithms of both sides of this equation are taken, it can be rewritten as:

$$\log[\text{metabolic rate}] = \log[a] + b(\log[\text{body mass}])$$

A log–log equation like this is plotted in Figure 2.5. From such a graph, a and b can be worked out, as a equals the metabolic rate when body mass is 1 and b is the slope of the line. Measurements on a number of different taxa show that the exponent, b, relating metabolic rate to body mass lies between about 0.5 and 0.9, irrespective of the units used for the measurements either of metabolic rate or of body mass (Kleiber, 1947; Peters, 1983; Reiss, 1985). Values for b tend to lie close to 0.75. The reasons for this are only now becoming understood. They are complicated but, at least in multicellular organisms, may be to do with the need to transport essential materials within individuals by means of space-filling fractal networks of branching tubes (West *et al.*, 1997). (Fractals are explained on page 181.) Because b is less than 1.0, larger species need *less* energy per day, *relative to their body mass*, than do smaller species. This means that small animals, such as shrews, have such a high metabolic rate relative to their body size that they may need to eat more than their body weight in food each day! An elephant, on the other hand, takes about three months to eat its own body weight of food.

2.4.2 Life style

Even among organisms of the same size, there is a great deal of variation in their metabolic rates. In a classic analysis Hemmingsen (1960) compared the basal metabolic rates of unicellular organisms, **poikilothermic animals** and **homoiothermic animals** (Figure 2.6). As you might expect, homoiotherms, which maintain a constantly high body temperature, need a lot more energy than poikilotherms. It is perhaps surprising just how much energy they need. As Figure 2.6 shows, a homoiotherm needs about 25–30 times as much energy as a poikilotherm of the same size. It is a lot cheaper to feed a pet snake weighing 5 kg than a pet dog weighing 5 kg!

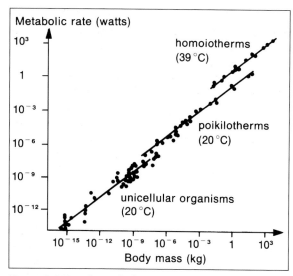

Figure 2.6 Basal metabolic rates of homoiotherms, poikilotherms and unicellular organisms as functions of their size. (From Peters, 1983.)

It is interesting to note, from Figure 2.6, that a poikilotherm has a metabolic rate some 8–10 times higher than a unicellular organism of the same size. Evidently there is quite an energetic cost to being multicellular.

If you look at Figure 2.6, you can see that much of the variation within the three groups – unicellular organisms, poikilotherms and homoiotherms – can be accounted for by body size, However, there is considerable scatter about the three lines. Attempts to identify the reasons for this scatter have mainly focused on birds and mammals. Essentially, two reasons for the variation can be suggested: life style and phylogeny. For example, mammals that feed on vertebrates have high basal metabolic rates relative to other mammals (Elgar & Harvey, 1987a). It might therefore be that a mammal that eats vertebrates needs a high basal metabolic rate because of its life style, perhaps because it must constantly be ready to rush after its prey. As Elgar and Harvey point out, however, the association might have a quite different explanation. It turns out that the correlation among mammals of high basal metabolic rate with the eating of vertebrates is mainly due to the possession of both of these traits by whales and dolphins (Cetacea) and seals (Pinnipedia). Perhaps the high basal metabolic rates of these animals reflect their marine life styles rather than their feeding habits. Of course, this too would be an ecological explanation for their relatively high basal metabolic rates. Alternatively, perhaps all cetaceans and pinnipeds have high basal metabolic rates irrespective of their ecology. This would mean that their high basal metabolic rates might be due to phylogeny rather than to life style. Another possible explanation arises from the fact that it is extremely difficult to measure the basal metabolic rates of seals and dolphins under laboratory conditions. They stress easily and often show abnormally high 'basal' metabolic rates. It might therefore be that the high values are an artefact.

2.5 Size determines more than metabolic rate

As we have seen, size is a very important determinant of an organism's metabolic rate. However, there are many other features of an organism's life that are strongly affected by size. Table 2.1 gives the values of the exponents relating various physiological, anatomical, ecological and behavioural measures to body mass. For instance, a value of 0.17 for incubation time in birds means that the length of time birds of different species spend sitting on their eggs is related to the mass of the parent bird by the equation:

Incubation time $\propto$ (body mass)$^{0.17}$

and this will be true irrespective of the units in which either incubation time or body mass are measured. Some exponents are negative. For example, the one relating heartbeat frequency in mammals to body mass is −0.25. This means that heartbeat frequency is related to the body mass of different mammals by the equation:

Heartbeat frequency $\propto$ (body mass)$^{-0.25}$

In other words, the larger a mammal is, the smaller the number of times its heart beats each minute. The smallest shrews have heartbeat frequencies of over 1200 times a minute.

We can conclude that an organism's size greatly influences its ecology. Knowing an individual's size immediately tells us something about how it interacts

Table 2.1 Exponents relating various anatomical, physiological, ecological and behavioural variables to body mass in the equation variable $\propto$ (body mass)^{b}. Taken from Peters (1983), Schmidt-Nielsen (1984) and Swihart *et al.* (1988).

Variable	Taxon	Exponent (*b*)
Home range size	Mammals	1.26
Skeletal mass	Rattlesnakes	1.17
Skeletal mass	Mammals	1.09
Skeletal mass	Fish	1.03
Lung volume	Mammals	1.02
Ingestion rate	Crustaceans	0.80
Brain mass	Mammals	0.70
Gestation length	Eutherian mammals	0.24
Age at maturity	Fish	0.20
Incubation time	Birds	0.17
Heartbeat frequency	Mammals	−0.25
Breathing rate	Mammals	−0.26

with other organisms and with the rest of its environment. In particular, the amount of energy an organism takes in each day is closely related to its body size. But what do organisms do with this energy? How much of it can they apportion to growth or to reproduction and how much do they need just to remain alive? These questions can be answered by looking at the energy budgets of organisms.

2.6 Energy budgets

Why do organisms take in food? We take eating so much for granted that this sounds like a silly question. However, it is worth emphasising that evolutionary success, for any organism, is judged by how successful it is at reproducing itself. It is an extraordinary thought that *every* individual alive today has an ancestry that goes back some 3500 million years, yet many of the individuals alive today will themselves fail to leave any offspring behind, the first in their line thus to fail. Organisms take in food so that energy and nutrients can be channelled towards the production of offspring.

Individuals cannot devote all their resources directly to reproduction. Some of their energy intake is needed to keep themselves alive and in good condition. In fact, around 10 to 30% of the energy absorbed from food ends up being used to digest food (Cossins & Roberts, 1996). When individuals are juvenile, they devote some of their energy to growth. Presumably, in most cases, a juvenile individual that tried to reproduce would not be very successful at it. For one thing, it would be too small to leave many offspring. It is more likely to maximise its **lifetime reproductive success** by waiting until it has grown more. Chapter 7 looks at the ecological pressures favouring growth or reproduction and at the conflict between reproduction and mortality: an individual that devotes too much of its resources to reproduction runs the risk of killing itself in the process. In this section we will look quantitatively at what happens to the energy that individuals ingest.

2.6.1 Assimilation efficiency

As we discussed at the beginning of this chapter, when an organism ingests food, only some of it is assimilated. The term **assimilation efficiency** refers to the percentage of the energy that an organism ingests that is assimilated rather than egested. As one might expect, organisms differ greatly in their assimilation efficiencies depending on the type of food they eat. Carnivores feeding on vertebrates may have assimilation efficiencies in excess of 90%; insectivores typically have assimilation efficiencies of 70–80%, while most herbivores have assimilation efficiencies of 30–60%, though zooplankton feeding on phyto-

plankton have assimilation efficiencies of 50–90% and the giant panda (*Ailuropoda melanoleuca*) has the lowest assimilation efficiency of any mammal yet measured at 20% (Ricklefs, 1980; Anon, 1982).

Assimilation efficiencies can be much lower than 20%. Cammen (1980) drew together data on the feeding habits of 19 species of invertebrate deposit feeders and detritivores found on the ocean bottom and which belonged to three phyla. The percentage of organic matter in the sediment on which these species fed ranged from 57% down to less than 1%. The crab *Scopimera globosa* fed on the poorest quality food, which contained only 0.19% organic matter. This means that the assimilation efficiency of *Scopimera globosa* must be less than 0.19%. Measurements on the contents of the foregut of this species show that there the food has on average an organic content of 12% (Ono, 1965). This means that a great deal of egestion must have taken place before the food reaches the foregut.

2.6.2 Production and respiration

Once a heterotroph has assimilated its food, the products of assimilation can either be respired to provide the energy needed to maintain existence and to repair old and damaged body tissue, or they can be diverted to growth and reproduction. Together, growth and reproduction are called **production**.

The percentage of the energy assimilated that an organism diverts to growth is called its **growth efficiency**. Growth efficiencies are economically very important to farmers. Juvenile pigs have growth efficiencies of up to 20%, which is very high for a farm animal. This means that pigs are very efficient at converting their feed into pig meat. The efficiency of this conversion is affected by many factors. As one might expect, their growth efficiency is highest when they do not have to expend any energy on thermoregulation and movement. Under intensive production some farmers therefore keep pigs warm and in almost total darkness – pigs move less when it is dark.

For any species, growth efficiencies are higher the smaller the individuals are, relative to their size at maturity. Figures 2.7a and 2.7b show growth efficiencies as functions of size for Leach's storm-petrel (*Oceanodroma leucorhoa*) and the fish *Ophiocephalus striatus*. In each case the growth efficiency decreases as the individuals mature. Because of this, it is rather difficult to make comparisons between species about their growth efficiencies. Nevertheless, as a general rule, juvenile poikilotherms have higher growth efficiencies than juvenile homoiotherms. However, if one looks at the growth efficiencies of *very* young organisms, one finds that homoiotherms may have growth efficiencies as high as 50–70%, almost the same as

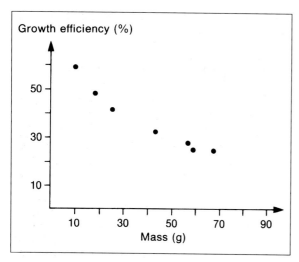

Figure 2.7a Growth efficiency as a function of size in juvenile Leach's storm-petrel (*Oceanodroma leucorhoa*). (Data from Ricklefs *et al.*, 1980.)

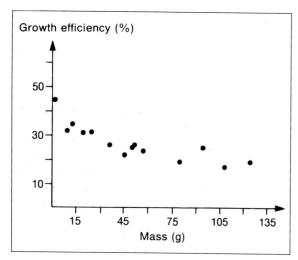

Figure 2.7b Growth efficiency as a function of size in juveniles of the fish *Ophiocephalus striatus*. (Data from Pandian, 1967.)

those of poikilotherms which may be as high as 50–80% (Calow, 1977). By comparison, bacteria, protoctists and metazoan cells in culture have growth efficiencies of about 60% (Calow, 1977).

2.6.3 Allocation to reproduction

Perhaps rather surprisingly, comparatively few studies appear to have been done which look at the allocation of assimilated food to reproduction. It would be very useful to have data on what might be termed **reproductive efficiencies** for different species. In animals it may be the case that homoiotherms allocate a smaller percentage of their assimilated food to reproduction than do poikilotherms.

Some plant ecologists have looked at the percentage

Table 2.2 The allocation of net assimilation to reproduction in different groups of flowering plants (Ogden, 1968; Harper, 1977).

Type of plant	Percentage of net assimilation devoted to reproduction
Herbaceous perennials (excluding vegetative reproduction)	1–15
Herbaceous perennials (including vegetative reproduction)	5–25
Annuals	15–30
Most grain crops	25–35
Maize, barley	35–40

of the **annual net assimilation**, that is photosynthesis minus respiration, allocated by a plant to reproduction. Table 2.2 shows that annual plants have the greatest reproductive efficiencies. This is as we might predict because annuals can devote as much of their assimilation as possible to reproduction; most perennials, unless they reproduce only once, have to hold back some of their assimilation to maintain a living plant capable of future reproduction. The extremely high values recorded in Table 2.2 for commercial grain crops are presumably a reflection partly of their annual nature and partly of the fact that such plants have been selected and grown under ideal conditions almost solely for the purpose of maximising yield. Wild plants probably allocate far more of their assimilation to other features such as anti-herbivore devices.

Measuring the proportion of an organism's **energy budget** that it devotes to reproduction may not be a true index of **reproductive effort** (Calow, 1977; Ronsheim, 1988). Energy may not be the resource that is limiting reproduction. For many plants, available nitrogen or phosphorus are more likely to be in short supply than sunlight. So far, however, most workers have looked at energy budgets rather than nutrient budgets. It may turn out to be the case that for most species the two are closely related anyway, so that measurements of energy allocations adequately reflect the evolutionary pressures facing organisms.

2.6. Drawing up a complete energy budget

A number of studies have been carried out to determine the energy budgets of individual species. Although they often take a long time to perform, and are difficult to carry out with great accuracy, energy budgets are valuable because they provide an insight into the ecological pressures operating on a species. We will look at one animal and two plant examples.

The energy budget of the crustacean *Idotea baltica*

The energy budget of the intertidal isopod *Idotea baltica* was studied by Strong & Daborn (1979). The following equation was used to represent the energy budget:

$$C = P + R + E_x + U + F$$

where:

C = energy ingested;
P = the sum of P_g (energy accumulated in growth) and P_r (energy shed as gametes);
R = respiratory or metabolic loss;
E_x = energy losses associated with ecdysis;
U = energy lost as soluble excretory products of all kinds;
F = energy passed through the gut as faeces.

The cumulative energy budgets are shown separately for males and females in Table 2.3.

Table 2.3 Cumulative energy budget for the isopod *Idotea baltica* calculated over the average lifespan: females, 415 days: males, 374 days (Strong & Daborn, 1979).

Variable	Symbol	Males			Females		
		J	%C	%A	J	%C	%A
Ingestion	C	2559.4	–	–	2528.4	–	–
Metabolism	R	959.2	37.5	56.7	673.7	26.6	57.7
Growth	P_g	565.6	22.1	33.4	309.0	12.2	26.5
Reproduction	P_r	–	–	–	99.0	3.9	8.5
Ecdysis	E_x	133.1	5.2	7.9	61.1	2.4	5.2
Non-faecal excretion	U	35.2	1.4	2.1	24.7	1.0	2.1
Assimilation*	A	1693.1	66.1	–	1167.5	46.2	–

*Calculated as sum of R, P_g, P_r, E_x, U.

Even the largest *I. baltica* individuals weigh less than 0.1 g, so measurements of the components of the energy budget were not easy: several assumptions and simplifications had to be made. For instance, U was not quantified directly and estimations of the rate of faecal production were possible on only two occasions. No measurements of sperm production were made which is why there is no entry for the value of P_r for males in Table 2.3.

Such energy budgets allow comparisons to be made between males and females of the way they allocate what they ingest to processes such as growth and reproduction. They can also be done for individuals of different ages and sizes allowing developmental changes to be seen. If carried out on several species, they allow interspecific comparisons to be made. At a community level they enable an investigator to see how the energy flowing through an entire community is partitioned.

The energy budgets of the annual plants groundsel (*Senecio vulgaris*) and corn marigold (*Chrysanthemum segetum*)

In two classic studies, Harper & Ogden (1970) and Howarth & Williams (1972) studied the allocation of dry weight to different stages of the life cycle in two annual plants in the same family, the Asteraceae. These were groundsel (*Senecio vulgaris*) and corn marigold (*Chrysanthemum segetum*). Large numbers of plants were grown in glasshouses and samples harvested at regular intervals throughout the growing season. At most harvests each plant was divided into roots, stems, leaves, receptacles and flowers. These parts were then weighed separately after drying to constant weight. Seeds were also collected and weighed throughout the seeding period.

The results for the two species are shown in Figure 2.8 in a form which allows the life cycles of the two species to be compared visually. These diagrams show how the compositions of the plants change as they grow. In both species, the root makes up almost all of the plant as it grows from the small seed. Then the first leaves develop and grow and a stem is produced to support these leaves. In both cases flowers start to be formed about half-way through the life cycle as the plants begin to shift their resources from leaf production to flower and then seed production.

It would be valuable to have this sort of information for a large number of plant species with differing life histories as such information tells us a great deal about how the life cycle of the plant has been shaped by natural selection to adapt it to the environment in which it lives (e.g. Silvertown & Dodd, 1996).

2.7 Distinguishing between growth and reproduction

So far we have assumed that growth and reproduction are two entirely separate processes. In many cases, however, this is not the case. Consider, for instance, a grass that produces tillers. In one sense this is growth; in another it is reproduction as the tillers soon grow roots and may even become detached from the parent plant (see Section 6.2.6). Perhaps the solution to the problem of what is growth and what is reproduction is that we should not try to divide all production into these two categories. Little is to be gained by forcing terms devised to describe higher animals to fit the life styles of all other organisms.

Many species show **colonial growth** – corals and bryozoans, for instance. Here, countless genetically identical multicellular subunits remain attached to one another. Essentially this is little different from the

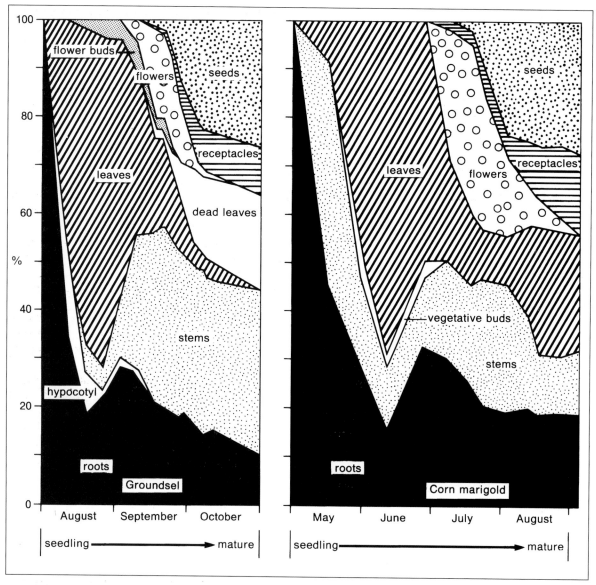

Figure 2.8 Allocation of dry weight to different structures throughout the life cycles of the plants groundsel (*Senecio vulgaris*) and corn marigold (*Chrysanthemum segetum*). (From Harper & Ogden, 1970.)

aggregation of genetically identical cells into a multi-cellular organism.

Fungi can also defy classification as individuals. Alan Rayner (1997) argues that fungi and many other microbes challenge the whole notion of the individual as the unit of reproduction. Some fungi exist as 'diffuse individuals' – there is one such 'individual' that extends over 15 hectares of forest in Montana, weighs more than 100 tonnes and is thought to be over a thousand years old. Other fungi exist not as discrete, bounded entities but as dynamic networks. We, as mammals, tend to think of discrete organisms as the norm, but in fact fungi have twice the total biomass of all the animals in the world. Perhaps discrete individuals are the odd life

form in a world of diffuse and colonial organisms.

Recently, many botanists have begun to think of plants as **modular organisms** where each unit such as a branch acts rather like a whole plant with leaves, flowers, seeds and so on. In most vascular plants growth is **indeterminate** in the sense that meristems (regions of actively dividing cells) continue to produce new shoots and roots until the opportunities for further local growth are exhausted (Waller, 1986). The individual and repeated sub-units of which a plant may be composed are sometimes referred to as modules. Growth is often decentralised in modular organisms, with each subunit capable of considerable regeneration. The population dynamics of modular organisms are considered in Section 4.7.

Summary

(1) Individual organisms are the products of natural selection in ways that aggregates of individuals are not.

(2) Autotrophs do not require a source of organic carbon; heterotrophs do.

(3) Heterotrophic nutrition involves ingestion, egestion, digestion, absorption, assimilation and excretion.

(4) Holozoons feed on relatively large pieces of dead organic material and include carnivores, herbivores and omnivores.

(5) There is a multiplicity of terms associated with heterotrophic nutrition. Herbivores, for instance, include granivores, frugivores, folivores, grazers and browsers.

(6) Parasites feed off hosts that are still alive. They may be endoparasitic or exoparasitic.

(7) Saprotrophs feed off dead organic matter which they either absorb in solution or ingest as very small pieces.

(8) The metabolic rates of organisms in the field are referred to as daily energy expenditures. They scale on body mass with exponents of between about 0.5 and 0.9 and typically exceed basal metabolic rates by factors of between 1.5 and 3.0.

(9) Relative to its body mass a large organism needs less energy per day than a smaller one. Large heterotrophs can therefore survive for longer between meals.

(10) Homoiotherms have basal metabolic rates some 25–30 times greater than poikilotherms of the same size which in turn have metabolic rates some 8–10 times greater than unicellular organisms of the same size.

(11) Many features of an organism's anatomy, physiology, ecology and behaviour are related to its mass by allometric equations.

(12) Assimilation efficiencies depend on the quality of the food a heterotroph ingests. They range from less than 0.2% to over 90%.

(13) Production equals growth plus reproduction.

(14) Within a species, growth efficiency decreases as individuals grow bigger.

(15) Energy may not always be the resource limiting reproduction.

(16) Energy budgets of individual species are difficult and time consuming to complete, but provide valuable information about the ecological pressures operating on a species.

(17) In many species it is not particularly useful to try to distinguish between growth and reproduction.

(18) Some species cannot easily be divided into separate individuals.

THREE

Autecology

3.1 The meaning of autecology

Autecology is the name given to ecological studies which concentrate on one species (*autos* is the Greek for self). An autecological study aims to answer the questions asked in Chapter 1 and to understand the processes described in Chapter 2 for a chosen species. To find out all there is to know about a species is an enormous task which requires considerable time, a great deal of observation and numerous experiments. Usually different ecologists are involved over many years in finding out about one species. Often they are working on specimens from different parts of the world. Several fields of biology are usually involved including genetics and biochemistry as well as the more traditional forms of ecology.

There are an estimated 30–40 million species in existence today. Very few of these have as yet been studied sufficiently for us to understand much of their autecology. This chapter gives two examples of species which have been investigated fairly extensively. One, the bracken fern, is an autotrophic plant, and the other, the European starling, is a heterotrophic vertebrate. These examples will give you an idea of how the results of many autecological studies can be combined to help us to understand the whole life cycle of a species. Much more is known about each of these species than is mentioned here, but you can get a flavour of the scope of autecology from these accounts.

3.2 The autecology of bracken

3.2.1 The importance of bracken

Bracken (*Pteridium aquilinum*, Phylum Filicinophyta) is the most successful fern in the world. In fact, it is one of the most successful plants in the world because it grows on every continent except Antarctica. The only places in which it seems unable to survive are on high mountains and within the Arctic and Antarctic circles; all these places have severely cold winters and this is probably what limits the bracken.

Bracken is a very suitable species for autecological study. Figure 3.1 shows an area covered with bracken. It is a very vigorous plant, and you can see from the photograph that the leaves cover a large area of ground. This seems to stop other plants growing with it and so very few other plant species are found in association with bracken.

Most of the bracken plant is **carcinogenic**, that is it can cause cancer in animals which eat it. Both the leaves (called fronds) and the dispersal units (called spores) contain carcinogenic substances. This can be a problem for people who regularly walk through areas of bracken, especially if spores are being released. Farmers who work where bracken is common have a higher risk of getting stomach cancers or leukaemia (I. A. Evans, 1986). The carcinogenic effects of bracken are especially apparent in the Japanese, many of whom eat the young curled leaf-fronds (called croziers) as a delicacy. This makes them three times as likely to get cancer (Hirayama, 1979).

Bracken is also poisonous to grazing animals. Pigs and horses which eat it develop symptoms associated

Figure 3.1 Fronds of bracken (*Pteridium aquilinum*) in an open woodland.

with lack of vitamin B_1 (thiamine). The fern contains an enzyme called thiaminase which destroys thiamine (W. C. Evans, 1986). Sheep and cattle do not usually need thiamine in their diets as they obtain it from microorganisms in the rumen (see Section 14.2.4), but if they eat enough bracken, they too suffer thiamine deficiency.

Although cattle are less likely than horses to show thiamine deficiency, they do develop another form of poisoning if fed on a diet containing bracken for a few weeks. This acute bracken poisoning is the result of decreased activity of the bone marrow which severely lowers the number of white blood cells and platelets in the blood (W. C. Evans, 1986).

Apart from the direct risk to the animals who eat bracken, the plant also provides shelter for the sheep tick (*Ixodes ricinus*). This is a blood-sucking arachnid (more closely related to spiders than to insects) capable of carrying several diseases which it may pass on to sheep and other vertebrates.

Despite all these fearful properties of bracken, it does have a few uses. It has been used as a source of potash fertiliser, as heating fuel, for roofing buildings and as bedding for animals, where the dried autumn leaves are considered more absorbent than straw and can be used as nutrient-rich manure afterwards. As already mentioned, the croziers are eaten by some people and the rhizomes have been used as a flour substitute in breadmaking and instead of hops to add bitterness in beermaking. Bracken was also used in the sixteenth and seventeenth centuries as a herbal cure for human infestations of worms (Rymer, 1976).

Because bracken is so poisonous, harbours disease-carrying ticks and is both widespread and vigorous, it is considered a pest in many countries (Barrow, 1991). The major ecological question which can be asked about bracken is: why is it so successful? Autecological studies have shown that there are many answers to this question. Bracken has several features which give it ecological advantages. Most of these features can be grouped under two headings: bracken is successful because of its life style and because of its anti-predator devices.

3.2.2 The life form of bracken

As has already been mentioned, bracken is a fern. It is unusual to find a fern which is an economic nuisance, because they are usually restricted to moist, shady areas and do not spread far. Bracken is unusual because it is much more vigorous than most ferns and because it seems able to survive in a wide range of conditions including well drained open hillsides. Being a fern, bracken reproduces by releasing tiny spores from underneath its leaves. These spores are wind dispersed and can survive for many months in a dry dust-like state. Because these spores are so tiny

and survive for so long, they can be blown very long distances before they settle. If it lands on a suitable site, a spore germinates and grows into a little flat green sheet of cells. This is called a **prothallus** and it produces male and female sex organs. The stationary female egg cell is fertilised by a motile sperm cell, which may well have swum from another prothallus, and so a new bracken plant is produced. The long-distance dispersal of spores in the winds which blow around the north and south hemispheres means that bracken can be carried to new areas where no adult plants exist. This widespread distribution of spores may help to explain the worldwide occurrence of the species.

Considering the abundance of bracken it is surprising that the establishment of a new adult plant from the germination of a spore is in fact an extremely rare occurrence. In Finland, Eino Oinonen (1967a) found that bracken only established from spores on sites which had been burned. Fire sterilises the soil and probably allows the tiny prothallus to survive without being attacked by fungi or soil-borne insects such as springtails (Collembola). The high-nutrient ash from the fire and the bare soil may also be important. In Britain bracken is also known to colonise bare sites such as demolished buildings, often growing on the mortar which held walls together. This mortar has similar properties to the ash resulting from fires.

Once established, the new bracken plant grows out in all directions by vegetative means to form a large circular colony. You can see how it does this in Figure 3.2. Only the leaves of the bracken are visible above the ground. Within the soil is a network of underground stems called **rhizomes**. Most of the plant is underground with only the leaves pushing into the light. This growth form where only individual leaves appear above the soil is shown in Figure 3.1. In fact all the leaves in the picture are probably from one plant and are joined under the soil by the rhizomes. Old rhizomes decay, and this may break the bracken plant up into several disconnected pieces, but each of these plants comes from one genetic individual, now divided up into a **clone** of plants with the same genetic material (see Chapter 6). Bracken requires soil deep enough for these rhizomes to grow, so it cannot inhabit the shallowest rocky soils, but it thrives almost everywhere else.

Oinonen found he could tell the age of clones by working out the date of the last fire, using data from the age of the trees in the area. He also measured the diameter of the circle of growth of the bracken clones. He obtained a remarkably good correlation between the size of clone and the estimated age of the clone. The results of this study can be seen in Figure 3.3. These data showed that the bracken increased the diameter of its clone by 10 m every 29 years. This

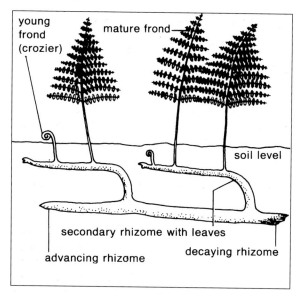

Figure 3.2 The life form of bracken. The deep underground rhizomes produce shallower rhizomes which bear leaves. The leaves push up through the soil as curled croziers which then expand into large fronds.

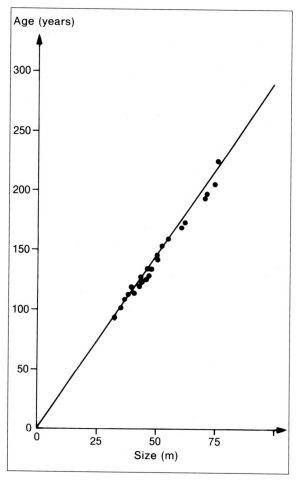

Figure 3.3 The relationship between the diameter of bracken clones and the estimated age of the clones. The data were collected from Finland by Oinonen (1967a).

allowed Oinonen to predict the age of any bracken clone he found. His results for bracken in Finland are fascinating. Many of the clones date to periods of war and occur in regions known to have been crossed by troops (Oinonen, 1967b). So the distribution of bracken in Finland is partly due to the camp-fires of soldiers, although natural fires must also have played a part. Some clones he found were over 450 m in diameter and he concluded that they had begun about AD 550–70. They occurred in areas known for archaeological finds of the Iron Age. It is tempting to imagine these clones as beginning on human camp-sites 1400 years ago.

The results of Oinonen's work show that bracken can be extremely long lived and persistent. Clones can live longer than most trees. So although the germination and establishment of spores is a rare occurrence, the death of established clones is probably also rare.

In temperate areas such as Britain and Finland the leaves of bracken turn red-brown as they die back for winter. The underground rhizome survives and produces new leaves the following spring. In more tropical regions, new leaves develop all year round.

The life form of bracken makes it a good **competitor**. It can out-compete other plants. The dense foliage of leaves prevents the seedlings of other species, including trees, from growing. Once bracken has become established, it is very difficult to eradicate because of its underground rhizomes. These lie deep in the soil defying efforts to dig or plough them up.

They are also protected from fire – either natural fires or attempts to burn off the bracken fronds. This is why bracken becomes a problem in tropical areas, such as the Amazon basin in Brazil, where burning cleared land which was once tropical forest encourages bracken. Bracken can be killed by some herbicides, but none are really successful and several applications are often needed. These herbicides are more likely to kill other fern species. As many of these are rare and occur in areas where bracken also grows there is a possible conflict with conservation in using such chemicals.

3.2.3 Anti-predator mechanisms

Some of the problems caused when mammals eat bracken have already been described. Although the poisons do not prevent it from being eaten, the effects are severe and often cause the death of the animal. This reduces the number of animals grazing on bracken.

Mature bracken fronds are tough to chew. They are full of tannins, lignins and silicates which make them unpalatable. The croziers and immature fronds are much more tender. If they were tough the frond would not be able to uncurl and expand properly. Because the frond is still growing and expanding it has a much higher protein concentration than the mature fronds which are extremely low in proteins. This makes young fronds a better food supply for herbivores. These tender young leaves need to be better protected than the tough old leaves.

The croziers are protected in several ways during the first month after they have emerged from the rhizome. They have high concentrations of poisons including cyanide and they contain substances which deter herbivores from eating them (called anti-feedants). They also possess ecdysones and related sterols, including α-ecdysone and 20-hydroxy-ecdysone. Ecdysones are insect moulting hormones. Bracken may affect the growth, moulting behaviour and egg-laying capacity of insects which eat the hormone compounds in the leaves but, as bracken contains so many anti-herbivore compounds, it is difficult to sort out which have an effect (Cooper-Driver, 1976; Lawton, 1976).

The developing fronds have glands which secrete nectar. These attract ants (Figure 3.4) which visit the fronds to drink the nectar. Once the ants know the nectar is there they defend the frond by attacking any other insects which visit it. This reduces the chance of insects eating the frond. So the bracken is protected from herbivores by the ants.

Despite all these poisons, anti-feedants and the defending ants, there are still over 100 species of insect which eat bracken leaves. They feed in several ways: chewing the leaf while sitting on its surface; burrowing into the leaf tissue; or sucking contents out of cells and the vascular system (Lawton, 1984). However, none of these insects ever becomes sufficiently numerous on a plant to damage much of the leaf area. Grazing and insect activity seem to have little effect on the distribution or vigour of bracken.

3.3 The autecology of the European starling

3.3.1 Appearance and distribution

The European starling (*Sturnus vulgaris*) is another widespread species, like the bracken fern just described. Unlike bracken, which is naturally widespread, the starling has a restricted geographical area where it occurs naturally, but it has been introduced into other continents by humans. Autecological studies include investigations of starlings in their natural environment and studies of how the birds have coped after introduction into new environments.

As the photograph in Figure 3.5 shows, the starling is a stocky bird with a strong beak. It is quite heavy for its size and usually weighs between 75 and 100 g (Feare, 1984). From a distance, the birds often appear a uniform dark brown or black. When looked at more closely fine patterns can often be seen on each feather. The feathers also have a metallic sheen of diffracted light seen as purple, green or bronze; so a clean, healthy starling is quite a handsome bird.

The European starling occurs naturally across central and southern Europe and in Asia as far east as India. This area is outlined in Figure 3.6 with a heavy line. Within this area many starlings **migrate** from one place to another and spend the winter in milder parts of their range. The arrows show the main directions of migration in autumn to areas where starlings go for the winter months. In summer they fly back to their breeding sites in Britain, northern Europe and Russia. Not all starlings migrate; many overwinter in places such as Britain, where they are also able to breed, and so do not have to travel such long distances.

Figure 3.6 also shows the areas where the European starling has spread after being introduced by humans. European starlings now occupy large

Figure 3.4 Nectaries on a developing bracken frond, being visited by *Myrmica* ants.

Figure 3.5 Two starlings (*Sturnus vulgaris*) in winter plumage.

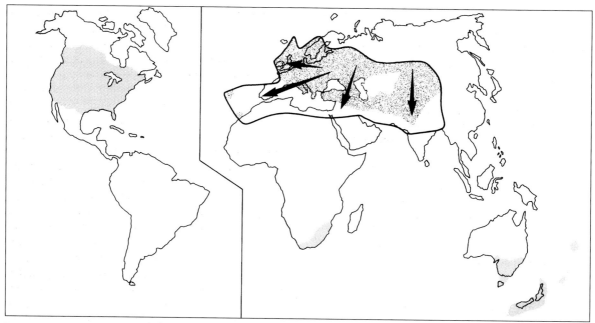

Figure 3.6 Map showing the world distribution of European starlings. Their natural distribution lies within the heavy line in Europe and Asia. The shaded area here indicates their breeding range and arrows show the direction of migration for winter. The other shaded areas indicate places where starlings have been introduced and subsequently spread. (Based on Feare, 1984.)

areas of North America, Australia, New Zealand and south-east Africa as well as several Pacific islands. Starlings were introduced into New Zealand in 1862 and since then they have become one of the most common birds except in mountainous regions and in very dense woodland (Falla, Sibson & Turbott, 1979).

In North America, 60 birds were released in Central Park, New York, in 1890 (Chapman, 1895). You can see from Figure 3.6 that their descendants have spread over most of North America. They are so numerous that they are a pest, both to farmers and to native wildlife. Starlings compete with other birds for nesting sites: the wood duck (*Aix sponsa*) and the eastern bluebird (*Sialia sialis*) seem to be affected particularly severely. In fact the eastern bluebird may become a threatened species because of the activity of starlings (Zeleny, 1969).

3.3.2 Feeding habits

As a heterotroph, the starling has to eat sufficient food to give it energy for all the activities of life. It needs this energy to maintain its basal metabolic rate, for maintenance of body temperature when the weather is cold, for flying and foraging, for migration, for regrowth of feathers after moulting and for breeding. Starlings vary their food intake depending on what they are doing (Feare, 1984). In winter, when their main need is for energy to keep warm, starlings eat a diet of seeds which are rich in carbohydrates and fats. During the breeding season, the birds switch to a high-protein diet of invertebrates upon

which they also feed their chicks. This diet is very variable, but includes insects, snails, worms and woodlice. Obviously the species eaten in different parts of the world vary as different species are available. In Europe, one of the main prey is leatherjackets (the larvae of the crane fly, *Tipula paludosa*). This fat grub, about 2.5 cm long, lives just below the soil surface and the starling feeds by probing the soil with its strong beak. In New Zealand, where the crane fly does not occur, the starlings eat other larvae which live in the soil.

A fascinating feature of this seasonal change of diet is that the digestive system of the starling also changes. The gut of captive birds which have been fed a diet of insects is shorter than the gut of birds who have been eating seeds. This change takes about two weeks in laboratory experiments. Wild birds which have been eating seeds also have longer guts (Al-Joborae, 1979).

Starlings are social birds: many of their activities are carried out in groups called flocks. If a few birds are seen to be feeding this quickly attracts more birds. In Britain starlings are often found feeding in fields in mixed flocks containing birds of other species. These include lapwings (*Vanellus vanellus*), golden plovers (*Pluvialis apricaria*), rooks (*Corvus frugilegus*) and jackdaws (*C. monedula*) (Feare, 1984).

What are the advantages of being in a flock? Figure 3.7 shows the feeding rate of starlings in different sized flocks. The observed birds were feeding at a trough where they were eating crushed barley supplied for cattle (Feare & Inglis, 1979). When the flock size was below five, the birds were nervous and spent a lot of time looking round, presumably watching for predators. When there were more than ten birds they became rather crowded around the trough, and the birds fought each other which interfered with feeding. So the birds' feeding rate went down when they were too close together, even though plenty of food was available. This observation shows that there are advantages to being in a flock in that it makes the birds safer from predators, but that too many birds too close together can be a disadvantage. Obviously the situation at the food trough is very different from that in open areas where the food source is scattered larvae and starlings can feed at a distance from each other. There have been many studies of flocking in birds and herding in animals; the findings and principles of such group behaviour are described and discussed in more detail in Section 8.2.

3.3.3 Roosting behaviour of starlings

As well as feeding in flocks, starlings sleep at night in large groups called **roosts**. Although they roost naturally in trees and sometimes in reed beds, starlings also now roost in cities, on window ledges and ornamental stonework. It is an experience to stand in the dusk at a roosting site and watch thousands of starlings darken the sky as they fly in for the night. Such city roosts have been discouraged because of the pollution caused by bird droppings to buildings and pavements.

Starlings may fly quite long distances from their

feeding sites to the roosting site. Maximum distances recorded for travel to night-time roosts are 38 km in Britain (Wynne-Edwards, 1929) and 80 km in America (Hamilton & Gilbert, 1969). Most birds fly much shorter distances, but probably still fly several kilometres. In Europe roosting sites can vary from summer to winter. In summer starlings favour deciduous trees for roosts; in winter, they prefer evergreen trees, either conifers or evergreen broadleaved trees (Marples, 1934; Delvingt, 1961).

So why do starlings roost together at night? They come together in much larger flocks than those in which they feed. Are they flocking, as they are when feeding, for protection from predators? This seems unlikely, and there is some evidence that predators are actually attracted to large winter roosts (Feare, 1984). Some species which sleep in a group huddle together for warmth, e.g. wrens (*Troglodytes troglodytes*). This does not seem to be the case with starlings as they appear to space themselves out on their perching sites. However, the preference for evergreen trees in winter suggests that shelter from cooling winds is probably important in a roost.

Roosts have a social organisation. There is a concentration of heavy males at the centre. Lighter birds, predominantly younger birds and females, are found towards the outer edges of the roost (Feare, 1984). The night-time roost may reflect social relationships between birds which have been formed during the daytime feeding activities. However, the congregation of so many birds at the roost may be an important opportunity for birds to sort out their social status.

Birds leave the roost in the morning in a very characteristic way. At approximate three-minute intervals waves of birds leave and fly in all directions in an expanding circle. Light intensity may act as a trigger for this movement if birds of one wave always react to the same intensity, so the same birds leave at a similar time each morning (Feare, 1984). So even though birds from different social groups mix in the roost, a group may all leave together in the morning. Feare suggested that the main value of such roosts may be the transfer of information about food sites found during the day. The largest roosts occur in winter when such information would be particularly important. It is not known how information passes from bird to bird. Perhaps a well fed bird is followed the next morning by hungry ones. Experiments on another bird species, the quelea (*Quelea quelea*), showed that hungry birds did indeed follow well fed ones to food, but mysteriously in some experiments hungry birds headed for the good food sources *before* the well fed birds had shown them the way (Groot, 1980). The roost has the appearance of an enormous spy ring with birds from each feeding flock distributed throughout the roost looking for well fed birds and trying to discover their secrets!

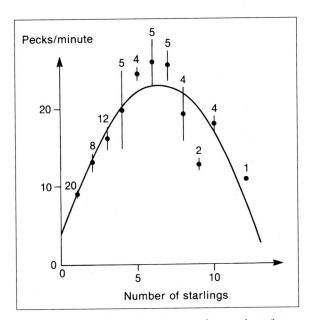

Figure 3.7 The relationship between the number of starlings feeding on crushed barley out of a cattle trough and their feeding rate measured as pecks per minute. The data are from Feare and Inglis (1979) and show that the starlings appear to feed best in flocks of 6–8 birds.

3.3.4 Reproduction

The European starling usually builds its nest in a hole in a tree. The process starts when the male starling finds and defends a suitable nest-hole. He then attracts a female by singing outside his nest-hole. When a potential mate arrives nearby, the male goes into his nest and sings from there. The male collects nest-making material of grass and twigs and puts them in the nest-hole. This also probably encourages a female to stay and start building the nest. The male also brings flowers to the nest as it is being built. Examples of nest decorations in Britain include daisy (*Bellis perennis*), lavender (*Lavendula* spp.), hawthorn (*Crataegus monogyna*) and daffodil (*Narcissus* spp.) (Feare, 1984). Once the pair are firmly attached to each other or **pair bonded**, the female finishes the nest. Starlings are facultatively polygynous: some males are monogamous, pairing with just one female; others mate with two females and so are polygynous (Bruun *et al.*, 1997).

Starling pairs usually copulate in secret, high up in the tree canopy. For the following few days, before the female lays the eggs, she is closely watched by her partner who thus prevents other males from copulating with her (Pinxten & Eens, 1997). The average number of eggs a starling lays in her nest is probably four to five. It is difficult to be certain about this as it has been found that sometimes another starling will lay an egg in the nest. Examination of blood proteins of chicks and the birds raising them has shown that this behaviour may be more common than was originally suspected (Evans, 1980). Birds are also known to remove eggs from nests and leave them on the ground (Feare, 1984).

These findings raise many questions. Do some females lay all their eggs in other birds' nests or do many females lay an extra egg somewhere else? Are the abandoned eggs removed by birds who have substituted their own egg in another bird's nest or are they eggs a nest owner has recognised as laid by another bird? Not enough is known about these incidents to answer such questions yet, but the subject is a fascinating one and may also shed light on the evolution of similar behaviour in birds such as the cuckoo (*Cuculus canorus*) which does not have a nest of its own, but lays all its eggs in the nests of other species of bird (see parasitism, Section 2.2.2).

3.3.5 Starlings and humans

Starlings have been associated with people for a long time. They are recorded as being kept as caged birds, presumably for their skill at mimicry, in Greece in the fifth century BC. Pliny (AD 23–79) was probably the first to conduct an autecological study when he described their behaviour of flying in flocks and feeding on crops.

Starlings seem to have increased in number as farming became more widespread. Starlings often feed in fields, for insects in summer and on crops in autumn and winter. As already mentioned, the birds in Britain, Australia and America also use human habitations as night-time roosting places. The first record of starlings roosting in a city comes from Dublin in Ireland in the 1840s (Potts, 1967). At first these birds slept in trees among the houses, but eventually they moved on to the buildings. The tall and ornate Victorian buildings were almost perfect as perches for starlings!

People have an ambivalent relationship with starlings. Because of their winter feeding habits they are considered a pest. They eat animal feeds and cereals and peck fruits such as apples and cherries, thus badly damaging crops. Their roosting is also discouraged close to habitation because of their droppings. However, their insect-eating habits in the summer have been seen as a benefit. In New Zealand hundreds of nest boxes are put out for starlings as they are thought to keep down insect pests in sheep-rearing areas (Feare, 1984).

Summary

(1) Autecology is the ecology of a single species in relation to its environment.
(2) Bracken (*Pteridium aquilinum*) is the most successful and widespread fern in the world.
(3) Bracken is a pest because it is poisonous to grazing farm animals and difficult to eradicate.
(4) Bracken is so successful because of its growth habit: its rhizomes are underground and so protected from fire, weedkillers and ploughing.
(5) Bracken is also successful because it has a number of chemical compounds which deter herbivores: cyanide, ecdysones, anti-feedants, thiaminase, tannins, lignins and silicates.
(6) Bracken has nectaries on its young leaves which attract ants. The ants then defend the leaves from other insects.
(7) The European starling (*Sturnus vulgaris*) is native to Europe and Asia, but has been introduced into several other continents.
(8) The starling is a communal bird which roosts and feeds in flocks.
(9) In autumn and winter, the starling eats seeds and fruits; in summer it switches to a high-protein insect diet to feed itself and its chicks.
(10) Starlings roost in trees and reeds, but also on buildings in cities where their noise and excreta are a considerable nuisance to people.

FOUR

Population dynamics

4.1 Populations and population change

Organisms do not live solitary lives in isolation from other members of their species. Usually they live in groups or, if normally single, they interact with other members of their species at various times in their lives. A group of organisms of the same species which live together in one geographical area at the same time is called a **population**. Within a population, all the individuals capable of reproduction have the opportunity to reproduce with other mature members of the group.

Although it is quite easy to define a population on paper, it is often difficult to study one in its natural environment (Crawley, 1990). First of all, the species to be investigated has to be identified. This is often easy, but may be difficult in some taxa or in particular circumstances. For example, insects often require microscopic scrutiny, one species in a mixed flock of birds may be difficult to pick out and count accurately, and plants sometimes hybridise with other species or occur as difficult taxonomic aggregates (see Section 6.2.5). The second problem is to decide what is an individual in the population. The bracken fronds in Figure 3.1, for example, are all members of one clone, that is one genetic individual, although they may look like a lot of separate plants (see also Section 4.7).

Perhaps the most difficult decision to make is what geographical area to study. What are the geographical limits of the population? Sometimes, if the population lives in a lake or woodland, or some other isolated habitat, then the decision may be relatively easy. Often, however, adjacent populations merge to a greater or lesser extent and it is not easy to decide on a study area.

Recently, ecologists have begun to talk of **metapopulations**. A metapopulation is a network of populations with occasional movement between them. The ecology and genetics of populations are viewed as a product of local population dynamics and the regional processes of migration, extinction and colonisation

(Husband & Barrett, 1996). From a conservation perspective, it can be an advantage if a species in an area exists as a metapopulation rather than as either a single large population or a number of small absolutely distinct populations. It means that if some of the populations go extinct, they may subsequently become re-established by occasional recolonisations from the surviving populations. This is now known to be the case for some butterfly species (Nève *et al.*, 1996). Theoretical models suggest that metapopulations will consist of a shifting mosaic of populations with only some of the available patches occupied at any one time. Perhaps more importantly, a threshold number of patches is needed: if the number of patches falls beneath a certain critical level, the metapopulation goes extinct because extinction exceeds colonisation.

Finally, the length of time required to investigate the population has to be decided. As the title of this chapter suggests, populations are always changing: they are dynamic. Often populations fluctuate considerably depending on the season, so studies at different times of year may give different results.

Figure 4.1 shows diagrammatically the four ways in which changes take place in the numbers of individuals in populations. These changes are what make populations dynamic. Populations gain individuals when young are born in the area (**birth, B**) or new members join the group from other populations (**immigration, I**). Immigration is most common in mobile species such as mammals, birds, fish, insects and so on. In plant species immigrants are usually seeds or spores. Populations lose organisms when they die (**death, D**), or when they leave an area to join another population (**emigration, E**).

If the gains due to the birth and immigration of individuals are equal to the losses due to death and emigration then the symbols representing these gains and losses can be used in the following equation:

$$B + I = D + E$$

This shows that the population is stable, neither increasing nor decreasing in number over a period of time. It is in **equilibrium**. If the population gains are

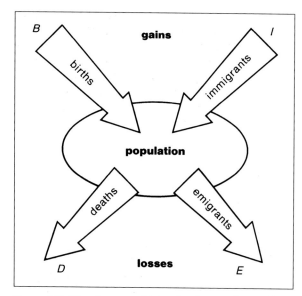

Figure 4.1 The changes which take place in the numbers of a population (the population dynamics) can be quantified using just four pieces of information. Gains in population numbers are due to births (*B*) and immigration (*I*): losses are due to deaths (*D*) and emigration (*E*).

greater than losses, that is if

$$B + I > D + E$$

then the population will increase in size, and if the losses are greater than the gains, that is if

$$B + I < D + E$$

then the population will decrease in size and may cease to exist.

The rate at which mature adults are replaced in a stable population which is in equilibrium will depend on the life span of the organisms. As we saw for bracken in Section 3.2.2. if an individual is capable of living several hundred years, then the death rate of adults will be low. It follows from this that the rate of successful replacement of adults will also be low. The birth rate may still be large, but in the case of a long-lived organism nearly all the young will die before they are mature. With bracken many spores are released but probably only a few prothalli produce established young ferns.

4.2 Dispersal of organisms

Organisms which leave one population and join another are said to have **dispersed**. This links *E* and *I* because if emigration from a population is high, this may mean that immigration (of organisms which have dispersed from other populations) is also high.

In plants the main dispersal units are seeds or spores. Fungi also produce spores as their main method of dispersal. Most seeds and spores do not travel very far and remain in the same population as their female parent. Some seeds, however, can travel very large distances with the help of special structures and can join other populations when they germinate. Such methods of seed dispersal include feathery attachments to catch the wind, sticky or hooked seeds to attach to animals and flotation devices. Examples of long-distance seed dispersal are discussed in Section 18.3.2. Fern and fungal spores are much lighter than seeds and are often carried very large distances by air currents. In rare situations whole mature plants can be moved to new sites, for example in aquatic habitats during floods. However, such situations are more likely to result in the transport of plant fragments which may root after deposition to start new plants.

Different animal species may disperse at different stages in their life cycle. Many marine organisms, such as crabs, corals and barnacles, have very small larval young which join the floating **plankton** and are carried by ocean currents until they mature into more stationary adults. In mammals and birds it is often the juveniles which emigrate as they begin to mature.

If a population has a high birth rate, due to favourable conditions, but a lower death rate, then the population will increase. It may become too large for the area to support so that increased emigration occurs. This happens regularly in populations of lemmings in arctic regions (there is more about population changes of lemmings in Section 5.3).

4.3 Dormancy

So far only dispersal in space (immigration and emigration) has been discussed. Dispersal can also occur through time. Plant seeds are a clear example of this as they can lie **dormant**, that is without germinating, for many years. When a seed does eventually germinate after a period of dormancy, many of the plants which were alive when the seed was produced may have died so that the population may be quite different. The store of seeds in the soil is called the **seed bank**. Often seeds will germinate on a patch of soil which has been disturbed. These seeds from the seed bank may have been dormant for many years but are stimulated to germinate when they come to the surface of the soil. Dormancy can be broken by a number of environmental features such as changes in temperature, carbon dioxide or water content of the area surrounding the seed. Different environmental stimuli will trigger germination in particular species. Annual weed seeds in fields of grain often germinate after the soil has been ploughed and seedlings may be found of species which have not grown in the area for many years.

Dormancy in seeds is very important to plant populations. Even if all the growing plants in a population die because of some disaster or adverse condition of the environment, seeds may survive in the soil and subsequently give rise to a new group of living plants. This is an extreme case, but a seed bank regulates the population size even under normal conditions. In good growing years many plants set seed and a large number of seeds are added to the seed bank. If a poor year follows, there will be few new seeds, but seeds germinating from the seed bank help to maintain the number of growing plants in the population.

4.4 The study of populations

4.4.1 The basic equation

There are many ways of studying populations and the changes taking place in them. All are concerned with the collection of information about numbers and ages of organisms. The study of population dynamics is called **demography**. The basic aim of any demographic study is to quantify the changes in a population by finding out the numbers of births, deaths, immigrants and emigrants. The changes in population size over a given time can be calculated by adding births and immigration to the original population number at time t (N_t) and subtracting the numbers of deaths and emigrants to give a new population size at time $t + 1$ (N_{t+1}). This sum is often represented by the equation:

$$N_{t+1} = N_t + B + I - D - E$$

Condensing population dynamics into a one-line equation like this can look rather bare because it does not indicate what is causing the changes. What affects the number of births? What causes deaths? Why do organisms leave or join the population and how? There is obviously a lot more to population dynamics than just counting individuals, but it is a good place to start!

4.4.2 Age structure in populations

One way to start a population study is to work out the age structure of the whole population at one point in time. To do this the size of the population and the ages of the individuals need to be discovered.

The number of individuals is easiest to estimate where the organisms are large and stationary, like trees. If the organisms are small and mobile, like insects or small mammals, then it will not be possible to count the population exactly. One method used to estimate populations under these latter circumstances, is called **mark**, **release**, **recapture**. First several individuals from the population are captured.

They are then marked in some way so they can be recognised again and released back into the population. The released organisms are given enough time to mix back with the unmarked ones, then more are captured. Hopefully some of the original marked individuals will be recaptured in the new group and the proportion of marked to unmarked individuals can be used to calculate the proportion of individuals in the

---

Mark

(1) A number of animals are caught and marked:
Number marked = M
(in this example $M = 10$)

Release

(2) M marked animals are released into the population and left for a period of time to mix with the unmarked animals.
Total population size = N
(N is the number we are trying to discover)

Recapture

(3) C animals are captured from the population:
Number captured = C
(in this example $C = 12$)

(4) Of these C animals, some will be marked. They are the recaptured animals. Number of recaptured animals = R
(in this example $R = 2$)

(5) The proportion of marked animals in the captured number is $\dfrac{R}{C}$ $\left(\text{in this example } \dfrac{2}{12}\right)$

(6) It is assumed that the proportion of marked animals in the whole population (N) is the same as the proportion in the captured animals.

so $\dfrac{R}{C} = \dfrac{M}{N}$ $\left(\text{in this example } \dfrac{2}{12} = \dfrac{10}{N}\right)$

so $N = \dfrac{CM}{R}$

(in this example $N = 12 \times 10 \div 2 = 60$)

Figure 4.2 A worked example showing how the mark, release, recapture method can be used to calculate an estimate of the population size (the Lincoln index).

whole population which were originally marked, and hence the population size. The estimate of the population size calculated by this method is called the **Lincoln index**. The stages in the calculation of the Lincoln index, with a simple example, are given in Figure 4.2.

For the mark, release, recapture method to give accurate results, certain assumptions which are made about the population and the effects of the investigation have to be true. The first assumption is that the marked animals, released after the first capture, mix in with the population after release, and have the same chance of being recaptured as any unmarked animal. For example, if the first set of captured animals are all active males, which are confident enough to enter traps for food, then the number of recaptured marked animals will be higher than it should be and the population size will be underestimated. Conversely, if most marked animals avoid entering a trap for the second time, only a few, or no, marked animals will be recaptured and the population size will be overestimated.

The population will also be overestimated if the markings wear off, or are destroyed, or if the marks lower the survival chances of the animal so that it is more likely to be spotted and captured by a predator before recapturing. In both these cases, fewer marked animals will be recaptured than should be.

It does not matter if animals die or emigrate from the population between the release and recapture, as long as the ratio of marked to unmarked animals leaving is the same as the ratio in the population as a whole. If this is the case, then the ratio in the remaining population stays constant and an accurate estimate of the size of the population at the time of release of marked animals will be obtained as long as no new unmarked animals enter the population.

New immigrants and newly born animals will all be unmarked and will lower the proportion of marked animals in the population. Their arrival increases the population size but, as long as no marked animals die or emigrate during the interval between release and recapture, this will not matter. The new arrivals will simply be included in the estimate which will be a measure of the population size at the time of the recapture.

Problems occur if both the number of marked animals decreases and the number of new arrivals increases during the study period. The ratio of marked to unmarked animals is decreased both by the loss of marked individuals from death and emigration and by the addition of new unmarked individuals. This will cause the population to be overestimated at the time of recapture. To minimise this effect, the second recapture should be carried out as soon after the first release as is possible (remember, the marked animals which are released have to mix

Figure 4.3 A felled larch tree (*Larix decidua*) showing annual growth rings.

evenly with the rest of the population before the second trapping occurs).

The other information needed to work out the population structure is the age of the individuals in the population. The age of some organisms can be determined fairly accurately because they possess some character which leaves a regular pattern in their structure representing some fixed time interval such as a year, lunar month or day. The best known example is annual growth rings in temperate trees. Each year as the sap rises in the tree in spring for the growth of new leaves, the tree grows a new layer of wood in a ring just beneath the bark. This spring wood has wide transport cells of tracheids or vessels. The diameter of new cells decreases during the following months until growth of wood stops in late summer. This produces a distinct tree ring for each year of growth and the rings can be counted, either in a felled stump or from a core of wood bored out of the living tree, so giving the age of the tree. Figure 4.3 shows a section cut across the trunk of a tree to show the annual growth rings. An autecological study using cores from black cherry trees to find out their ages is given later in this section.

Tropical trees do not show this distinct growth pattern because of the lack of winter dormancy followed by a spring burst of growth. They may show no variation in wood structure or have rings which correspond with flowering or leaf emergence. However these rings may be misleading if used to age the tree. For example some trees flower more than once a year (e.g. certain figs, *Ficus*) while others may miss one or more years, and leaf production can also vary seasonally with several bursts of growth a year (Creber, 1977). Leaf bases may also be useful for dating plants such as palms if they have a regular yearly leaf production as the scars left by the leaves can be counted.

Structures which can be used to determine the age of animals include annual growth rings in teeth, in fish scales, in shells of molluscs and in the otoliths of mammals. (Otoliths are small calcareous grains found in the inner ear which are used for balancing.)

It is usually impossible to determine the exact age of the organisms being studied, even to the nearest year. This is because most organisms do not have a structure which records yearly cycles or because study of such structures can only be done by killing or harming the organisms. In such cases the age of individuals has to be estimated in some way. For example the bracken (*Pteridium aquilinum*) mentioned in Section 3.2.2 was aged using the size of the clone based on a historic study of fire sites. Such estimates, however carefully they are done, only give approximate ages, and in many studies the organisms are grouped into broad categories based on the stage in the life cycle each individual has reached. This system of using stages rather than age has been studied and refined by Russian ecologists. They investigated many forms of plants and found that several stages of life can be recognised. These are seed, seedling, juvenile, immature, virginile, reproductive (young, mature and old), subsenile and senile (Gatsuk *et al.*, 1980). The same sort of categories can be applied to other organisms such as animals and fungi. Many ecologists argue that the stage an individual has reached is more important than its exact age. This is probably just as well as, in most cases, stage is much more easily determined.

Auclair and Cottam (1971) carried out an age structure study on the black cherry (*Prunus serotina*) populations found in oak forests in southern Wisconsin, North America. The Wisconsin forests are a relatively recent vegetation type which developed from savannah with standard trees in the mid-nineteenth century when the regular burning which had maintained the savannah ceased (see Section 16.4.3). Black cherry does not grow well in the shade of other trees (it is a **shade-intolerant species**) so it is most abundant in the gaps left by disturbance where trees have died.

Forty areas of oak forest were chosen for the study where black cherry was common but where different oak species (*Quercus rubra*, *Q. alba* and *Q. velutina*) were dominant. 854 cherry trees were cored one foot above ground level and the ages of the trees determined by counting the annual rings. The oldest cherry dated was established in 1876, but most of the trees (77%) were 25–50 years old at the time of the study. The results of this study can be seen in Figure 4.4. You can see that the age distribution of the trees is very uneven, and can be divided into four sections. There is a scatter of trees established before 1916. Then from 1916 to 1930 a steady number of trees were recruited into the forest populations. The most interesting period is the high, but rather variable, number of trees established each year between 1931 and 1941. After this fewer new trees seem to have established. The authors suggest that the two high records for 1955 and 1956 are an artefact of sampling some recently cut forests where black cherry had become abundant.

It might be expected that the age structure of the trees would follow a pattern which reflected age-related survival with many young trees, fewer middle-aged trees and fewer still older trees as most would have already died before the time of the study. The pattern seen in Figure 4.4 does not illustrate such age-related survival. Perhaps several older trees have died from shading or disease over the years, but this does not explain why the number of trees germinating in the 1930s was so high, nor why additions since then have been rather low.

Auclair and Cottam investigated many features of the forests and the autecology of the black cherry in order to decide what factors could have caused the patterns seen in Figure 4.4. They concluded that the increase of establishment of black cherries from 1916 onwards correlated with use of the wooded areas for cattle grazing. The sudden further increase in tree numbers in the 1930s could be due to germination of a seed bank. Black cherry seeds are distributed widely by birds and can lie dormant for some time. The very variable pattern for 1930 to 1940 seems to have been an effect of fluctuating weather conditions. There were several years of drought in that decade when fodder was in short supply, so that the woodlands were, as a consequence, rather heavily grazed. In some years, especially 1934 and again in 1936, farmers even had to cut foliage from the trees for their animals to eat. This considerable and repeated disturbance of the forests was probably responsible for the increase in black cherry numbers. In the early 1940s crop land increased and, probably as a consequence of this, grazing of woodland by cattle and horses decreased. The return to a regime with less woodland disturbance probably accounts for the marked drop in the number of trees which

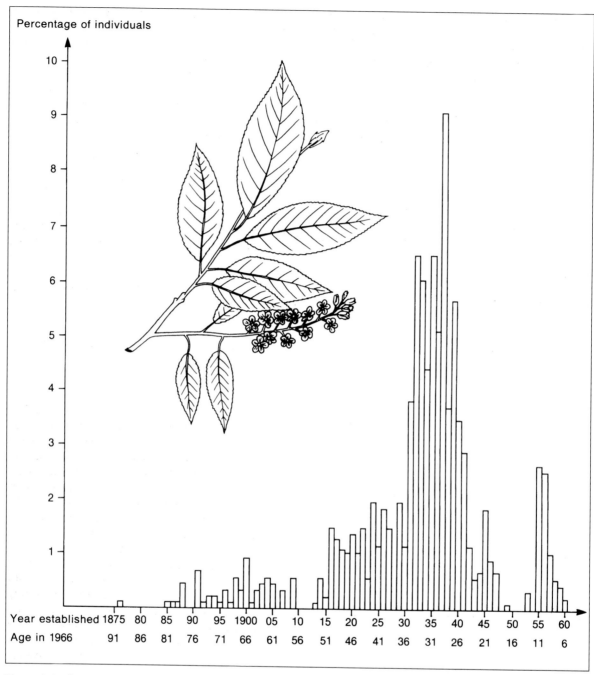

Figure 4.4 The age structure of black cherry (*Prunus serotina*) populations found in oak forests in Wisconsin. The age of each tree was obtained by boring a core from the trunk and counting the annual rings. (Data from Auclair & Cottam, 1971.)

established after 1941.

To determine the age or stage structure of an existing population at one point in time, as in the cherry example given above, is relatively easy and quick. This can be a valuable start towards the understanding of a particular population. Sometimes it may be the only information which can be gathered if time for study is short or if the population is difficult to reach or work with.

However, just to look at the population structure at one time is a limited approach. The conclusions reached about population structure are made as a static survey. The dynamics of the population, such as changes in births, deaths, immigration and emigration and the risks or advantages of different plant stages are not being investigated. A survey at one point in time of a population, like the *Prunus* trees, is looking at individuals which joined the population in many different years. Each plant will have been subject to different environmental effects. A very old tree could have germinated and developed under very different conditions from the younger specimens. It may be unwise to extrapolate present features of populations back into the past. However, one advantage of such studies is that, for long-lived organisms, a range of ages far greater than the life span of the ecologist can be studied in a short time. This is something not really possible using the next type of study.

4.4.3 The fate of a cohort

A different method of population study is to follow one group of individuals from the beginning of their life through to their death. Such a group of organisms from one population which are all roughly the same age is called a **cohort**. Cohort studies have the advantage over age structure studies in that they follow through the dynamics of part of the population. Such data are easier to use when forming ideas about major causes of mortality, reproductive success and so on.

As mentioned earlier, cohort studies are not suitable for very long-lived species as they would take too long to complete. For long-lived species a mixture of age structure and cohort studies is best.

4.4.4 Age at death

A third way to study population structure is to look at the remains of organisms. This is the only method available to palaeoecologists who have only fossilised remains to work on. Any organism which leaves a durable part that can be aged, such as shell or bone, can be investigated. This method provided one of the first life tables (see Section 4.5.1): for the Dall mountain sheep (*Ovis dalli*) in Alaska (Deevey. 1947). Murie (1944) collected 608 skulls and estimated their age at death using growth rings on the horns. The Dall mountain sheep data are used in Figure 4.7 and Section 4.5.3.

4.4.5 Long-term population studies

Some of the most valuable results in population dynamics have come from long-term projects. If several cohorts from successive years are followed through their life span, then the differences in mortality and movements between populations from year to year become more obvious. Individuals can be recognised either from natural marks (e.g. a tiger's pattern of stripes) or from marks or tags added by researchers.

An example of such a long-term project is a study of the red deer (*Cervus elaphus*) living on an isolated island called Rhum, off the west coast of Scotland. The study started in 1957 and by now most of the

deer in part of the island are known as individuals. They can be recognised using various features, such as ear shape, coat patterns and (in stags in summer) antler shapes. Several animals have also been marked with ear tags and collars.

The red deer has a short breeding season, called a **rut**, during which matings can be observed. By careful observation the most likely father of each calf can be determined by calculating back from its date of birth to the probable day of conception. The stag whose harem the mother was in at that time is most likely to be the father. As lineages and relationships are discovered, so the lives of parents and their offspring can be compared. By now extensive data have been collected about the deer population. This information provides the background for a detailed autecological study of the behaviour, reproductive success and population dynamics of the deer. You can find out more about the Rhum red deer in Clutton-Brock *et al.*, 1982.

4.5 Presentation of demographic data

4.5.1 Life tables

The numerical data collected during a population study can be presented as a table of figures known as a **life table**. Life tables usually represent data for a cohort (called **cohort life tables**) but can also be produced using age structure data. If age structure data are used, as for the black cherry data shown in Figure 4.4, it is assumed that rates of birth, death, immigration and emigration, measured at the time of the study, have been similar in the past.

Life tables usually include the numbers of organisms surviving or dying in a given time for various age classes and the corresponding survival and death rates. It is usually difficult to tell quickly by looking at a life table what is important in terms of population dynamics. Often, therefore, the data are also represented in graphic forms such as population pyramids (Section 4.5.2) and survivorship curves (Section 4.5.3).

One way of representing the life table information more diagrammatically is shown in Figure 4.5. The number of individuals at each stage of the life cycle is recorded in boxes connected by arrows. The figures in the triangles are the probabilities of any individual in the box representing one life stage surviving to the next stage. This example is the field grasshopper (*Chorthippus brunneus*). This species is an annual, so the probability of an adult (top box) surviving into the next year (lowest box) is nil, hence the figure on the arrow is zero. The workers who investigated this species did not have to consider emigration and immigration because the population was isolated. On

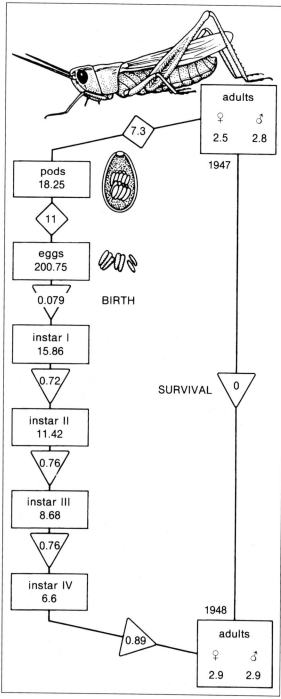

Figure 4.5 Diagrammatic life table of the field grass-hopper *Chorthippus brunneus*. Numbers in triangles are probabilities of survival to next stage: numbers in boxes are individuals per 10 m² surviving at that stage. (Adapted from Begon & Mortimer, 1986; data from Richards & Waloff, 1954.)

average each female laid 80.3 eggs so the 2.5 females per 10 m² laid 200.75 eggs per 10 m². As you can see egg mortality was very high; only 79 per 1000 survived to be first instars, hence the probability in

the triangle is 0.079. By the next year the 200 eggs had produced only about 6 mature adults.

4.5.2 *Population pyramids*

Another way of representing demographic data is as **population pyramids**. These are often used in human demography as they have the advantage over bar graphs such as Figure 4.4 in that males and females can be shown separately. The pyramid is really just two bar graphs back to back, with males to the left and females to the right of a central line. For humans, where population pyramids usually represent numbers for a whole country, the shape of the pyramid reflects birth and death rates. Bottom-heavy pyramids show increasing annual birth rates or high birth rates and high infant mortality. Top-heavy pyramids with more old people than expected show a decreasing birth rate and good adult survival. Figure 4.6 shows a very irregular or **gashed pyramid** from Russia. The indentations in the year groups 10–19 and 35–44 years of age are probably due to the Second World War. The older cohort was of fighting age during the war so the low numbers are probably due to increased mortality. The younger cohorts were born during 1939–49 when the birth rate was low because of the war when so many young men died.

4.5.3 *Survivorship curves*

Cohort life table data are often shown as a **survivorship curve** for a particular population. This is a graph showing the number of individuals which survive per thousand of population through each phase of life. In most studies, of course, there are fewer than a thousand individuals in the population, but this is standardised so that life tables can easily be compared. Using actual numbers would make this more difficult. The sort of survivorship curve this produces can be seen in Figure 4.7a. A second way of presenting survivorship curves is to use a log scale for the numbers of individuals. This type of semi-log plot is shown in Figure 4.7b. The same data has been used as in the normal plot in Figure 4.7a, so that you can compare the two curves obtained. The main advantage of using a semi-log plot is that any population where the proportion of individuals dying each unit of time (such as each week or each year) is a constant will plot as a straight line. On a normal plot, it would be difficult to tell this situation, which produces an exponential curve, from any other curve which is not exponential.

The general shape of the semi-log survivorship curve is therefore quite useful as it can indicate the population dynamics of the organism. Figure 4.8 shows the three basic semi-log plot survivorship curves found in populations. A straight line, like (i), shows that the mortality is constant whatever the

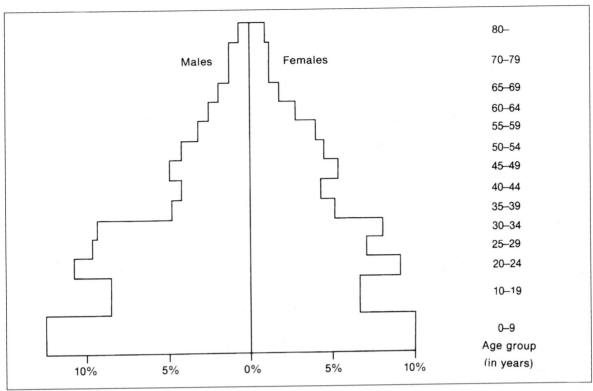

Figure 4.6 A population pyramid of Russian people taken in 1959. The data are divided into males and females. This pyramid is unusually irregular in overall shape and the data for males and females are not symmetrical. (From Hollingsworth, 1969.)

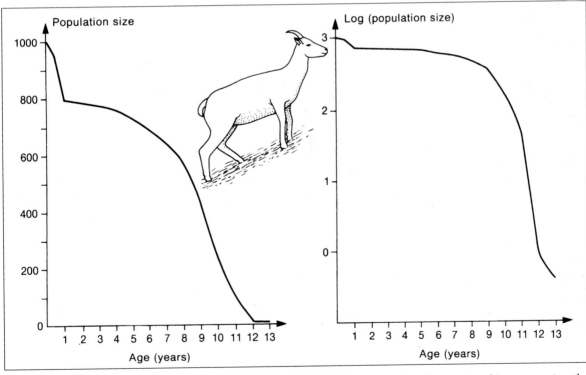

Figure 4.7a Normal plot survivorship curve for a cohort of Dall mountain sheep of survivors per thousand population against age in years.

Figure 4.7b Semi-log plot survivorship curve using the same data as Figure 4.7a, but a log plot of survivors per thousand population against age in years. (Data from Deevey, 1947.)

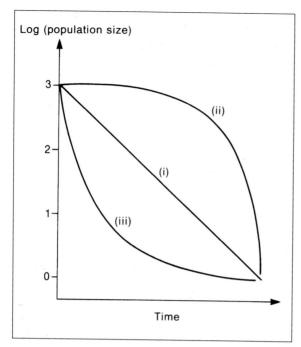

Figure 4.8 The three general shapes of survivorship curves on a semi-log plot which illustrate different patterns of population dynamics. (i) shows a steady mortality rate throughout life; (ii) shows good survival of young with high death rates only in old age; (iii) shows high mortality in the young with very little in adults.

age of organisms. A curve like (ii) shows that the highest mortality rate occurs in older individuals. Conversely, if the curve is like (iii), then mortality rates are very high in young individuals and relatively low in older ones. Populations rarely show such extremes as (ii) or (iii); in fact most populations show a wavy curve, like the one in Figure 4.7b.

The shape of survivorship curves can be used to tell at a glance what are the critical periods in the lives of individual organisms. Wherever the curve becomes steeper there is an increase in mortality which indicates some environmental or developmental effect on the population. The normal plot curves, such as Figure 4.7a, show the early pattern of mortality in the young most clearly, while the semi-log curves, like Figure 4.7b, show the pattern of mortality in the older organisms more clearly.

Often, as in the Dall mountain sheep graph (Figure 4.7b), the curve is steep to begin with, showing a high rate of mortality in the young. Another common feature is the steepening of the curve at the high age range as older animals also show a high death rate. In the case of the mountain sheep it has been shown that these steeper curves are due to predation by wolves which are more likely to catch immature and old animals. The middle-aged sheep which are fully grown and healthy can avoid the wolves (Murie, 1944). Less obvious kinks in the

curve may occur and these are not so easy to explain. They may represent mortality changes at the age the organism matures (possibly indicating increased death rate as animals have to disperse or learn to fight for mates) or other hazards such as mating, giving birth or poor environmental conditions.

Whatever the causes of death, drawing a survivorship curve can help to pinpoint critical periods in the life of the organism. This will then indicate areas worth further investigation in the population study.

4.6 Evolutionary strategies

4.6.1 Strategies as shown by survivorship curves

The general forms of survivorship curves shown in Figure 4.8 can indicate what evolutionary strategy a species has and how population numbers are maintained. Individuals in populations with curves like (ii) usually have few offspring. These offspring are well cared for so that their chance of survival is high. Such strategies occur in birds and large mammals including humans. Species with curves like (iii) usually have large numbers of offspring, most of which die before they reach maturity. Most plants, fungi, fish, amphibians and invertebrates have this strategy.

4.6.2 r- and K-strategies

Another way of classifying evolutionary strategies was suggested by MacArthur and Wilson (1967). They applied the terms **r-selected** and **K-selected** to populations and this system has been widely used since then. The initials r and K come from the **logistic equation** for describing the actual rate of growth of populations (R):

$$R = \frac{dN}{dt} = rN\left(1 - \frac{N}{K}\right)$$

where:

r = the maximum rate of increase of the population;

K = the number of organisms able to live in the population when it is in equilibrium, that is the **carrying capacity** of the population;

N = the number of organisms in the population at time t.

The importance of the logistic equation in describing population growth and regulation is discussed in Section 5.1.2.

From the equation you can see that an r-selected population is one in which the maximum rate of increase (r) is important. An r-selected population can take advantage of a favourable situation by having the ability to increase population size rapidly. This means having many offspring which under normal circumstances die before reaching maturity but which may survive if circumstances change. Hence r-selection is associated with the type (iii) survivorship curve of Figure 4.8.

Similarly, a K-selected population is associated with a steady carrying capacity (K). K-selected populations are less able to take advantage of particular opportunities to expand than are r-selected populations. They are in general more stable and less likely to suffer high mortality rates of immature individuals. Usually K-selected organisms have few, well cared for young and tend to be associated with type (i) and (ii) survivorship curves.

The two classes r- and K-selection described above are really the extreme ends of a continuum. It was soon realised (Pianka, 1970) that each end is associated with a whole group of characteristics of life which fit together into a particular evolutionary strategy. Some of the more important ecological characters associated with the r- and K-selection strategies are listed in Table 4.1.

Table 4.1 The characteristics of populations which fall at the extremes of the evolutionary continuum known as r- and K-strategy.

| Character | Continuum | |
	r-strategy	K-strategy
Population size	Variable Usually below the maximum the environment can support	Constant In equilibrium Near the maximum the environment can support
	Emigration common Recolonisation high	Recolonisation uncommon
Mortality	Often high Variable Not density dependent*	Often regular Density dependent*
Survivorship curve (semi-log plot)		
Competition	Poor competitor	Good competitor
Life span	Short	Usually more than one year to very long
Environment	Variable and unpredictable	Constant or variable but predictable

*see Section 5.2.1

Table 4.2 The percentages of the four genetic lines of *Taraxacum officinale* found in three habitats with differing amounts of disturbance. (Data from Solbrig & Simpson, 1974.)

| Habitat | Percentage of genetic lines present | | | |
	A	B	C	D
I Trampled, dry grassland, mown every week, bare ground present (high disturbance)	73	13	14	0
II Moderately trampled grassland mown every week, in shade (medium disturbance)	53	32	14	1
III Wet, shaded meadowland mown once a year in spring (low disturbance)	17	8	11	64

In ecological systems it is probable that populations are constantly undergoing r- or K-selection. Their position on the r–K continuum depends on the strength of selection pressures and where they balance out. If some of the population enter new conditions suitable to a more r- or more K-oriented strategy then a new balance will be set up. It is difficult to show that this happens in natural situations. There is some evidence for r- and K-selection in the dandelion (*Taraxacum officinale*) which is a perennial herb which grows in grasslands. Solbrig and Simpson (1974) studied this species in Michigan, North America. The dandelion usually produces seed without the need for sexual reproduction (it is **apomictic** – see Section 6.2.5) so whole lineages are really clones of one genetic type. Production of new genetic combinations by sexual reproduction, to form a new genetic line, is a rare event. Thus there are usually only a few such lines in any one locality. Solbrig and Simpson identified just four in their study and they labelled them A, B, C and D. They found these dandelion types in three grassy areas which were close together, but had different amounts of disturbance. The habitats are listed in the first column of Table 4.2. You can also see from the table that they found different proportions of the four genetic lines at each site.

It can be predicted that the forms found most frequently in the trampled, highly disturbed land (mainly type A) would be r-selected, while those in the almost undisturbed meadowland (mainly type D) were likely to be K-selected. To investigate the differences between genetic lines, Solbrig and Simpson took 100 plants from each site and grew them in a garden experiment. They also collected seed from the four lines and grew the progeny in a number of situations to investigate their competitive ability.

In the experiments dandelions were grown under a number of different conditions. These included plants in gardens in open weed-free soil, plants grown individually in pots and several plants grown together in pots and therefore in competition with the same and different genetic lines. The resulting growth of the plants differed considerably depending on the situation. For example, when grown in competition, type A produced an average of 0.24 flowers per plant per year, while in open, competition-free soil it produced up to 35.4 flowers per plant per year! However, in each situation type A produced more flowers than did type D in the same conditions. A summary of their results for types A and D is shown in Table 4.3.

Table 4.3 The results of various experiments on the *Taraxacum officinale* lines A and D. (Data adapted from Solbrig & Simpson, 1974.)

Character	A	D
Time of first flowering	In first year	In second year
Flowers per plant (average)	More than D (0.24–35.4)	Fewer than A (0.04–27.7)
Seeds per flower head (average)	About 100	About 200
Mean weight of seed	0.32 mg	0.44 mg
Total numbers of seeds produced	4 to 5 times as many as D	

If you compare the results in Table 4.3 with the characters associated with the *r*- to *K*-strategy continuum in Table 4.1, you can see that type D, which has delayed flowering and a higher competitive ability than type A, falls at the *K*-strategy end of the continuum, while type A falls at the *r*-strategy end. This fits the prediction made from the disturbance levels in the two habitats.

4.7 Modular organisms

It was mentioned in Section 4.1 that it is sometimes difficult to decide where one individual ends and another begins. For the bracken fern in Figure 3.1 each frond could be counted as an individual unit, yet they are all part of a single clone. Many organisms are easy to recognise as single individuals because they are limited in their growth. This is the case for organisms such as arthropods, vertebrates and molluscs. However, some organisms do not have such a regulated form; they are capable of increasing in size by growing an unspecified number of similar additional parts. These organisms which consist in total, or in part, of many repeatable units are called **modular organisms**. Most plants and a variety of animals show some form of modular growth. Figure 4.9 shows a coral which is a modular organism. Each

Figure 4.9 Part of a coral (*Gonopora*) showing the individual polyps.

Figure 4.10 A solitary oak tree (*Quercus robur*).

of the little bodies or polyps has its own mouth, tentacles and gut, and the polyps are connected by living tissue and have all grown from a single larva. Although the coral has a very precise structural form, the number of polyps can vary in a single coral

Box 4
Human population dynamics

The longest detailed record of population changes is for humans (*Homo sapiens*). Using a mixture of archaeology, historic records, census data and the anthropology of recent societies, demographers have pieced together human population changes since Palaeolithic times two million years ago.

The term 'population' is used rather differently in human demography from its strict ecological definition. It usually represents the number of individuals in a country, or sometimes in the world. Many countries probably contain several ecological populations, though this was more likely when travel was harder than it is now. However, the social structure of humans and the patterns of world trade (even in prehistoric times flints were traded) and travel make the ecological definition of population difficult to apply.

Our estimates of world population are most accurate for the last few hundred years because detailed documentation exists. Before this, estimates have to be based on interpretation of fragments of evidence. Before written records the demographer has little to go on except archaeology and estimates based on recent anthropology. The study of recent hunter-gatherer and settled agricultural societies can be used to indicate the density of population which was likely to have occurred in such societies in the past.

From Lower Palaeolithic times the growth rate for the population has been estimated as only about 0.001%. About 10 000 years ago the number of people in the world was approximately 10 million. At the beginning of the Neolithic period, there seems to have been a rapid increase in the population up to about 50 million. This increase in population growth rate to 0.1% coincides with a changeover from hunter-gatherer societies to more stable agricultural ones (see Cohen, 1980; Hassan, 1980). Figure 4.11a shows the world population for the last 10 000 years on a conventional semi-log plot. If the data are represented on a log-log scale like Figure 4.11b which covers the last million years, then an unusual graph is obtained. The graph consists of three steps which indicate three main periods of population increase. The first is the suggested increase in population caused by the development of tools by Palaeolithic humans. The second corresponds with the change from hunter-gatherer societies to settled agriculture (the Neolithic 'agricultural revolution'). The third burst in population growth is well documented as it is associated with the recent industrialisation of the eighteenth and nineteenth centuries (Deevey, 1960).

Attempts to answer the question 'How many people can the Earth support?' range from just one billion to over one thousand billion (Cohen, 1995). Producing a more precise answer is difficult as it relies on estimates about increasing food production, on the one hand, and increasing loss of agricultural land through soil erosion and other causes on the other (Olsson, 1993; Biswas, 1994). In 1996 there were approxi-

mately 5.8 billion people, increasing by 80 million a year – that's more than 200 000 extra people every 24 hours. A more ethical question is 'How many people should there be?'. From an ecological perspective there is much to be said for having significantly fewer people than there are now. Even if the world population stabilises at its present level the pressure humans exert on the rest of the biosphere is likely to intensify as more countries make the shift from agricultural to industrial or post-industrial economies. A third question is 'How many people will there be?'. The 1996 United Nations median prediction was 8 billion people by the year 2020. An independent prediction by demographers at the International Institute for Applied Systems Analysis in Laxenburg, Austria suggests that the world's population will peak at 10.8 billion in 2080 (Coghlan, 1996). However, previous attempts to predict future populations have often proved wide of the mark.

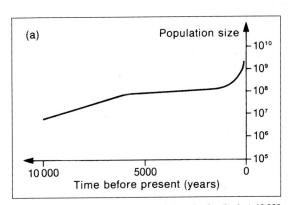

Figure 4.11a The world human population size for the last 10 000 years on a semi-log plot.

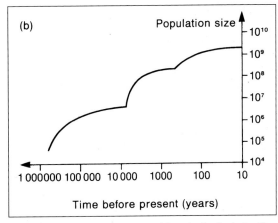

Figure 4.11b The world human population size for the last million years on a log-log plot. (Data from Deevey, 1960.)

as it grows and from one coral to the next. A similar situation occurs in plants which grow by sending out runners or root suckers to form a clump of connected stems.

A tree, like the oak tree in Figure 4.10, is easily identified as a single individual. It has a single trunk and is not surrounded by other stems of the same species. However, the tree shows modular growth because of its many branches and twigs. The tree has a large number of repeated 'units' of structure: branches; twigs; twiglets with leaves; twigs with flowers or fruit. The number of branches, twigs, flowers and fruits on the tree will vary with age, stage of growth, environment and so on. It may be possible to estimate the numbers of the various parts, but it will not be possible to predict them accurately just by knowing the species.

In organisms with modular growth, it is possible to consider these repeated units as if they were individuals. The twigs, leaves or flowers can be counted and treated as a population. The birth (= growth) and death (= shedding) of such units can be studied to show the changes in the tree during the year or throughout the life of the tree.

Summary

(1) A population is a group of organisms of the same species which live together in space and time and so have the potential to interbreed.

(2) The numbers of organisms in a population are increased by birth or germination and immigration.

(3) The numbers of organisms in a population are decreased by death and emigration.

(4) Plants emigrate or disperse as seeds or spores; in animals the young or juveniles usually disperse.

(5) Plants can disperse in time if seeds join the dormant seed bank.

(6) Demography is the study of population dynamics.

(7) Population structure can be described in terms of the age of organisms or the stage of development they have reached.

(8) A population can be studied at one point in time or the fate of a cohort can be followed.

(9) Demographic data can be shown as life tables, population pyramids or survivorship curves.

(10) The shape of a survivorship curve indicates the evolutionary strategy of a population.

(11) Some populations are r-selected: they have a high rate of population increase close to the maximum rate of increase (r) and can thus take advantage of changing conditions.

(12) Some populations are K-selected: the population is maintained close to the carrying capacity (K) and does best in stable or predictable conditions.

(13) r- and K-strategies lie at either end of a continuum.

(14) Different populations of the same species can be more r- or more K-selected, depending on their environment.

(15) Many organisms, including plants and corals, show modular growth; they possess repeated structures, the number of which varies with the age or stage of the organism and the season.

(16) Over the last million years, the human population has shown three major bursts in the rate of population increase: the first accompanied the development of Palaeolithic tools; the second marked the Neolithic settlement and shift to agriculture; and the third coincided with industrialisation.

(17) From an ecological perspective there is much to be said for having significantly fewer people on the Earth than there already are.

FIVE

Population regulation

5.1 Population growth

5.1.1 Population growth without regulation – exponential growth

As mentioned in Chapter 4, populations are dynamic: they are always changing. Organisms are born, die, immigrate or emigrate. But what would happen if a population grew in size without stopping? Darwin considered this situation and wrote 'The elephant is reckoned to be the slowest breeder of all known animals . . . it breeds when thirty years old, and goes on breeding till 90 years old, bringing forth three pairs of young in this interval; if this be so, at the end of the fifth century there would be alive fifteen million elephants, descended from the first pair.' (Darwin, 1859, Chapter 3.) But this does not happen: we are not crowded out by elephants, mice or any other species. In other words populations do not usually increase to such an extent that they are out of balance with the rest of the environment. There are exceptions, where species suddenly increase in numbers and a plague follows, but numbers usually return to lower levels quite quickly (see Sections 5.3 and 19.5).

What stops elephant populations, or any other species, from increasing continuously? The sizes of populations are often **regulated** by environmental factors. The main factors known to be involved in population regulation are described in Section 5.2. To understand the effects of population regulation we must first see what happens to a population when no controlling influence exists.

To understand what can happen in an unregulated population let us consider a very simple example of a hypothetical population. We can take as a start one breeding pair of animals which gives birth to two offspring every year, and each offspring can reproduce when it is one year old. From this we can draw up a table of population numbers as shown in Table 5.1.

The data for the growth of this hypothetical population can be shown graphically as in Figure

Table 5.1 The increase of population size in a hypothetical population which starts from a single breeding pair that have two offspring each year. The offspring are capable of breeding after one year and also have two young each year.

Year	0	1	2	3	4	5	6	. . . n
Breeding pairs	1	1	2	4	8	16	32	2^{n-1}
Offspring	0	2	4	8	16	32	64	2^n
Population	2	4	8	16	32	64	128	2^{n+1}
Log$_{10}$ population	0.3	0.6	0.9	1.2	1.5	1.8	2.1	$0.3(n+1)$

5.1a. As you can see the curve of population size quickly begins to rise very steeply. This population is undergoing **exponential growth**. If each pair of animals had given rise to four individuals each year, then the population growth would have been even faster. If we display the population increase as a semi-log plot (shown in Figure 5.1b) then the data lie on a straight line. This is because the **intrinsic rate of increase** (the maximum number of offspring born per individual in the population) is a constant. In the example above, the intrinsic rate of increase is one per year. If four offspring were born for each pair of animals each year then the line would still be straight on a semi-log plot, but it would be steeper: the intrinsic rate of increase would be two per year.

Real populations seldom show exponential growth rates for many generations. A species introduced into a new and favourable area (like the introductions described in Section 19.5) may well show exponential growth for a while, but some environmental factor usually begins to regulate the population when it reaches a certain size.

If you look back at Figure 4.11a (in the box on p. 35) you will see what appears to be an exponential curve for human world population. However, this is a semi-log plot! The population is plotted as a log to the base 10, just as in Figure 5.1b. In other words, in the human population as a whole, even the *rate* of increase is increasing. This incredible rate of growth

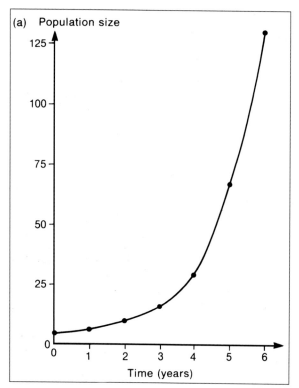

Figure 5.1a The increase in size of a hypothetical population through time using the data in Table 5.1. The population shows exponential growth.

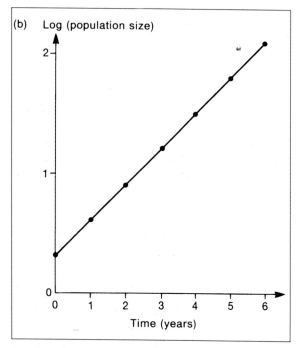

Figure 5.1b A semi-log plot of the data from Table 5.1 with the population size plotted as $\log_{10}$ population size.

in the human population cannot continue indefinitely. As already mentioned in Chapter 4, we will run out of food and the space to grow it in. In other

words at some stage population regulation will occur, either because we impose it ourselves, or because other environmental factors take over. Even the solution most favoured by science fiction writers – colonisation of other planets – would require unattainable efforts. Even now about 200 000 people would have to be launched into space *each day* just to keep the population steady!

5.1.2 *Simple population regulation – the logistic growth curve*

Often a population invading a new area where space and food are plentiful will undergo exponential growth to begin with. Soon, however, the external constraints of the environment influence the rate of increase which begins to decline. Eventually, in most cases, the population size stabilises. The number in the population at this stage is called the **carrying capacity** of the environment.

Populations of protoctists kept in laboratory conditions show this type of population increase and levelling off. A classic experiment of this sort was conducted by Gause (1934). He put 20 individuals of one species, either *Paramecium aurelia* or the larger *P. caudatum*, into a glass tube with a standard amount of bacteria as food. Every day the *Paramecium* were given food and every other day the mix was given an extra rinse to make sure toxins did not build up in the tube.

The population in each test tube grew in size during the experiment and the resulting population growth curves are shown in Figure 5.2. The *Paramecium* multiplied by **binary fission**, that is each organism divided into two without sexual reproduction. At first the shape of the curves in Figure 5.2 are like that of the exponential curve of Figure 5.1a: the *Paramecium* were dividing close to their maximum population growth rate. As the *Paramecium* numbers increased the rate of population increase slowed down. After four or five days the number of *Paramecium* in the tube no longer increased: the population had reached its carrying capacity. If the larger species, *P. caudatum*, was in the tube then the carrying capacity was smaller (about 400 *Paramecium*/cm³) than if *P. aurelia* was present (about 1000 *Paramecium*/cm³). However, the volume of protoctists in each case was approximately the same.

The shape of the curves for population growth in *Paramecium* seen in Figure 5.2 are typical of logistic growth. For *Paramecium* the rate of increase of the populations rises to a maximum after about two days and then decreases to zero when the carrying capacity is reached. It is this rise then fall in the rate of increase of population numbers which is predicted by

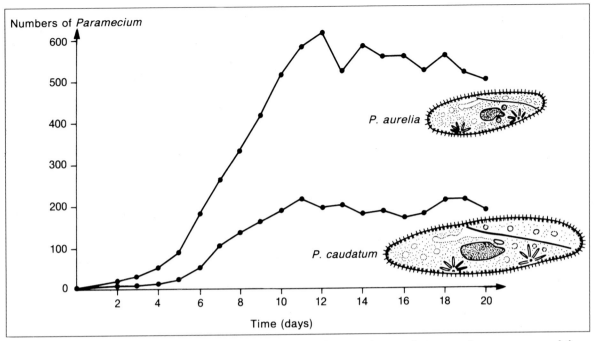

Figure 5.2 Population growth curves for two species of *Paramecium*. The population numbers are means of three replicate experiments for each species. (Data from Gause, 1934.)

the **logistic equation** (see also Section 4.6.2):

$$\frac{\mathrm{d}N}{\mathrm{d}t} = rN\left(1 - \frac{N}{K}\right)$$

where: $\dfrac{\mathrm{d}N}{\mathrm{d}t}$ = rate of growth of population;

    r = intrinsic rate of increase per individual of population;

    K = carrying capacity of population;

    N = number of organisms in population at time t.

This equation can be represented graphically in two ways. If the number of organisms in the population N is plotted against time t (as in Figure 5.3a) then a curve very similar in shape to the population increase in Gause's *Paramecium* is obtained (compare Figure 5.3a with Figure 5.2). This curve is sometimes called a **sigmoid curve** as it is roughly S-shaped (sigma is the Greek letter S). Most populations undergoing logistic growth tend to overshoot the carrying capacity (K), after which numbers oscillate about this number. This feature can be seen in Gause's experiments in Figure 5.2.

If the rate of increase dN/dt, not the increase itself, is plotted against time a curve like Figure 5.3b is obtained. When N is very small (as at the beginning of Gause's experiments or during colonisation) N/K is very small so $(1 - N/K)$ is approximately one and the logistic equation approximates to:

$$\frac{\mathrm{d}N}{\mathrm{d}t} = rN$$

This is the equation for exponential growth. As the carrying capacity is reached, N/K tends to one so the logistic equation becomes:

$$\frac{\mathrm{d}N}{\mathrm{d}t} = rN\,(1 - 1) = 0$$

In other words, the rate of population growth becomes zero.

Both these curves, Figures 5.3a and 5.3b, can be referred to as 'the logistic curve' which may be rather confusing, but as already mentioned, both represent different features of the logistic equation.

5.2 Factors which regulate population size

5.2.1 Types of regulation

As has been already noted, few species show exponential growth for many generations. Populations seem to fail to achieve their maximum potential for growth. The regulation of population size limits population growth. The position of a population on the $r - K$ continuum (see Section 4.6.2) will be close to the optimum of potential population increase and carrying capacity for that species in the environment where it lives.

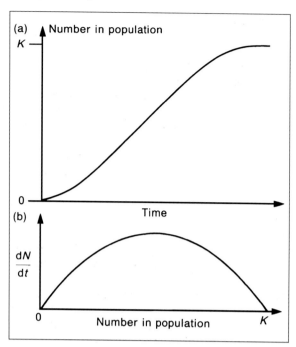

Figure 5.3 Two ways of illustrating the changes in a population undergoing logistic growth where $dN/dt = rN(1 - N/K)$. Figure 5.3a is the change in population size with time. Figure 5.3b is the rate of change of population growth with population size.

The environmental factors which are responsible for regulating populations are often the ones that affect the growth of the individual. This is because populations are made up of individuals. Population increase depends on the reproductive fitness and life span of these individuals. Factors such as energy and nutrient availability, flood, drought, predators and disease will all affect population size.

Environmental factors can affect populations in various ways depending on whether the factor is variable or not, and how predictable the variation is. It is worth classifying these environmental factors here so they can be borne in mind for the rest of the section.

Constantly limiting factors

These are always in short supply, but are relatively constant so that a population is limited to a certain fairly constant size by the factor. Individuals may have to compete for the resource. For example, plants compete for space and light, birds for nesting territories and many heterotrophs for food. These factors do not usually produce large changes in population.

Variably limiting factors

These are variable but predictable, like seasonal drought and cold, or variation in food availability. They are only an influence at certain times of year

when they may cause the population to crash. Some species avoid such factors by taking evasive action. For example, some birds migrate and deciduous trees drop their leaves.

Unpredictable factors

These change in an irregular way so population dynamics may vary considerably over time. Unpredictable factors include weather effects, grazing or predation pressure and disease levels.

Unpredictable factors tend to alter population size in a fairly haphazard way: a volcanic eruption, fire, or virulent disease epidemic may kill most, or even all of a population. If the population has died out completely, then new individuals may migrate into the area and start a new population. It can be argued that such limiting effects are not really population regulators. This is because such factors are not constantly or repeatedly limiting population size, but merely damaging it in a one-off event, after which the population will recover up to a level where real regulating effects take over. **Population regulation** implies that one or more factors are always involved in controlling population size, whether they are constant or variable in operation.

Many factors have an increasing effect on a population the more closely spaced the individuals in the population become. Such factors are said to be **density dependent**. Density-dependent influences often include resources that are in limited supply such as space, water and nutrients. The more individuals there are competing for the resources, the higher the mortality will be in the population. Disease can also be density dependent as epidemics spread more rapidly if organisms are close together and in frequent contact rather than sparsely distributed. Factors which do not show this effect are said to be **density independent** and include fire, volcanic eruption and other catastrophes, extreme weather conditions like heavy frosts and high winds, and toxic pollution.

Some of the major factors which alter population size, both unpredictable hazards and population regulators, are discussed in this section. Spatial heterogeneity may mean that more than one factor regulates a single population and that density-independent factors can be important in population regulation. Experiments or observations of natural populations are included which give examples of how these factors can regulate populations. Although these are examples where one main factor is involved in regulation, the size of most populations is influenced by a whole complex of interactive factors. Different factors may be important at different times of year, in different years, or on individuals of different ages and in different parts of the geographical range of the population.

5.2.2 Space

The most basic requirement an organism has is space. Protoctists and most animals, except for stationary ones like scale insects, corals or barnacles, are not restricted by lack of physical space. They may have problems when finding the right place to breed, but this is a form of population restriction which falls under the heading of territories (see Section 5.2.4).

Plants, on the other hand, are often crowded closely together. As each individual grows it spreads out roots and branches and takes up more space. Unless it is an isolated plant, at some point in growth it starts to interfere with or be interfered with by neighbouring plants of the same or different species. Thus a mature plant will probably have displaced several others.

Tree plantations are good examples of groups of plants of the same age and species growing up and competing for space. The results of several studies of Ponderosa pine (*Pinus ponderosa*) in plantations (Meyer, 1938) are recorded diagrammatically in Figure 5.4. The data are a combination from plantations of several ages from seedlings to adults, but we can go back a stage further to imagine the situation for seeds too. To consider how a stand of trees might develop, start at the right-hand side of the graph in Figure 5.4 with the seeds which have a very low mean plant weight and are small and well spaced out. At first the germinating seeds are not close enough to one another to compete for space. The seedlings can therefore increase in volume (and weight, which is easier to measure and so is recorded here) while remaining at the same density. This is represented by the dotted vertical line on the lower right of Figure 5.4.

As the seedlings get larger, however, they begin to compete for space. Some plants are out-competed and die as others take over their space. Because plants are dying the density of plants (that is the number of plants per unit area) falls and the line begins to slope upwards to the left. This process of plant death and takeover is called **self-thinning**. As the pines grow, each one covers more ground area at the expense of other individuals which die. As each successful individual grows it increases the surface area of the ground it covers. The increase in ground surface area (measured in units squared, e.g. m²) covered by each tree means fewer trees per unit area.

As each tree increases its area, it also increases the volume it fills (measured in cubic units). The plant matter in the tree is proportional to its volume and therefore the weight of the tree is also proportional to the volume. This means that if the data on increase in weight and decrease in density are represented on a log–log plot (as in Figure 5.4), then the values are predicted to fall on a straight line with a gradient of $-\frac{3}{2}$. The $\frac{3}{2}$ figure is the ratio of the power of volume (units cubed) to area (units squared). The minus occurs because the density is falling as volume and mass are increasing, so the slope is a negative one. This gradient of slope is known to occur in about 80 plant species (White, 1980). Because it seems to be a common occurrence and follows a physical character of volume to area, it is called the $-\frac{3}{2}$ **power law**.

The more crowded seedlings are, the larger the number of deaths in a given area. If there are only a few seedlings, then those individuals can grow very large before any self-thinning occurs. The amount of self-thinning where space is the limiting factor is related to the initial number of organisms present and is therefore density dependent.

Plants can sometimes avoid the consequences of the $-\frac{3}{2}$ law by not increasing their mass with age. If they remain small when space is limiting they avoid the intense competition for space. Often a plantation stagnates and no more timber growth occurs until the trees are thinned by selective felling. The floor of

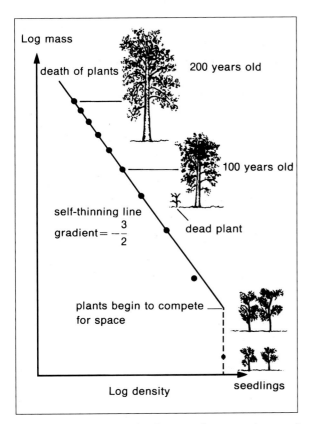

Figure 5.4 The relationship between density and mass of plants in a population of Ponderosa pine (*Pinus ponderosa*). The dotted line shows the phase when seedlings are too small to compete with each other for space at the density they are planted. The solid line shows density-dependent growth, where the plants can grow in size only at the expense of others which die, hence the density of plants decreases. (Data are based on nearly 850 sample plots, Meyer, 1938: and the drawing based on White, 1980.)

a deciduous woodland often has many small oaks which look two or three years old but which are actually much older. When a space appears in the canopy above, due to the death of a mature tree, the seedlings close to the gap begin to grow upwards. One or more of the waiting seedlings may succeed in reaching the canopy space above.

5.2.3 Food and water

As was mentioned in Section 2.2, both heterotrophs and autotrophs require energy, nutrients and water to survive, grow and reproduce. Food for a photo-autotroph means mineral nutrients, and carbon dioxide and the presence of enough light to syn-thesise organic molecules. Lack of food or scarcity of particular nutrients can reduce the growth rate of an organism, or even cause failure to reproduce or premature death. Drought conditions can cause similar hardships. In England, the beech tree (*Fagus sylvatica*) is sometimes badly affected by drought. Hot dry summers prevent the developing seeds from ripening fully, so that they do not germinate. In long dry periods, like the 18-month drought of 1975–6 which ended after a very hot summer, many beech trees died outright or were weakened so that they subsequently caught fungal diseases or blew over in strong winds.

It is not easy to see the direct effects of shortage or abundance of food and water on plant population dynamics. As with the beech example, the influence of a shortage, such as drought, may only be appar-ent years later when weakened plants are damaged by some other factor. Most plants are autotrophic, so they will always be able to fix energy from the Sun. Sunlight does vary in intensity through the year in many latitudes (see Section 9.2.4), but the changes are very predictable, so rather than regulating popu-lations sunlight affects the whole life style of species. What is more important is the space required for plant leaves to catch the light and this was discussed in Section 5.2.2. Carbon dioxide is also unimportant in population regulation. The gas can limit plant growth, in fact some commercial crop growers create a high carbon dioxide atmosphere in their green-houses to increase growth rates. However, the levels of carbon dioxide in the atmosphere will limit all plants to the same extent, and other, less predictable factors are likely to be more important. The role of carbon dioxide and its effect on climate and plant growth are considered in Sections 13.2, 13.3 and 21.4.

The vital nutrients for plant growth, especially nitrogen, potassium and phosphorus, are often in short supply in soils. Species are known to be affect-ed differently by various soils. Many species of low-growing plants favour poor soils where they are not overgrown by the lush foliage of the larger plants

that grow well in fertile soil. Hence soil nutrient content may limit the population size of certain species, but population changes from year to year will be regulated by other factors. One of the few times when soil nutrient content changes sufficiently to affect population dynamics is after a fire. The fire clears vegetation and converts it into high-nutrient ash and smoke.

For some plants, lack of nutrients is less important than lack of water. Some plant populations are com-pletely regulated by rainfall or lack of it: many desert or semi-desert areas 'bloom' with vivid carpets of flowers after heavy rain which may only occur every ten years or so. The dormant seeds in the dry soils germinate and grow rapidly after the rain, then flower, set seed and die as the area again dries out. In other plant communities rainfall is greater and more frequent, but occasional dry periods can be a consid-erable problem. The effects of drought conditions are obvious when plants wilt or die outright, but more subtle effects of water shortage include failure to ripen seed (as in the beech) or loss of leaves which weakens the plant as it has to replace them later.

The direct effects of food and water on the health of animals, and on their potential to raise young, are more obvious than for plants and it is easy to see how these influence population size. In fact the lack of sufficient food to maximise **reproductive potential** may be the most important regulator of population size in animals. If food is rare, animals have to spend considerable time finding it. If it is of poor nutrient quality then there may not be enough time available to eat a sufficient amount. Food availability is often seasonal and unpredictable. In temperate regions, food may abound in summer and autumn as young shoots and fruits, but foliage becomes rare and of low quality in winter or during summer drought. Hence herbivores alter their life style to migrate, hibernate, survive as a dormant phase or live as best they can, often with considerable loss of life. Because there is a lack of herbivore prey in winter, insectivores and carnivores also have difficulties at these times.

It is often difficult to measure quantitively the relative abundance of the food available for a popula-tion. Many observations show that when food is scarce, animals reproduce less successfully than when food is more abundant, but the details of the way food reg-ulates a population are difficult to disentangle. In some species, the mode of life of the animal makes it a suitable candidate for experimental study. In such cases, the data comes from laboratory or field experiments rather than natural situations. However, such experiments can give us a better idea of how populations are regulated. An example of such experimentation is given below.

The Australian bushfly (*Musca vetustissima*) lays its eggs in the fresh dung of grazing herbivores such as cattle, horses and sheep. After a short time, the dung hardens on the surface, with the consequence that no more eggs can be laid in it. When those eggs which have been laid in the dung hatch, the larvae feed on the dung until they are ready to pupate in the nearby soil.

An experiment was set up in which a fixed volume of cow dung (2 dm³ in each case) was populated with different numbers of eggs ranging from 1400 to 4250 (Hughes & Walker, 1970). Figure 5.5 shows the percentage of eggs which survived to emerge as adults for the different densities of eggs. Each dot represents a 2 dm³ 'cowpat'. You can see from the figure that when around 1500 eggs were placed in the cowpat, about 50% survived to adulthood, that is about 750 adults, whereas when 4000 eggs were present, only about 25% survived, that is about 1000 adults. The average number of adults emerging for each cowpat was 915. The curve on the graph is not drawn to fit the actual data, but marks the percentage survival for any original egg number which will give exactly 915 adults. It is, in fact, quite a good fit to the data. This indicates that for a cowpat of 2 dm³, however many eggs are laid in it, the pat will only support the development of about 915 adults.

Presumably different sized cowpats, with different amounts of food and space, would support different numbers of eggs through to adulthood. The mortality of young in Figure 5.5 shows density dependence. The more eggs there are in the cowpat, the higher the percentage mortality.

Many herbivores are food specific. Many insects, for example, may only live on one species of plant. In these cases the abundance of the host plant will affect population size. If the plant population crashes for some reason then so will the herbivore popula-

tion. Similarly, for carnivores, a good year for their prey species will allow the predator to catch ample food and raise more offspring.

5.2.4 Territories

A **territory** is an area which contains a resource of sufficient value to an animal that it defends the area. The animal holds or defends the territory by marking the boundaries in some way and challenging strangers if they approach too closely. Territories may vary considerably in size depending on what resource they contain. Also they may be held for a very short time or for longer, depending on how long the resource is useful to the animal. Many species are known to hold territories, including mammals, birds, reptiles, amphibians, fish, insects and crustaceans.

Territories are usually held because they contain a food source, or because they provide some site needed in breeding, such as an area to attract mates or a nest site. Often these two reasons are connected and a breeding territory will contain the nest site and sufficient area to supply food to the growing offspring.

Food territories may be held by a solitary animal. A good example is the tiger which hunts alone and has a very large territory. Often the territory of a male tiger will overlap that of one or more females. Some territories support whole social groups of animals like the meerkats (*Suricata suricatta*) where a few individuals keep watch for danger or strangers, and the whole adult troop goes to defend the area if necessary.

Territories which are not held for food are usually very small. The male starling, as mentioned in Section 3.3.4, defends potential nest sites where he sings to attract a female. Animals which have a **lek**, that is an area where males gather and display to attract females, have minute territories. The best males hold tiny areas at the very centre of the display area, less good males have larger territories around the edges of the lek. Many birds, like the European ruff (*Philomachus pugnax*) and the American sage grouse (*Centrocercus urophrasianus*) have leks.

The size and number of territories in an area affects how many breeding individuals can live in the area and so influences population size. The small breeding territories do not usually regulate populations, unless actual nesting sites are limiting, as may be the case in cliff-dwelling seabirds where food is abundant, but safe ledges for nests may not be. It is the larger food territories which probably have most effect on population size: so what is it that regulates the size of these territories?

In a territory defended for food the amount the animal needs and the amount and quality of the food in the defended area will all affect territory size. Observation of the rufus hummingbird (*Selasphorus*

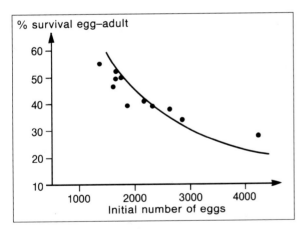

Figure 5.5 The relationship between the density of eggs and emergent adults of the Australian bushfly (*Musca vetustissima*) in 2 dm³ of cow dung. Experimental data are recorded as dots. The continuous line is the percentage of eggs required to survive from each egg density to give exactly 915 adults. (Data from Hughes & Walker, 1970.)

rufus) which feeds on nectar from flowers, a relatively easy resource to quantify, has given interesting insights into territory size. Rufus hummingbirds breed in northern America, even as far north as Alaska. For the winter they migrate south to Mexico and Guatemala. On migration the hummingbirds stop in mountain meadows to rest and feed. They defend feeding territories for short periods of one to seven days before moving on to another meadow further south.

In the meadows of California, Gass *et al.* (1976) found that the hummingbirds held territories where they fed mainly from two species of nectar-producing flower, the red columbine (*Aquilegia formosa*) and the Indian paintbrush (*Castilleja miniata*). They measured the nectar production and sugar concentration in the nectar of the two flowers to obtain the value of each flower to the hummingbirds; they then converted their results into a measure called a floral unit. They also observed the birds to see what territories they defended and how many flowers these territories contained. For example, a territory of 250 floral units contained 250 columbine flowers or 780 paintbrush flowers or a mixture of the two. Figure 5.6 shows the relationship Gass *et al.* found between territory size and floral unit density. They showed that territories were maintained to contain a certain number of floral units. If flowers were sparse, then the

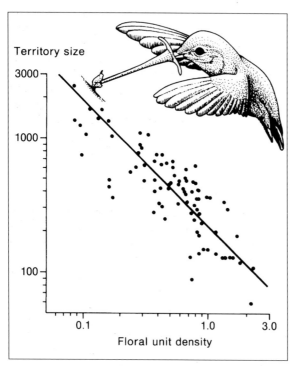

Figure 5.6 The territory size defended by rufus hummingbirds (*Selasphorus rufus*) plotted against the density of floral units in the territory. The more dense the floral units, the smaller the territory defended by the hummingbirds. (From Gass *et al.*, 1976.)

hummingbirds held large territories; if the flowers were abundant, territory size was much smaller.

There are times when it is not worth the hummingbirds defending a territory. If flowers and nectar are very plentiful in the meadows, then the birds abandon their territories as there is no need to defend food. If nectar is very sparse, then the birds again abandon territories as the areas they need to defend to include enough flowers is far too large (Kodric-Brown & Brown, 1978).

The most obvious way in which territories can regulate population size is in species which hold dual purpose territories for food and nesting. This is the case in many bird species. For example in great tits (*Parus major*) Krebs (1971) found that if he removed territory holders from good sites in Wytham Wood in Oxfordshire, the territories did not remain vacant. After a re-shuffle of areas by remaining birds, new birds moved in to take over the remaining vacant territories. These new birds had come from hedgerows outside the wood where they had been holding poor territories. The reproductive success of birds in the wood was higher than those in the hedgerow.

The work on great tits shows that both number and quality of territories in an area can regulate population size. The number of territories in the wood remained fairly stable, indicating that there was a limited number of territories available to the population. This limited the total number of breeding pairs in the area, and hence the potential population growth. Each year the number of breeding adults remains about the same.

5.2.5 *Herbivores and predators*

Predators obviously affect the population size of their prey, but the effect of herbivory on the population dynamics of vegetation is less apparent. If an organism eats seeds or seedlings, so killing plants, this directly affects the total plant population size. But there are many seeds, and few seedlings survive to maturity because of the $-\frac{3}{2}$ self-thinning effect (Section 5.2.2), So the effect of the herbivory on adult plant populations is lessened. Many herbivores do not even kill the plant they feed off. The effect this has on population dynamics is even harder to determine. Flower spikes may be removed so reducing seed production, or defoliation or root grazing may weaken the plant and prevent it reproducing. Defoliation experiments show that plants respond differently to this depending on species, stage of reproduction reached and amount of leaf area lost. Leaf removal can delay flowering by several weeks, reduce seed numbers to zero or reduce seed size.

Herbivory can have a positive effect on some plant populations by providing more opportunity for new individuals to establish. For example, if rabbits are excluded from chalk grassland, the ungrazed vegetation grows taller and there is an overall reduction in the number of individual plants and in the diversity of species. A very rich, heavily grazed grassland full of flowering plants changed to one dominated by species of grass with very few herbs (Tansley & Adamson, 1925). Rabbits evidently eat grasses in preference to herbs, so that when rabbits are present, the herbs can flower, set seed and increase. Once rabbits are excluded, the grass crowds out the other species.

Predators, despite their obvious effect on prey numbers, may not always have a long-term effect on population regulation. If numbers of prey are influenced mainly by food availability, then the predator may not be regulating the population at all. When prey numbers are high, because of abundant food, any prey the predator takes is likely to have died of starvation when food was no longer plentiful. When food is short, so that the prey population is small, each death from predation leaves a little more food for survivors and the population will increase in size once food is again plentiful. Only if predators are very efficient, or the prey population is slow to reproduce, is predation likely to regulate the prey population.

One example of an efficient predator is the protoctist *Didinium nasutum* which feeds on the species of *Paramecium* used by Gause in his experiments on population growth described in Section 5.1.2. Gause (1934) added small numbers of *Didinium* to populations of *Paramecium*. The results were always the same. The *Didinium* population increased until all the paramecia had been eaten; then it died out in turn, unless more paramecia were added. If a refuge was added to the glass tube for the paramecia to hide in, the situation became more complicated. In some cases the paramecia still disappeared, eaten by *Didinium*, but in other experiments some of the paramecia hid in the refuge and were still there after *Didinium* had died of starvation. In other words, in the case of these protoctists, features of the environment were important in the predator–prey relationship.

5.2.6 *Weather and climate*

The climate of an area is usually seasonal, but this seasonality is always predictable. There may be wet and dry seasons or hot and cold ones, but they occur regularly each year. Within this predictable climate there is a shorter term unpredictability: the **weather**. There is considerable variation from day to day in temperature, wind speed, precipitation (rain or snow), and so on. Such changeable features occur most noticeably in temperate zones. In the tropics the weather is more stable, but there are often extreme diurnal variations such as cold nights and hot days (see also Section 18.3.5).

Both the annual variations in the climate and the shorter, more rapid changes in weather can have a regulating effect on populations. How much a population is affected depends very much on the life span of the species concerned. A short-lived species like an insect or small annual plant may complete the whole of its life cycle in one part of the annual cycle of variation in climate. Many annual plants germinate in spring and flower and set seed before the following winter, or grow only through the wet season. They survive in the seed bank during the cold or dry part of the year. If the length of time for growth is short, as it is in the tundra or in a brief rainy season, then the size to which the plant can grow in the time available will be limited. The amount of energy it can store for reproduction will also be limited. The same is true for organisms living in shallow ponds which dry up for part of the year.

For longer-lived species, the annual cycles may provide a regular period of hardship such as drought or cold when water and food are in short supply. The effect of this seasonality on food availability and therefore on population size has already been mentioned in Section 5.2.3. Mortality may be particularly high during this period, especially among the immature or old. On top of the annual cycles of climate, the extremes of weather will also affect the longer-lived species. Extremes of temperature, high rainfall or drought, mild winters or very cold spells all alter the ability of individuals to survive or reproduce. In years when the weather is unfavourable at a critical point in the life cycle of a species, population numbers may crash or reproduction may fail. Adverse weather conditions can affect flower development, pollination and germination in plants or birth, rearing of young and adult mortality in animals. Climate change and the variability of weather, as it affects population regulation, is discussed further in Sections 9.2.5 and 9.4.2.

5.2.7 *Parasites and diseases*

A **parasite** is an organism which lives on or in another organism. The organism on which the parasite lives is called the **host**. Parasites depend on their hosts for food. A heavy infection of parasites, called the **parasite load**, may eventually cause the death of the host, directly or indirectly, by weakening it so that it succumbs to a predator or disease.

But how does a parasite differ from a predator or herbivore? The definition of a parasite given above could fit members of both these groups. A parasite is recognised as being much smaller than its host, so that often many parasites can live on a single host. Parasites usually reproduce much faster than their host and have a much faster rate of population increase (Pianka, 1988). However, as with many other definitions in ecology, there is a boundary area between these three groups where it is difficult to decide which category to use. For example, European cuckoos (*Cuculus canorus*) are much larger than their host birds, yet the baby cuckoo in the nest of a meadow pipit is considered as a parasite. On the other hand a colony of monkeys eating the fruit or leaves of a tree are much smaller than their host yet are considered to be herbivores, not parasites.

If an organism causes disease in its host, often leading to the death of the host, it is called a **pathogen**. These organisms are usually very small and include viruses, bacteria, fungi and protoctists. An alternative name for such small organisms is **microparasites**.

Both plant and animal populations can be infected with parasites and pathogens. The population dynamics of the host and the disease organism are linked. If the pathogen or parasite population is very small, then it usually has little effect on the host population. If the parasite population increases in numbers, so that the host organism has a heavy parasitic load, then this is likely to weaken or even kill the host. Usually a heavy parasite load affects population dynamics by reducing the birth rate and by increasing deaths from secondary causes.

Most populations have low levels of disease present, but sometimes a disease will multiply and spread rapidly in a population causing an epidemic. An epidemic can be caused by a particularly vigorous or virulent strain of microparasite, such as the fungus *Ceratocystis ulmi* which killed the majority of elm trees (*Ulmus* spp.) in Britain in the late 1970s. It may be helped by other environmental conditions.

Parasite infections or disease epidemics are more likely to spread if the host species is very common or closely spaced together. This makes regulation of disease and parasites important, especially in agriculture and among human populations where such conditions frequently occur. Because of the importance of such factors in crop growing and animal husbandry, there has been much investigation of the effects of pathogens and parasites on population dynamics. Several different models have been produced to explain the way hosts and parasites or pathogens regulate each other. For reviews of this and further references, see Burdon (1987) and Brown (1996).

5.2.8 *Natural disasters*

Disasters are uncommon and unpredictable events which can occur in any area of the world. Such events include fires, floods, winds of various destructive kinds from gales to hurricanes, tidal waves, mud flows, earthquakes, volcanic eruptions and meteor impacts.

Some of these disasters are the result of geological processes such as plate tectonics, or the impact of extraterrestrial debris. Such events may be very rare, but have large-scale effects. A huge meteor crashing to earth is thought to have caused the extinction of the dinosaurs and many other species at the end of the Cretaceous period, 65 million years ago. Other geological events such as volcanos may be rather local in effect, covering only a few hundred square kilometres with lava or thick dust; however, atmospheric dust from volcanos can block sunlight over a much larger area.

Some disasters are caused by extreme weather conditions. These include high winds, floods and fires (influenced by hot dry periods and caused in nature by lightning strikes). They tend to occur more often than geological disasters. Their effects may be apparent over a large area, but are not global in influence. The effects of disasters on populations are considerable. These events do not regulate populations in the continuous and fairly stable way like other factors discussed in this chapter. A fire, flood, hurricane or volcanic eruption (Figure 5.7) may wipe out many species in the vicinity of the disaster. New populations may then be established by immigrants from neighbouring populations. Survivors in seed banks may provide the next generation for the plant populations. If these have also been destroyed, fresh incoming seed is required to colonise the area. In many cases new species will colonise the devastated area. It may then be some time before populations of the original species re-invade the site.

Fires and high winds may be extremely important factors in the variation in age structure of plant

Figure 5.7 The devastation which can be caused by natural disasters: about 800 square kilometres of forest on and around the slopes of Mount Saint Helens, Washington State, USA, was completely flattened by the volcanic eruption of 18 May 1980.

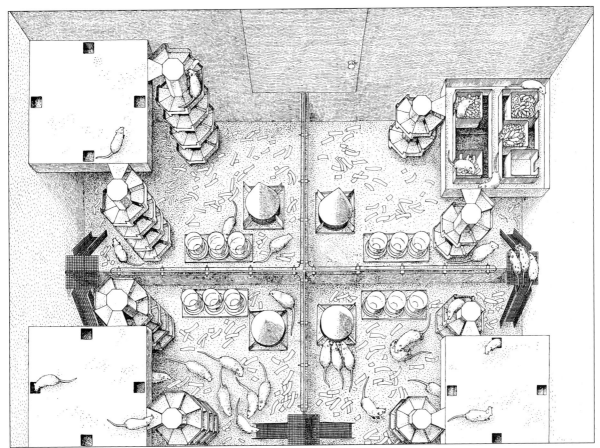

Figure 5.8 The design of the cage system used to investigate the effect of overcrowding on a population of laboratory rats. (From Calhoun, 1962.)

populations. Young trees and bushes are most likely to suffer fire damage, while mature trees are most likely to be blown down. In grasslands, frequent fires are caused by the presence of dry grass leaves. These fires are low in temperature and do not kill the growing points of the grass in the soil, but damage any young trees or bushes. Thus, grassland fires tend to prevent establishment of woody species. This maintains the grassland and prevents it being invaded by scrub.

In tropical and temperate forests, large gaps are occasionally created by wind damage. This provides the opportunity for young forest trees to grow and produces patchiness in the forest. Diversity in tree age and forest structure is therefore maintained. Forest fires are much hotter than grassland fires and may kill all the trees and animals in a large area. The fires cause nutrient loss to the atmosphere and release some nutrients into the soil in ash. These fires may devastate the tree populations, but some tree species will be favoured by fire. Several pine (*Pinus*) species and, in Australia, *Eucalyptus* species have seeds whose germination is triggered by fire. The seedlings thus grow in a nutrient-rich soil and experience little competition from other species while they are growing.

Fire and high winds are probably extremely important factors in maintaining diversity in species and the structure of vegetation in an area. Because they are infrequent they do not tend to be studied as often as the other more continuous factors which regulate populations.

5.2.9 *Self-regulation and stress*

Often there is no obvious reason why a population does not increase in numbers. It is quite likely that one or more of the factors mentioned in Sections 5.2.2–7, or some other environmental influence, is responsible. However, natural populations are usually difficult to study, so it may be hard to determine the exact causes of hardship or death or reasons for emigration.

Laboratory experiments can have a much tighter control over environmental factors and because the population is small and confined it is easier to study. Sometimes such experiments give extremely interesting results. One such study was carried out on rats by Calhoun (1962). Calhoun created his population by putting 16 young female and 16 young male rats into an enclosure. They were given all the food and water they needed and left to reproduce. After one year the number of rats in the population had risen to 80 adults. The enclosure, which was made up of four areas each with a nesting unit, was designed to hold about 50 rats, so it was overcrowded with 80. The design of the population box can be seen in Figure 5.8.

The population was kept at 80 by the removal of young rats after they had been weaned from their mothers' milk. This represented emigration from the population. The population structure at this time consisted of a couple of dominant males who held territories in the box, each with a harem of about five female rats. These females were good mothers and successfully raised about half their young. The other females, not in harems, were not so successful and about 80% of their offspring died before they were fully weaned. This was because the non-harem mothers provided inadequate care for their young. Outside the territories of the two males the females were more crowded and failed to build proper nests for their young. Often they abandoned their young around the compound. The young often left the nests before they were capable of surviving alone, presumably because the poor nests offered little comfort. The behaviour of male rats also became unusual for caged animals. There was increased fighting among some males, while others showed almost total inactivity and did not appear to notice the existence of other animals. Some rats showed homosexual or hypersexual activity and there was cannibalism of the young.

Calhoun's rat experiment shows how completely the normal social behaviour of many of the rats broke down after the population became overcrowded.

Other experiments have shown that such overcrowding can cause physical as well as behavioural problems. Mice developed ulcers or a lowered resistance to disease when kept in similar conditions. In natural conditions there would be more space for the population and overcrowding would be reduced by increased emigration. However, overcrowding does occur in nature, for example on islands, or during widespread favourable periods when densities may reach high levels, especially in small mammals (Chitty, 1977).

Can we extrapolate from such experiments to human societies? Can we predict our response to future crowding, or even detect symptoms of overcrowding in some existing societies?

It is tempting to answer yes. Overcrowding in industrialised cultures can lead many humans into the aptly named 'rat race' of competitive behaviour. Being successful means having a good job and earning money. Many people in employment show stress-related conditions such as ulcers and heart attacks. At the other end of the scale, when unemployment is persistently high, there are people without an active role in society who may become over-aggressive or lethargic. Most would argue that this sort of interpretation is too simplistic. After all, human societies are much more complex and varied than the social structures developed by rats in a box.

5.3 Patterns in population dynamics

Some populations have regular patterns of increases and decreases in population numbers. These **population cycles** are found in several mammals including

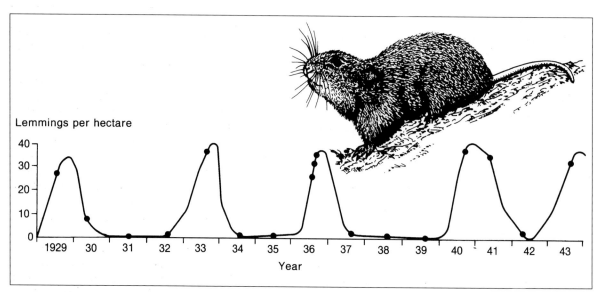

Figure 5.9 Changes in the number of collared lemmings (*Dicrostonyx*) in northern Manitoba, Canada, from field estimates and trapping data. (From Shelford, 1943, 1945.)

Box 5
Weeds

Most people use the word 'weed' about some plant they know. Everybody knows what they mean by a weed, but the word can mean different things to different people. To the gardener a weed is simply a plant in the wrong place. To a farmer a weed is any plant which is a nuisance or lowers the economic value of the farm's produce. It may compete with crops or poison animals.

Weeds tend to be associated with cultivation. Often they are rapid invaders of freshly dug land. Weeds tend to be fast growing and abundant in an area. They are difficult to get rid of, because they build up a seed bank which replaces the adult population even if this is destroyed. They often have seeds which can be dispersed over long distances, produce lots of seed and have seed which can remain dormant for many years. Often the adult plant is short lived, usually an annual.

So why do weeds exist?

We have to consider where weeds lived before humans created the cultivated ground with which they are now associated. For example, towards the end of the last ice age in Britain, about 12 000 years ago, there was a period when the glacial ice was retreating. This left large areas of bare, open, treeless soils. Pollen and seed fossils show that many species now called weeds were present at the time (Godwin, 1960). We do not know much about their ecology, but they must have been growing on nutrient-rich, open soils left by the retreating glaciers and the melting permafrost. Such soils were probably quite similar to the bare cultivated land they now inhabit.

By about 10 000 years ago, Britain was invaded by trees and soon most of the country was covered in forest. This, of course, meant the end of the vast areas of ideal habitats for

weeds. So where did all the weeds go between then and the reopening of the landscape by human agriculture? For about 5000 years, these species must have survived in the few treeless spaces left amongst the woodlands – places like seashores, river gravels, marshes, dried-up lakes, steep unstable slopes like screes, cliffs, areas of landslip and the region above the tree line on mountains.

Most of these possible sites are small in area and many are ephemeral. Open soils caused by landslips, uprooted trees, drought or fire are quickly overgrown or re-flooded and then no longer offer opportunity for colonisation. The weeds would have to invade and grow quickly in the available space so that they could flower and set seed before the gap closed. The seed would need to be widely dispersed and able to remain dormant for long periods so that it could take advantage of the next landslip or bare river bank.

Thus, for many thousands of years, these weed species grew in situations where *r*-selection was likely to be very strong. Then along came humans who destroyed the forest and cultivated the land. By doing this, humans provided the weed species with abundant areas to colonise. From a precarious existence, with small, short-lived populations, the weeds could increase greatly in population number and size. Hence, by constantly renewing the availability of prime sites for these species in close proximity to our crops, we have allowed a group of species which were probably very rare in the natural vegetation to become a human nuisance. In the last 5000 years or so, the genotypes of many of these weed species will have been further selected to thrive under arable practices (Cousens & Mortimer, 1995). Thus they have become even more successful in habitats such as cornfields and gardens: hence our present-day use of the term 'weed'.

small mammals such as voles and lemmings and larger carnivores such as foxes and lynx. Figure 5.9 shows how the fluctuations in lemming numbers over several years follow a regular pattern. The lemming populations increase rapidly from only one or two lemmings per hectare to about 30 or 40 lemmings per hectare. This increase occurs in only a few months. Shortly afterwards the numbers fall back to the earlier low levels and remain low for the next few years.

As you can see from Figure 5.9, the peaks and troughs of lemming numbers are regularly spaced about four years apart. This four-year cycle occurs not only in lemmings, but also in mice (Muridae) and foxes (*Alopex* and *Vulpes* spp.). Other species such as the snowshoe hare (*Lepus americanus*), muskrat (*Ondatra zibethica*), some foxes and the Canadian lynx (*Lynx canadensis*) have ten-year population cycles (Finerty, 1980).

So what makes these populations increase and decrease with such regularity, and why should the cycles be four or ten years long?

In the case of the carnivores such as foxes and lynx the answer may be fairly simple. These species feed on lemmings, mice and hares, so when the small mammals are abundant the carnivores have plenty of food. This means they can raise and feed more young in the peak years of the cycle. When the small mammal populations crash then the carnivores starve and many die taking population levels back to the low numbers between peak years.

This still does not explain why the herbivorous mammals have regular cycles. Lemmings can raise two or three broods of young a year. They are towards the *r*-selected end of the *r*–*K* continuum (see Section 4.6.2) and can rapidly increase their population size in favourable conditions. The sharp rise in population size in the small mammals is probably due to some favourable environmental factor which results in an increase in the birth rate.

The equally rapid fall in the population in the lemmings seems to result from a high death rate and lowering of the birth rate. The population crash does not seem to be due to disease or to starvation because of lack of food. There are many stories of emigrating lemmings going mad and committing mass suicide by plunging into rivers. The phrase 'to rush like lemmings to their fate' comes from this idea. So are lemmings showing behavioural stress like the overcrowded rats in Section 5.2.9? The tales of madness appear to be unfounded. Lemmings often swim rivers during migration and inevitably some will be washed away and drowned in the process. The more lemmings there are on the move, the more accidental deaths there will be. Krebs (1964) looked at 4000 dead lemmings collected after a population peak when the death rate was high. He could find no evidence of stress in the animals and they all seemed quite well fed. Lemmings are short-lived animals and do not seem to survive for more than a year in the wild. The rapid fall in numbers seems to be due to a failure to breed coupled with the dying off of the older animals.

It was once suggested that the increase in predators such as the foxes caused the lemming population to crash. This was the case with the efficient predator *Didinium* on the laboratory populations of *Paramecium* mentioned in Section 5.2.5, but foxes are not such efficient predators. They do not eat sufficient lemmings to cause the crash.

The fact is that we are still not sure why lemming populations cycle with such regularity. Anything related to weather is not so regular: peaks and slumps in conditions might occur, but not on such a regular four-year pattern. The most likely suggestion is that the period between population peaks is the shortest time needed by the population to recover from the last collapse, but it does not really explain why the population collapses in the first place.

As mentioned earlier, the snowshoe hare cycles regularly too, on an 8–11-year cycle. Its predator, the Canadian lynx, as in the case of foxes feeding on lemmings, increases in numbers when the hares do, and many starve when the hare population crashes. The hares eat twigs on shrubs and may exhaust the food supply, as many hares show the characteristics of starvation. The bushes respond to the heavy browsing by increasing toxins in their new shoots for two or three years after a 'hare high'. This extended period, during which food is in short supply, keeps the hare populations low and is thought to explain why the cycling is as long as ten years. If you want to read more about population cycles look at Colinvaux (1993) and Begon, Harper and Townsend (1996).

Summary

(1) A population that grows in size without any regulation will show an exponential increase in numbers.
(2) Populations seldom grow exponentially for many generations.
(3) Population size often stabilises at the carrying capacity of the environment.
(4) The idealised growth of a population from founders to carrying capacity follows a pattern predicted by the logistic growth equation. Under these circumstances a sigmoid curve results when population size against time is plotted.
(5) Populations do not usually show unlimited exponential growth because they are affected by environmental factors. Such factors can be regulating, i.e. constantly or variably limiting, or they can be haphazard and unpredictable.
(6) Some environmental factors are density dependent: they have an increasing effect with increasing density of organisms in the population.
(7) Environmental factors which alter population size include space, food and water, the possession of territories, the presence of predators for animals or herbivores for plants, the weather, parasites, diseases and natural catastrophes.
(8) Some animals, if overcrowded, show signs of stress which may regulate the population, but this has mainly been shown to occur in unnatural conditions such as captivity.
(9) Some populations, notably small mammals and their predators, show regular cycles of population numbers.
(10) Population sizes of some predators seem to cycle because of changes in their food supply, but it is not understood why the small mammal populations cycle in the first place.

SIX

Ecological genetics

6.1 The importance of genetics to ecology

6.1.1 The source of variation

Every organism has a collection of genes consisting of lengths of **DNA** (deoxyribonucleic acid). This DNA can be read like a book and is used by the organism to construct proteins including enzymes. Each cell has a copy of this genetic material stored in the nucleus. The genes are arranged in a set order on several **chromosomes**. Most cells have two sets of chromosomes and, usually, one set comes from each of the two parents of the individual. The genes are responsible for the growth, development, maturation and reproduction of the individual.

In this book we are not concerned with how genes are constructed or exactly how the cell uses them. Ecological genetics is the study of how the genes an organism possesses affect its life and the way it interacts with its environment. The study is also concerned with how genes are passed on to new individuals in the population.

A single gene can have many different forms called **alleles**. If a gene has more than one allele, it is said to be **polymorphic**. Because many genes are polymorphic considerable variation is possible between different organisms of the same species. The set of alleles an individual possesses on its chromosomes is called its **genotype**. Different genotypes result in individuals with different appearances. An obvious instance of this is hair colour and pattern in cats, or shapes and sizes of dogs. In wild populations genetic differences are not usually as obvious as these examples. Many alleles affect the organism in ways which are not immediately apparent, for example differences in blood groups. The genotype of each individual causes the organism to interact with its environment. Together the genotype and the environment are responsible for the physical appearance of the individual. This outward expression of the genotype is called its **phenotype**. It is the variation in genotypes and the expression of this in different phenotypes which is of interest to the ecological geneticist.

6.1.2 Genetic and environmental variation

Most populations contain individuals with different genotypes. This variation in genotypes is responsible for some of the variation seen in the population. The environment also has an effect on the phenotype of each individual. For example, if a plant grows in a deep rich soil with a good supply of rain, it will be larger and more vigorous than if it had grown in a poor, dry soil. This kind of environmental effect can be seen in Figure 6.1. The plants in the photographs were originally split off from the same plant so that they each have the same genotype: they are **clones**. They were then grown in different environments and this is reflected in the different appearances of the plants.

So some of the variation in a population will be caused by the presence of different genotypes; some will be the result of the environment producing additional variation in the phenotypes.

6.1.3 The role of variation in natural selection

The term **natural selection** was used by Charles Darwin (1859) to describe the process by which environmental conditions favoured some phenotypes in a population and not others. He explained how organisms which did particularly well in their environment were more likely to survive, reproduce and leave offspring than other, less able organisms in the same species. Since then, the phrase 'survival of the fittest', coined by Herbert Spencer, a contemporary of Darwin, has been used extensively to describe this feature of genotype selection. Those most likely to survive are the healthy individuals which fit their environment best. Only those individuals which manage to reproduce in some way will perpetuate their genes in the population.

Variation within populations provides the basis on

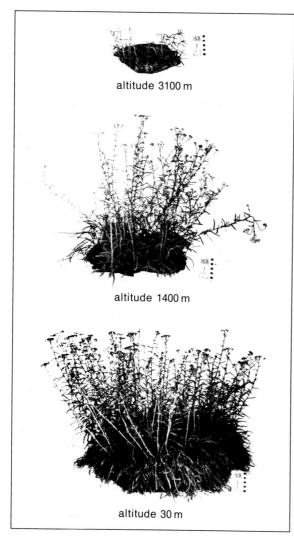

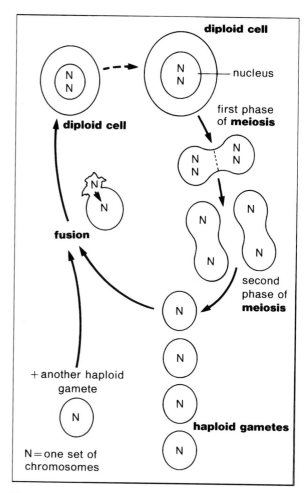

Figure 6.2 A simplified sketch to show the stages under-gone by a diploid cell dividing by meiosis to produce haploid gametes.

Figure 6.1 Clones of common yarrow (*Achillea mille-folium*) grown at different altitudes. They were photo-graphed in 1937 as part of a series of classic experiments on phenotypic variation carried out by Clausen, Keck and Hiesey (1940).

which natural selection can act. Ecology includes the study of natural selection in action when it investi-gates why some organisms die without offspring while others go on to reproduce successfully. Organisms within a species have been selected by the environment, so that they maximise the advantages they obtain from it, yet some populations and even species become extinct. Hence natural selection is at the interface between genetics and ecology.

6.2 Reproductive systems

6.2.1 *Formation of genetic variation*

An organism has no say in the genotype it inherits from its parents. It can, however, affect the genes its offspring inherit by the method it uses to pass on its

genes. There are several ways in which an organism can pass on genetic material in such a way as to maximise or decrease genetic variation in its progeny. Although the system used by an organism is aimed at increasing the success or fitness of its offspring, it also has consequences for the population in which the organisms live.

The various reproductive systems used by organ-isms fall into two main categories. One is **sexual reproduction** which involves the organism producing special cells which contain only half the normal number of chromosomes. These cells are called **gametes**, and because they have only half the genetic material they are described as being **haploid** as opposed to normal cells which are **diploid**. (The terminology for this is standard and is shown in the summary diagram in Figure 6.2.) The two haploid gametes can come from the same organism (**self-fertilisation**) or from different organisms (**cross-fertilisation** or **outcrossing**).

The second main category is **asexual reproduction** where no haploid cells are involved in producing the new offspring. There are several forms of asexual reproduction including simple outgrowth of part of the adult to form a new organism. This is called **vegetative reproduction** in plants, and budding in invertebrates such as *Hydra*: both are forms of **cloning**. Protoctists often simply divide in two (**binary fission**, as noted for the *Paramecium* in Section 5.1.2), but also have quite complex sexual cycles. Another form of asexual reproduction can occur when the sex cells which would have formed gametes fail to undergo meiosis, but start to develop into an embryo without fertilisation. This is called **seed apomixis** in plants and **parthenogenesis** in animals. The aphid group of insects often reproduce by parthenogenesis where an unfertilised female gives birth to more females.

6.2.2 Obligate cross-fertilisation

This form of sexual reproduction is universal in vertebrates but more uncommon in plants and invertebrates. In plants only about 2% of British angiosperm species have separate sexes with only male flowers on some plants and only female flowers on others. These are described as **dioecious** and include the holly (*Ilex aquifolium*) (Figure 6.3). Some areas have higher percentages of dioecy, for example New Zealand's flora has about 12% dioecious species, although the reasons for such differences between floras are not known. (See also Section 7.7.)

6.2.3 Facultative cross-fertilisation

The majority of angiosperms and probably many ferns and invertebrates show facultative cross-fertilisation. That is, although each organism can produce both male and female gametes, they tend to outbreed most of the time. In fact they often have mechanisms which prevent self-fertilisation occurring, although under certain circumstances this may happen. Some of the mechanisms found in angiosperm plants which promote outcrossing by making self-fertilisation difficult are illustrated in Figure 6.5.

If by the end of the life of a flower or the flowering season, a plant has not set enough seed by outcrossing alone, many incompatibility mechanisms begin to break down. The plant will then be able to self-fertilise as a last resort. Setting some self-fertilised seed is better than failing to reproduce.

6.2.4 Self-fertilisation

Many individuals which can produce both male and female gametes (**hermaphrodites**) have the ability to fertilise their own eggs. Some isolated animals, such as the tapeworm gut parasite (*Taenia*) often do this, and so do many plants. Some plants produce special flowers which never open to insect visitors, but remain closed and hidden among the foliage. These

Figure 6.3 Twig of a female holly tree (*Ilex aquifolium*) with berries.

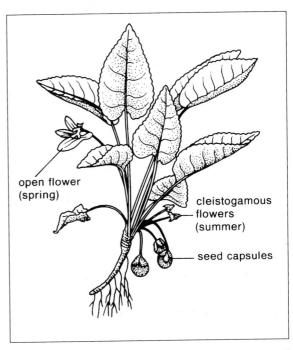

Figure 6.4 The hairy violet (*Viola hirta*) showing typical heart-shaped leaves, a typical open spring flower, large round seed capsules from the spring flowers and small cleistogamous flowers which develop in summer.

53

are called **cleistogamous** flowers and self-fertilisation occurs within the enclosed bud-like structure. An example of this is seen in violets (*Viola* spp.). In spring the violet has normal open flowers which attract insect visitors and promote cross-fertilisation. Later in the year, the violet produces cleistogamous flowers which are self-fertilised (Figure 6.4).

Many species which do not have special closed flowers also produce self-fertilised seed either by chance pollination, or as mentioned in Section 6.2.3, after failing to produce outcrossed seed.

6.2.5 *Seed apomixis*

When a diploid cell in or near the ovary of a plant develops without undergoing meiosis to form a normal looking seed, the process is called **seed apomixis**. It is very difficult to study exactly what happens in the flower. Sometimes pollen from another plant is needed to trigger development, even though it plays no genetic part in production of the embryo. Often a plant will produce some seeds by apomixis and some by normal sexual means. This makes it hard to detect

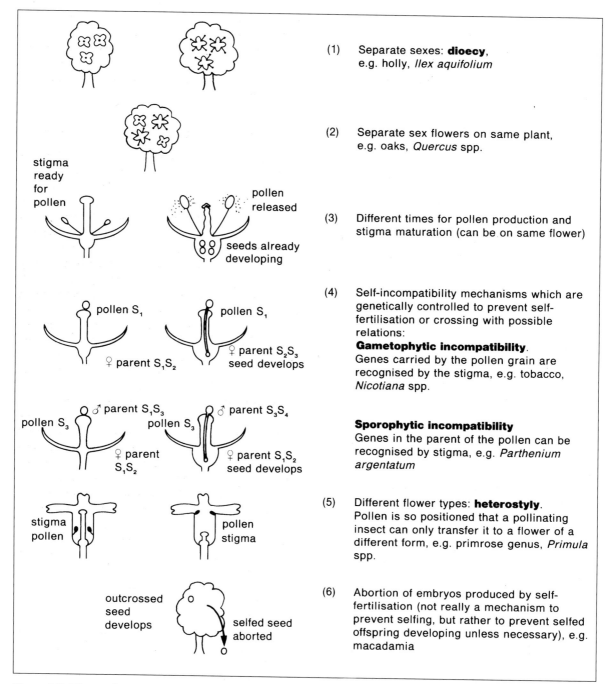

(1)   Separate sexes: **dioecy**, e.g. holly, *Ilex aquifolium*

(2)   Separate sex flowers on same plant, e.g. oaks, *Quercus* spp.

(3)   Different times for pollen production and stigma maturation (can be on same flower)

(4)   Self-incompatibility mechanisms which are genetically controlled to prevent self-fertilisation or crossing with possible relations:
Gametophytic incompatibility.
Genes carried by the pollen grain are recognised by the stigma, e.g. tobacco, *Nicotiana* spp.

Sporophytic incompatibility
Genes in the parent of the pollen can be recognised by stigma, e.g. *Parthenium argentatum*

(5)   Different flower types: **heterostyly**.
Pollen is so positioned that a pollinating insect can only transfer it to a flower of a different form, e.g. primrose genus, *Primula* spp.

(6)   Abortion of embryos produced by self-fertilisation (not really a mechanism to prevent selfing, but rather to prevent selfed offspring developing unless necessary), e.g. macadamia

Figure 6.5 The mechanisms which enable flowering plants to prevent self-fertilisation.

whether apomixis is occurring and to determine the frequency with which apomixis occurs in a population. A good example of an apomictic plant in which it *is* easy to tell that no pollen is needed is the dandelion (*Taraxacum officinale* aggregate). If you cut off all the stigmas from a dandelion bud before it opens so that self-pollination is prevented, then keep the cut area free of other pollen, it still sets a full head of seed.

The seed produced by apomixis has the same genotype as the parent plant. This leads to lines of individuals which all look very similar. If sexual reproduction also takes place occasionally, genes will be recombined and a new plant form will arise. This results in a large number of similar looking, but non-interbreeding, lines which produce offspring genetically identical to themselves. This is why plants which reproduce by seed apomixis are often considered difficult groups to identify. They tend to be lumped together under a single species name followed by the word 'aggregate' or 'agg.'. Examples, in addition to the dandelion, include the blackberry (*Rubus fruticosa* agg.) and the briar rose (*Rosa rugosa* agg.).

6.2.6 *Vegetative reproduction*

Many plant populations can increase in number without setting seed at all. They do this by **vegetative reproduction**. This can take many forms. Some plants, like elms (*Ulmus* spp.) send up new shoots from their roots, while species like potatoes (*Solanum*) regrow from root tubers. Often root suckers take many years to separate off from the parent plant. Some plants send out shoots which run along the surface of the ground, then root and produce a new plant. Examples include the strawberry (*Fragaria vesca*) and creeping buttercup (*Ranunculus repens*). Some plants produce tiny plants on the leaves or in the axils of leaves. These plantlets drop off or, if the parent leaf touches the ground, put out roots and grow into new individuals. The fern *Asplenium bulbiferum* is an example of a plant which reproduces this way. All these forms of vegetative reproduction produce new individuals with the same genotype as the parent. They are clones of the parent.

6.3 Genetic consequences of different reproductive systems

6.3.1 *The source of inherited chromosomes*

The different reproductive systems considered in Section 6.2 all result in new individuals entering the population. Each individual will be diploid in the majority of animals and in most plants (ferns, gymnosperms and angiosperms). Differences between the genotypes of individuals depend on the origins of their chromosomes. Outcrossing results in individuals obtaining half their genes from one parent, half from the other. Individuals produced by self-fertilisation obviously get all their chromosomes from the single parent. An individual produced asexually (whether by seed apomixis, parthenogenesis or vegetative reproduction) will have an exact copy of its parent's genes.

At first glance, the source of the chromosomes may not seem important, as long as the individual inherits two sets. However, the source of the chromosomes and the way they are inherited can have a considerable effect on the genotype of the individuals produced, and on the overall genetic range of the population. This in turn of course affects the phenotype of the individual, and the amount of variation in the population.

6.3.2 *The consequences of outcrossing*

To understand the consequences of **outbreeding**, let us first consider a single polymorphic gene **A** which has four alleles A_1, A_2, A_3 and A_4. If we set up a cross between two parents, each of which has two of the alleles, as shown in Figure 6.6 stage (i), then we can work out the possible genotypes of their progeny. Each parent will produce four gametes from each meiotic division. These gametes will each have only one allele from the pair possessed by the parent, Figure 6.6 stage (ii). The process of outcrossing will bring together one haploid gamete from each parent. Chance will play a large part in which two gametes fuse to form a new individual. If the two parents shown in Figure 6.6. produced many offspring, then roughly equal numbers of the four new combinations shown in stage (iii) would be found.

Now we can look at the consequences of outcrossing.

First, you can see that all the allele combinations in stage (iii) are different from either parent. In other words outcrossing produces genotypes in the offspring which are different from those in the parents.

Second, there are many different possibilities for offspring genotypes. This example has produced four new ones in stage (iii), yet here we have only considered one gene. If you consider that some genes have more than four alleles, while there are thousands of polymorphic genes, then the number of possible genotypes is almost endless!

Third, the two alleles inherited by the offspring may be different. The example given above shows such an example. Often, of course, both parents have a copy of the same allele and the offspring may inherit them both. In general, however, outcrossing results in a high degree of variation in alleles. If an organism has two different alleles for the same gene (for example A_1 and A_3) it is said to be **heterozygous**. If

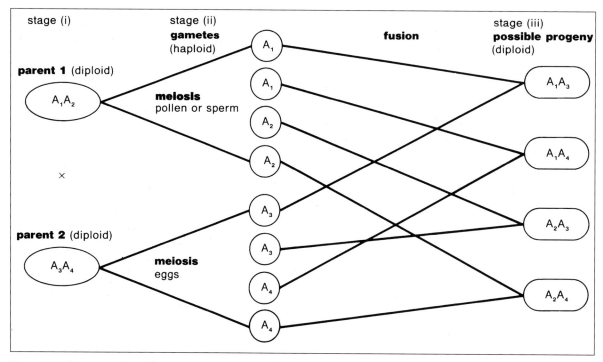

Figure 6.6 Representation of outcrossing using a single polymorphic gene to show the resulting genotypes of progeny.

the two alleles are the same (for example A_1 and A_1) it is said to be **homozygous** for that gene.

Now it is interesting to note that organisms with a high degree of heterozygosity are usually very vigorous and healthy. They show what is called hybrid vigour or **heterosis**. Another advantage of being heterozygous is that if one of the inherited alleles is disadvantageous (**deleterious**) or fatal (**lethal**), then the other allele may be expressed instead. In other words, the heterozygote has two chances of having an allele that will be to its advantage.

So the advantages to a plant or animal of outcrossing are that it is likely to produce vigorous, healthy offspring. There is a disadvantage: that if one parent has a wonderful combination of alleles, it will only be able to pass half of them to its offspring. So the offspring may not have such successful genotypes. If the parent has several offspring this problem is offset by the fact that the offspring will be variable and therefore at least some will be well suited to the environment in which they find themselves.

6.3.3   The consequences of self-fertilisation

If we look at a similar example to that in Section 6.3.2 but consider a self-fertilising organism we see a different result in the genotypes of the progeny in Figure 6.7. As all the gametes from stage (ii) are from the same organism, they can only contain one of two different alleles. The consequence of this is that half

the resulting offspring in stage (iii) are homozygous. Over many generations of selfing, heterozygosity is lost, so that the genes are almost all homozygous.

If we self-pollinate a normally vigorous, outcrossing plant so that it is self-fertilised and try to grow the seed produced, we find some quite serious problems. Many seeds will be empty with no embryo. Of those with an embryo present, some may fail to germinate, or when the seedlings emerge some will grow poorly and die before maturity. The plants that survive will probably be much less vigorous than their parent. This phenomenon is called **inbreeding depression**. It is the result of increased homozygosity.

The surviving offspring will show a decrease in heterosis, the vigour of variability lost due to the decrease of heterozygosity. The high mortality is the result of the previously hidden deleterious and lethal alleles being expressed because they have been brought together in the homozygous state. As these lethal alleles cause death before the plants reach maturity, they are selected out of the plant line, so that after several generations of selfing the plants, or crossing them with close relatives (**inbreeding**), the lethal and many of the deleterious alleles will have disappeared.

We can hypothesise from this that if we look at a wild population of plants which are normally self-fertile or inbred with close relatives, we will find a high degree of homozygosity and a low number of deleterious alleles. There would be very little genotypic variation in the population. If the plants lived in

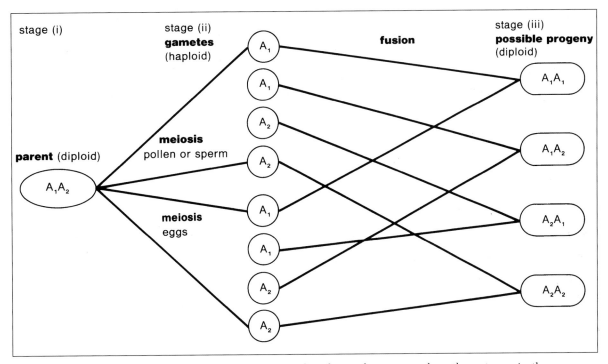

Figure 6.7 Representation of self-fertilisation using a single polymorphic gene to show the outcome in the progeny.

a fairly uniform or predictable environment, similarity would not be a disadvantage. If the plants are well suited to their environment, then their progeny will be also. If an outcrossing event occurs with a different inbred population from another site, then some degree of heterozygosity will be found in the offspring: they will show hybrid vigour. A small number of these individuals may have the opportunity to colonise a new area. While the new population is still small, inbreeding will occur and the heterosis will again be lost. This will be different from the situation where a normal outcrosser is selfed: the high number of fatalities in offspring seen in that case will not occur. This is because there are no lethal or deleterious alleles in the parents as they have themselves come from previously inbred lines.

So inbreeding is successful if the organism is not normally outbreeding and lives in a predictable environment where it is an advantage to be like your parent. As we saw with the violet, many plants both outcross and self. This is obviously a **bet-hedging** plan. The outcrossing maintains variation and a degree of heterosis, while the selfing at the end of the season guarantees that progeny are produced and eliminates some of the more deleterious alleles. Where organisms are isolated or likely to live only with relatives, inbreeding may be the normal method of sexual reproduction. If the species is dioecious, then obviously an individual in that species has no option but to outbreed, but it runs the risk of failing to reproduce if it is isolated from plants of the opposite sex.

6.3.4 The consequences of asexual reproduction

Asexual reproduction, except in the case of seed apomixis, does not usually involve gamete production, exchange and fusion, so offspring with the same genotype as their parent (in the absence of mutation) are the result. As mentioned in Section 6.2.5 seed apomixis can occur in a plant which also reproduces by outcrossing so that, occasionally, a new genotype is produced. New asexual lines are often the result of outcrossing events so the new combination of genes passed on by the asexual parent may well have a high degree of heterozygosity and therefore heterosis. Such a population reproduced by asexual means will be genetically uniform and probably well suited to the environment in which it has expanded. Occasional outcrossing followed by asexual reproduction seems to be the best genetic and reproductive plan of the three: it has the advantages of assured reproduction, heterosis and suitability to the environment.

In fact most plants, many marine invertebrates and some common insects (such as aphids, Aphidoidea) reproduce asexually most of the time with occasional sexual recombinations. There may be drawbacks to widespread asexual reproduction however, as it can produce a population with little or no genetic variation, though the advantages of sexual reproduction are still incompletely understood (Zeyl & Bell, 1997). All the individuals in an area will have the same genotype. They will thus be highly vulnerable to attack by disease: if one organism is susceptible, all are likely to catch it. Individuals in a uniform population are also likely to respond similarly to a change in environment. This could be a change in climate or the occurrence of another species such as a pollinator or predator. A population of just one genotype is not open to natural selection and no individual has a 'fitter' genotype. (See Section 6.4.2 and Box 6, p.62 for further discussion of the consequences of population uniformity.)

6.4   Patterns of genetic variation

6.4.1   *External influences on genetic variation*

Apart from the effects of the various reproductive systems used by organisms (as discussed in Sections 6.2 and 6.3), there are other conditions which influence the genotypes within populations. One obvious factor is the genetic nature of the individuals who founded the population, and any subsequent changes due to migration and mortality. A second cause of change and pattern is the process of natural selection. Different environments favour particular alleles which result in a higher concentration of these in the population. These influences on genotypes are discussed in more detail below.

6.4.2   *Founder effects and bottlenecks*

Populations which have been founded by only a few individuals will have a limited amount of genetic variation. Such **founder effects** are probably important on islands, especially those which lie a long way from other land. The arrival of organisms by sea drift or carried by wind or flying animals will be a rare occurrence. Some organisms such as plants and insects are likely to reach islands, while others, such as small mammals are absent from many islands, even those only a few miles offshore. Within a species, which organisms arrive on an island will depend largely on chance. The frequency with which a species arrives on an island and the genotypes of the arrivals influence the early population size and the genetic diversity. (See Section 18.3.2 for more on island biota.) Other places where founder effects are important are in ponds and lakes, in grassy clearings on the forest floor and any other isolated habitat where migrants are rare.

A nice example of the founder effect is provided by a recent study on a plant called *Lactuca muralis* in the Asteraceae (daisy family). This plant was studied on some 240 islands off the Pacific Coast of Canada over the course of a decade (Cody & Overton, 1996). *Lactuca muralis* is wind dispersed. It turns out that islands are likely to be colonised by seeds that are

significantly lighter than those on the mainland – by some 15% on average. The reason, as shown by experiments, is simply that the lighter the seed the further it disperses. Interestingly, there is then natural selection for larger achenes so that within about eight years the seeds have the same mass as those on the mainland. (See Section 18.3.4 for more on this effect.)

Sometimes an existing population suffers a crash in numbers due to a disaster of some kind (see Sections 5.2.6, 5.2.7 and 5.2.8). Disease epidemics are probably the most frequent cause of large population crashes, but drought, fire, flood, predation or lack of food may also be involved. Such times when a population is unusually small are known as **bottlenecks**. During such a population crash, some alleles will decrease in frequency, and may be lost altogether due to the excessive mortality of individuals. If only a few individuals survive in the population, then inbreeding will occur after the event producing increased homozygosity in the resulting genotypes.

An extreme example of the results of a genetic bottleneck may be provided by the cheetah (*Acinonyx jubatus*) (Figure 6.8). When 55 cheetahs from two isolated populations were studied, they were found to be homozygous at every gene examined (50 genes were looked at). When an even more extensive study was undertaken, only 3.2% of the genes studied in

Figure 6.8 A female cheetah (*Acinonyx jubatus*) with her cubs.

the cheetahs were found to be polymorphic. In various other animals, including humans, other cats and insects, the number of polymorphic genes is usually between 10 and 43%. Only 20 000 years ago the cheetah lived in Asia, Europe and North America; now it is restricted to parts of Africa. Stephen O'Brien has calculated that a bottleneck in the cheetah population occurred somewhere between 3000 and 13 000 years ago (De Smet, 1993).

The cheetah is obviously now a vulnerable species with fewer than 20 000 individuals. There is evidence of inbreeding depression as male fertility is ten times lower than in other cat species. Also, infant mortality may be as high as 70% in the wild; this is high for a *K*-selected mammal. Increased levels of homozygosity mean that if a disease attacks one animal because it has no resistance, then it is likely to be able to attack all the population. This vulnerability was shown in 1983 among captive animals in an American wildlife park. Within the colony, 90% developed an infectious disease which attacks cats; despite intensive medical treatment 18 animals died (O'Brien *et al.*, 1983, 1985).

Other biologists have questioned O'Brien's conclusions about inbreeding (Laurenson *et al.*, 1995; Lewin, 1996). Indeed, the debate has become increasingly acrimonious. O'Brien's work has been described as 'a loosely strung chain of poorly established findings' and as 'not meeting the minimum standards of evidence'. O'Brien has retaliated, describing some of the work of his opponents as 'an embarrassment' (to science, not O'Brien) that 'should never have been published' (Lewin, 1996 p.14). O'Brien's version of the cheetah story has not greatly changed. However, his opponents claim that re-analysis of his data shows that the cheetah is not as genetically impoverished as previously thought and that the high infant mortality rates are simply due to predation. On one point everybody is agreed, namely that habitat loss is the main threat to the species' future.

6.4.3 Isolation of populations

After the foundation of a population, the amount of isolation it then experiences may affect the genetic variability. Immigration from other populations will probably introduce new alleles into the population. The frequencies of the different alleles for each polymorphic gene may change randomly (**genetic drift**) or certain alleles may be favoured by natural selection.

Isolated populations often show many phenotypic differences from other populations of the same species. This is the case, for example, in many populations found on islands. Small mammals on offshore islands around the British Isles show this phenomenon. Berry (1964, 1977) studied the house mouse

(*Mus musculus*) on various islands. He looked at skeletal variations and protein differences in island populations and compared these with populations on mainland Britain. Quantifying the genetic differences between populations is no easy matter, and several ways have been suggested to calculate the 'genetic distance' between populations (Williamson, 1981). Berry found that there was 5–15% difference between populations on the mainland, but 25–30% difference between island and mainland populations.

It is in fact quite difficult to interpret these results. The differences could be due to founder effects or to the process of natural selection in island environments. Most of the differences seem to be due to different frequencies of alleles, rather than to the absence of various polymorphisms from islands. The founder effect would predict a lower number of alleles in island populations. Berry's data suggests that the differences between mainland and island populations are the result of natural selection (Williamson, 1981).

Sometimes the effect of immigration on the introduction of new genotypes into a population can be traced if the alleles involved have very obvious effects on phenotype. An example of this is the introduction of the deleterious allele causing Huntington's disease in humans. Normally, 2–6 people per 100 000 develop the genetic disease. In Maracaibo, Venezuela, the figure is 52 people per 100 000. The introduction of the deleterious allele can be traced back to a sailor on a German trade ship in the 1860s. Similarly, the same allele was introduced into Tasmania in 1842 by a female carrier who had 13 children, of whom 9 inherited the allele. Huntington's disease is thought to have spread to the rest of Australia through this single family (Phillips, 1982).

6.4.4 Ecotypes and ecoclines

It was mentioned for the island mice in Section 6.4.3 that the variation seen in island populations may have more to do with natural selection to suit life on islands than to founder or bottleneck effects. Many populations show a series of phenotypic characters which suggest they are adapted to live in the environment where they are found. These characters usually remain even if the organisms are put in a different environment, and are inherited by their offspring, indicating that the characters are genetically rather than environmentally determined.

An example of a plant species which shows these variations between populations is the harebell (*Campanula rotundifolia*). This species was studied by the Swedish ecologist Turesson (1925). The results of his study are summarised in Figure 6.9. He found that mountain populations were shorter, flowered earlier and had stronger rosettes of leaves than the

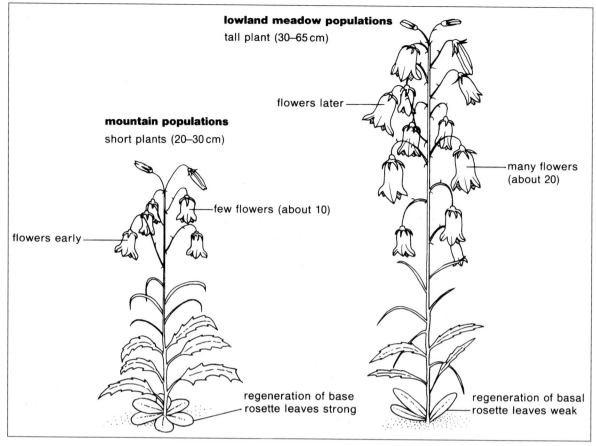

Figure 6.9 Some of the differences between populations of mountain and lowland harebells (*Campanula rotundifolia*) as described by Turesson (1925).

lowland forms. This suited their growth in short mountain turf and their adaptation for rapid flowering and seed production in the shorter growing season in high altitude mountain habitats.

Turesson called these different forms **ecotypes**. Many different ecotypes of plants have been recognised including ecotypes from coastal areas, cliffs, mountains, fields and so on. All these populations tend to grow some distance from the other ecotypes of the same species. The different ecotypes of a species, therefore, have a degree of genetic isolation from each other, so that natural selection within the populations could have occurred to produce these easily recognisable forms.

Much of the recognisable variation in plants is not between widely separated and distinct populations such as the ecotypic forms described by Turesson. Variation often occurs within populations. For example, a species with a continuous distribution up the side of a mountain may have lowland forms at the bottom, grading into more upland forms at higher altitude. The population cannot be divided at any one point into two ecotypes. For this pattern of distribution the term **ecocline** has been applied. An ecocline can be used to describe any gradual change across a

population or series of adjacent populations. Often the changes observed appear to have an ecological explanation, although the exact environmental causes of the variation along the ecocline may not be known. In fact some clines may simply be caused by chance effects of allele distribution rather than natural selection, but this is extremely difficult to prove.

The presence of polymorphisms in clover is a good example of an ecocline stretching right across Europe. White clover (*Trifolium repens*) has a gene for the production of a cyanogenic glucoside with two alleles (**G** = glucoside produced; **g** = no glucoside produced). It has another gene which produces an enzyme, a lysozyme, that breaks down the glucoside to produce free cyanide (**L** = lysozyme present; **l** = lysozyme absent). In plants with the alleles **G** and **L**, when a leaf is crushed, the lysozyme comes into contact with the glucoside and cyanide is released. Plants which do not have either **G** or **L** do not produce cyanide when crushed. When the distribution of cyanide and non-cyanide producers was investigated, it was found that plants which produce cyanide are found mainly in the south and west of Europe, while plants which do not release cyanide occur mainly in the north-east of Europe (Figure

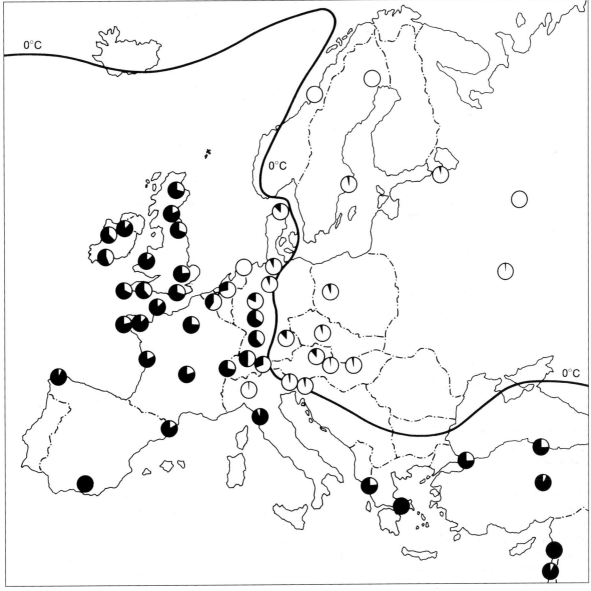

Figure 6.10 An ecocline of frequency of cyanide production within populations of white clover (*Trifolium repens*) across Europe. The black section shows the frequency of the cyanide-producing plants. The line shows the January isotherm of 0°C. (From Briggs & Walters, 1984.)

6.10). The ratio of non-cyanide to cyanide producers changes in the populations of clover producing a SW–NE cline overall (Daday, 1954).

The production of cyanide is thought to be an anti-grazing device to deter slugs and snails, and perhaps larger grazers too, from causing leaf damage. So why do northern plants not have this cyanide protection? What factors cause the ecocline shown in Figure 6.10?

In the south-west, grazers such as slugs and snails are abundant because the winters are quite mild, and summers fairly wet, consequently cyanogenesis is an advantage to the plants. Towards the north-east, harder winters cut down the number of slugs

surviving the winter, so grazing damage may not be so extensive. More important may be the fact that if leaves freeze, cell membranes are damaged by ice formation. This releases the lysozyme within the leaf, so that it comes into contact with the glucoside. Release of cyanide within the leaf is probably not beneficial to the clover.

The genetics of the cline show that in the south-west the alleles **GL** are homozygous in most populations, while in the north-east alleles **gl** are homozygous. The degree of homozygosity decreases towards the centre of the cline and increases to nearly 100% at either end. The same cline can be found in ascending mountains, where heavy frosts in winter seem to

Box 6
The farming revolution

Great changes have taken place in farming over the last 50 years or so. With improved breeding techniques and a much clearer knowledge of how genes operate, crop and livestock breeders have been able to manipulate the production of plants and animals to suit the requirements of the modern farmer. Plant breeding has produced much more uniform crops which ripen synchronously or are shaped so that they are easily harvested with modern equipment. For example, modern wheat has a shorter, stronger stalk than old varieties, and is less likely to blow over before harvesting. There are even varieties of pea now available which do not have leaves; the mass of tendrils which grow in their place hold the plants up, so they do not need expensive staking. Selection of animals has led to increased growth rates for meat or higher milk yields.

The production of such improved crops and animals has led to their widespread use by farmers. But problems can arise from the extensive use of such genetically uniform populations. Densely packed, genetically similar monocultures are susceptible to outbreaks of disease or attack by pests. One possible response to these problems is to counteract infections by the application of chemicals. This solution is widely used by farmers. Crops are sprayed with pesticides, fungicides and selective herbicides, animals are injected routinely with antibiotics, even though they may not have a detectable disease.

Such mass applications of chemicals have strong selective effects on the populations of diseases and pests. Any individuals with some resistance to the chemical substances being applied will have a greater chance of survival and therefore reproduction than susceptible organisms. Thus the genetic balance in the pest populations will swing towards increased resistance. This in turn will lead to the pesticide or fungicide becoming ineffective, and alternatives will have to be found.

An alternative solution to this heavy use of chemicals is to find naturally occurring genes which confer resistance within the crop or animal to attack by the disease or pest. These genes can then be introduced into a farmed population by selective breeding. The sources of these new genes are often old farmyard varieties which have been rejected in the past for some other reason, or wild relatives of crops. Sometimes the resistant gene is in a very different species, which makes introduction of the useful gene very difficult as the two species cannot be crossed.

The importance of wild species and old varieties as a resource of varied and useful genes has not been appreciated

for very long. There is still the risk that good genotypes will be abandoned and lost without their value being appreciated. This is probably especially important at the moment in countries where European domestic animals and crops are being imported to 'improve' farming practice. The endemic crops or livestock are much better adapted to the different environmental conditions where they are.

In Britain, many breeds of farm animals and crops which were once used went out of fashion a few decades ago and reached the verge of extinction. Some have gone for ever, others were rescued just in time like the sheep in Figure 6.11, and organisations like the Rare Breeds Survival Trust now monitor the rescue progress. Many breeds of sheep, pigs, chickens and cattle are now maintained, although their populations are often small, and many breeds have gone through bottlenecks. It is easier to maintain the genetic variability of plants than of domestic animals. This is because large numbers of seed can be kept viable for many years under correct conditions. However, even stored seed gets old and needs to be regrown every 10–100 years or so (depending on the species). These rare breed and seed collections are extremely important as they represent a vast gene pool of variability. We never know when some genetic character will become important or some minor crop species relevant to our food producing programme.

Figure 6.11 A Manx Loghtan ram. By 1913 this breed had almost become extinct but has now increased in size to several hundred animals.

have the same selective effect against cyanide production (Briggs & Walters, 1984, 1997).

Whether variation in population patterns form ecotypes, ecoclines or a non-directional mosaic distribution will depend on how far apart populations are from each other, the amount of genetic exchange between populations, the distribution of environmental features and the degree of natural selection acting on individuals. As all these features can vary with geographical location and species so the expression of genotypic and phenotypic pattern within species also varies. Sometimes distinct ecotypes are found; in other cases clines occur; but in most populations the patterns revealed will be complex and the causes of the patterns may be difficult to determine.

6.5 Genetic variation within an organism

After two gametes fuse to form a new organism the genetic material also fuses to create a nucleus. From then on the cell will divide many times to form a multicellular organism. The nucleus will also divide and the genetic material is copied again and again for each new cell. In principle every cell will have the same genotype, a copy of the genotype in the original zygote. The only way this can alter is if a mutation occurs during replication. A mutation may affect only a single gene, or many genes may be influenced if a chromosome breaks and rejoins incorrectly, or gets omitted altogether so that part of it is left as an extra in the other cell. Any cells which then develop from division of this cell will carry the mutation. If a single allele is miscopied, it may still function adequately but produce a slightly different protein, or it may become useless or even harmful.

Some organisms, especially vegetatively reproductive plants, can live for hundreds or thousands of years. Over this time they may well accumulate many such mutations in various parts of their modular structure so that the genotype of the single organism is not uniform but patchy for one or more genes. Sometimes it is very obvious that a modular unit has such a mutation. The plant in Figure 6.12 has a branch of leaves which are unable to make chlorophyll. This type of mutation would obviously be fatal if it occurred in the main growing shoot before the plant had any leaves, but because most of the plant has chlorophyll the branch can survive

Figure 6.12 Part of a horse chestnut tree (*Aesculus hippocastanum*). The group of leaves in the centre of the photograph lack chlorophyll.

even though it has lost the mechanism for autotrophism.

Other genetic changes can occur which produce individual branches on a plant with differences in, for example, flowering times or flower form or resistance to insect attack. Such phenotypes may well have advantages over branches with other characteristics. Because plants are modular organisms these different branches could compete for nutrients, pollinators, light and so on. The more successful branches may well grow larger and produce more seed. In such cases, even a clone may not behave as one uniform plant with one genotype.

Summary

(1) The units of inheritable material called genes are arranged on chromosomes in the cells of organisms.
(2) Different forms of a single gene found at the same position on homologous chromosomes are called alleles.
(3) Most organisms have at least two sets of chromosomes making up their genotype.
(4) Organisms can pass genes on to their offspring by different breeding systems: by outcrossing, self-fertilisation, seed apomixis, parthenogenesis and vegetative (or asexual) reproduction.
(5) The degree of heterozygosity within an individual depends on the breeding systems of its parents.
(6) In individuals with a high degree of heterozygosity, many of the genes are represented by different alleles (polymorphism); the individuals are often vigorous (heterotic).
(7) After several generations of inbreeding, individuals are usually homozygous at most loci.
(8) Genotypic variation within a population depends on the breeding systems of the organisms present, founder effects, immigration, bottlenecks and natural selection.
(9) The genotypic variation in populations may produce patterns of variation called ecotypes or ecoclines, although often the variation appears random.
(10) Ecotypes are isolated populations with distinct features which indicate the population is genetically adapted to its environment.
(11) Ecoclines occur where a gradual change in phenotype is seen across a population or connected group of populations.

SEVEN

Behavioural ecology

7.1 What is behavioural ecology?

In the past ecologists often took little account of the behaviour of organisms they studied. Increasingly it has been realised that just as knowledge of genetics is important for an understanding of ecology (see Chapter 6), so is a knowledge of behaviour. An individual needs to do many things in order to survive and reproduce. Some of these are simply bodily functions, like those described in Chapter 2, such as ingestion and excretion, but around these functions is a pattern of activity, such as food searching, resource defence, mate location and parental care, which can be described as **behaviour**.

Behavioural ecology investigates how the behaviour organisms show is related to their ecology. For example, if we want to understand how an individual's feeding behaviour helps it to survive and reproduce, we need to know a lot about its ecology. We need information about the habitat in which it lives and the food on which it feeds. We also need to know something about the other organisms, if there are any, that feed on the same food.

Much of the research done on behavioural ecology looks at three problems that organisms face: obtaining food; avoiding being eaten; and reproducing (Krebs & Davies, 1993). In recent years some ecologists have begun to use the language of behavioural ecology to see how plants solve these problems. In this sense even plants and fungi can be said to behave.

7.2 Optimisation theory

The central importance of reproduction in an organism's life has been examined in previous chapters, so why do animals not spend all their time reproducing or looking after their offspring? Presumably because, in the real world, animals also need to spend time feeding, avoiding predators, sheltering from inclement weather and so on. Clearly there are constraints on behaviour and organisms have to make the best of a bad job in an environment not ideally suited to their single goal of reproduction.

The notion that, for any situation, there is an optimum solution which produces the best possible result has long been familiar to engineers and economists. Such **optimisation theory** has recently been used in a variety of biological cases (Alexander, 1982; Sibly & Calow, 1986). The idea that optimisation theory can be applied to ecology is very attractive, because the optimum solution can be calculated *before* the situation is studied, to predict what an organism should be doing. Such predictions can then be tested. In other words, the theory can produce a testable hypothesis, although in practice it is often difficult to identify correctly the constraints under which the organism lives. The sorts of questions tackled by optimisation theory include 'Should an organism reproduce once in its life or on several occasions?' and 'What proportion of its energy intake should an organism devote to growth as opposed to reproduction?'. We will examine optimisation theory in the context of **optimal foraging**. This is a field in which biologists have used optimisation theory to make quantitative predictions about feeding behaviour which can then be tested by observation and experiment.

7.3 Optimal foraging

In order to survive, grow and reproduce an animal needs to feed. In feeding, an animal obtains the raw materials and the energy it needs for growth, repair and reproduction. We are so used to seeing animals feed that it is easy to take their feeding behaviour for granted. It has only been within the last 20 years that biologists have begun to investigate the extent to which animals can be said to feed, or forage, optimally. Many behavioural ecologists have looked at the extent to which animals maximise their energy intake. We will look at optimal foraging in two very different animals and in a plant.

7.3.1 Optimal foraging in crows

In many of the coastal areas of Canada, crows (*Corvus caurinus*) feed on shellfish. On the west coast, they hunt for whelks (*Thais lamellosa*) at low tide. Having found a whelk, they fly with it in their beak to above a nearby rock. Here they stall and drop the whelk from the air so as to smash its shell on the rock and so expose the flesh inside. Zach (1979) noted that crows always drop the shell from a height of about 5 m. However, the whelk does not usually break when it is first dropped. Typically a crow has to drop each whelk several times in order to break it open.

A crow obviously has to expend energy in order to fly upwards. Zach hypothesised that crows might drop whelks from that height at which the crows would minimise the *total* upward vertical flight required per whelk eaten. Experiments showed, as one would expect, that if whelks are dropped from near the ground, many drops are required to break open the shell. At greater heights, fewer drops are needed (Figure 7.1). As a result of these experiments, Zach was able to calculate the total upward vertical

flight necessary to break open an average whelk dropped from different heights. As you can see from Figure 7.2, the calculated dropping height that minimises the total upward vertical flight is very close to the crows' average of 5.2 m.

Inspection of Figure 7.2 shows that although dropping the shell from a height of less than 4 m is so unlikely to break it that too many drops are needed, any height from 5 m to 11 m would fulfil the optimisation criterion 'minimisation of upward vertical flight per whelk broken'. Zach suggested that the greater the height from which a whelk is dropped, the more the chance that the whelk will fragment into many small fragments, some of which are too small to retrieve. This may be the reason why the crows usually only fly to a height of about 5–6 m, rather than 10–11 m. If enough data were available this hypothesis of Zach's could be tested. The new optimisation criterion might be 'minimisation of upward vertical flight per gram of whelk flesh eaten'. One of the advantages of making quantitative predictions is that it enables one's predictions to be tested precisely and forces one to be rigorous.

7.3.2 Foraging in African elephants

The African elephant (*Loxodonta africana*) is the largest living land mammal. Adult males weigh up to about 5500 kg, adult females up to about 2800 kg. As one might expect, these animals face very different problems in satisfying their nutritional requirements than do the crows considered in Section 7.3.1. For a start, African elephants are herbivores. This is a characteristic they share with Asian elephants and all the other very large land mammals: giraffes, hippopotamuses and the various Asian and African rhinoceroses are all herbivores. It is difficult to

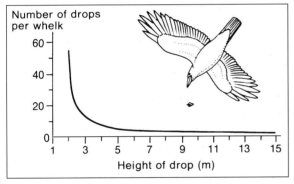

Figure 7.1 Mean number of drops required to break open whelks (*Thais lamellosa*) when dropped by crows (*Corvus caurinus*) as a function of the height from which the whelks were dropped. (After Zach, 1979.)

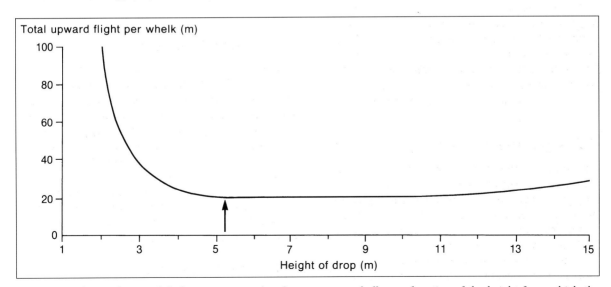

Figure 7.2 The total vertical flight necessary to break an average whelk as a function of the height from which the whelks are dropped. Arrow indicates mean height used by crows. (After Zach, 1979.)

imagine such large animals being agile enough to hunt their prey as lions and tigers do.

Although elephants eat a great deal of food each day, their daily energy requirements are not, as was discussed in Chapter 2, simply proportional to their body weight. Relative to their body weight, large animals need *less* food per day. The daily food intake of an adult male is about 60 kg dry matter (Owen-Smith, 1988). Males and non-lactating females eat about 1.0–1.2% of their body mass per day: lactating females about 1.2–1.5% (expressed as dry mass of food/live mass of elephant).

Elephants consume both grass (Figure 7.3) and browse (Figure 7.4). Browse, that is the leaves and young shoots of shrubs and trees, is usually more nutritious than grass. However, browse is less abundant than grass. It probably takes an elephant longer to find and consume, say, 10 kg of browse than 10 kg of grass. It is only small herbivores that can feed entirely off browse. One advantage to an elephant of being very big is that food remains in the digestive system for quite a long time. This means that more of the nutrients can be extracted. Consequently, large herbivores are able to survive on lower quality food than can small herbivores.

7.3.3 *Optimal foraging in plants*

It may seem rather odd to talk about foraging in plants, but on reflection the concept is a useful one which allows us to look at the capture of light by plants in a new way. Consider first of all the arrangement of leaves on a tree. They are carefully positioned so as to intercept as much of the sunlight that falls on the tree as possible. Consequently, the best arrangement for the plant is one which ensures that leaf overlap is as even as possible. It is better for a plant to position its leaves so that light always has to pass through, say, four leaves before reaching the ground, rather than having its leaves so that some light is never intercepted, while some has to pass through eight layers of leaves to reach the ground (Figure 7.5). In such a way the tree can be said to be optimising its foraging for sunlight.

What is true of leaves is also true of roots. Roots obtain minerals and water. It is important therefore that they ramify to penetrate as much soil as possible. Equally, the roots of an individual should compete with each other as little as possible. That is, roots should be evenly spaced. An exception to this prediction would be where there is a regular gradient of nutrients and/or water. In this case one would obviously expect roots to cluster near the relevant sources.

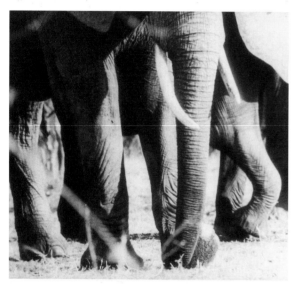

Figure 7.3 African elephants (*Loxodonta africana*) obtaining grass by kicking it from the ground with their feet (Luangwa Valley, Zambia).

Figure 7.4 African elephant using its trunk to browse (Luangwa Valley, Zambia).

Figure 7.5 The overlapping arrangement of maple leaves showing that much of the sunlight is intercepted as it passes through the tree canopy.

Ground ivy (*Glechoma hederacea*) is a perennial herb, native to Europe and Asia, commonly found growing in woods and grasslands and on waste ground (Figure 7.6). A plant produces **stolons** which are thin horizontal stems that grow roots at nodes where leaves arise. At the end of a stolon a plant produces a **ramet**. This is a morphological unit with the potential for independent existence. Ramets are therefore produced vegetatively and are genetically identical (see Section 6.2.6). A plant, of whatever size and however divided into ramets, which originates from a single seed is called a **genet**, as all the parts share exactly the same genes.

A genet of ground ivy can be thought of as sending out stolons which produce ramets which gather in water, nutrients, carbon dioxide and sunlight. The effects of soil nutrient availability on the development of genetically identical ground ivy clones were investigated by Slade and Hutchings (1987). Ramets were planted individually in a greenhouse into pots containing either nutrient-poor sand or sand mixed with a commercial plant food. Clones growing in the nutrient-rich sand had short stolons, copious branching and a rapid accumulation of many large ramets with large leaf areas. Proportional allocation of dry weight to leaves and petioles was high, whereas allocation to stolons and roots was low. The reverse was the case for the nutrient-poor clones.

These had long stolons, less frequent branching and a few small ramets with small leaf areas. Proportional allocation of dry weight to leaves and petioles was low, whereas allocation to stolons and roots was high.

The results of this study of growth in ground ivy can be interpreted if one considers that the plants are foraging for nutrients. When a single clone hits a patch that is poor in nutrients, it makes sense to produce long stolons and branch little, so as to maximise the chances of escaping from that patch to a better area. When a patch rich in nutrients is found, the plant behaves adaptively by producing short stolons and branching copiously, thus foraging intensively. In these nutrient-rich patches, plants can afford to produce many leaves. In nutrient-poor patches, plants do best to hurry on as quickly as possible in search of a better patch.

More recent work has investigated the ability of *Glechoma hederacea* to forage in patchy environments (Wijesinghe & Hutchings, 1997). Plants were grown in eight experimental environments, each containing a patchwork of nutrient-poor and nutrient-rich soil. When the patches are large (25 x 25 cm) the plants, as one would expect from Slade and Hutchings (1987), do well: they seek out the richer patches. Interestingly, when the patches get too small (6.25 x 6.25 cm), the plants do as badly as if they were

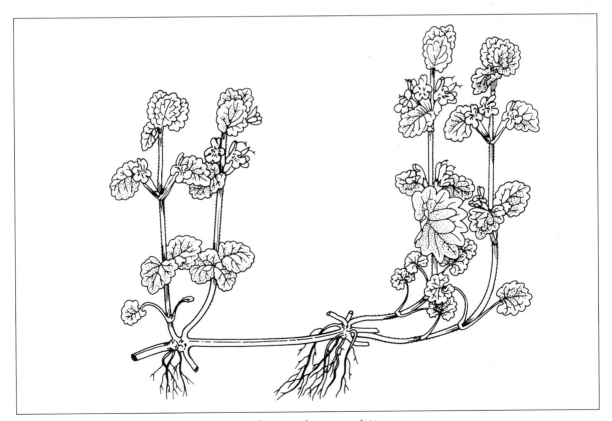

Figure 7.6 Ground ivy (*Glechoma hederacea*) sends out stolons to exploit new areas.

grown in a uniformly poor environment. It seems that the environmental variation is on too small a scale for them to respond to it.

7.4 Growth versus reproduction

Organisms use the nutrients and energy that they obtain from the environment to grow and reproduce but the functions of growth and reproduction are in conflict. If an organism puts a lot of its resources into growth, it will have less to put into reproduction and vice versa. In many trees the width of their annual rings provides an index of how much a tree is growing. In Douglas firs (*Pseudotsuga menziesii*) there is an inverse relationship between the number of seed cones produced and tree ring growth (Figure 7.7).

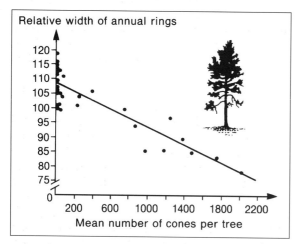

Figure 7.7 The relationship between the number of cones produced by individual trees of Douglas fir (*Pseudotsuga menziesii*) and that year's growth. (After Eis *et al.*, 1965, from Silvertown, 1987.)

Reproduction may have other costs as well as reducing growth. In red deer (*Cervus elaphus*) adult females produce a single calf most years, but not every year (Clutton-Brock *et al.*, 1982). Unless the adult females are very old, they rarely die in the years when they have not got a calf. However, in years when they have got a calf, and therefore have all the costs of pregnancy and milk production to bear, the adult females have a significantly greater risk of dying (Figure 7.8). Most deaths occur at the very end of winter as a result of starvation, just before the grass starts growing again in spring.

7.5 Reproducing only once versus reproducing several times

Section 7.4 considered the trade-off between reproduction and growth and between reproduction and survival. There are many other ecological trade-offs subject to natural selection. When an organism does reproduce, it has the choice of how many offspring to produce, whether or not to provide parental care for them and whether to reproduce again or not. (It must be emphasised that organisms do not make these decisions consciously! Depending on the ecology of the species, natural selection operates to ensure that certain strategies are favoured in some situations but others in different ones.)

A species is said to be **semelparous** if individuals reproduce only once in their lives and **iteroparous** if they reproduce on two or more occasions. Salmon and most bamboos are examples of species which are semelparous. They grow throughout their lives and then put all their energy into just one massive episode of reproduction. We are more familiar with iteroparity. This is what happens, for example, in

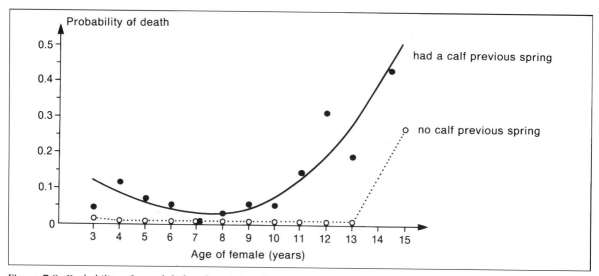

Figure 7.8 Probability of an adult female red deer dying as a function of her age and whether or not she gave birth to a calf the previous spring. (After Clutton-Brock *et al.*, 1982.)

most trees and nearly all mammals and birds. Theoretical ecologists have tried to predict which species should be semelparous and which iteroparous. Unfortunately, as is often the case, it is easier to produce such theoretical models than to obtain the data to test them.

One study which investigated why some species are semelparous and others iteroparous was carried out by Schaffer and Schaffer (1977). Over three summers they studied species of *Yucca* and *Agave* in western North America. *Yucca* and *Agave* are semi-succulent plants and both genera contain semelparous and iteroparous species. Yuccas are pollinated exclusively by moths; agaves mainly by large bees. When Schaffer and Schaffer looked at the behaviour of the pollinators of the semelparous species, they found that the pollinators, whether bees or moths, preferred the individual plants with the largest flowering spikes. However, the pollinators of the iteroparous species showed no preference for individual plants with the largest spikes.

Apparently individual plants in the semelparous species are driven by pollinator behaviour to produce larger and larger spikes. In these species, an individual plant that produced only a small flowering spike would attract few pollinators and thus fail to get many of its flowers fertilised. However, although producing a massive flowering spike has the advantage that most of the flowers on it are fertilised, it has the disadvantage that the parent plant puts so much of its resources into this spike that it has insufficient reserves left to enable it to survive to reproduce again. In other words, it is semelparous. Conversely, in those yuccas and agaves where the pollinators show no preference for massive flowering spikes, individual plants produce smaller spikes and thus have sufficient reserves to enable them to survive to reproduce again. As is so often the case in ecology, answering one question only gives rise to another. It is not known for certain why some pollinators prefer massive flowering spikes while others do not.

7.6 Parental care

Whether an individual reproduces on one occasion in its life or on several occasions, the females producing the offspring have to decide how much to invest in each offspring. (As discussed in Section 7.5, these decisions are not made consciously, but are the result of millions of years of natural selection.) Just as there is a trade-off between an individual investing a lot in growth and its investing a lot in reproduction, so a female cannot both produce a large number of offspring and invest a great deal in each one. In animals, males too have the option of providing care for their offspring.

7.6.1 *Offspring size*

There are many options open to organisms about how they invest the energy available to them for reproduction. The female can have a few large offspring, or she can produce a large number of small offspring. Equally, the offspring may be abandoned soon after fertilisation to fend for themselves or they may be given considerable parental care. Species such as ferns and corals abandon their young having invested very little in each one, while mammals and many seed-bearing plants invest substantially in each offspring. The variation in offspring size is perhaps most apparent in plants. The seeds of the orchid *Goodyera repens* each weigh as little as one microgram (10^{-6}g), while that of the double coconut (*Lodoicea maldivica*) weighs in at about 20 kg (2 x 10^4g) (Harper *et al.*, 1970). A classic study by Black (1958) on the clover *Trifolium subterraneum* showed that larger seeds gave rise to clover seedlings which did much better in intraspecific competition. This is hardly surprising; larger seeds have greater food reserves, enabling the seedlings they produce to grow vigorously and have the greatest chance of outcompeting their neighbours for space and sunlight.

If large seeds do so well, why are some seeds still as small as those of *Goodyera*? Are there disadvantages to being a large seed? From the point of view of the seed, there are not usually many. However, from the point of view of the parent plant which produced that seed, the larger each seed is, the fewer it can produce in total. There are other disadvantages to producing large seeds. Large seeds disperse over smaller distances and may attract herbivores which see them as valuable food sources.

Overall, there is an evolutionary balance between the number and size of offspring that a parent has. Consider, for instance, the development of marine benthic invertebrates. Here there are some species which produce a small number of large eggs; other species produce large numbers of smaller eggs. The former strategy is more common in poor environments. It seems probable that in poor environments small eggs would stand almost no chance of developing to adulthood. This means that a parent does better to produce a small number of large eggs, each of which has a reasonable chance of survival (Christiansen & Fenchel, 1979; Sibley & Calow, 1986).

7.6.2 *Which sex looks after the offspring?*

In animals, once an offspring has been born or an egg laid, there are four patterns of parental care possible:

Neither parent provides any parental care;
Mother alone provides parental care;
Father alone provides parental care;
Both parents provide parental care.

To try to explain why any one of these systems is found in a particular situation, behavioural ecologists have found it most valuable to focus on the **strategies** open to each parent. Consider, for instance, the situation in mammals. Here internal fertilisation is a physiological fact. The newly fertilised egg is retained and nourished in the mother's body for some time before birth. After birth, the mother is therefore literally left holding the baby. The father can do one of two things: either he can leave the mother to bring up the offspring on her own or he can remain and help. Help can be provided by males in a variety of ways in mammals. The male may provide the offspring with food; he may bring food to the mother either before or after she produces their offspring; or he may provide some sort of defence for the mother and/or their offspring. One thing males do not seem to be able to do is to provide milk. This may explain why polygyny is the rule in most mammals, in which a successful male mates with several females. A male's best strategy is usually to try to mate with as many females as possible and leave each to look after the offspring by herself.

In fish all four patterns of parental care are found (Gross & Shine, 1981; Krebs & Davies, 1987). Most fish show external fertilisation and in these species it is usually the case that neither parent provides any parental care. However, when parental care is provided, it is more likely to be provided by the male than by the female. This seems to be because in such species it is typical for the males to be territorial. A female comes to the male's territory, lays her eggs and then leaves. The male may then continue to defend his territory, hoping to attract further females. Unless the eggs float out of the male's territory, such continued defence of his territory includes defence of the fertilised eggs. In those fish species which show internal fertilisation, female parental care is the norm. This is hardly surprising. Unless the male remains with the female from fertilisation to egg-laying or birth, he will have little opportunity to look after the offspring or eggs.

7.7   Breeding systems in plants

The first angiosperm flowers were probably hermaphrodite: each flower containing both ovules and stamens. Over 70% of today's angiosperms still have flowers that are hermaphrodite, but almost 30% do not and show various alternative breeding systems. This is shown in Table 7.1 which summarises data collected by Yampolsky and Yampolsky (1922) on over 120 000 angiosperm species.

Why all this variation? Freeman *et al.* (1980) studied plants which sometimes change sex during an individual plant's lifetime. For instance, a plant which

Table 7.1 Classification of breeding systems in over 120 000 species of angiosperms. Hermaphrodites have male and female parts in each flower; monoecious plants have separate male and female flowers on the same individual; dioecious plants have individuals which are either male or female; andromonoecious plants are individuals with flowers some of which are male and some of which are hermaphroditic; gynomonoecious plants are individuals with flowers some of which are female and some of which are hermaphroditic. (Data from Yampolsky & Yampolsky, 1922.)

Breeding system	Percentage of species
Hermaphrodite	72%
Monoecious	5%
Dioecious	4%
Andromonoecious	2%
Gynomonoecious	3%
Hermaphrodites and individuals of one or other sex	7%
Some individuals monoecious and some dioecious	4%
Other combinations of the above	4%

has borne only female flowers may subsequently start to produce only male flowers. In the western USA male plants are typically associated with drier soils, females with moister soils. Pollen production obviously takes place earlier in the year than seed and fruit ripening. It has been suggested that this means that a plant which found itself growing a long way from permanent water would be unlikely to obtain sufficient moisture for any seeds and fruits to ripen. Under these circumstances an individual plant that finds itself in a dry patch does best to specialise as a male. Plants in damper patches then do best to specialise as females.

So why do the majority of angiosperms have hermaphroditic flowers when the data above suggest it would be better for individual flowers to specialise as male or female? If this were the case we would expect plants to be either dioecious (with each plant being either male or female) or monoecious (with each plant bearing male flowers and female ones). Charnov (1984) points out that in the gymnosperms (cone-bearing plants) this is precisely what is found. Almost all gymnosperms are either monoecious or dioecious. It may be significant that nearly all gymnosperms are wind pollinated, whilst the majority of angiosperms use animal vectors to carry their pollen.

Imagine what would happen to a female animal-pollinated angiosperm visited by only a very few pollinators. Unless those pollinators bring pollen from a plant of the same species, the female will fail to set any seed. On the other hand, a plant with hermaphroditic flowers will at least achieve some self-fertilisation.

7.8 Alternative strategies

In the example of crow foraging investigated by Zach and discussed in Section 7.3.1, there was a single height from which all the crows preferred to drop their whelks. Often, however, it is found that two or more **alternative strategies** are adopted by members of a species in response to a problem. In North American bullfrogs (*Rana catesbeiana*), for example, males defend territories in ponds to which females come and lay their eggs (Howard, 1978). Females prefer to lay their eggs in warm territories and in territories where the vegetation is not too dense. Both factors reduce predation on the eggs by leeches. When the water is warm, the eggs develop more quickly and are therefore exposed to predation for fewer days; if the vegetation is not too dense, the eggs form a spherical mass, which leeches find more difficult to attack than the thin film of eggs that develops on top of the plants in territories with a dense mat of vegetation. As might be expected, males fight for access to those territories in which females prefer to mate and lay their eggs. The heaviest, strongest males end up in the best areas, where they croak loudly to advertise their presence to other males.

Rana catesbeiana, however, shows the existence of alternative male mating strategies (Figure 7.9). Small young males are too weak to defend territories against larger, stronger males. Instead they behave as **satellites**, sitting silently within the territory of a stronger male and attempting to intercept and mate with the females he attracts. In this they are not very

successful. Only about 3% of matings are by such satellite males. Nevertheless this is perhaps the best chance a small male bullfrog has of obtaining matings. Satellites make the best of a bad job by parasitising the largest calling males on the best territories where most females are attracted.

In the green treefrog (*Hyla cinerea*) males show the same two alternative strategies – calling and behaving as satellites. In this species, however, a satellite appears to obtain about as many matings as a calling male (13 versus 17 in the one study done), even when females attempt to reject satellite males (Perrill *et al.*, 1982). In this case it seems that the two strategies may be equally successful. This accords with the observation that individual males frequently switch strategies and that there is no tendency for calling males to be larger.

As young bullfrogs age and grow, they may switch from being satellites to being callers. Behaving as a satellite when young, therefore, has the advantage, in addition to securing the occasional mating, of giving first-hand experience of the breeding system. In other species, mating strategies may persist for life. In gelada baboons (*Theropithecus gelada*) males have two ways of obtaining a harem of females (Dunbar, 1984). The demography of gelada society is such that almost every male who survives to adulthood will obtain a harem at some point in his lifetime. Some males pursue a low-cost/low-gain strategy. They quietly join a relatively large harem belonging to another male as a submissive follower, and some years later bud off a harem of their own. Other males opt for the high-cost/high-gain strategy of challenging a harem-holder directly. If the challenger wins the fights that ensue, he takes over the harem and immediately begins mating with females, many of

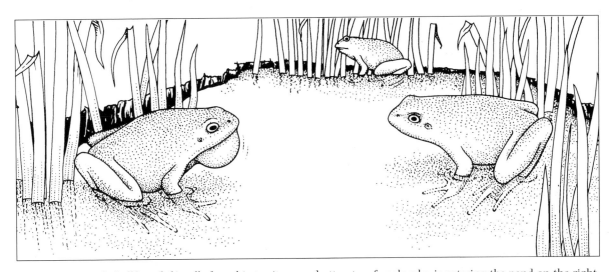

Figure 7.9 A male bullfrog (left) calls from his territory and attracts a female who is entering the pond on the right. In the middle of the picture is a small male who sits silently in the larger male's territory as a satellite and attempts to intercept the female on her way to the caller. (After Krebs & Davies, 1987.)

whom come into oestrus prematurely. Meanwhile, the defeated former harem-holder retires to follower status in the same harem and thereafter essentially spends his time investing in his immature offspring. Dunbar showed that the two strategies are equally effective in producing offspring.

In other species alternative strategies may persist for a lifetime, yet result in very different payoffs. In the bee *Centris pallida* studied by Alcock (1979), males practise one of two very different mate-location behaviours: patrolling or hovering. Patrolling involves hunting over a large home range in an emergence area, searching for sites where buried female virgins are about to emerge from pupation. Patrollers try to find these females, excavate them and mate with them. Other males hover in and around emergence areas or near flowering trees, waiting for receptive virgin females to fly by them. Such virgin females are ones that have emerged from the ground without being mated by a patrolling male. This happens either because the female emerges from an isolated nest or because when she emerges a group of males is so busy fighting that

none succeeds in mating with her before she flies off to a flowering tree to feed. Such females constitute a pool of potential mates and are more likely to be detected by males that are hovering and scanning the air overhead than by patrollers searching close to the ground for excavation sites. When a hovering male sees a female he pursues her and tries to capture and mate with her.

Male head widths range from 4.6 mm to 6.2 mm, which means that some males weigh three times more than others. Size is strongly correlated with the mating tactics that males employ. As can be seen from Figure 7.10, large males patrol for pre-emergent females, while small males specialise in hovering. Large males tend to patrol because of aggression between males at digging sites. When a male, using his sense of smell, finds a spot containing a pre-emergent female, there is no guarantee that he will ultimately succeed in mating with her. Excavating a female takes six minutes on average, during which time other males are likely to arrive. Violent fights subsequently occur, usually won by the largest male. This means that small males are better advised to hover rather than patrol, even though large males do most of the mating.

7.9 Games theory

For the crows considered in Section 7.3.1, the optimal height from which a particular crow does best when it drops its whelk is independent of the heights from which other whelks are being dropped. The strategy adopted by each individual does not depend on what the other individuals are doing. This is not always the case in biology. In many situations the benefit or **payoff** of the behaviour of an individual depends on what other individuals are doing. The theory which investigates the strategies that individuals should adopt in such situations is called **games theory**.

Consider, for example, the case of the green treefrog described in Section 7.8, where males could either hold territories and call or behave as silent satellites, and where the two strategies appear equally effective at securing matings. Imagine a population of treefrogs where all the males are calling. The first male to behave as a satellite will probably do much better than the other males. Croaking requires a surprisingly large amount of energy, so that satellites should be able to last longer on their stored food reserves. Additionally, croaking may carry some other disadvantage. It is known that in some amphibians calling attracts predatory bats (Figure 7.11). In such situations behaving as a satellite would be better than croaking, provided that there were still croakers around to attract females in the first place! We can envisage that at equilibrium a population

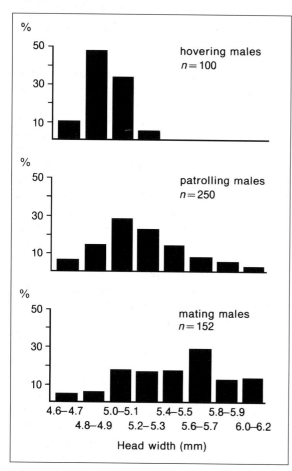

Figure 7.10 The distribution of males of different head widths in the bee *Centris pallida* that were captured while hovering, patrolling and copulating. (After Alcock, 1979.)

Figure 7.11 A fringe-lipped bat (*Trachops cirrhosus*) catching a male puddle frog which had been calling at a Panamanian pond.

might consist of both callers and satellites.

The notion that alternative strategies might co-exist in equilibrium was presented by Maynard Smith and Price (1973) who introduced the notion of an **evolutionarily stable strategy** (ESS). An evolutionarily stable strategy is one which, when most members of a population adopt or play it, cannot be bettered by any other strategy. The term was first used in an analysis of fighting behaviour. Suppose, for the moment, that an animal may adopt only one of two strategies in a conflict – to behave as either a **hawk** or a **dove**. Hawks fight much harder than doves and consequently win in conflicts with doves. The disadvantage of being a hawk is the risk of being hurt in a fight with another hawk. Imagine the following situation. When a dove fights a dove, one of them eventually gives up and the other wins the contest. Over a number of conflicts each gets on average +5 points. (A point might be a unit of food or a mating.) When a hawk and a dove fight, the hawk always wins, gaining +10 points, the dove scoring 0 points. When two hawks meet they get on average −5 points as one of them gets hurt. It can be shown (Box 7, page 75) that neither hawk nor dove alone is an ESS. For a population to be in equilibrium it transpires that half the individuals must be hawks and half doves.

Games theory has now been applied to a huge number of cases in evolutionary biology, and we will refer to it again. It is true to say that the theory of the subject is now understood in some detail although field studies that can test predictions of the theory are rare. This is not altogether surprising. Even the simplest models usually require the costs and benefits of strategies to be measured in units of **lifetime individual reproductive success**. Yet there are only a small number of species for which we have good data on lifetime individual reproductive success (Clutton-Brock, 1988). Another problem is that it can never be proved that a superior strategy does not exist. For instance, in the hawk–dove example, it can be shown that a population of hawks and doves can be invaded by the **bourgeois** strategy of 'Play hawk when at home and dove when away from home'.

All this may sound rather theoretical. Yet an example where the bourgeois strategy in nature seems to settle conflicts is found in the speckled wood butterfly (*Pararge aegeria*) (Davies, 1978). In speckled wood butterflies males compete for patches of sunlight on the ground layer of woods. Females are attracted to these spots, and it is here that nearly all courtship occurs. In Davies' study only about 60% of the males held these temporary territories at any one time; the remainder patrolled for females up in the canopy top. Such patrolling males continually flew down from the canopy and rapidly took over vacant sunspots. If, however, the sunspot was already occupied, the intruder and territory owner went into a short spiral flight lasting a few seconds, at the end of

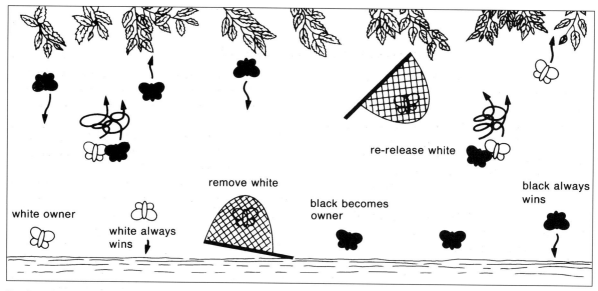

Figure 7.12 An experiment which shows that the rule for settling contests for territories in the speckled wood butterfly is 'resident always wins'. One male is arbitrarily represented as white, the other as black. (After Davies, 1978.)

which, in every one of 210 observed cases, the intruder retreated even if he was in prime condition and the territory owner had tatty wings.

Apparently the males are using the bourgeois strategy to settle their contests. Further evidence for this came from removal experiments (Figure 7.12). Davies' final set of experiments consisted of trying to release a second male into an occupied territory without either butterfly noticing the other's presence, so that both butterflies believed themselves to be the territory holder. This was rather difficult, but success was achieved on five occasions. A spiral flight soon ensued which lasted, as Table 7.2 indicates, over 10 times as long as normal. Evidently each butterfly thought that it was the territory holder, and so neither backed down. It now appears that sitting in a sunny patch warms up the butterfly, making it more active (Stutt & Willmer, 1998) and thus more likely to win a territory fight. As usual in ecology, the story may be more complex than it seems at first.

Table 7.2 The spiral flight between two male speckled wood butterflies is much longer when they both believe they are the territory holder than when one is definitely the territory holder and the other definitely an intruder. (After Davies, 1978.)

Male behaviour	Spiral flight duration(s) Mean ± 1 SE
Unambiguous territory ownership by one male	3.65 ± 0.23
Both males behaving as territory holders	39.60 ± 7.35

7.10 Constraints on adaptation

Suppose a biologist, having set out to investigate the function of a structure or a behaviour, produced a mathematical model which predicted, say, the length of a bird's wing or the rate at which its offspring should be supplied with food, but then found that the predictions disagreed with the facts of the situation. What could our biologist conclude? There are two possibilities; either the model is at fault, whether in its overall hypothesis or in some particular details, or the structure or behaviour under investigation is **suboptimal**.

It is hardly surprising that optimisation models are often at fault. For one thing, they are almost bound to be oversimplified. A preliminary model to predict the length of a bird's wing might concentrate initially on the length most economic for flying. Subsequent modifications of the model could incorporate other functions of birds' wings, such as thermoregulation and display. Quantifying the precise costs of a structure or a behaviour is likely to be extremely difficult. Time, energy and other resources devoted to one item cannot be used for others. In general, optimisation models may be at fault by failing to identify correctly what is being optimised, by failing to consider all the options open to the organism, by failing to identify all the constraints involved (e.g. birds' wings may be used for thermoregulation as well as for locomotion) or by making invalid assumptions about the hereditary nature of the trait in question (e.g. selection for alleles for larger wings might also lead to larger birds overall).

A number of biologists (e.g. Gould & Lewontin, 1979) have pointed out that it is easy to fall into the

---

Box 7
Using games theory to analyse fighting behaviour

Consider two possible strategies for an animal faced with the prospect of a fight: either behave as a 'hawk'; or behave as a 'dove'.

Hawks are aggressive and fight to win. In the process, however, they risk injury to themselves. Doves merely display. Their strategy is to hope the other individual gives up after a while. Being a dove has the disadvantage that one may waste a lot of time displaying. On the other hand, doves, unlike hawks, never get injured.

Let us now construct what is called a **payoff matrix**. A payoff matrix shows the results of a contest. Typical values for the payoff matrix might be as follows:

Contestants	Change in contestant fitness after encounter
Hawk vs hawk	On average each loses −5 fitness units
Dove vs dove	On average each gains +5 fitness units
Hawk vs dove	Hawk gains +10 fitness units; dove gains 0

Now imagine a population consisting entirely of doves. The result of an encounter between two doves is that each, on average, gains +5 fitness units. However, consider what would happen if just a few doves start to behave as hawks, or if a few hawks arrive in this population from elsewhere. These hawks almost never encounter each other, as there are so few of them. Almost all of their encounters, therefore, are with doves. But when a hawk meets a dove, the hawk gains +10

fitness points. This is better than doves do even when they meet each other, let alone when they meet a hawk. This means that a population consisting entirely of doves is open to invasion by hawks.

In exactly the same way imagine a population consisting entirely of hawks. A fight between hawks leads to each, on average, losing 5 fitness units. This population is open to invasion by doves. Doves do not do very well in encounters with hawks – they 'gain' 0 fitness points. However, 'gaining' 0 is better than losing 5, which is what a hawk can expect to get as the result of an encounter with another hawk.

We can determine what the ratio of hawkish to dovish behaviour needs to be for the population to be stable so that it cannot be invaded either by more hawks or by more doves. Suppose that a proportion p of the population are doves and therefore a proportion $1 - p$ are hawks. Then, for a reasonably large population, the average payoff to a dove of a contest is $+5p$. The average payoff to a hawk is $+10p - 5(1 - p)$. For a population to be in stable equilibrium, the two payoffs must be equal. Therefore we have:

$$10p - 5(1 - p) = 5p$$
i.e.
$$p = 0.5$$

This means that for the particular set of values we have chosen, either the population should consist of equal numbers of hawks and doves, or each individual should play a **mixed strategy**. 'Play dove with probability 0.5 and hawk with probability 0.5'.

---

mistake of believing that all organisms, and all parts of all organisms, are perfectly adapted. This is, after all, highly unlikely. After all, natural selection presumably still operates, and natural selection obviously works by weeding out imperfections, which clearly implies the existence of such imperfections. A number of reasons can be identified for expecting sub-optimality (Dawkins, 1982), namely time lags, historical constraints and the lack of available genetic variation.

7.10.1 Time lags

Van Valen (1973) argued that organisms are like the **Red Queen** in Lewis Carroll's book *Through The Looking Glass* – perpetually running to stay in the same place (see Section 19.3.3). Species are built under the influence of genes selected in some earlier era when conditions may have been different. Moths often fly into lights. This is thought to be because in the days before candles and lamps were invented, small sources of bright light, as perceived by moths, were sure to be celestial bodies such as stars. Flying

at a fixed angle to such an object enables the moth to fly in a straight line. Flying at a fixed angle to a candle, however, results in a spiral to oblivion.

We would expect the problems of time lags to be especially acute for our own species. Clearly our life styles have altered by so much and so recently that the rest of our biological make-up cannot be expected yet to have caught up. This is why so many people suffer from tooth decay, atherosclerosis, bowel cancer and other 'modern' diseases.

7.10.2 Historical constraints

Question: 'What do giraffes have that no other animals have?' Answer: 'Baby giraffes' (or 'Parent giraffes'). In other words, organisms evolve from what was there before. A giraffe cannot suddenly evolve a trunk or tusks. The arrangement of the neurones in the vertebrate eye provides an example of the significance of historical constraints. In vertebrate eyes the retina appears to be back to front. The light-sensitive cells are at the back of the retina, and light has to pass through the intervening cells, with some inevitable distortion, before it reaches them. In cephalopods, however, the retina is the 'right way round' and this is presumably more efficient. The present vertebrate condition is the result of some bit of evolution that happened millions of years ago, with which vertebrates are now stuck.

7.10.3 Lack of genetic variation

Cat breeders have not yet bred a domestic cat with true spots, valuable though such an animal would be. A fortune awaits the first person to breed a yellow African violet. In these and other cases the necessary genetic variation simply has not yet manifested itself. Whether there are many cases where natural, as opposed to artificial, selection has thus been foiled by the absence of genetic variation is less certain. One instance where natural selection does often seem to lack the necessary genetic variation is in the determination of an offspring's sex. In most species parents appear unable to control this, even though such control would often be adaptive. For example, in most mammals any particular female is equally likely to produce sons or daughters, even though there are cases where it would benefit a mother to be able to determine the sex of her offspring. There simply does not appear to be any genetic variation in the ability of mothers to determine the sex of their offspring. However, there are species, such as ants, bees and wasps, where mothers can control the sex of their offspring. In these species females are diploid and males haploid. Because of this, mothers can determine the sex of their offspring by controlling whether or not a sperm is allowed to fuse with each of her eggs. Fertilised eggs develop into females and unfertilised ones into males.

Summary

(1) Behavioural ecology is the study of how the behaviour of individuals enables them to survive and reproduce in their environments.

(2) Animals need to feed in order to obtain nutrients and energy. Optimisation theory can be used to identify the evolutionary pressures acting on the behaviour of animals.

(3) Crows and African elephants appear to forage in ways which maximise their rates of energy and nutrient gain.

(4) Plants can also be said to forage. The pattern of growth in *Glechoma hederacea* can be interpreted in terms of the plant seeking out patches rich in nutrients.

(5) Growth and reproduction are in conflict. More of one means less of the other.

(6) Reproduction may also conflict with survival.

(7) In some species individuals reproduce only once in their lives; in other species individuals reproduce on two or more occasions. The reasons for these differences probably lie in the ecology of the species.

(8) The balance between the number of offspring an individual produces and their size is largely determined by the ecology of the species.

(9) Patterns of parental care are best understood by looking at the evolutionary options open to each parent.

(10) Flowering plants show a variety of breeding systems. Most are hermaphrodite, but in some circumstances it pays flowers to specialise.

(11) Sometimes two or more alternative strategies are found within a species. For instance, in a number of frogs some adult males behave as callers, while others silently try to intercept females.

(12) Games theory can be used to interpret behaviour when the benefits of an individual behaving in a certain way depend on what other individuals are doing. One behaviour which games theory has helped to explain is the territorial behaviour of male speckled wood butterflies.

(13) When the predictions of an optimisation model disagree with what is found in nature, there are two possible explanations. One is that the model is at fault. For example, some important constraints may not have been identified. Another reason is that the structure or behaviour under investigation is suboptimal.

EIGHT

Sociobiology

8.1 Living in groups

Many animals are social, that is, they associate in a group with other individuals, usually of the same species. For example, starlings feed in small flocks and join large communal roosts at night, although during the mating season they go off in pairs (see Section 3.3). Other species, such as most termites, ants and bees, have much more rigidly organised social units than the starling. Whether an animal lives in a loose and changing association or in a tight family unit, it has to live with and relate to the other animals around it. The study of the origins and biological basis for such social behaviour is called **sociobiology**.

Many species are social, while others are solitary. This suggests that for some there are advantages to group living, while for others the disadvantages are such that a solitary existence is preferred by natural selection. Once a species evolves group living there often follow important consequences for the individuals in the group in terms of inter-individual relationships and group structure. This chapter looks at the advantages, disadvantages and consequences of social behaviour. The evolutionary pressures favouring helping behaviour are investigated as is the way in which social behaviour plays a vital role in the lives of many species including termites, army ants, lions and naked mole rats.

8.2 The advantages of group living

8.2.1 Less risk of predation

One frequent advantage of living in a group is that it reduces an organism's risk of being eaten. For example, woodpigeons (*Columba palumbus*) are safer in larger flocks. Kenward (1978) used a trained goshawk (*Accipiter gentilis*) to attack feeding flocks of woodpigeons and found that the predator's attacks were less successful when directed at larger groups of pigeons (Figure 8.1). Kenward found that the reason for this is that the larger a flock of woodpigeons, the

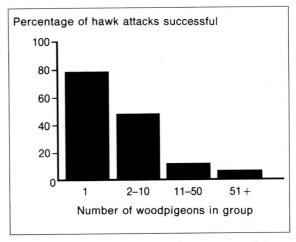

Figure 8.1 Goshawks (*Accipiter gentilis*) are less successful when they attack larger flocks of woodpigeons (*Columba palumbus*). (After Kenward, 1978.)

greater the distance at which they tend to detect predators. This is because the more birds there are in a flock, the greater the chance that one of them will spot a predator when it is still some distance off.

Another advantage of being in a group is that together the group may be able to defend itself from predators. The social insects are particularly good at this. Indeed, many social insects have soldiers, which are specialised for the defence of the group (Figure 8.2). Similarly, a group of musk oxen (*Ovibos moschatus*) is less vulnerable to attacks by grey wolves (*Canis lupus*) than are solitary individuals because a group of musk oxen can gather into a defensive circle when attacked by wolves. They stand facing outwards towards the wolves, their large horns forming a defensive wall to protect their young which hide within the circle of adults.

If a predator does succeed in attacking a group of potential prey, it is faced with the question of which individual to attack. In some cases it has been shown that a predator gets confused if there are several potential prey individuals to choose from. Neill and Cullen (1974) investigated the hunting behaviour of squid (*Loligo vulgaris*), cuttlefish (*Sepia officinalis*), pike

77

Figure 8.2 Red ants attacking a larger wood ant.

(*Esox lucius*) and perch (*Perca fluviatilis*) attempting to catch small fish in laboratory tanks. All four species were more successful when hunting individuals than when their prey was in shoals. Further, they were more successful hunting small shoals than larger ones.

Even in cases where predators are as successful in attacking large groups as attacking small groups or individuals, it still pays individuals to aggregate into groups (Hamilton, 1971). Although in this case the group as a whole does not benefit, each individual does better from being in a group. Consider two groups of potential prey, one with 10 individuals, the other with 100. If the two groups are equally likely to be attacked and have a single individual removed from them and eaten, then the individuals in the larger group are 10 times less likely to be eaten. This means that solitary individuals are more vulnerable to predation with the result that individuals are likely to aggregate into groups. Hamilton called this phenomenon the 'selfish herd' effect. Only if larger groups receive proportionately more attacks will the selfish herd phenomenon *not* favour the aggregation of individuals into groups under the threat of predation.

8.2.2 More chance of obtaining food

Another reason for living in a group is that food may be easier to obtain. A number of studies on the Serengeti plains of Africa have shown that lions (*Panthera leo*), spotted hyenas (*Crocuta crocuta*) and African wild dogs (*Lycaon pictus*) are all able to catch larger prey when hunting in groups (Kruuk, 1972; Schaller, 1972; Malcolm & van Lawick, 1975). Ward and Zahavi (1973) suggest that when some bird species gather into large flocks, they may be able to exchange information about food sources. A bird which observes that another individual appears

better fed than itself might do well to try and follow that individual in future. This was suggested as a benefit of the mass roosting behaviour of starlings in Section 3.3.3.

In fact, the suggestion that group living favours food acquisition has little evidence to support it. For example, although groups of lions can catch larger prey than individuals, the whole group has to share the meat, so there always seems to be less for each individual group member! This suggests that group living probably evolved for other reasons. The hypothesis that birds may benefit from living in groups by exchanging information about food sources has so far only been convincingly demonstrated in one species, the cliff swallow (*Hirundo pyrrhonota*) (Brown, 1986; Elgar & Harvey, 1987b).

8.2.3 Other advantages of group living

Many of the advantages of group living considered so far apply even if the group members belong to different species. Mixed-species groups do occur and sometimes may be favoured over single-species groups. Titmice (Paridae) often forage in mixed-species flocks. Presumably this is because the different species forage on different foods or in different ways and so minimise the level of food competition while maintaining the anti-predator advantages of group living. It may even be that mixed-species groups are *better* at detecting predators. Olive baboons (*Papio anubis*) and impala (*Aepyceros melampus*) are often found in mixed herds. This may be because together they stand the best chance of detecting predators. Impala have very sensitive noses and ears, while baboons, unlike impala, possess colour vision and often sit high up in trees looking out for predators. So between them they can smell, hear and see any approaching danger (Bertram, 1978a).

If individuals huddle together, they effectively decrease their surface area to volume ratio. This may help them to reduce the rate at which they lose water and warmth. Nestling great tits (*Parus major*) huddle together for warmth and lose less weight the more of them there are (Mertens, 1969). Woodlice (Isopoda) may huddle together to reduce dehydration.

The costs of locomotion may be reduced by moving in groups. It is known that fish in shoals sometimes show far greater endurance than fish of the same species and size swimming on their own. Weihs (1973) suggested that this might be because of the trail of vortices set up by a fish as it moves its body through the water. In Figure 8.3 a fish swimming directly behind fish A can be calculated to experience approximately 30% more resistance to movement than a fish swimming in position B, that is, midway between fish A and fish C. The effort required by fish B can be calculated to be less than that needed by a

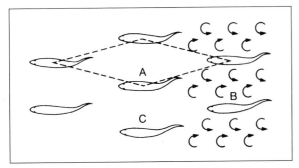

Figure 8.3 Diagrammatic representation of part of a fish shoal as seen from above. Lateral movements of the fish set up vortices shown as curved arrows. These mean that a fish swimming in position B, behind and mid-way between fish A and fish C, experiences less drag than if it was swimming on its own or directly behind either fish A or fish C. The dashed lines show the characteristic diamond shaped arrangement of fish in shoals. (From Weihs, 1973.)

Figure 8.4 An adult male red deer (*Cervus elaphus*) roaring in the autumn as he defends a harem of females against other males.

solitary individual swimming at the same speed. Photographs of fish shoals show the characteristic diamond patterns predicted by Weihs' calculations and shown by the dashed lines in Figure 8.3. Similar aerodynamic advantages have been suggested for geese and other large birds which sometimes fly in characteristic V-shaped formations (Gould & Heppner, 1974). It has been reported that the birds take it in turn to lead, rather like team pursuit cyclists.

Once a species does show permanent group living, a number of benefits may follow. Honeybees (*Apis mellifera*), like many other of the social insects, have taken full advantage of sociality. The 60 000 or so bees in a typical healthy midsummer honeybee colony have the 'beepower' to ventilate the nest and keep it at an optimum temperature for larval growth. The developing bees are given great protection and individuals can specialise in the type of food they collect and may learn from each other.

8.3 Disadvantages of group living

So far we have mainly looked at the advantages of living in a group, but there are disadvantages. For example, as mentioned for lions in Section 8.2.2, although belonging to a group may mean that more food can be obtained by the group, that food may have to be shared out among the group members. Being in a group thus means that there is potential competition for food. Just how important this competition is depends on the ecology of the species. Individuals in a large colony of rabbits (*Oryctolagus cuniculus*) may have to travel further each day for their food than if there were fewer individuals in the group. Among herbivores, such feeding competition is likely to be most acute for species such as rabbits which have a fixed home from which they go out to

obtain food and to which they frequently return. Competition for food is likely to be particularly severe in the social carnivores. For a carnivore at a carcass the presence of other individuals means a reduction in the amount of food available per individual. Such competition is probably less of a problem for herbivores grazing in a field or browsing among the leaves of a tree.

Within a group it may be easier for a few individuals to monopolise opportunities for breeding. In many species competition among adult males results in a **dominance hierarchy** among the males. The individuals at the top of the hierarchy obtain most of the matings. In some species competition between the males for females is so intense that some adult males never breed. For example, in the autumn, adult male red deer (*Cervus elaphus*) fight among themselves for the right to hold groups or harems of females (Figure 8.4). A successful male waits for the adult females in his harem to come into **oestrus** when they permit him to mate with them. Unsuccessful males live in bachelor groups or remain as solitary individuals, trying occasionally, but almost never with any success, to steal a female from a harem.

Females, too, may compete with one another for opportunities to reproduce. In grey wolves (*Canis lupus*), wild dogs (*Lycaon pictus*) and dwarf mongooses (*Helogale parvula*) each pack usually has an alpha, or dominant, female who does most or all of the reproducing. Cases have even been reported in wild dogs where breeding by a subordinate female has led to the killing of her pups by the dominant female.

For any one species, the balance between the advantages and disadvantages of group living may shift throughout the year or from place to place. In

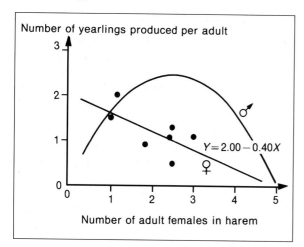

Figure 8.5 The effect of harem size in the yellow-bellied marmot (*Marmota flaviventris*) on the reproductive success of the single resident male and of the females in the harem. The straight line shows the regression of female reproductive success, measured in the number of yearlings produced per female, as a function of the number of adult females in the harem. The equation for the regression is given. The curve shows the effect of harem size on the reproductive success of the resident male. (From Downhower & Armitage, 1971.)

many species, group size varies with the supply of food or intensity of predation. It is also worth noting that natural selection may operate differently on males and females with respect to **optimal group size**.

Downhower and Armitage (1971) studied groups of yellow-bellied marmots (*Marmota flaviventris*) in Colorado, USA. This species forms stable harems, consisting of a single adult male, one to three adult females and their young. Downhower and Armitage studied a total of seven harems, five of them for seven years, two of them for five years. The straight line in Figure 8.5 shows how the average number of yearlings reared by each adult female depends on the number of adult females in the harem. It is apparent that the more adult females there are in a group, the fewer offspring each female marmot rears. This appears to be because the more females there are in a harem, the less food there is per female. Evidently, the optimal group for each adult female consists of one adult female (herself), one adult male and their young. However, an adult male's **reproductive success** equals the *sum* of the reproductive successes of the females in his harem. Another way of expressing this is that male reproductive success equals the average number of adult females in the harem multiplied by the average number of offspring surviving to each female. The curve in Figure 8.5 shows how male reproductive success depends on the number of females in the harem. Whereas females prefer a harem of one male and one female, males prefer a harem of one male and two or three females.

8.4 Optimal group size

We know a great deal about the advantages and disadvantages of animals leading social rather than solitary lives, or vice versa, but remarkably little about why animals feed or sleep or socialise in groups of precisely the size they do. Consider the woodpigeon example discussed in Section 8.2.1. Here a major advantage of flocking is to reduce predation. In order, however, to predict the optimal group size for woodpigeons, it is necessary to know the disadvantages, and any other advantages, of not being solitary. The consequences of living in groups of different sizes would have to be quantified as functions of group size in comparable units. It would not be adequate just to measure *benefits* of increased group sizes in terms of greater survival, but also *costs* in terms, say, of less food eaten per day due to competition.

Even if this difficulty could be overcome, one would need not just to consider the optimal group size of the group as a whole, but the optima for different classes of individuals. It is known that subordinate woodpigeons feed at the edges of large groups. As a result they are more likely to be killed by predators than are the central dominant birds, despite the fact that subordinates interrupt their feeding more often to look for such predators.

8.5 Evolution of helping behaviour

The existence of social behaviour is often accompanied by instances of helping behaviour or **altruism**. Darwin's (1859) theory of natural selection is frequently taken to mean that natural selection operates on individuals to ensure they maximise the number of their descendants. But Darwin himself raised the objection 'What about the social insects?'. These appear to raise a crushing blow to Darwin's theory. In a typical honeybee colony, for instance, there are tens of thousands of adult females. Yet only one of these ever reproduces! How can natural selection result in the evolution of such **worker sterility**?

8.5.1 Kin selection

Darwin pointed out that sterility in such circumstances could evolve by a process which he termed 'family selection' and which is nowadays known as 'kin selection'. **Kin selection** is the name given to the type of altruism where the decrease in an individual's fitness as a result of its spending time and energy helping its relatives is more than compensated for by the increased fitness of its relatives.

A convenient way to understand the principle of kin selection is to consider the options open to a worker ant, bee or wasp. She could produce her own offspring, or help the colony produce reproductives. Now if she adopts the latter course, the reproductives will be the offspring of her own mother, as in most social insects all the workers are offspring of a single queen. Consequently, the reproductives produced will be either her full or half sibs. (They will be her full sibs if their father is her own father; half sibs if they are fathered by another male.) So she will be contributing to the production of close relatives whether she tries to have offspring herself or helps the queen, her mother, to produce them. It is true that she might be more closely related to her own offspring than to those produced by her mother, and this would certainly be the case if those produced by her mother are only her half sibs. Yet set against this is the fact that the entire colony is geared up to the production of reproductives by a single queen who is carefully protected by the colony. For many social insects it seems virtually certain that for an individual worker to attempt to have her own offspring would be tantamount to suicide. The chances of them surviving would be very slim. Whatever the evolutionary origins of social behaviour in the social insects, and kin selection was undoubtedly important in that, there is no doubt that kin selection is a powerful force preserving worker sterility.

We will look at several examples of kin selection in the accounts below of social behaviour in a number of species. An early report of kin selection was provided by Blest (1963) in an elegant study of saturniid moths on Barro Colorado Island in the Panama Canal Zone. Blest looked at different species of moths and compared the activities of the females after they had laid their eggs. In some species the females died soon after producing their offspring, typically within 24 hours. In other species, females usually lived for several days after they had laid their eggs. Furthermore, these females became increasingly excitable with age and spent the last 12 hours of life in unrestrained flight activity.

Blest noticed that the females that lived for several days after having laid their eggs, and that flew conspicuously, were distasteful to predators and brightly coloured. On the other hand, the females that died soon after laying their eggs were not brightly coloured and normally relied on being cryptic to avoid predation. Blest suggested that post-reproductive survival could affect the survival chances of other members of the species. In the cryptic palatable moths, the longer a female survives after egg-laying, the greater the opportunity for predators to learn to recognise the colour and pattern of such palatable moths. So it is best for such females to die quickly after reproduction. However, the brightly coloured and distasteful moths help their relatives by prolonging their post-reproductive life span, thus giving predators the opportunity to learn that moths with such a distinctive patterning belong to a distasteful species which the predator would do best to avoid in future.

8.5.2 Reciprocal altruism

Kin selection does not involve reciprocation: the beneficiary of the altruism does not repay the altruist for the altruism. However, another way in which altruism can evolve is by **reciprocal altruism** (Trivers, 1971). In essence, reciprocal altruism is 'You scratch my back, I'll scratch yours'. One individual provides another with help, and the second subsequently reciprocates or pays back the first. A difficulty with reciprocal altruism is the risk of **cheating**. Cheating occurs when an individual fails to reciprocate. The benefits of cheating are low if there are many opportunities for reciprocal altruism, if individuals can recognise each other and if they have good memories. In such situations, cheats are liable to be found out and outlawed or ostracised.

There are only a few cases where reciprocal altruism has convincingly been shown to occur in nature. One of them is in the vampire bat (*Desmodus rotundus*) (Wilkinson, 1984). Vampire bats are colonial mammals. During the day they roost together, often in hollow trees (Figure 8.6). At night they feed on the blood of domestic stock, such as cattle, horses, goats and pigs. Compared to other bats they have good eyesight and a keen sense of smell, which probably helps them to locate their hosts. A bat will often land on the ground near potential prey and then silently move to the animal. The bat then painlessly inflicts a small wound with its needle-sharp incisors, secretes some anticoagulants in its saliva and laps up a meal which may take up to 20 minutes to ingest.

If an individual fails to find a meal, it is in trouble. Wilkinson found, by accident, that after about 50–60 hours without blood, a vampire bat starves to death. At his two study sites fully 18% of the individuals failed to obtain a meal on any one night. If these

Figure 8.6 A colony of vampire bats (*Desmodus rotundus*) at their daytime roost.

Box 8
The mathematics of kin selection

Common sense tells us that the more distantly related two individuals are, the less likely kin selection is to be involved. To take the most extreme case, kin selection cannot be the reason for altruism between unrelated individuals. The first full quantitative treatment of kin selection was given by Hamilton (1964). Hamilton was able to show that the condition for the spread of altruism through kin selection could be expressed by a single mathematical relationship:

$$\frac{b}{c} > \frac{1}{r} \tag{8.1}$$

where b is the benefit (in terms of Darwinian individual fitness) to the beneficiary of the altruism, c is the cost (again in terms of Darwinian individual fitness) that the altruist suffers and r is the degree of relatedness between the two individuals. (To give some examples of r, in an outbred population of a sexually reproducing diploid species such as ourselves, the degree of relatedness between a parent and his or her offspring is a half, between an uncle or aunt and a niece or nephew is a quarter, and between two first cousins is an eighth.)

For example, let us consider a pack of wolves in which only one male and one female breed while the other individuals in the pack provide these two with food and help protect the pack against other wolves. One can ask the question: 'Why doesn't a subordinate pair of the wolves go off and breed on their own?' Hamilton would argue that they should *unless* Equation 8.1 is satisfied. Suppose that if our pair of wolves went off and reproduced together, they could expect to rear N cubs. Then by *not* going off and rearing N cubs, but rather remaining in the pack and helping the dominant pair to breed, they each experience a cost c equal to N. Now suppose that by remaining in the pack and helping, the subordinate pair enable the dominant pair to rear n more cubs than they would have done had the two subordinates left the pack. Then the condition for the altruism of the subordinate female to be favoured by kin selection is for:

$$\frac{n}{N} > \frac{1}{r_f} \tag{8.2}$$

where r_f is the *average* of the degree of relatedness of the subordinate female to the dominant female and the degree of relatedness of the subordinate female to the dominant male.

Similarly, the condition for the altruism of the subordinate male to be favoured by kin selection is for:

$$\frac{n}{N} > \frac{1}{r_m} \tag{8.3}$$

where r_m is the average of the degree of relatedness of the subordinate male to the dominant female and the degree of relatedness of the subordinate male to the dominant male.

It is perhaps unsurprising that it has proved rather difficult to provide quantitative tests of kin selection! It is worth noting from Equations 8.2 and 8.3 that altruism may be favoured by one sex but not by the other. In most species of mammals males in a group are less closely related to the other males than females are to the other females. This is because a male usually leaves the group in which he was born before he reproduces and thus finds himself in a group of unrelated individuals. Females, on the other hand, tend to remain in their **natal groups**. Because of this, we would expect the females of most mammals to show more kin-selected altruism than the males.

It is worth giving an introduction to the population genetics of kin selection in order to show why Equation 8.1 holds. Imagine a population of organisms in which the only sort of helping behaviour that goes on is that individuals look after their own offspring for a time. Now imagine a mutation, giving rise to an allele **A**, such that individuals with the genotype **Aa** are altruistic to their full sibs. (This means that we have assumed that **A** is dominant to **a**. It can be shown that exactly the same conclusions result if this is not the case.) **aa** individuals, of course, are not altruistic to their sibs. To determine whether the allele **A** will start to spread through the population we have to consider what happens when **A** is rare. (Once **A** becomes common, a separate argument, which we will not give, is necessary.) Individuals with the genotype **Aa** are almost bound to mate with individuals with the genotype **aa**. (This is because, as **A** is rare, the chances of two **Aa** individuals mating with one another is very small.)

Consider an **Aa** offspring resulting from the mating of an **Aa** individual with an **aa** individual. On average, half of its sibs will be **Aa** and half **aa**. As only half of its sibs are **Aa**, the **Aa** individual will need to help a sib twice as much as it harms itself for **Aa** individuals to be fitter than **aa** individuals, which is the requirement for **A** to spread at the expense of **a**. In other words, for **A** to spread, we require:

$$\frac{b}{c} > 2 \tag{8.4}$$

as, in this case, $r = 0.5$. This is why the geneticist J. B. S. Haldane once remarked that he was prepared to lay down his life for two of his brothers!

individuals fail to feed within the next 24 hours, they are likely to die. What Wilkinson found was that a bat that had failed to obtain a meal was usually provided with regurgitated blood by a roost-mate which had successfully fed the previous night. Often such altruism was provided by a mother for her offspring.

However, on a number of occasions, the bat receiving regurgitated food was either unrelated or related only distantly to the bat providing the food. Further, experiments on unrelated captive bats showed that reciprocation takes place. It looks as though individuals remember from which individuals they have received blood and subsequently reciprocate if an opportunity presents itself.

8.5.3 *Group selection*

A third way in which it has been suggested that evolution may favour altruism is by **group selection**. The first really detailed arguments in favour of group selection were put forward by Wynne-Edwards (1962). Wynne-Edwards was primarily concerned with attempting to explain how population density is regulated. It seemed that nature provides many examples where individuals do not appear to be trying to maximise their offspring number. Rather, many species seem to be surrounded by an abundance of food. Wynne-Edwards argued that animal populations should restrict their population density and rate of reproduction rather than endanger their food supply through the over-exploitation typical, for instance, of many human fisheries.

But how could such careful groups evolve? Wynne-Edwards suggested that groups containing individuals who reproduced flat out, maximising the number of their offspring, would exterminate themselves by exhausting the local food supplies. On the other hand, prudent groups, where altruistic individuals restrained their reproduction, would outlive such selfish groups and come to predominate.

Most biologists, with some notable exceptions, have not reacted enthusiastically to the theory of group selection. This has been partly because of the data collected by field workers which show that individuals *do* seem to be maximising the number of their offspring. Another reason why few biologists consider group selection important is that it faces a fundamental theoretical problem (Maynard Smith, 1964). Imagine a group consisting of individuals reproducing sub-maximally as required by group selection. Now suppose that a mutation arises which causes its holder to maximise the number of its offspring. *Within* the group individual Darwinian selection will favour the mutation which will spread rapidly as the numerous descendants of the mutant take over.

Only if two strict conditions are met is group selection likely to be important. First, that prudent groups do indeed persist longer than groups with individuals maximising the number of their offspring. Second, that the movement of individuals between groups is very rare. If such movement is at all common, individuals which maximise their reproductive success are likely to move groups before their group goes extinct. The available data, which are admittedly still pretty incomplete, suggest that group selection is usually likely to be outweighed by the opposing pressure of individual selection (Maynard Smith, 1976), though there are a few species where group selection may be important (e.g. Nichol, 1996).

8.6 The unit of selection and social behaviour

A great deal of sociobiological argument has become dominated by thinking at the level not of the species, nor of intraspecific groups, nor even of individuals, but of genes. Essentially this is because genes replicate themselves very faithfully and last far longer than groups or individuals (Dawkins, 1976). For most purposes in ecology, this gene's eye view of life can adequately be approximated by thinking at the level of the individual. We will now look at the behaviour and ecology of four groups of social organisms, and then briefly at humans, to examine the causes and consequences of their social behaviour. It will become apparent when we need to consider selection at gene level, and when individual selection is an adequate theoretical framework.

8.6.1 *Termites*

Termites are insects belonging to the order Isoptera and are closely related to the cockroaches (order Dictyoptera). Although some species attack living trees, most termites are ecologically important as decomposers. Three-quarters of all the known species belong to a single family, the Termitidae.

In many species of termites their hind guts hold huge numbers of symbiotic bacteria and specialised flagellate protozoa. Just as in the mammalian gut (Section 14.2.4), these microorganisms enable the termites to utilise normally indigestible substances. In some species it is known that they can digest over 90% of the cellulose and about 50% of the lignin ingested (Brian, 1983). These microorganisms are lost when the cuticular lining of the hind gut is shed at ecdysis; however, a newly moulted individual will be fed special anal secretions rich in these microorganisms. Different termite species specialise on different foods. Most eat wood, but the most primitive termite known, the Australian *Mastotermes darwiniensis*, has the broadest diet of any termite species (Wilson, 1975). Workers may eat crop plants, rubber, ivory, sugar, dung, the plastic lining of electric cables and wooden buildings. Unattended houses have occasionally been reduced to dust within a couple of years!

Permanent termite mounds are a characteristic feature of the landscape of many countries. These afford protection for the colony, which may number up to a million individuals in some species. Nests can be a variety of shapes and sizes depending on the species which lives in them. Some are flat-topped like mushrooms so that the rain bounces off and keeps the nest dry; others have chimneys and deep tunnels down to the water table so that the internal microclimate can be regulated. In *Macrotermes natalensis* circulation of air within the nest keeps the air cool and the level of carbon dioxide low (Figure 8.7).

All of the 2200 or so termite species are **eusocial**. This means that each species satisfies three conditions:

There is co-operation in looking after the young;
Some individuals are permanently sterile;
There is an overlap of at least two generations contributing to colony labour.

Eusociality can be seen as the pinnacle of altruism. After all, a eusocial species contains large numbers of individuals which never reproduce, but dedicate their lives to helping other individuals to reproduce.

Termites, as is the case for all the social insects, have **castes**. A caste is a collection of individuals within the colony that are morphologically distinct from individuals in other castes and perform specific tasks. Castes allow colony members to specialise. In a colony, there are always at least three castes – a queen, a king and lots of workers. Furthermore, in many species the workers themselves may be sub-divided into two or three castes. For instance in *Trinervitermes*, an advanced termite, the adult males may end up as either small soldiers or large soldiers (Figure 8.8). These have different functions and are physically distinct. Not only that, but as termites pass through their moults, and gradually develop into adults, the juveniles may specialise and do different jobs from the adults. This is a bit like an assembly line in a factory where different people, or robots, are specialised to perform different jobs with the greatest efficiency.

Termites, in common with most other organisms, but unlike species in the order Hymenoptera (ants, bees and wasps), are diploid, so that males and females have the same number of chromosomes. Unlike workers in the Hymenoptera, termite workers include both males and females. Termites are also distinctive in that the primary reproductive male (the king) stays with the queen after the nuptial flight, helps her to construct the first nest and fertilises her intermittently as the colony develops (Wilson, 1975). Figure 8.9 shows how huge the queen may be. She ends up as a highly specialised egg-laying machine.

8.6.2 Army ants

Ecologically the ants (family Formicidae) are extremely diverse. Although one ant looks pretty much like another, the range of foods they eat is enormous (Wilson, 1975). There are species that feed only on isopods, others that specialise on centipedes, and some even eat other ants. Others rely exclusively on the carbohydrate-rich excretions of aphids or other insects; some have fungus gardens in which symbiotic fungi are grown on organic material collected and tended by the ants. In *Cyphomyrmex*

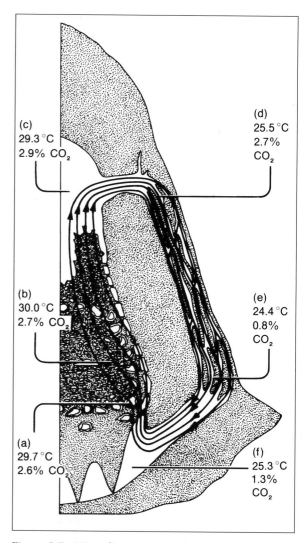

Figure 8.7 Microclimate regulation in the nest of the termite *Macrotermes natalensis*. The mounds in this species may be 5 m high and can have a basal diameter of 5 m. This diagram shows a section through a medium-sized nest which might contain two million individuals. Narrow channels (d to e) allow the radiation of heat, outward diffusion of carbon dioxide and inward diffusion of oxygen. The temperatures and carbon dioxide concentrations of various parts of the nest are shown. (Based on Lüscher, 1961.)

rimosus the fungi are grown exclusively on caterpillar faeces! Better known are the leaf-cutter ants which provide their fungi with fresh leaves. The remarkable relationship that some ants have with acacia trees is discussed in Section 19.2.2.

In common with the other Hymenoptera, the ants are **haplodiploid**. This means that while females are diploid, males develop from unfertilised eggs produced by **parthenogenesis** and are haploid with only half the chromosome complement of females. An important consequence of this is that in the Hymenoptera a reproductive female can control the

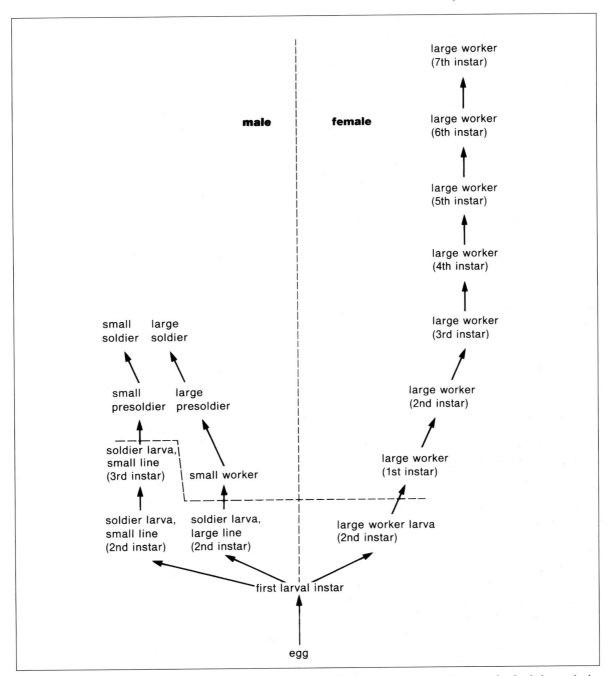

Figure 8.8 Castes in the termite *Trinervitermes*. Each time an individual moults it gets bigger. Individuals beneath the dashed line are helpless; those above it are capable of independent action and serve the colony. (From Oster & Wilson, 1978.)

sex of her offspring. If she wants to produce a female, she allows one of the sperm she has stored in her spermatheca to fertilise one of her eggs. If she wants a male, she simply lays an unfertilised egg.

In all the social Hymenoptera, the workers are female. Why might reproductive females want all their workers to be females? One suggestion is that in species where the queen mates only once, so that all the workers are full sisters, workers are more closely related to each other than they are to their own offspring. This remarkable truth follows from the fact

that any offspring a female has are related to her by just a half. However, full sisters share *all* their father's genes in common (remember the father is haploid), and half of their mother's. This means that in outbred haplodiploid species the coefficient of relatedness between full sisters is three-quarters.

The army (or driver) ants in the subfamilies Dorylinae and Ecitoninae are remarkable creatures. Colonies may number up to 40 million individuals. One of their most fascinating features is that they do not have permanent nests; rather they have bivouacs

(Schneirla, 1971). After a day out raiding, towards dusk a few of the workers anchor themselves by their leg hooks to some raised object like a log and dangle. Newcomers are attracted and attach themselves to these living ropes by their legs or mandibles. Strands may be dozens of ants long and coalesce into ropes. These bivouacs allow the colony to rest at night and protect the queen at the centre.

The next morning the colony goes out raiding (Figure 8.10). Their behaviour is legendary, though they do not consume 'everything in their path' as movie makers sometimes maintain. Nevertheless, they may kill and eat small vertebrates, and if your house is invaded by them the best thing to do is to get out of it for a few hours and be thankful that when you return it will be clean of cockroaches and other pests!

Why do army ants not have a permanent home? Wilson (1975) points out that the huge colonies would eat their home range bare of food. With their relatively slow rate of movement they would not be able to cover the large area necessary to provide them with their food on a sustained basis. Their strategy, rather, is to move around reducing the amount of animal matter in one area and then returning to it later once the level of available food

has built up again. Because of this, the queen cannot be the permanently huge immobile creature characteristic of many termites with their fixed nests that may last for decades. So, army ant colonies alternate between nomadic phases, when the colony moves to a new bivouac site every day, and stationary phases when it remains at one bivouac site for two to three weeks. During this stationary phase the queen has a burst of reproduction producing hundreds of thousands of eggs. These soon develop into new workers and the colony resumes its nomadic life.

8.6.3 Lions

Lions (*Panthera leo*) have been studied in the Serengeti and Ngorongoro Crater in East Africa continuously since 1966 (Schaller, 1972; Bertram, 1978b; Packer *et al.*, 1988). Despite this intensive study, it is still not known for certain why lions are the most social of the cat family (Packer, 1986). Although the adult females, the lionesses, hunt in groups, lions seem to get the most food *per individual* when they are solitary, or possibly when they hunt in pairs. However, there are other advantages to being social. For example, groups of lions can defend their kills against other animals and frequently take the

Figure 8.9 A termite queen (*Macrotermes subhyalinus*) with the king and a retinue of workers. The king is the larger termite close to the female's side.

Figure 8.10 A colony of army ants (*Dorylus* sp.) out raiding.

Figure 8.11 Two of the lionesses in a pride chasing a Thomson's gazelle in the Ngorongoro Crater.

carcasses killed by other carnivores, such as cheetahs.

The lionesses benefit in other ways from being in a group. Females remain in their natal groups, called **prides**. A pride typically consists of 4–12 adult females, 1–6 adult males and their offspring. The females in a pride are quite closely related to one another and lions are one of the few species to practise **communal suckling**. Cubs may suck from any adult female with milk. This means that, to some extent, females can take it in turns to go hunting (Figure 8.11) and to look after and feed the young.

Males benefit from the presence of other adult males in the pride. This is because the adult males in a pride run the risk of being ousted by other males. As one might expect, the more adult males there are in a pride, the longer they tend to be able to fight off other males. This means that they can sire more offspring. It used to be thought that the adult males in a pride were full or half brothers, and kin selection was invoked to explain the rather surprising observation that there is no dominance hierarchy among the males within a pride; they all share fairly equally in the numerous acts of mating. Now it is known that almost half of the breeding coalitions of known origins contain non-relatives (Packer & Pusey, 1982). In fact fights do occur between the males in a pride and males can be badly hurt in these. Individual selection may be what is responsible for the relatively harmonious relationships between the males in a pride. After all, if the length of time a male can expect to remain with a pride depends strongly and positively on the number of adult males, it may not be worth their while for males to fight too much among themselves; co-operation may be the better strategy.

When a pride is taken over by a new group of adult males, the new males seek out and kill as many as possible of the young cubs. About a quarter of all cubs die from such **infanticide**. The function to the males of this infanticide is that it serves to bring the adult females back into oestrus. In common with many mammals, including ourselves, frequent lactation prevents ovulation. Infanticide appears to be a strategy used by males to ensure that they begin to reproduce as quickly as possible after taking over a pride. Infanticide is now known to occur in several other species of mammals for similar reasons.

One remarkable feature of lion reproductive physiology is that during oestrus copulation occurs on average every 25 minutes. As oestrus lasts, typically, four days, and females often take more than one oestrus to conceive, an awful lot of copulations are required to ensure one conception. A number of attempts have been made to explain this. One is that by requiring so many copulations for a single fertilisation, females are reducing the *value* of each copulation to a male, and thus making it not worthwhile for the males in a pride to fight one another for access to oestrous females. Such fighting among the males, so the argument continues, is to the disadvantage of *females* as, should one or more of the resident adult males be injured, the pride is more likely to be taken over by new males which would, as we have described, kill many of the existing cubs. Another, and almost exactly opposite, argument is that this prolonged sexual activity by females, which is particularly apparent just after a pride has been taken over, attracts *other* males to the pride and ensures that the largest and most pugnacious coalition of males takes over the pride. Such a coalition, this argument goes, will defend the pride for longest against other males, which is to the advantage of the females. Sociobiologists are rarely short of ideas!

8.6.4 Naked mole rats

Why should termites, bees, ants and wasps be the only animals to have evolved sterile castes? In the 1970s, the zoologist Richard Alexander gave a lecture at a number of universities. In his lecture Alexander raised the question 'Why aren't any birds or mammals eusocial?'. (The definition of eusociality was given in Section 8.6.1. Essentially, a eusocial species is one in which some individuals are permanently sterile, there is co-operation over reproduction and there are overlapping generations.) At first sight birds and mammals would appear prime candidates for eusociality; after all, they all show considerable parental care and in a number of species there are individuals, often juveniles, who help some of the adults to rear young. Alexander argued that eusociality would be most likely to develop in a situation where a nest could be very securely defended. The evolution of sterile workers might then be favoured by natural selection as the workers might gain genetically by sacrificing themselves in the defense of the nest. Alexander suggested a subterranean mammal living in exceptionally hard impenetrable soil. In such a habitat a nest would be inaccessible to predators except via a nest entrance which could be defended by the sterile workers. A colony of such animals would need a lot of food, so Alexander predicted that they might eat large, nutritious tubers produced by plants in an arid region with irregular rainfall and bush fires – the sort of habitat that favours the evolution of plants with large underground reserves of food (Gamlin, 1987).

In 1975, one of the people in the audience listening to Alexander was a mammalogist called Terry Vaughan. Vaughan had spent a year in Kenya and one of the mammals he had collected was a rodent, the naked mole rat (*Heterocephalus glaber*). As Vaughan listened to the lecture he realised that the naked mole rat lived in exactly the sort of habitat that Alexander was describing. After the lecture he told Alexander that the only person he knew who was studying naked mole rats was Jennifer Jarvis. Alexander wrote to her, asking about the animal's behaviour. Alexander's letter arrived at an opportune moment (Gamlin, 1987). Jarvis had been studying naked mole rats for several years and had tried to set up laboratory colonies. Her attempts were frustrated by the fact that she had never caught a breeding female and never seemed able to get more than one female to come into oestrus in her laboratory colony.

The naked mole rat story is a fine example, cynics would say a rare one, of the predictive value of ecology. The details of the naked mole rat's social organisation and ecology conform closely to Alexander's predictions (Jarvis, 1981, 1985; Gamlin, 1987).

The most detailed studies on the naked mole rats have come from entire colonies dug up and transported to zoos where they have been allowed to establish themselves in artificial burrow systems which permit detailed observation (Figure 8.12). Within a colony, which in the wild may contain up to 300 individuals, only a single pair breeds. The breeding female may live for over 20 years and does little work other than suckle the young. She frequently leaves her nest chamber (Figure 8.13) and patrols the colony. **Pheromones**, airborne chemicals that alter the behaviour of animals in a group are important in other rodent species. They are secreted by a breeding female in her urine which she leaves in special toilet areas. However, experiments have shown that pheromones are less important in naked mole rats and that direct contact with the queen, including scent and behavioural interactions seem the main factors (Faulkes & Abbott, 1993). When a queen dies fighting often breaks out and females may kill each other. Colonies hardly ever contain more than one breeding female.

In other species of mole rats (family Bathyergidae) females usually have litters with between two and five young. In the naked mole rat the average litter size is 12 and litters of up to 27, the greatest number for any mammal, have been recorded. Naked mole rats are also unusual in that breeding is nonseasonal. The reproductive female of a naked mole rat colony produces a litter as often as every 80 days.

Once a naked mole rat is about three months old it joins the 'frequent worker' caste. This caste contains the smallest adults in the colony. These do most of the day-to-day work: digging, foraging, transporting soil, keeping the tunnels clear and building nests for the breeding female. Some individuals in this

Figure 8.12 A group of naked mole rats huddled in a nest chamber. These individuals belong to a captive colony at the Institute of Zoology, London Zoo.

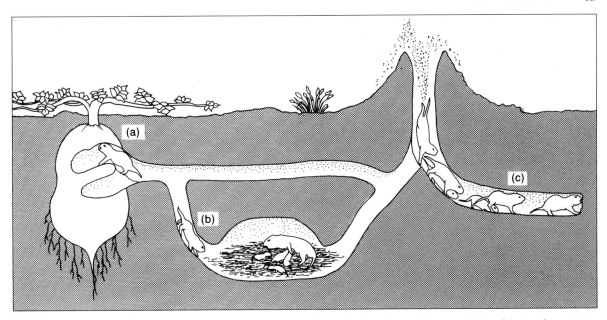

Figure 8.13 Cross-section of part of the burrow system of a naked mole rat (*Heterocephalus glaber*) colony: (a) a frequent worker eating part of a growing tuber; (b) the main chamber occupied by the breeding female, subsidiary adults and young; (c) a digging chain of frequent workers enlarging the burrow system. (From Jarvis, 1985.)

caste, which includes both males and females, grow very slowly and remain as frequent workers throughout their lives. Other individuals, again both males and females, grow into 'infrequent workers'. These spend most of their time asleep but are important in defence. The main predators of naked mole rats are certain carnivorous birds, which may grab the mole rats as they kick soil out of their burrows, and snakes which are the only predators able to pursue mole rats underground.

Mole rat colonies are most vulnerable when soil is being ejected. This tends to occur at night or early in the morning when snakes are least active. The individuals most likely to be caught are the frequent workers because they are the ones sometimes found near the surface ejecting soil. If one of them is captured by a predator, its alarm grunts stimulate nest-mates to block off the burrow just behind it, abandoning the small worker to its fate. Should a snake manage to gain entry to the colony, the large infrequent workers spring into action. These are the soldiers of the colony and attack invading snakes by biting them again and again.

As Alexander predicted, naked mole rats feed on the underground tubers of plants. Some of these tubers weigh up to 50 kg; by comparison, the average mole rat weighs only 40 g. These tubers are not consumed by a colony in one go; rather, the mole rats leave sufficient of a tuber to allow it to regrow, thus ensuring a continuous food supply.

One problem with living underground in a sealed burrow is the build-up of poisonous carbon dioxide (see Figure 8.7). Naked mole rats have the lowest metabolic rates of any mammals, thus minimising the amount of oxygen they use and carbon dioxide they produce. They are also the nearest thing yet discovered to a poikilothermic mammal. The temperature in a naked mole rat colony remains a steady 29–30 °C. In the laboratory, naked mole rats cannot survive ambient temperatures outside the range 15–38 °C. At 15 °C a naked mole rat is so sluggish it can hardly move, and its body fats begin to solidify. Their low metabolic rate may be the reason why they have lost most of their hair. Loss of hair may also reduce the risks of harmful ectoparasites which might be a particular danger to a highly social species living in a confined area.

8.7 Human sociobiology

Humans, wherever they live, have a highly complex society. Does sociobiology apply to humans? Some writers on human behaviour and social organisation have been very enthusiastic about sociobiology (e.g. Wilson, 1978). Others have dismissed sociobiology as being of no relevance to human biology whatsoever (e.g. Sahlins, 1976). In recent years, careful field studies have begun to show us whether sociobiology can make any useful predictions about our behaviour and society (Chagnon & Irons, 1979; Betzig *et al.*, 1988).

The central assumption of sociobiology is that behaviours have evolved which maximise the survival and replication either of genes or of the individuals which carry them. To many anthropologists it seems laughable to suppose that humans are maximising their reproductive success. What of the many adults who choose not to reproduce, for instance? Sociobiologists argue that many humans who fail to reproduce have little if any choice about the matter because of the society in which they live. For example, in polygynous societies, where males may have two or more wives, some men never get the opportunity to reproduce.

8.7.1 Parental investment in the later mediaeval Portuguese nobility

One specific instance where a human society has been interpreted using sociobiological principles is a historical one. It involved an investigation of some of the factors affecting reproductive success in the Portuguese nobility of the fifteenth and sixteenth centuries (Boone, 1988). We are fortunate that there exists a seventeenth-century manuscript containing the detailed genealogies of several hundred Portuguese noble lineages. These are just the sort of data that a sociobiologist needs.

Four status groups can be defined within the Portuguese nobility:

Primary titles: the royal family, dukes, counts, marquises, viscounts and barons;
Royal bureaucracy: overseers of the royal and ducal houses, secretaries of the treasury, key judges, etc.;
Senhorial class: landed aristocrats holding hereditary titles to land and associated labour;
Military/untitled: men holding at best a small lifetime ecclesiastical military pension. Many spent their careers in the army; many were members of religious military orders.

Table 8.1 gives the average number of surviving offspring for males depending on their natal status (i.e. the social class into which they were born). It is apparent that the higher their natal status, the more children they tended to have. There are several reasons for this. One is that sons were significantly less

Table 8.1 Reproductive performance, measured as number of offspring that survived to adulthood, and the probability of dying in warfare, in males as a function of their social class. Data for Portuguese nobility in the fifteenth and sixteenth centuries (Boone, 1988).

Natal status of males	Average number of surviving offspring	Probability of dying in warfare
Primary titles	2.93	0.16
Royal bureaucracy	2.55	0.22
Senhorial	2.54	0.19
Military/untitled	1.52	0.24

likely to die in wars the higher their natal status, as is also shown in Table 8.1. Males of higher status were also more likely to marry more than once. It is clear that high natal status is biologically advantageous for males. The position for females is more complicated. Women could change their status by marrying, as once married a woman acquired the status of her husband. Table 8.2 shows that *natal* status had no significant effect on the average reproductive performance of women who married. However, Table 8.2 also shows that women of higher *married* status had more surviving offspring.

Table 8.2 Reproductive performance, measured as number of offspring that survive to adulthood, in females as a function both of their natal status and of their married status. Data for Portuguese nobility in the fifteenth and sixteenth centuries (Boone, 1988).

Class	Reproductive performance	
	Natal status	Married status
Primary titles	3.78	3.62
Royal bureaucracy	3.20	3.95
Senhorial	3.60	3.79
Military/untitled	3.32	2.61

When a woman married, her family had to give a dowry or bride payment to the family of the man she was marrying. Dowries were expensive. For women of high social class they could be huge. Further, in common with many societies, women could not marry a man of lower status: women might marry 'up', but never 'down'. This inevitably meant that in a society which forbad polygyny there were more women in the highest classes than there were men to marry. This may account for the fact that the higher a woman's social class, the more likely she was to enter a convent and become a nun. Forty per cent of the women born into families with primary titles became nuns!

Boone's reasoning may help to explain why so many women in the Portuguese nobility became nuns, and why they were more likely to do so the higher their social class, but why did so many men go off and die in wars (Table 8.1)? The answer seems to be the existence of a system of inheritance in which only the first-born son inherited the real estate (Boone, 1988). This inheritance system appeared in Portugal in the thirteenth and fourteenth centuries and resulted in younger sons often becoming landless. Their only options were to join the Church, to try and succeed in the army or to try and succeed in business by joining the rising *burguesas* (commercial class).

We cannot be certain that Boone's interpretations are correct. The attractiveness of his explanation is that a single chain of reasoning, beginning with the maximisation of individual reproductive success, appears to link together phenomena as apparently separate as the death of men in battle, the entry of women to convents, and the way in which real estate was inherited. On the other hand, it can be argued that Boone's analysis raises as many questions as it answers. For instance, would it not make more biological sense for women, whatever their social class, to raise at least some children, rather than become nuns?

Many people doubt that the subtleties of human behaviour can adequately be explained by sociobiology. There are, after all, many ways in which humans differ from all other species, including the possession of a moral code, widespread religious belief and the existence of property.

8.7.2 *Helping behaviour in humans*

In Section 8.7.1 we looked at a sociobiological analysis of one particular historical situation. Sociobiology also claims to explain everyday helping behaviour in humans using theories of kin selection and reciprocal altruism.

It can be argued that kin selection is responsible for much of the altruism we direct at our relatives. Unfortunately, it has proved difficult to provide quantitative tests for the predictions of kin selection in the social insects, let alone in ourselves. The evidence for the importance of kin selection in humans is quite strong in the few societies where fathers direct their paternal care towards the children of their sisters. This seems odd at first as a father is, of course, more closely related to his own children than to those of his sister. But consider the situation of a man in a society where there is a good deal of uncertainty about paternity. His sister, who has the same mother as he does even if her father might have been different, is definitely related to him, and so, therefore, are her children. However, the man may not be the father of his wife's children and so may not be related to them at all. The evidence is quite good that it is in societies which have the greatest incidence of adultery that a man is most likely to invest in his sister's children rather than in his wife's (Alexander, 1974; Kurland, 1979).

Whatever the importance of kin selection in human society, there is little doubt that reciprocal altruism is of great importance. Indeed, human society is largely based around reciprocation. Humans are extremely good at remembering people who fail to reciprocate. Most of us feel guilty if we fail to give a birthday present to someone who gives us one, unless they happen to be an older relative (kin selection). The entire money system only works because of reciprocation. An English five-pound note, for example, is just a piece of paper which carries the words 'I promise to pay the bearer on demand the sum of five pounds'. Promises only mean something if reciprocation occurs.

Even the simplest human society is tremendously complex and there are many who doubt whether sociobiology will be able to tell us much that cannot be explained in some other way. Time alone will tell how useful a sociobiological framework is for an understanding of ourselves.

Summary

(1) Sociobiology is the study of the biological basis of social behaviour.

(2) There are advantages and disadvantages to being in a group. Group living often decreases the risks of predation but may increase competition for food or mates.

(3) The evolution of altruism in animals can often be explained by kin selection or reciprocation.

(4) Kin selection occurs when the decrease in an individual's fitness as a result of its altruism is more than compensated for by the increased fitness of its relatives.

(5) Reciprocal altruism occurs when an individual pays back a previous act of altruism.

(6) Group selection is probably unimportant in the evolution of animal behaviour.

(7) Sociobiologists attempt to explain behaviours by looking at how natural selection works at the level of genes.

(8) Termites are eusocial species with worker castes of both males and females.

(9) Ants, bees and wasps are haplodiploid and only have females as workers.

(10) Lions are the most social of the cats. The females co-operate in hunting and exhibit communal suckling. The males practise infanticide when they take over a pride.

(11) Naked mole rats are the nearest thing known to a eusocial vertebrate. Only one pair breeds in a colony. Their social system is a consequence of the nature of their food supply and habitat.

(12) There is considerable academic controversy as to whether human societies can validly be analysed using sociobiological reasoning.

(13) Reciprocation is certainly of great importance in human behaviour.

NINE

The environment

9.1 What is the environment?

The astronaut in Figure 9.1 is surrounded by very hostile conditions on the Moon. There is no breathable atmosphere, no running water, very high temperatures in the glare of the Sun and very low ones on the dark side of the Moon, no soil, just dust and rocks and the risk of meteors. In fact the Moon is probably unable to sustain any form of life, although biologists live in hope that some microorganisms may be found there. This is why the astronaut has to wear a spacesuit to carry his oxygen supply and maintain his temperature correctly.

All these characteristics of temperature, light, aridity and ground structure make up the **environment** on the Moon. It is a very unusual environment, by Earth standards, because there are no living organisms; in other words the environment on the Moon is **abiotic**: it is made up only of **physical** conditions. On Earth the abiotic environment of an organism is composed of physical variables such as temperature, rain- or snowfall, nutrient and toxic content of the soil, the power of wave action and wind speed. Unlike on the Moon, on Earth an organism also experiences the influence of other organisms, for example through competition, predation, herbivory, pollination and seed dispersal. The effects of such organisms forms the **biotic** part of the environment.

Although the abiotic and biotic components of the environment can be treated separately, as they are here in Sections 9.2 and 9.3, the relationships between them are complex. For example, the amount of incident light falling on a leaf may depend not only upon the position of the Sun and the amount of cloud cover, but also on the angle at which the leaf is held, and the shading effect of the surrounding vegetation. A nutrient-poor soil may be colonised by plants with root nodules containing nitrogen-fixing bacteria, but the presence of such plants alters the soil by increasing nitrogen availability, thus allowing subsequent colonisation by other plant species (see Section 16.3.1). Fire is a physical phenomenon which depends on the weather, but a fire can only burn because of the build-up of organic matter from the biotic environment, such as occurs in savannah grasslands.

For any organism within a community (see Section 14.1), the environment which it experiences will be the result of these complex interactions between biotic and abiotic factors. Some factors will be more important than others for limiting an individual's growth, development, survival or repro-

Figure 9.1 Commander David R. Scott from Apollo 15. The spacesuit contains a portable environment to support him.

ductive success. Different factors maybe important at different times in the life of the organism: a caterpillar will have different requirements from an adult butterfly; seed germination will be affected by different aspects of the environment than those which affect a mature plant. Whether or not an organism can survive at all stages of its life in a particular environment is therefore of considerable importance in determining the distribution within habitats and the overall global range of individual species. Figure 9.2 summarises the complex interactions between the abiotic and biotic environments of an organism.

Investigations of the components of the environment and the responses of organisms play a crucial part in ecological study as they add to our understanding of both the distribution of species and the structure of communities.

9.2   The physical environment

9.2.1   The composition of the physical environment

The physical or abiotic environment experienced by an organism depends on several factors: geology (rock and soil types); topography (landscape); world location (latitudinal light and temperature variations); climate and weather; and catastrophes (fire, earthquakes etc.). Some of these factors such as the geology and topography of an area are relatively stable; they may be different at different places, but at any one site they will remain constant for periods of time much longer than the life of the organisms living there. Other factors, such as atmospheric conditions including humidity, wind speed, temperature

and sunlight, will be very variable at one locality from one day or year to the next. Such abiotic factors will also change throughout the day and night, so that an organism, however short its life span, will have to live through changes in the environment.

9.2.2   Geology and soil

The different rock types which form the geology of an area are the product of many long and complicated processes. These include the effects of the movements of whole continents by a process called **plate tectonics** (see Section 18.2), igneous activity such as volcanos, the accumulation of sediments and erosion of rocks.

Figure 9.3 shows how complex the pattern produced by these processes can be. It is actually a simplified geological map of Britain which shows only a few of the many different rock types which occur. Geologically Britain can be divided into two parts by a line running from the south west to the north east. Scotland, Wales and the north and west of England are composed of old Palaeozoic rocks with scattered granites and volcanics indicating past igneous activity. To the south and east of England are lower lying, younger, Mesozoic and Tertiary sedimentary rocks.

The geology of an area also influences the topography of the land (see Section 9.2.3): in many places rocks are exposed as cliffs or smaller outcrops. However, over much of England, the rock is covered in drift deposits of boulder clays or tills which were carried there by glaciers and left when they retreated. The boulder clays and tills are covered with a fairly shallow skin of fertile soil.

The inorganic components of soil are derived from rocks by **weathering**. Hence it is important which

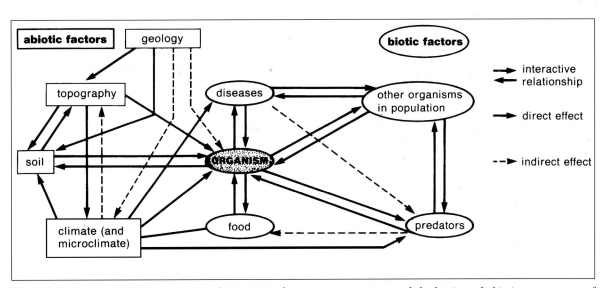

Figure 9.2 Summary of the complexity of interaction between an organism and the biotic and abiotic components of its environment.

rock types are available in an area to produce the soil. Large fragments of rock are broken into smaller pieces by **mechanical weathering**. Water fills cracks in the rock and the water expands when it freezes. pushing the rock apart. Heat can also fragment rock. In desert areas, which often experience intense heat by day and cold at night, the outer layers of rock gradually peel off to leave rounded outcrops.

Rocks are also broken down by **chemical weathering**. Carbon dioxide dissolves in rainwater to form a mild acid, even in unpolluted rain. This acidity reacts with the rock surface. Limestones dissolve to form calcium hydrogencarbonate; feldspars, in igneous rocks like granites, break down into clay particles and potassium carbonate. The potassium carbonate accumulates in the colloidal clay particles and organic matter in the soil where it becomes available for plant use. Various other minerals such as iron and magnesium hydrogencarbonates, colloidal aluminium hydroxide and silicic acid are products of

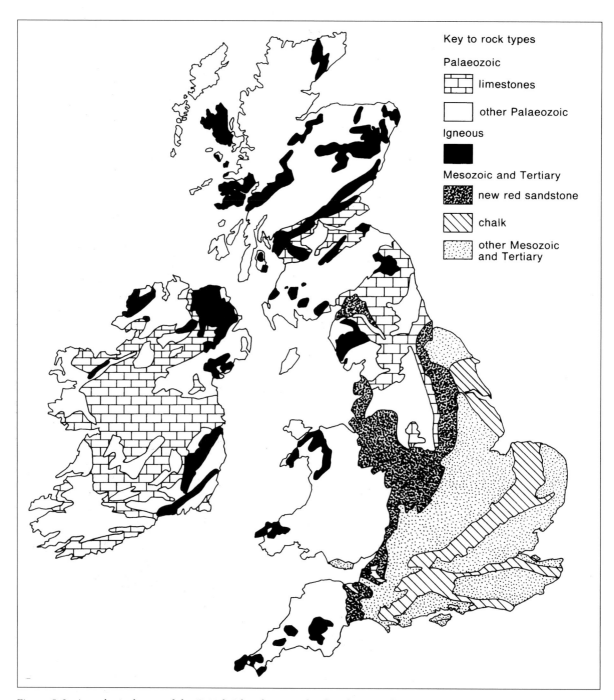

Key to rock types

Palaeozoic

limestones

other Palaeozoic

Igneous

Mesozoic and Tertiary

new red sandstone

chalk

other Mesozoic and Tertiary

Figure 9.3 A geological map of the British Isles showing the distribution of some major rock types.

the weathering of various rock types. Quartz is very resistant to weathering and remains as pebbles or sand.

Often the weathered products of rock types remain close to the site of their formation. The soils of central England are sandy, and the same red as the new red sandstones of the area shown in Figure 9.3. Iron and aluminium hydroxides are concentrated in some tropical regions where there is a very wet season followed by a dry period when evaporation occurs. As the surface water evaporates in the dry season, the deeper groundwaters are drawn up by plant roots

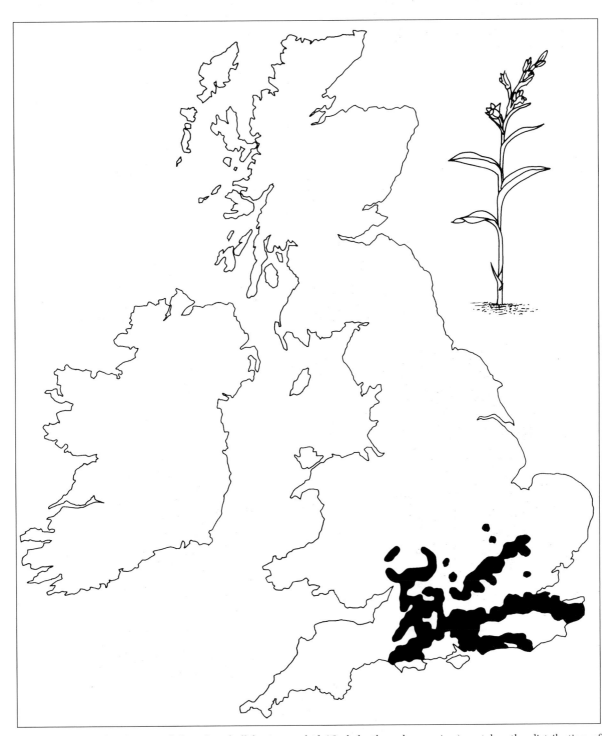

Figure 9.4 The distribution of the white helleborine orchid (*Cephalanthera damasonium*) matches the distribution of chalk in southern England. Compare the pattern with the geological map in Figure 9.3. (Redrawn from Perring & Walters, 1962.)

and capillary action, thus concentrating the minerals which were leached down in the wet season. This weathering regime results in a build-up of a reddish material called **laterite** which is extremely nutrient and humus poor. Lateritic soils often develop an impervious hardened surface which cause them to become infertile and once the vegetation on them has died, erosion quickly follows during the rainy season. This type of deposit, although bad news for farmers, does have some uses: it can be cut into bricks which harden on exposure to air, and the aluminium-rich laterites are mined for bauxite, an ore of aluminium (Holmes, 1978).

The distribution of plant species is often quite strikingly similar to the distribution of a major rock type. Compare, for example, the distribution of an orchid, the white helleborine (*Cephalanthera damasonium*) plotted in Figure 9.4 with the outcropping of chalk in Figure 9.3. In fact a whole habitat type, the species-rich turf known as chalk grassland, is associated with the chalk exposures.

For more detailed descriptions of the structure of soils which develop under various conditions, see Section 15.2.

9.2.3 Topography

Topography, the height and shape of the land, can play an important part in the distribution of organisms. Temperature decreases with altitude, so that snow lies on high mountains all year round, even at the equator as on Mt Kilimanjaro in Tanzania (see Section 18.3.5). Slopes may be north or south facing, and much better drained than an area of flat land or a hollow. The angle of the slope may be so severe that landslips occur, or the hollow may collect cold air (a frost hollow).

The distance below the surface of a sea or lake is important for aquatic organisms. Water absorbs light quite rapidly, especially longer wavelengths of visible light. Red and orange are absorbed by about 30 m depth; green and blue light penetrate furthest, down to about 140 m in the clearest water. Even at this depth, where only about 1% of the sunlight penetrates, green algae may still survive (Colinvaux, 1993). The topography of the shore and the amount of exposure it receives during low tides will affect the distribution of marine organisms. Some of the organisms of the shore are described in Sections 9.4.3 and 14.2.3 and the abiotic environment in lakes in Section 15.3.7. Figure 9.5 summarises some of the features and effects of topography.

Even very small changes in topography can be important. In woods on heavy clay soils the shallow depressions in the woodland floor become waterlogged after heavy rain and sometimes contain standing water, while the sides and tops of ridges remain relatively dry. In Cambridgeshire woods this produces distinct patterns of vegetation. The plant dog's mercury (*Mercurialis perennis*) is intolerant of waterlogging and only grows on the tops of ridges. Ferrous iron is associated with waterlogged soil where ferric iron (Fe^{3+}) is reduced to the more soluble ferrous iron (Fe^{2+}) by soil microorganisms. *Mercurialis perennis* is killed in about three weeks by only 5 parts per million of ferrous iron. The oxlip (*Primula elatior*) which also grows in these woods is more tolerant of ferrous iron: it can survive concentrations of 20 p.p.m. It is found in the wetter, lower lying areas of woodland floor. It appears to be excluded from the higher, well drained areas by strong competition from the dog's mercury which actively spreads by rhizomes (Martin, 1968). Figure 9.6 shows these two species growing close together: note that the dog's mercury is on the slightly higher ground.

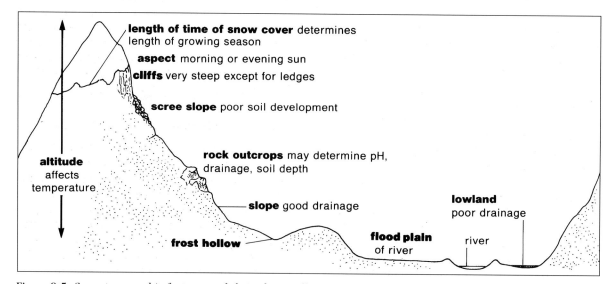

Figure 9.5 Some topographic features and their abiotic effects.

Figure 9.6 Dog's mercury (*Mercurialis perennis*) and oxlip (*Primula elatior*). The oxlips are the plants with the nodding clusters of flowers. These two species are not usually found so close together.

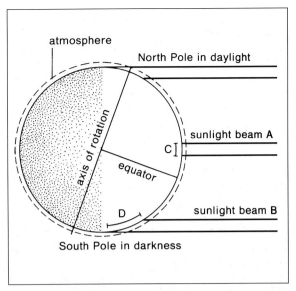

Figure 9.7 A diagram of the Earth showing how energy from the Sun is distributed differently at the poles and the equator. The energy falling via beam A on to a square metre of surface C is greater per unit time than that falling on to the same area of D via beam B because of the difference in distance travelled through the atmosphere and the angle of the beam to the surface at the two sites.

9.2.4 Latitudinal light and temperature variation

The Sun is the major source of light and heat on Earth. It is the variation of this solar energy which creates day and night and the great changes in climate from the hot tropical forests and baking desert to the freezing poles. The Earth spins on its axis of rotation every 24 hours (see Figure 9.7), so that most of the world has 24-hour cycles of day and night. Figure 9.7 also shows why the differences between equator and poles occur. As the energy passes through the atmosphere, about 20–40% of it is absorbed by the gases, dust and water vapour in the atmosphere. The energy which reaches the outer atmosphere near the poles (represented by beam B on the diagram) has more atmosphere to travel through before it reaches the Earth's surface than does the energy reaching the outer atmosphere near the equator (beam A). More energy is therefore lost in the atmosphere at higher latitudes. Also, because of the curve of the Earth's surface, the energy which has passed through the atmosphere at the poles falls obliquely on the Earth's surface. It is spread over a larger surface area (D on the diagram) than a beam of the same size shining on the Earth near the equator (C). This means that, per unit time, the light and heat reaching the Earth at the equator is higher than at the poles. Over a year the equator gets twice as much energy as either pole (Woodward, 1987).

Variation of light and temperature with latitude is, however, much more complicated than this indicates. As already mentioned, the Earth spins on its axis of rotation every 24 hours. This axis, as is shown in Figure 9.7, is tilted at an angle to the Sun. This means that, as the Earth orbits the Sun during each year, the part of the Earth directly facing the Sun changes. The Sun thus appears to be directly over-head at the equator at the two equinoxes, but over the Tropic of Cancer at the northern hemisphere's summer solstice (the position shown in the diagram) and over the Tropic of Capricorn at the winter solstice. This affects the relative lengths of day and night in any 24-hour period.

At the equator, day and night are of almost equal length all year. Seasonality is minimal, although in many parts of the tropics there is a rainy season when the sun is hidden by clouds for months on end and this causes lower ground temperatures. At high latitudes, inside the Arctic and Antarctic Circles, seasonality is extreme. During the summer the Sun never sets and total daylight irradiance over 24 hours is actually higher than it is in the tropics (where there are only 12 hours light). However, during the winter months the Sun hardly rises above the horizon and total irradiance is very low.

9.2.5 Climate and weather

Some of the features of climate in different parts of the world, including seasonality in temperate regions, have been explained earlier. However, within any climate, there are considerable differences in the conditions experienced from day to day or year to year. These fluctuations are due to a complex inter-action of factors including the movement of areas of low and high pressure in the atmosphere which influences cloud cover, rainfall and temperature. A long-lived organism may experience a considerable

Figure 9.8 *Hunters in the Snow* by Pieter Bruegel the Elder. Painted in February 1565 during the first really cold winter of the little ice age. (Painting in the Kunsthistorisches Museum, Vienna.)

variation in temperature, drought length and wind speed during its life, a shorter lived one may have its whole life cycle disrupted by 'freak' weather conditions.

The variation which can occur in weather from year to year is shown by the temperature changes in Europe over the last 500 years. The past climate has been painstakingly reconstructed from many sources of historical writings and records, most notably by Gordon Manley (1974) and Hubert Lamb (1977, 1982). A cooling occurred in the climate from 1540 to 1700. In the winter of 1564–5 the River Thames froze over. This was the first of many hard winters and during it (February 1565) Pieter Bruegel the Elder painted his famous *Hunters in the Snow* (Figure 9.8). This marked the beginning of a whole series of winter snow and skating paintings to be produced over the next 200 years, especially by Flemish artists.

This whole period of climatic deterioration is called the **little ice age**, but it was the sequence of particularly cool summers and long cold winters of the worst decade, 1690–1700, which had the most serious effect. During this time, the Alpine glaciers are reported to have advanced considerably. This extremely cold weather caused chaos in Swiss farming areas. The grain rotted under the snow and the

hay for animals ran out before the snow had gone. Cows had to be fed on pine branches or slaughtered. The same decade in Scotland is recorded as a time of famine which resulted in the emigration of many people to Ireland: in many parishes 30–65% of the population starved to death (Lamb, 1982).

These are fairly short term, if rather disastrous, fluctuations in what has been a fairly stable climate during the past 500 years. There is evidence, however, that the climate has changed markedly and several times in the last two million years. The study of microfossils, like pollen and spores, which were deposited in bogs and lakes as these became infilled with sediment, can be used to reconstruct past vegetation. The flora and fauna of Britain have changed considerably in the last 100 000 years. Fossil-bearing deposits from London, many from the site of Trafalgar Square, indicate that the climate about 100 000 years ago was rather warmer than now, and the flora was grazed by elephants, hippopotamuses and rhinoceroses. However, between then and about 10 000 years ago, Britain was treeless except for scattered junipers (*Juniperus*) and dwarf birch (*Betula nana*). Half of Britain, down to The Wash of East Anglia, was covered in huge ice sheets, and the rest had a tundra type vegetation of grasses, sedges and herbaceous angiosperms (Pennington, 1974). This was the true **ice age** after which the poor weather of the seventeenth and eighteenth centuries was named.

The temperature and amount of ice at the poles can be determined from the ratio of two oxygen isotopes ($^{16}O/^{18}O$) found in fossils or sediments. If the isotope ratios are analysed for several chronologically ordered samples, then they can be used to indicate climatic changes. Figure 9.9 is a diagram of oxygen isotope ratios from small fossil shells in two deep ocean cores. The fossils provide a temperature record for the last 450 000 years. The diagram cannot give an actual temperature, but indicates whether the climate was warm or cold. It shows the warm phase about 100 000 years ago when elephants roamed along the banks of the Thames, and that this was followed by the long glacial period we know as the ice age. However, as you can see from the diagram, there have been several earlier ice ages, followed by relatively short warm **interglacials**. This cycling appears to be caused by very long-term changes in the orbit of the Earth around the Sun (Hays *et al.*, 1976). These temperature oscillations seem to have started fairly gradually about two million years ago and have been getting more pronounced.

The oxygen isotope record shows that about 10 000 years ago there was a rapid increase in temperature marking the end of the previous ice age. Britain, which was still attached to Europe across the south-west part of the English Channel and the Dogger Bank, was invaded by birch and pine forest and then by deciduous forest. Because the written history of humans has fallen in this interglacial, our climate and the vegetation which goes with it seem 'normal' to us, but the warm phases are actually much shorter than the glacial periods. About 90% of the last one million years in Britain has been glaciated (Lamb, 1982). There seems to be every reason to suppose that this interglacial will not continue indefinitely: glacials have followed the last ten or so warm spells and unless humans alter the climate (see Section 13.3), a glacial will probably follow this one, but not for perhaps another 4000 years!

Within fluctuating temperature cycles, an individual organism will be affected most by the extremes of intense cold or heat. Both these may be associated with water shortages: in the cold because the water has frozen and is not available; in the heat because of drought or excessive evapotranspiration. Another factor which may affect plants is the length of the growing season. Plants do not seem to be able to grow below 0 °C, and only slowly at temperatures just above freezing point.

Cooling to low temperatures, even above freezing, may have deleterious effects on an organism. Plasma membrane lipids may solidify, enzymes may operate only slowly, proteins may denature by spontaneous unfolding or disassociation and rates of biochemical reactions may be changed so that reaction pathways become out of phase, causing a build-up of intermediate substances in the sequence (Franks, 1983).

Freezing also causes problems. Ice crystals may form in cells causing mechanical damage to the membranes or cell walls. Such damage can result in leakages and cell death. Freezing temperatures are overcome in two ways, either by tolerance or by prevention of ice formation. In tolerant species freezing occurs in most of the cell but does not cause damage. This happens in many plants, for example fruit trees like apples and pears, but at very low temperatures of about −40 to −50 °C the tree is killed. Freezing can be prevented either by using an antifreeze such as sugar or alcohol concentrated in the cell, or by preventing **nucleation** of ice crystals, a mechanism known to occur in arctic fish, but not fully understood. The supercooled cells will not freeze until they reach temperatures as low as −50 °C, when freezing suddenly occurs and the cells are killed (Franks, 1983). Some plant species can withstand even lower temperatures. Sakai and Weiser (1973) found that

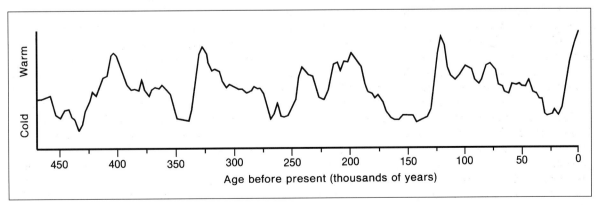

Figure 9.9 Oxygen isotope data from marine Foraminiferae shells showing the different warm and cold periods representing interglacials and glacials over the last 500 000 years. (Redrawn from Hays *et al.*, 1976.)

some twigs of species of poplar (*Populus*), birch (*Betula*) and larch (*Larix*) even survived being immersed in liquid nitrogen at −196 °C for two hours!

One might expect plants growing in areas where winter or night-time temperatures fall below freezing to be more tolerant of low temperatures than plants which live in hot climates. This is indeed the case.

Even within species, resistance to low temperatures varies depending on where the population lives. The western hemlock (*Tsuga heterophylla*) from the coast of North America cannot survive temperatures lower than −20 °C, while from the colder mountains inland and in Alaska, the same species survives temperatures of −35 to −50 °C (Sakal & Weiser, 1973). These differences within the species reflect different **ecotypes** (see Section 6.4.4): intraspecific differences between populations which are genetically based and which usually, as in this case, are adaptations to particular features of the environment.

Temperature cycles are important to organisms which experience winter or summer dormancy. For example, the dormouse (*Muscardinus avellanarius*) pictured in Figure 9.10 is on the extreme western

Figure 9.10 A dormouse (*Muscardinus avellanarius*) on a hazel branch.

edge of its range in Britain. Here the summers are less hot and the winters usually milder than over much of its range on the continent. The more reliable continental summers assure a plentiful supply of nuts in autumn. The dormouse hibernates during the winter. If the temperature is consistently low, it sleeps right through until spring, but in the warmer and more erratic British winters, the dormouse may wake up too often in warm spells, thus using up its fat reserves which it needs to survive and warm up on leaving hibernation (Bright, 1987). In addition, if the weather gets too cold, its body temperature falls too low and it must raise its metabolic rate by waking up or risk freezing to death.

Seed dormancy and germination are often also dependent on temperature cycles. Many plant seeds will germinate immediately if sown fresh in correct conditions. However, once they have become dormant, they may need periods of low temperature, equivalent to one, two or more winters, before germination will occur. This is a common feature of many alpine plants and temperate trees.

Precipitation is another important variable of climate. It may be evenly spread throughout the year or concentrated into a rainy season. Where temperatures are relatively high all year round, with a hot drought period, plants may drop their leaves in summer and grow them again in the cooler, wetter winter months. In cold areas leaves may drop in autumn, as in the temperate deciduous forest of Northern Europe. Where there is permanently frozen ground (**permafrost**) the surface layer of soil may become solid in winter and very waterlogged in summer as drainage is impeded. The effect of precipitation as snow will also affect organisms. Snow acts as an insulator protecting ground-dwelling or low-lying organisms from extremes of cold and ice-blast from high winds blowing sharp ice crystals.

One interesting example of an unusual climatic environment comes from the fossil plants in Alaska. During the Cretaceous period, about 100 million years ago, Alaska was situated at an even higher latitude than it is at present, at 75° N. It moved further north to about 85° N due to the movement of the continent by plate tectonics. The floras seem to have consisted mostly of deciduous plants including ferns, cycads, conifers and angiosperms. The angiosperms were often temperate deciduous trees related to the plane (*Platanus*) (Spicer & Parrish, 1986). Similar fossil forests of gymnosperms and angiosperms related to recent southern hemisphere taxa such as the Podocarpaceae and *Nothofagus* are found in Antarctica, close to the Cretaceous south pole. These are floras which would be impossible in similar latitudes today. The Cretaceous climate near the poles is thought to have been rather similar to that of present-day Britain. The very high number of deciduous species, greater than in our temperate climate, is probably due not so much to low temperatures and lack of unfrozen water as to lack of light during the winter months which makes photosynthesis impossible. Unfortunately, no environmental conditions like this remain today and any reconstruction of these fossil communities is just conjecture.

9.2.6 Catastrophes

There are several abiotic events which can have a considerable effect on organisms, but which occur infrequently and are not regular. These are catastrophic occurrences like volcanic eruptions, earthquakes, tidal waves and floods, hurricanes and freak storms, fires, landslides, bogbursts and impacting meteors. All tend to cause loss of life to a greater or lesser extent and clearance of the existing vegetation to some degree. Most of these are so infrequent and devastating that there is little an organism can do to survive them.

Fire is, perhaps, the most frequently occurring catastrophe and, if it is not too fierce, some animals can shelter underground during the blaze and regeneration of vegetation from root systems can occur. Some plants which grow where fire is fairly frequent have seeds which lie dormant until a fire has passed and then germinate, so that even if the adults are killed, seedlings can take their place in the cleared land. *Eucalyptus* and some pines are examples of this. In one pine forest in Minnesota, both fire and high winds were found to be important in clearing patches of forest which then regenerated, maintaining species richness and age range. A major fire had occurred in the wood about once every 10 years between 1650 and 1922. Since then a lack of fire and overgrazing by deer has almost completely halted regeneration of the pines and maples in the forest (Peet, 1984). Fire can be much more devastating if it is infrequent as there is far more organic matter to burn (see Sections 16.4.3 and 20.5.3).

Because they are fairly infrequent, the importance of catastrophes in communities tends to be overlooked. Fire and high winds are probably both very important causes of plant death which allows subsequent regeneration. Fire tends to clear ground vegetation, wind is more likely to affect trees. In tropical rainforest, hurricanes may be far more important in creating gaps for growth of young trees than the death of individual old trees in the canopy. A consideration of the effect of catastrophes on populations of organisms can be found in Section 5.2.8.

9.3 The biotic environment

9.3.1 Types of interaction

The biotic environment is experienced by an individual as interactions with other organisms. These will include individuals of the same species (**intraspecific effects**) and individuals of many other species (**interspecific effects**). Throughout its life an individual will come into contact with many other organisms. Within any community there will be a complex interplay of relationships both within and between species.

The way in which organisms fit together in a community and react to each other is fundamental to the understanding of the community and therefore of basic importance in ecology. Many examples of relationships between organisms, which constitute the biotic environment, are to be found throughout this book. The major categories of interactions between organisms are given below. These relationships, in many subtle and varying forms, can be discovered in any community of organisms. A few examples of each type of interaction are given here with one or more references to other sections where a more detailed description can be found. Not all the interactions and not all references have been given: the reader will find many more while reading other chapters.

9.3.2 Intraspecific relationships (within species)

Reproduction
One of the first phases in sexual reproduction is the location of a mate. This may occur, for example, by the advertisement by one sex of its whereabouts using scent (the pheromones of female moths) or sound (croaking in some species of frog) or sight (lights in glow-worms). The selection of a mate often includes competition between individuals of the same sex, either males for females or more rarely females for males. In red deer, for example, males fight for harems of females. In other species males compete without fighting and females choose the one which they see as most attractive; this is common in bird species where the males are colourful, for example peacocks and birds of paradise. Competition between females for males occurs in the jacanas, which are small water birds.

Care of offspring
Offspring are usually cared for by their parents: by the female only, for example in mammals like the red deer; or by both parents, for example in many birds where both parents may hunt for food to feed the nestlings. More rarely offspring are cared for by older siblings. This occurs among colonial insects such as

termites (Section 8.6.1) or in a creche system, where one or two adults in a social group remain with the offspring, for example in the naked mole rat (Section 8.6.4). Care of offspring includes feeding, guarding, keeping them warm and transporting immature individuals.

Social behaviour

Many social interactions are altruistic and centre around finding food and defence. Guarding and defence of territories, young and more vulnerable members of the group are also common in social animals (the whole of Chapter 8 is devoted to such social interactions).

Competition

This occurs between individuals within a species for environmental resources such as food, space, light, water and mineral nutrients.

In both animals and plants, competition can occur at any time during the life cycle: between sperm or actively growing pollen tubes for the chance to fertilise an egg; between embryos in the womb for parental nutrients; between seedlings for light and space (Section 5.2.2); between offspring in the same litter or in different litters in the same social group (Section 8.3); between adults (e.g. Sections 3.3.2, 5.2.1, 5.2.3 and 8.3); and between adults and young: for example between trees and the seedlings on the woodland floor (Section 14.2.2); or between male lions taking over a pride and the young cubs of the previous male (Section 8.6.3).

9.3.3   *Interspecific relationships (between species)*

Reproduction

Pollination in many angiosperms and a few gymnosperms involves a more mobile species as pollen transporter. Such carriers include insects, birds, slugs, bats and a few other mammals. The pollinator benefits by receiving nectar or pollen; the plant benefits by being pollinated with pollen from another individual (Section 19.4).

Propagule dispersal

Many plants use animals to disperse their propagules, for example birds which eat fleshy fruits and excrete the seeds, animals which carry seeds attached to their fur, and flies which visit some fungi and disperse the spores (Section 18.3.2).

Care of offspring

Care of offspring by another species is rare. In one strange example illustrated in Figure 9.11, the larva of the large blue butterfly (*Maculinea arion*), after feeding on thyme (*Thymus*) drops to the ground where it is found by *Myrmica* ants. The ants take a sweet substance from a honey gland on the larva and

the larva is then carried by the ants to their nest. Here the butterfly larva eats ant larvae and hibernates in the safety of the nest. Another more familiar example are the European cuckoos (*Cuculus* spp.) which lay their eggs in the nests of small birds such as the meadow pipit, hedge sparrow and reed warbler. The offspring are then reared by the owners of the nest. This is a parasitic interaction as the foster parent usually loses all its own fledglings and gains nothing.

Mutualism

Where both the organisms gain benefits by long-term association with each other the relationship is said to be mutualistic. Often one of the organisms gains food and the other protection, for example some ant species live in the specially swollen bases of thorns on some *Acacia* trees. The ants obtain a home and defend the tree by attacking insects which they find on the tree; the tree gains protection for its leaves (Section 19.2.2).

Trees such as beech, oak and pine gain amino acids from fungal associations; the fungi in return receive carbohydrates and vitamins from the tree (Section 13.5, see also nitrogen-fixing bacteria in root nodules mentioned in Section 13.4).

Parasitism and disease

Parasitism occurs where individuals of one species live or reproduce using the food and other resources of an individual of another species, often with a noticeable deleterious effect on the host. The relationships between parasites and hosts are many and varied (Sections 2.2.2, 5.2.7 and 9.4.2). Some parasites live within the host (**endoparasites**) such as gut parasites (Section 14.2.4); others, like the disease-carrying fleas on the black rat (Section 9.4.2), live outside the host (**ectoparasites**).

Predation

Apart from autotrophs (Section 2.2), all organisms have to feed on other organisms. Herbivores, insecti-

Figure 9.11 Large blue butterfly larva (*Maculinea arion*) rearing up ready for adoption by a *Myrmica* ant.

vores and carnivores are all involved in inter-specific interactions. The evolutionary pressures on predators to be good at locating and consuming their prey, and on potential victims to be good at avoiding being eaten, causes a constant struggle between individuals of both species for survival, resulting in dynamic co-evolution (Section 19.3.2).

Protection

Many organisms attempting to avoid predation use other organisms for protection, for example insects hide under tree bark; birds nest in holes in plants to protect their young; some moths have wing patterns that imitate bark or leaves; and some insects and frogs look like twigs or leaves. Some organisms gain protection from predation by using the defence mechanisms of another species. The cinnabar moth caterpillar eats the poisonous ragwort plant (*Senecio jacobaea*) and concentrates the plant's toxins in its own tissues. The caterpillars are brightly coloured with black and yellow stripes to warn birds that they are unpleasant to eat. Some edible species even copy the warning colorations of harmful ones to gain protection (Section 19.2.1).

Competition

Competition between individuals of different species is probably extremely important for determining the abundance, health, reproductive capacity and distribution of species within a community. Competition probably occurs for all the limited environmental resources such as food, space, light, pollinators, water and minerals. Competition can occur between similar or related species (Section 9.4.3) or between quite different species which need the same resource. Species have evolved to minimise competition (Section 10.3).

Defence

Organisms respond to predation, parasitism and competition by using defence mechanisms. These include mechanical defences such as sharp spines on cacti (protecting the plant, which contains a store of water, from thirsty desert animals) and hedgehogs, and toxic substances which kill or deter grazing animals, such as the toxins in the ragwort also used by the cinnabar moth caterpillar, or tannins in leaves (Section 14.2.2). Sometimes toxins are produced by plants to kill other plants in the surrounding vegetation.

9.4 Biotic and abiotic interactions

9.4.1 The complexity of the environment

The interlocking effects of the abiotic and biotic environment are many and complex. The physical environment will always exert an influence directly on the organisms in a community. Day, night, rain, drought, cold, high winds: all these factors will affect the individual and, because of this, they will also affect its interactions with other organisms, and therefore the biotic environment. Any biotic factor which alters the distribution or abundance of a particular species will also affect the abundance and distribution of its predators and so on throughout the food web (see Section 11.7). Situations may also exist where the food source of a predator is still present, but the environment has rendered it unobtainable. For example, in Britain, the number of herons dropped considerably in the cold winter of 1962–3 because so many lakes and ponds were completely frozen over that the birds could not reach enough fish (Stafford, 1971).

All communities will be affected by complex interactions of the biotic and abiotic factors in the environment. A few examples of the interrelationships where physical effects cause biological reactions are given below.

9.4.2 Pathogens and climate

Many outbreaks of disease seem to be associated with climatic conditions. Changes in the abiotic environment can influence the severity of outbreaks of diseases in three main ways: by weakening the host, by providing favourable conditions for the pathogen and by affecting the behaviour of the transmitter of the disease if one exists.

An example of host weakness increasing the effectiveness of the pathogen occurred during the unusually hot summer of 1976. The very high temperatures of the summer of 1976 came at the end of a long period of 16 months' drought which killed many plants outright. Many which survived were severely weakened, most notably trees. Far more elms than usual that year were infected with the fatal fungal infection known as Dutch elm disease (*Ceratocystis ulmi*) and other fungal diseases of trees were especially prevalent. A second abiotic factor – soil type – was also involved as the trees on the more freely draining soils (notably the beeches on chalk) were especially badly hit by bark disease (Ford, 1982).

In some cases a disease organism is transmitted from one species to another or within a species from one organism to another by a third organism known as a vector. Many vectors are insects. A change in the number or behaviour of vectors caused by environmental conditions will alter the occurrence of the disease it carries. The Black Death or bubonic plague, much feared throughout mediaeval Europe, often occurred in hot summers in England. A famous documented example is the plague of 1665–6 in London. The summers of 1665 and 1666 were notably hot and were probably also responsible for drying out the London timber sufficiently to cause the fire of London that year. The disease is caused by a bacterium, *Pasturella pestis*, which is carried by black rats and can be transmitted to humans via an insect vector: the rat flea. The flea is most active at high temperatures (between 20 and 32 °C). Hence in the hot summers of 1665 and 1666, the chances of rat fleas carrying infection to humans was increased considerably (Ford, 1982).

9.4.3 *Abiotic effects on competition*

Any physical factor such as waterlogging (Section 9.2.3), soil pH, temperature, drought and light which affects the growth of an organism will alter its competitive ability within a community. Competition is one of the most important biotic effects of the environment and any factor, either biotic or abiotic, which alters the competitive ability of an organism, either beneficially or deleteriously, will be extremely important in determining the distribution of individuals in the community.

In many cases a species does not grow or range over the whole area where the physical environment is suitable for survival: it may be that the species is being excluded from part of its range by another species (**competitive exclusion**: see Section 10.3 for a more detailed discussion).

The occurrence and causes of competitive exclusion are not easy to prove. In the marine shore environment, where there is constant disturbance and battering by waves, competition for space on rocks to which organisms can attach themselves to prevent being washed away is intense. Many mobile molluscs like limpets and whelks compete with more permanently fixed barnacles and seaweeds for available space. On British shores the distribution of two barnacle species overlaps. *Semibalanus balanoides* is a northern species almost at its southern limit, and *Chthamalus montagui* is a southern species at its northern limit.

The relative numbers of each species fluctuate considerably over periods of many years depending on climatic changes (warm or cold periods) and the relationship of the two species has been unravelled by a large number of researchers (see Ford, 1982). The feeding and reproductive behaviour of the two species is very dependent on temperature. *Semibalanus* can only breed after a cold season (below 10 °C) and will feed in temperatures as low as 1.8 °C, although its optimal temperature is 17 °C. *Chthamalus* will not breed in temperatures lower than 15 °C and although it can feed at 5 °C, its optimal temperature is 30 °C. Where the *Semibalanus* is at its optimum temperature, *Chthamalus* is too cold, and can be outcompeted physically by the rapidly growing *Semibalanus* which pushes it off the rock, or grows over it. In warmer periods when the sea water temperature rises above 17 °C, the *Chthamalus* can compete much better, especially as *Semibalanus* may cease to breed.

Figure 9.12 illustrates an example of how the environment can affect competition between two species of succulent angiosperm, *Sedum rosea* (the common pink roseroot) and *S. telephium*. Both species

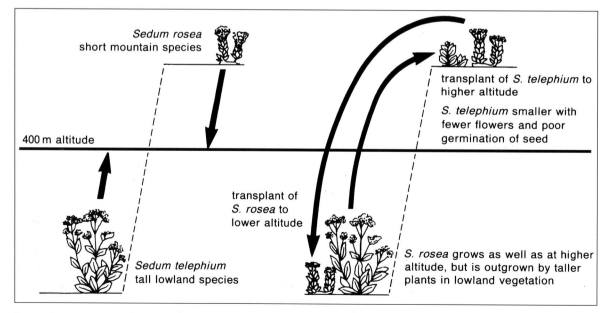

Figure 9.12 Diagram showing the result of transplanting two species of *Sedum* to different altitudes. *S. telephium* transplanted above its normal upper altitude limit of 400 m grows smaller with fewer flowers and poor seed germination. There is no alteration in the growth habit of *S. rosea* when transplanted below its normal lower limit of 400 m.

occur naturally in the Lake District of England, where *S. rosea* never grows below an altitude of 400 m and *S. telephium* does not grow above this altitude. Transplant experiments show that the upper limit in *S. telephium* is due to sensitivity to low temperatures – at colder, higher altitudes the plant grows poorly, seldom flowers and seed germination is low, but it can survive as an adult for several years. The other species, *S. rosea*, is totally tolerant of temperature changes, and flowers and seeds well at this altitude. If it is transplanted to a lower altitude, below 400 m, it still flourishes, so it is not being excluded from lower communities for abiotic reasons. At this lower altitude, however, it continues to grow to its normal height of about 20 cm, so it is outcompeted for light by the more vigorous, taller *S. telephium* and the other taller plants found at lower altitudes (Woodward, 1987).

Summary

(1) The environment of an organism has a physical (abiotic) and a biological (biotic) component.

(2) The abiotic environment includes stable characteristics such as geology and topography, variable factors like climate, nutrient availability and wave action, and occasional catastrophes such as fire and volcanic eruptions.

(3) The biotic environment involves the interaction of organisms and includes aspects of competition, predation, herbivory, reproduction and dispersal.

(4) The abiotic and biotic components of the environment frequently interact in complex ways.

(5) Mechanical and chemical weathering of rocks produces the inorganic components of soil; rock type therefore influences soil type as well as topography.

(6) The global climate has fluctuated considerably in a cyclical way during the last 2 million years as several ice ages have pushed icecaps and therefore vegetation types to much lower latitudes than at present.

(7) Low temperatures can kill organisms which are not adapted to withstand them. Different organisms can withstand different degrees of cold; some plants can tolerate temperatures well below $-60\,°C$.

(8) Pathogens are influenced by the climate: suitable conditions such as mild winters or hot summers may considerably increase the severity of disease or parasite attack.

TEN

Habitats and niches

10.1 Habitats

The word **habitat** is used extensively in ecology when describing where an organism lives. Unfortunately, it is difficult to give a precise definition of the term habitat. The word is a Latin one and literally means 'it inhabits' or 'it dwells'. It was first used in eighteenth-century floras or faunas to describe the natural place of growth or occurrence of a species. These guides to the plants or animals of a region used always to be written in Latin, hence the Latin word *habitat*. When floras and faunas began to be written in modern languages, the term 'habitat' remained untranslated and began to be used as a technical term.

It is easy to give the habitats of some species. For instance, the lowland gorilla (*Gorilla gorilla*) has as its habitat lowland tropical secondary forest; the fungus *Hericium abietis* is found on coniferous logs and trees in the Pacific north-west of the USA (Figure 10.1). Some species, though, have several habitats. Those of

Figure 10.1 The fungus *Hericium abietis*. This species is found on coniferous logs and trees in north-west America.

the tiger (*Panthera tigris*) include tropical rainforest, snow-covered coniferous and deciduous forests and mangrove swamps (Sunquist, 1985).

Ecologists soon realised that for smaller organisms, especially if they lived in a very restricted area such as on a particular plant or animal or in a specific region of the soil, it was useful to be more precise about where they lived. Consequently the term **microhabitat** was coined. Any one environment is divided up into many, possibly thousands of, micro-habitats. Figure 10.2 shows the range of micro-habitats available to insect herbivores and fungal parasites on a typical flowering plant. Of course, few plants are so unfortunate as to be attacked by all these organisms at the same time!

10.2 Niches

The distinction between habitat and **niche** is an important one in ecology. The word niche originally meant a shallow ornamental recess formed in the wall of a building, usually for the purpose of containing a statue or other decorative object. In ecology it came to stand for the precise way in which a species fits into its environment. A habitat is a description of where an organism is found, but its niche is a complete description of how the organism relates to its physical and biological environment. The concept of the niche has been very important in ecology, but it takes a great deal of field work to determine the niche of an organism in any detail.

10.2.1 Determining niches

In his classic book *Animal Ecology* Charles Elton pointed out that the word niche should be used to indicate what an animal was actually *doing* in its community. 'When an ecologist says "there goes a badger" he should include in his thoughts some definite idea of the animal's place in the community to which it belongs, just as if he had said "there goes the vicar"'(Elton, 1927, p. 64). A complete account of the niche of a badger would therefore include some

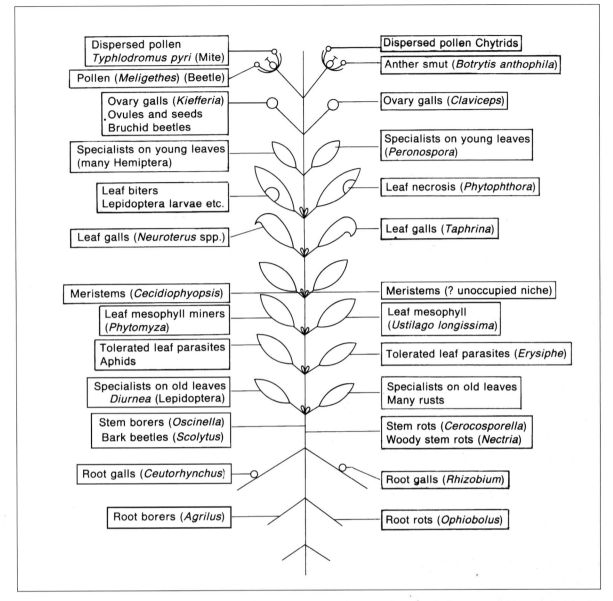

Figure 10.2 A generalised diagram of a typical flowering plant showing the microhabitats available to insect herbivores and fungal parasites. Examples are given of insect herbivores (left) and fungal parasites (right) which can attack particular parts of a plant. (After Harper, 1977.)

sort of a description of what badgers eat, what, if anything, eats them, what sort of a physical environment they need in terms of temperature, soil type, and so on.

Providing a quantitative description of a niche is difficult. Figure 10.3 shows an attempt to represent on a two-dimensional graph the feeding niche of the bird *Polioptila caerulea*, the blue-grey gnatcatcher (Root, 1967; Whittaker *et al.*, 1973). Along the horizontal axis is drawn the length of its prey. The vertical axis shows the height above ground at which blue-grey gnatcatchers feed. Also shown on the figure are contour lines showing the frequency with which the birds fed at a particular height and on a

particular length of prey. We can see that the birds concentrated on prey about 4 mm in length which they caught about 10 to 15 feet off the ground.

An attempt to represent the data of Figure 10.3 as a three-dimensional volume is given in Figure 10.4. Here the feeding niche of the birds is represented as a hill. The species can only be found on the surface of or inside the hill; the peak of the hill shows where the birds are most likely to be found. Hutchinson (1958) proposed that the environmental variables that are relevant to a species could be conceived of as a set of independent axes. In Figures 10.3 and 10.4 only two axes are plotted. In principle, other axes could be added. Ambient temperature might be a third, risk

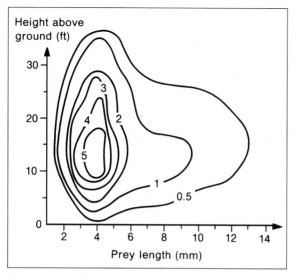

Figure 10.3 A two-dimensional representation of the feeding niche of the blue-grey gnatcatcher (*Polioptila caerulea*). The curved contours show the feeding frequencies (equal to percentage of total diet) for adult gnatcatchers during the incubation period in July and August in oak woodlands in California. (From Whittaker *et al.*, 1973).

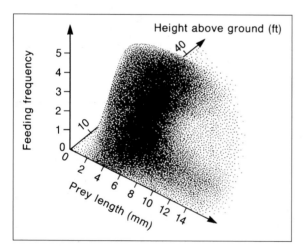

Figure 10.4 The feeding niche of the blue-grey gnatcatcher plotted as a three-dimensional hill. (Data as for Figure 10.3.)

from predation at different times of the year a fourth, and so on. It becomes almost impossible graphically to represent niches when more than two environmental variables are involved, but computers can easily hold and analyse the data. Hutchinson suggested that the niche of a species could be defined as an **n-dimensional hypervolume**, where the number of different environmental variables equals $n - 1$.

10.2.2 Each species has its own unique niche

Ecology has few laws. One of them is that each species has its own unique niche (Grinnell, 1924).

Suppose that two species occupied the same niche. It is almost inconceivable that they could occupy exactly the same niche *and* coexist. One of them, so the theory goes, would be bound to out-compete the other, driving it to extinction.

Two species may occupy separate *n*-dimensional hypervolumes, yet overlap considerably on each of the separate axes representing the environmental variables. When these axes show environmental variables such as prey length or protein concentration of food, they are often referred to as **resource axes**. Figure 10.5a shows how two species may overlap considerably on each of two resource axes, yet occupy almost distinct three dimensional niche volumes. Figure 10.5b shows, in a different way, how the data for several species can be represented on one graph.

The idea that each species occupies its own niche is all very well in theory, but is it true in practice? MacArthur (1958) conducted a detailed study of the population ecology of five species of warblers in Vermont and Maine, USA. These five species are very similar in size and beak length, are all mainly insectivorous and all belong to the same genus, *Dendroica*. Table 10.1 shows that within the five species, average beak lengths differ by only a few per cent. MacArthur chose to study these species precisely because earlier ecologists studying them had been unable to find any differences in their requirements.

Table 10.1 Mean beak lengths of the five species of warblers studied by MacArthur (1958).

Species	Mean beak length (mm)
Dendroica coronata	12.47
D. virens	12.58
D. tigrina	12.82
D. fusca	12.97
D. castanea	13.04

MacArthur's warblers, as they are now known, feed on firs (*Abies*) and spruces (*Picea*). Most of the trees in which the birds fed were 50-60 feet tall. MacArthur classified the trees into six vertical zones, each approximately 10 feet in height. He then divided each branch into three horizontal zones: a bare or lichen-covered base (B), a middle zone of old needles (M) and a terminal zone of needles and buds less than 1.5 years old (T). The number of seconds each bird spent in each of the 16 possible zones was recorded and the data are presented in Figure 10.6.

It is apparent from Figure 10.6 that the five warblers occupy feeding niches which, to a considerable extent, are distinct. Furthermore, MacArthur noticed that even when feeding in the same zone of a tree, the species appear to use different feeding

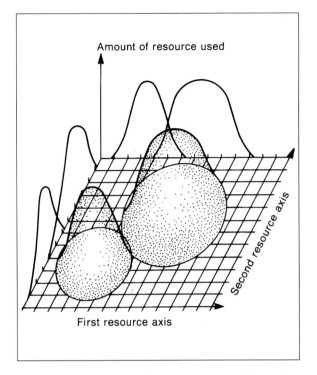

Figure 10.5a Diagrammatic representation of the niches of two species. Note how on each of the two resource axes there is considerable overlap between the two species. Despite this, the two species occupy three-dimensional niche volumes which overlap only slightly. (From Pianka, 1976.)

Figure 10.5b Diagrammatic representation of the niches of several species, showing how in two dimensions niches may overlap only slightly, despite considerable overlap along each of the one-dimensional resource axes. (From Pianka, 1976.)

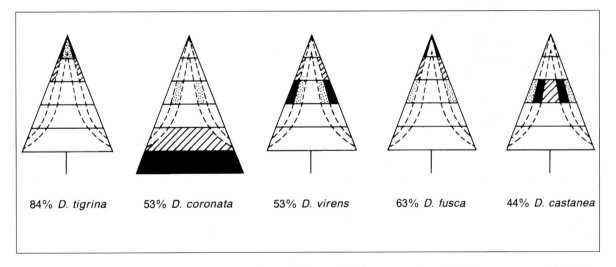

Figure 10.6 The zones of the trees where the five warblers studied by MacArthur (1958) spent most of their time feeding. For each of the five species, the solid black areas show the zone where the birds were most likely to be found, the hatched areas show the second most popular zone, and the dotted areas show the third most popular zone. The figures at the bases of the trees show the percentage of the time that the birds spent in these three zones when feeding in trees.

techniques. For example, *D. virens* obtains more of its food from stationary prey hidden among the foliage than do the other species which, to a significant extent, all search out flying prey. The five warblers also show some differentiation in their nest positions within the trees (Figure 10.7). Observations by MacArthur on the winter feeding behaviour of the birds also showed differences between the species.

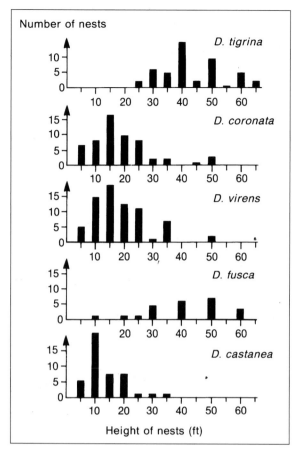

Figure 10.7 The heights at which the five species of warblers studied by MacArthur (1958) made their nests.

10.3 Gause's competitive exclusion principle

The notion that no two species can coexist if they occupy the same niche is now known as **Gause's competitive exclusion principle**, even though earlier workers such as Grinnell (1924) and even Darwin

(1859) had expressed essentially this idea. Gause himself never formulated the principle so succinctly as this. However, he carried out some careful and detailed laboratory experiments on coexistence among closely related species of yeast and protoctists and concluded that 'as a result of competition two similar species scarcely ever occupy similar niches, but displace each other in such a manner that each takes possession of certain kinds of food and modes of life in which it has an advantage over its competitor' (Gause, 1934, p. 19).

Gause first grew the protoctists *Paramecium caudatum* and *P. aurelia* in separate cultures which contained bacteria on which the *Paramecium* fed. The results displayed in Figures 10.8a and 10.8b show that each species grew in numbers according to the logistic equation as discussed in Chapter 5. It is clear from Figure 10.8 that *P. aurelia* grows in numbers more quickly than *P. caudatum* and ends up with more individuals in the same volume of culture medium. In other words, the carrying capacity of the culture medium is greater for *P. aurelia* than it is for *P. caudatum*, and *P. aurelia* has the greater rate of natural increase of the two species. These differences make sense as *P. caudatum* is larger than *P. aurelia*.

Gause then grew the two species together in the same small 5 cm³ culture volume. A new culture medium was inoculated with a mixture of both species and the numbers of both species were counted each day. The average results from three experiments are plotted in Figure 10.9. Each of the three experiments followed the same course. Initially both species grew in numbers, but eventually *P. caudatum* declined and became extinct. Gause attributed the inevitable extinctions to the existence of 'but a single niche in the conditions of the experiment' (Gause, 1934, p. 98).

In Gause's experiments, *P. aurelia* always won the **competition** between the two species. Park (1948,

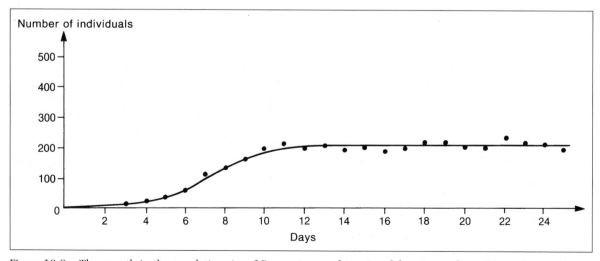

Figure 10.8a The growth in the population size of *Paramecium caudatum* in a laboratory culture. (From Gause, 1934.)

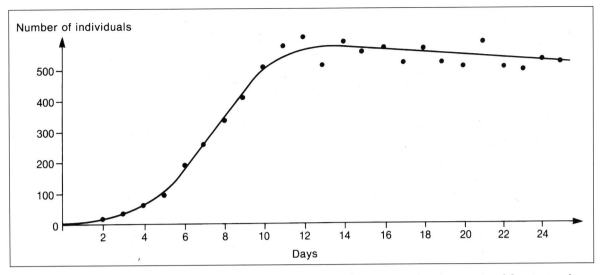

Figure 10.8b The growth in the population size of *Paramecium aurelia* maintained on its own in a laboratory culture. (From Gause, 1934.)

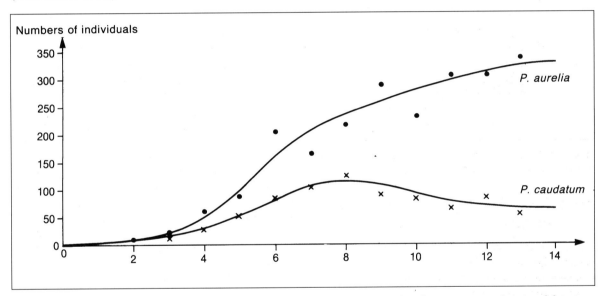

Figure 10.9 The changes in the numbers of *Paramecium caudatum* and *P. aurelia* when grown together in a laboratory culture. At first, both species grow in numbers. Eventually, however, *P. caudatum* declines in numbers while *P. aurelia* continues to increase its population size. (Data from Gause, 1934.)

1954) carried out a long series of laboratory experiments on two species of flour beetle, *Tribolium confusum* and *T. castaneum*, and found that in mixed cultures which of the two species survived and which became extinct depended on the environmental conditions. When conditions were hot (34 °C) and humid (70% relative humidity), *T. confusum* always went extinct. However, when conditions were cooler (24 °C) and drier (30% relative humidity), *T. castaneum* always became extinct. Of additional interest was the observation that under some environmental conditions, it could not always be predicted which species would go extinct (Table 10.2). Sometimes one species survived, sometimes the other.

Table 10.2 The outcome of competition between the two species of flour beetle *Tribolium confusum* and *T. castaneum* averaged over many replicates under different laboratory conditions. (Taken from Krebs, 1985, using data from Park, 1954.)

Temperature	Relative humidity	% of times *T. confusum* wins	% of times *T. castaneum* wins
24 °C	30%	100	0
24 °C	70%	71	29
29 °C	30%	87	13
29 °C	70%	14	86
34 °C	30%	90	10
34 °C	70%	0	100

10.4 Species coexistence

Subsequent to the work of Gause, Park and MacArthur, a great many studies of species coexistence were conducted in the field. It soon became apparent that however closely related a group of organisms was, niche differences could be found. But this does not provide a test of Gause's competitive exclusion principle. By the very fact that two species are different species, they are almost inevitably going to occupy different niches. The important question is 'How different do the niches of species have to be to allow species coexistence?'

10.4.1 Size ratios in closely related species

During the summer of 1958 the American zoologist G. E. Hutchinson collected some water bugs in the genus *Corixa* in a small pond on a hill to the west of Palermo in Italy. There were vast numbers of the water bugs in the pond which on closer investigation all proved to belong to just two species, *C. punctata* and the smaller *C. affinis*. Hutchinson found himself asking why there should be two and not 20 or 200 species of the genus in the pond (Hutchinson, 1959). More generally, Hutchinson wondered why there are about a million species alive today (we now think the figure is nearer 30 million) rather than, say, 10 000 species or 100 million species.

Hutchinson found that individuals of the larger *C. punctata* were about 46% longer than individuals of *C. affinis*. He began to wonder whether such size differences might permit the two species to coexist by enabling them to feed on different prey. Hutchinson extended his studies to birds and mammals and from a few measurements found that when pairs of closely related species were examined which fed at the same level of a food web, linear measurements of skeletal characters relevant to feeding (such as skull length in mammals) differed by an average of 28%. For instance, in Britain, skull lengths of the stoat (*Mustela erminea*) are on average 31% longer than those of the weasel (*M. nivalis*) (Hutchinson, 1959).

As closely related species are approximately the same shape, a difference of 28% in linear measurements (such as skull length) between two species means that the larger species is approximately $(1.28)^3 = 2.1$ times heavier than the lighter. Subsequent to Hutchinson's paper, a great many ecologists rushed out into the field, or went to museums, and weighed closely related species found in the same area. Often it turned out that they differed in mass by a factor of about two. Hutchinson's tentative suggestion became elevated to the level of an immutable law. An example of the sort of data used to support Hutchinson's suggestion is given in Figure 10.10. These data, on the geographical ecology of desert

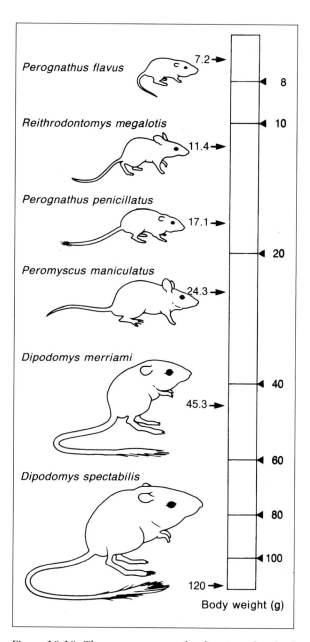

Figure 10.10 The mass, name and a drawing of each of the six species of rodents in the Sonoran Desert. (From Brown, 1975.)

rodents, were published in a book dedicated to the memory of Robert MacArthur who died at the age of just 42 (Brown, 1975).

Brown concluded that 'communities of coexisting species have distributions of body sizes with remarkably regular spacing' (Brown, 1975, p. 322) and at first sight the data in Figure 10.10 appear to support this assertion. However, if one calculates the ratios of the body weights of these six rodents arranged in pairs, starting with the lightest and finishing with the heaviest, one obtains the numbers 1.58, 1.50, 1.42, 1.86 and 2.65. At the very least, these figures show considerable scatter about the 'predicted' value of 2.1. Furthermore, the value of 2.1 is itself, of course, *not* a prediction. It simply follows from the value of 1.28 that Hutchinson measured. The most serious criticism of the many studies such as Brown's was provided by Simberloff and Boecklen (1981).

Simberloff and Boecklen pointed out that, inevitably, if one arranges a number of closely related species in order of body size and then calculates the ratios of the weights of neighbouring species starting at the lightest and ascending to the heaviest, one will get a collection of numbers greater than one. What Simberloff and Boecklen then did was to analyse, using appropriate and quite sophisticated statistical techniques, whether these numbers were *clustered* around a particular value or whether they varied as much as would be expected if the weights of the species were independent of each other, and merely arranged at random between a lower limit (equal to the weight of the lightest species) and an upper limit (equal to the weight of the heaviest species). They found almost no evidence whatsoever that the numbers were clustered around a particular value. In other words, analysis of the size ratios of pairs of closely related species found in the same area provides very little evidence to suggest that their sizes have been affected by **interspecific competition**. Subsequent work supports the conclusions of Simberloff and Boecklen (Pagel & Greenough. 1987).

10.4.2 Niche overlap and species coexistence

A different approach to the question of how much niche overlap can be tolerated between species before one of them goes extinct was taken by May and MacArthur (1972). If one makes, for example, the extreme assumption that animal species never vary in their diets, then obviously, however small the difference in their feeding niches, they will be able to coexist indefinitely because each species has access to foods which the other will not take. In real life, of course, animal diets are not so fixed. May and MacArthur produced a mathematical model which showed that when fluctuations in feeding niches were taken into account, the *average* food size for species adjacent on a resource axis to coexist indefinitely must differ by an amount roughly equal to the standard deviation in the food size taken by either species. This result is shown diagrammatically in Figure 10.11. This is an important result because for the first time it allowed a genuine prediction of the amount of niche overlap that could be tolerated in nature. Subsequent theoretical extensions of May and MacArthur's work have reinforced their conclusions (Maiorana, 1978; Rappoldt & Hogeweg, 1980), though few ecologists seem yet to have collected the data for a rigorous testing of their conclusions.

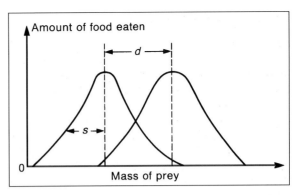

Figure 10.11 Diagrammatic representations of the feeding niches of two species on a one-dimensional resource axis, such as prey mass. The prediction of May and MacArthur (1972) is that in order for the two species to coexist in the face of interspecific competition, the difference, *d*, between the means of the prey masses taken by each species must be at least as great as the standard deviation, *s*, of the prey mass taken by either species.

10.5 Fundamental and realised niches

The red squirrel *(Sciurus vulgaris)* is native throughout wooded parts of Eurasia. The larger grey squirrel *(S. carolinensis)* is one of three species in the genus native to North America where it is found in hardwood forests. Repeated attempts to introduce the species to Britain were made from as early as 1828 and the last recorded introduction was in 1929, by when the species was already firmly established (Lever, 1977). While the grey squirrel has spread through much of Britain, the red squirrel has at the same time had its range reduced, particularly since the 1940s. Red squirrels are nowadays most abundant in Britain in large areas of mature conifer forest. The reasons for the decline of the red squirrel are not known for certain, but it does seem clear that the introduction of the grey squirrel has, at the very least, resulted in the red squirrel being unable to recolonise areas from which it has become absent through habitat destruction or disease (Tittensor, 1977a; Reynolds, 1985). A careful attempt to reintroduce red squirrels to Regent's Park, London, failed despite the use of selective feeders which provided supplementary food for red squirrels, but excluded grey squirrels (Bertram & Moltu, n.d.).

The red squirrel in Britain is still declining but it is unknown if it is in danger of becoming extinct. Grey squirrels predominate in areas of mature hardwood or mixed woodland, provided at least 25% of the trees are hardwood (Tittensor, 1977b). Red squirrels predominate in conifer woods, mixed woods dominated by conifers and places such as Brownsea Island and the Isle of Wight where the grey squirrel has never reached.

Hutchinson (1958) introduced the terms 'fundamental niche' and 'realised niche'. He suggested that the niche an organism could occupy in the absence of competitors and predators be called its **fundamental niche**, while the role it actually played in the community should be called its **realised niche**. In real life, therefore, the realised niche of an organism is smaller than its fundamental niche. In the case of the two squirrels in Britain, the realised niche of the red squirrel has contracted with the establishment and spread of the grey squirrel.

MacArthur and Pianka (1966) considered what might happen when several competitors invaded a habitat. They argued that in the presence of a greater number of competing species, each species should concentrate its search efforts for food on a smaller part of the total available habitat – so-called **ecological compression**. On the other hand, within the reduced habitat searched by each species, a full range of food items may still be present. Consequently, MacArthur and Pianka argued, the range of food taken by a species should be independent of the number of competing species present. Similarly when interspecific competition is relaxed, species should increase the range of habitats in which they obtain food, but not the actual range of food taken, This idea is illustrated pictorially in Figure 10.12.

Pianka (1975) found that in communities with several lizard species, the more lizard species there were, the *less* overlap there was between the feeding niches of the lizards. This held true for lizards in North America, Australia and Kalahari. Perhaps if evolution allows the coexistence of many species with very similar life styles, these species must have relatively narrow feeding niches. Another way of interpreting these results is to suppose that, as argued in the previous paragraph, the presence of more species of competitors leads to ecological compression. This

ecological compression results in species specialising as to which habitats they occupy. Consequently, they might inevitably specialise more in terms of the food they eat. This increased specialisation might be enough to reduce niche overlap.

10.6 Resource partitioning

Where two or more species divide out a resource such as food or nesting sites **resource partitioning** is said to occur. For instance, Figure 10.13 shows how the root systems of some common Mojave Desert plants partition out the water supply (Cody, 1986). Here the partitioning is not that different plant species take different sorts of water – there is, after all, only one sort of water. Rather, the plants are dividing up *access* to the water. Some specialise on ephemeral sources of water, which may be trapped by surface roots; others rely on relatively permanent sources of deep water.

A classic instance of resource partitioning among animals was provided by Bell (1971) in his studies of grazing mammals on the Serengeti plain in Tanzania. Careful analyses of the stomach contents of most of the principle large herbivores on the Serengeti plain – zebra (*Equus burchelli*), wildebeest (*Connochaetes taurinus*), topi (*Damaliscus korrigum*) and Thomson's gazelle (*Gazella thomsoni*) – revealed considerable resource partitioning. Zebra were able to take the diet highest in cell wall content. This is the poorest diet and zebra were able to live on it for two reasons. First, they are the heaviest of the four species, with an average adult female having a body mass of 220 kg. As discussed in Chapter 2, the larger a species of animal, the poorer the quality of diet on which it can subsist because of its low metabolic rate, relative to its body size. The second reason why zebra can survive

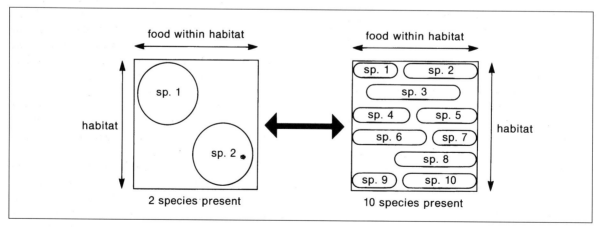

Figure 10.12 The compression hypothesis. As more species make use of a habitat, the habitat used by each species shrinks. However, the range of food items taken by each species remains constant. The actual diet may become more concentrated, but the range of items should not change. The hypothesis only applies to short-term non-evolutionary changes. (From MacArthur & Wilson, 1967.)

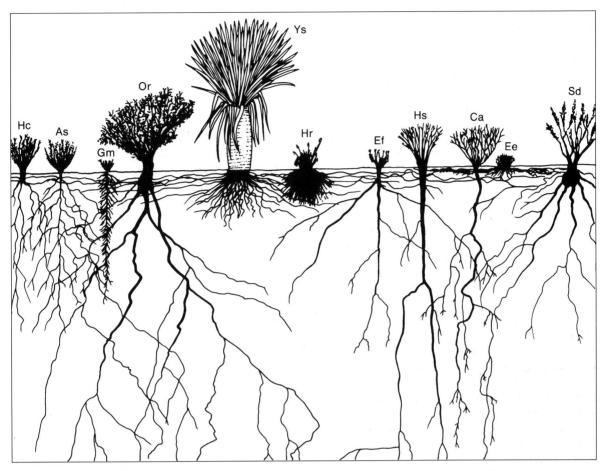

Figure 10.13 The root systems of some common Mojave Desert plants. Species abbreviations, from left to right, are: Hc, *Haplopappus cooperi*; As, *Acamptopappus sphaerocephalus*; Gm, *Gutierrezia microcephala*; Or, *Opuntia ramosa*; Ys, *Yucca schidigera*; Hr, *Hilaria rigida*; Ef, *Eriogonum fasciculatum*; Hs, *Hymenoclea salsola*; Ca, *Cassia armata*; Ee, *Echinocereus engelmannii*; Sd, *Salvia dorrii*. (From Cody. 1986.)

on low quality forage is that, unlike the other three species, they lack a rumen. The ruminant strategy is to obtain maximum nourishment by thorough digestion. The equid strategy, adopted by zebras and horses, is to pass food through the gut rapidly, getting what can easily be obtained from it, but not attempting to maximise digestibility. Zebras thus need more food for their size than do ruminants. However, this evolutionary strategy means that zebras can survive off the poorest quality herbage, unlike the other Serengeti herbivores.

The species that ate the highest quality diet was the smallest ruminant, the Thomson's gazelle. Adult female Thomson's gazelles weigh about 16 kg. Of the four herbivores, Thomson's gazelles were the only ones to take a large amount of dicotyledonous material: 40%, chiefly fruit. This high-protein high-energy diet is necessary for the small ruminant with its high metabolic rate, relative to its size. Topi (110 kg) and wildebeest (160 kg) took food intermediate in quality between that taken by Thomson's gazelles and zebra.

An elegant study of resource partitioning in a large group of species which at first appear morphologically very similar was provided by Gilbert (1985). Gilbert studied adult hoverflies (Diptera, Syrphidae) of which there are about 250 species in Britain. Gilbert was able to identify the hoverflies to species level when they were feeding without disturbing them. He found that a few species fed almost entirely on leaves, but most species fed on nectar or on pollen or on a mixture of the two. Gilbert showed that a tight correlation existed between the shape and size of the mouthparts in the hoverflies and the nature of their food source. Considerable niche separation existed between all the species, though the more similar two species were morphologically, the greater the overlap in their feeding niches.

Perhaps the most clear-cut instances of resource partitioning come from studies on host-specific parasites. Often, any one parasite can parasitise only one, or at most a few, host species. Furthermore, host species are rarely attacked by two closely related parasite species.

10.7 **Character displacement**

In 1835 Charles Darwin collected some dull-looking finches on the Galapagos Islands. They have since come to be known as 'Darwin's finches', for they were cited by him as a classic instance of the evolution of several species from a common ancestor. Just before the Second World War, David Lack was able to spend five months on the Galapagos Islands. He argued that the differences between the birds, which Darwin had observed, were *adaptive* (Lack, 1947). In particular, he maintained that interspecific competition was an important force structuring the Galapagos communities and that **ecological isolation** is as necessary for the coexistence of closely related species as is reproductive isolation.

Particularly interesting were the results of Lack's careful measurements on beak depth in the three ground-finches: the large ground-finch (*Geospiza magnirostris*), the medium ground-finch (*G. fortis*) and the small ground-finch (*G. fuliginosa*). Some of these results are plotted in Figure 10.14. From this figure it can be seen that on six islands where all three ground-finches exist, complete separation occurs between the beak lengths of the three species. Similarly, on Charles and Chatham, from which *G. magnirostris* is absent, the beak depths of the other two species are still distinct. Most interesting, however, is the observation that on Daphne, which has just the one species, *G. fortis*, and on Crossman, which also has just one species, this time *G. fuliginosa*, the birds have beaks of about the same depth. On the islands which hold both *G. fortis* and *G. fuliginosa*, **character displacement** has occurred. Over evolutionary time, natural selection has caused the beaks of the two species to diverge as the birds specialised on different foods. Character displacement is distinct from ecological compression which was described in Section 10.5. Ecological compression takes place on a much shorter time-scale and does not involve heritable change.

10.8 **Interspecific competition in natural communities**

Over the last 15–20 years ecologists have come to change their views on the importance of competition between different species (interspecific competition) as a force shaping natural communities and driving evolutionary change. After all, natural selection is fundamentally about competition between the members of a species (intraspecific competition). Botanists, in particular, have long been aware that intraspecific competition is typically stronger than interspecific competition (Black, 1960; Harper & McNaughton, 1962). This does not mean, however, that competition between species never occurs. As so often is the

case in ecology, what is at issue is the relative importance of an effect. Competition between species for shared resources undoubtedly occurs, but this does not necessarily mean that it is very important. Furthermore, it is perfectly possible that interspecific competition is more important in some communities than in others.

An early review of the evidence for and against the theory that interspecific competition is important in natural communities was provided by Connell (1975). Connell pointed out that three general methods have been used to detect interspecific competition under natural conditions. The first is to look and see whether nature conforms to the predictions that follow from this theory. MacArthur's observations on the *Dendroica* warblers (Section 10.2.2) fit into this category. He showed that the feeding and nesting habits of the warblers were consistent with the hypothesis that interspecific competition was important in the structuring of that community.

The second approach described by Connell is to look for 'natural experiments'. For instance, the decline in red squirrel numbers in Britain as grey squirrel numbers increased (Section 10.5) is consistent with the theory's predictions.

The third approach, and the one which Connell favours, is to conduct controlled experiments in the field. This alone, Connell argues, allows one to ensure that everything varies in the same way between treatment and control except for the factor being tested. Now, insistence on controlled field experiments is a very strict requirement. After all, we do not have controlled field experiments to test the hypothesis that hunting by humans has led to a reduction in the numbers of African elephants, but no one seriously doubts this. Nevertheless, Connell's insistence on careful experimentation is helpful. It might, for example, be the case that red squirrel numbers have declined in Britain due to the effects of disease rather than competition by grey squirrels.

It is perhaps unsurprising that the most convincing evidence, using Connell's criteria, for the importance of interspecific competition comes from plants. This is because experiments can be conducted on plants more easily than on animals. For instance, many early experiments showed the importance of competition between the roots of canopy trees and those in the understorey in temperate woods (reviewed by Connell, 1975). Trenches were dug around plots containing understorey trees, cutting the roots of neighbouring adult trees and so reducing competition for water and nutrients. In every case the smaller trees inside the trenched plots survived and grew much better than those in nearby control plots. It should be mentioned that many of these studies did not distinguish between intraspecific and interspecific competition.

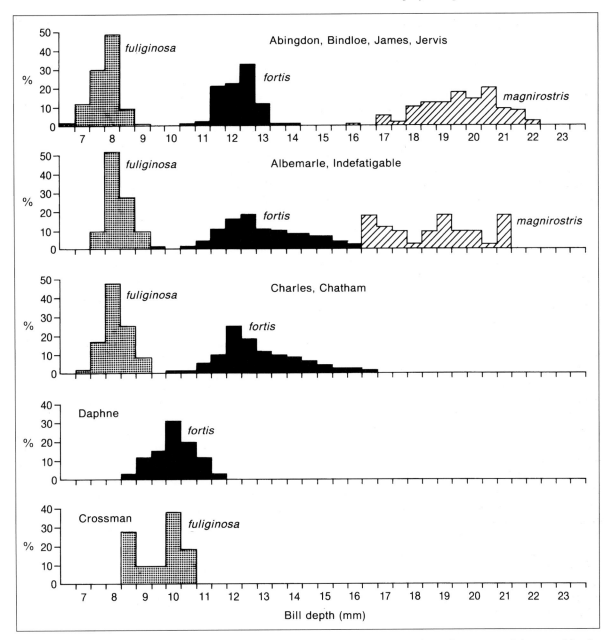

Figure 10.14 The bill depths of the three species of ground-finch (*Geospiza* spp.) on the various Galapagos Islands. (From Lack, 1947.)

From Connell's review it looks as though many species seldom reach the population densities necessary for interspecific competition to be important. It seems that their numbers are kept down by intraspecific factors, by physical conditions or by interspecific factors, such as predation and parasitism, that are not normally included within the definition of interspecific competition.

A more recent paper (den Boer, 1986) reaches the same conclusion as Connell but even concludes that *intraspecific* competition is not of great importance in structuring natural communities! 'As a future trend we can expect a further depreciation of competition,

both intra- and interspecific, as being a major force in ecology and evolution. The manifold influences of weather, climate and other physical factors will assume greater importance, and the claim of universality of competition will probably be replaced by a gradual re-evaluation of the role of predation' (den Boer, 1986, p. 28).

Some ecologists feel that den Boer has gone too far (e.g. Abrams, 1986; Giller, 1986) but there have certainly been some ecologists who have changed their minds and now feel that interspecific competition is less important than they previously thought. For instance, Sinclair (1979) had extended the work of Bell (1971) on grazing in the Serengeti, and had concluded that interspecific competition was the dominant structuring process in the Serengeti community, producing the observed separation and coexistence of herbivores. However, Sinclair (1985), as a result of 'natural experiments' including variations in the numbers of wildebeest, concluded that predation appeared to play as much of a role as interspecific competition in structuring the community. In particular, most of the larger herbivores are found *nearer* to the wildebeest than expected from a random distribution. This is contrary to the 'interspecific competition hypothesis', but agrees with the 'risk of predation hypothesis' (see Section 8.2.1).

One example where interspecific feeding competition is still believed to be important in community structure is provided by Gilbert *et al.* (1985). Gilbert and his co-workers studied the ten most common hoverflies in an ancient woodland near Cambridge. They found that the ratios of the proboscis lengths of these species were significantly more constant than expected by chance. This agrees with Hutchinson's (1959) suggestion that closely related species should reduce interspecific feeding competition by having dissimilar feeding apparatuses. There is also strong evidence for character displacement in hyaenas and other carnivores as a result of interspecific competition (Werdelin, 1996).

As ecologists have demanded better evidence for the importance of interspecific competition, so a number of the cases of apparent character displacement have been shown to be due to other causes (Arthur, 1982). However, it still looks as though Lack's example of character displacement in the ground-finches *Geospiza fortis* and *G. fuliginosa* is valid (Grant, 1986). It could be argued, as an alternative hypothesis to that of character displacement, that the intermediate beak depths seen on the islands where only one of the ground-finches is found (Figure 10.14) simply reflect adaptation to intermediate seed sizes found on those islands. On the islands where the two species coexist, the bill depths might differ because their food supplies differ due to effects *other* than interspecific competition. This alternative explanation to character displacement appears incorrect in this case, as Schluter, Price and Grant (1985) showed that the intermediate beak depths seen on the islands where only one of the ground-finches is found was not due to the presence of intermediate seed sizes on these islands.

10.9 Do plants need niches?

Temperate chalk grasslands, like the one in Figure 10.15, may contain as many as 40 species of vascular plants in each square metre. Over 150 tree species

can be found in a single hectare of some tropical forests. Yet most plants require much the same basic resources: light, water, carbon dioxide, oxygen, nitrate, phosphate, a dozen or so minerals in solution, and space. Can it really be that different plant species have different niches? Silvertown and Law (1987) reviewed advances in plant community ecology and questioned whether Gause's competitive exclusion principle held for plants.

Rather surprisingly, Silvertown and Law argue that competition between species may be less important in plants than in animals. They maintain that terrestrial plants are often surrounded by other members of the same species, and that this means that competition between different plant species is unimportant. This argument is surprising because the problem with traditional niche theory in plants is that each plant seems to be competing with individuals from many other species, as, for example, in chalk grasslands and tropical forests.

A more convincing argument for the maintenance of high species richness in many plant communities is that occasional episodes of density-independent mortality can *delay* and greatly reduce the effects of interspecific competition. These **non-equilibrium** theories date back at least as far as Andrewartha and Birch (1954) who argued that many insect species were affected far more by such occasional density-independent disasters than by density-dependent mortality. The crucial point about some of these non-equilibrium models is that calculations suggest

Figure 10.15 A species-rich temperate short grassland on a shallow chalky soil.

that the average time required for interspecific competition to eliminate species may be *longer* than the average time between speciation events. In other words, by the time one species has become extinct through interspecific competition, another one or two may have evolved! The larger the community, the more important this effect becomes. Tropical rainforests are places where such non-equilibrium coexistence may be particularly important.

Predators and parasites may also stop a single species from increasing in numbers to a level where it can out-compete other species. Janzen (1975) has argued that in the tropics, whenever a single plant species begins to increase substantially in numbers, host-specific herbivores congregate. These stop and even reverse the growth in the numbers of that particular plant species and prevent it from seriously competing with other species.

Some botanists have maintained that plant species may conform to the predictions of traditional niche theory and occupy distinct niches. Grubb (1977) introduced the notion of the **regeneration niche**. He argued that different plant species often need specific conditions for their seeds to germinate and for their seedlings to survive and grow. Grime (1979) has further examined the huge variety of ways in which plants regenerate. The fact that adult plants may not occupy distinct niches does not mean that this must be the case for *all* stages of the life cycle. The niche of an organism needs to be considered over the entire life cycle of the organism.

10.10 Community structure of fish on coral reefs

Individual coral reefs may hold an astonishing diversity of fish species. The debate in recent years about what allows the coexistence of so many species has mirrored the same debate about plant communities (Section 10.9). Essentially, two reasons have been suggested as to why coral reefs have so many fish species (Mapstone & Fowler, 1988). First, that coral reefs are communities at equilibrium showing precise resource partitioning in response to the competition between the various fish species. Second, that the number of fish species on coral reefs is kept high largely by **stochastic** processes. Stochastic processes are unpredictable and operate in a relatively density-independent fashion. This is the opposite of the traditional, equilibrium hypothesis which emphasises density-dependent competition between species. Advocates of the stochastic hypothesis even refer to 'assemblages' of reef fish, denying that the fish really coexist as a community where the actions of one member affects other members.

Although there are not yet enough data to rule out either of these two explanations, experiments on the recruitment of new fish to the reefs may provide the answer. The traditional equilibrium model assumes that the probability of an individual fish larva surviving to reproduce is limited in a density-dependent manner by the abundance of the adult fish. The alternative stochastic model predicts that recruitment to the adult phase is independent of the density of the adults.

Summary

(1) A habitat is the place where an organism is found in nature.

(2) Any one environment contains many micro-habitats.

(3) The niche of an organism is a complete description of how the organism fits into its physical and biological environment.

(4) A great deal of field work is needed to determine the niche of a species in any detail.

(5) Hutchinson proposed that the niche of an organism could be defined as an *n*-dimensional hypervolume with separate environmental variables plotted on the axes.

(6) Gause's competitive exclusion principle states that no two species can coexist if they occupy the same niche.

(7) The size ratios of closely related species provide little evidence that interspecific competition has been important in evolution.

(8) Species can coexist even in the face of interspecific competition provided that their niches do not overlap too much.

(9) The fundamental niche of an organism is the niche it could occupy in the absence of competitors and predators.

(10) The realised niche of an organism is the role actually played by it in the community.

(11) Resource partitioning occurs when two or more species in a community divide out a resource such as food.

(12) Character displacement occurs when interspecific competition results in natural selection causing the morphologies of two closely related species to diverge.

(13) There is little rigorous experimental evidence to demonstrate that interspecific competition is important in nature.

(14) A number of theories suggest that interspecific competition may be of little importance in communities as different as tropical rainforests, temperate grasslands and coral reefs.

(15) Much more field work is needed before the importance of traditional niche theory can be evaluated.

ELEVEN

Trophic levels

11.1 Why study trophic levels?

There are many approaches to studying the ecology of organisms in their natural surroundings. We have already seen several of these: changes in population size, the occupation of niches, and so on. Another way is for ecologists to study organisms from the point of view of their feeding relationships – what they feed on and what feeds on them. There are a number of reasons why this is a good approach. First, one of the major problems faced by all organisms is how to obtain enough energy and nutrients to survive, grow and reproduce. Second, if the feeding relationships of a group of organisms can be unravelled, then we get a clearer understanding of how those organisms interact. Finally, there simply is not enough time, nor sufficient funds, to study every species in detail. Ecologists therefore often cannot always look at each and every species, one by one. Rather, they study the autotrophs, the herbivores, the carnivores, the decomposers and so on, lumping together the species that have a similar role in the community.

In this chapter we will look at the arrangement of organisms into these community groups or **trophic levels.** Trophic literally means feeding, so trophic levels are the levels or positions at which species feed. Examples of trophic levels include 'herbivores' and 'decomposers'. In this chapter we will look at the characteristics of organisms in each of the trophic levels. We will then examine approaches to the study of the feeding relationships of organisms. Finally, we will see whether there are any rules governing the trophic levels and feeding relationships observed in nature.

11.2 Autotrophs

Autotrophs, as we briefly discussed in Chapter 2, are organisms that literally 'feed themselves'. Unlike **heterotrophs**, they do not require organic compounds as their source of energy. Autotrophs can be divided into two groups, the **photoautotrophs** and the **chemoautotrophs**, and we will look at each of these groups in turn. The autotrophs are sometimes also referred to as **producers**. This is because they manufacture or produce the organic molecules on which all other organisms depend for their source of energy.

11.2.1 Photoautotrophs

The autotrophs with which we are most familiar are the **photoautotrophs** or photosynthetic organisms. As their name suggests, photoautotrophs obtain their energy from light: the Sun. The organisms best known to us as photoautotrophs are, of course, plants. Most ecologists accept a five kingdom classification of life (Margulis & Schwartz, 1982). Under this classification, from which viruses are excluded, organisms are classified into the following five kingdoms: Prokaryotae (unicellular), Protoctista (eukaryotes that do not fit into the other kingdoms: often unicellular), Plantae (multicellular eukaryotes with photosynthetic nutrition), Fungi (multicellular eukaryotes with hyphae and living saprotrophically or parasitically) and Animalia (multicellular eukaryotes lacking cell walls and living heterotrophically).

Of these five kingdoms, three show photoautotrophic nutrition. Almost all species of plants are photoautotrophs, the exceptions being parasitic plants such as broomrapes (*Orobanche* spp.) and dodders (*Cuscuta* spp.) (see Box 2, p. 8). It is not appropriate to consider plant nutrition in detail here, but the essentials are that carbon dioxide is fixed by chloroplast enzymes (ribulose bisphosphate carboxylase in C_3 plants, phosphoenolpyruvate carboxylase in C_4 plants) while inorganic substances such as nitrate (NO_3^-) and phosphate (PO_4^{3-}) along with water are taken up into the plant, typically by root hairs. The plant's chloroplasts, with their photosynthetic pigments, are then able to use the energy from sunlight (or almost any other source of visible light) to split water into oxygen and hydrogen. The oxygen diffuses away from the plant as a waste product, unless it is used in respiration. The hydrogen is used to reduce carbon dioxide (CO_2) to the fundamental building block of all carbohydrates, the three-carbon sugar phosphoglyceraldehyde.

The other eukaryotes that show photoautotrophic nutrition are some protoctists including *Euglena*, green algae, red algae and brown algae. These organisms are able to photosynthesise in essentially the same way as plants.

The third kingdom with members that show photosynthesis is the Prokaryotae. This group is divided into the Cyanobacteria and the Bacteria. All cyanobacteria show photosynthesis. Indeed, they look rather like chloroplasts and there is now very strong evidence that the chloroplasts possessed by eukaryotes once were cyanobacteria which somehow became permanent parts of eukaryotic cells. This is part of the **endosymbiotic theory** of the origin of the eukaryotic cell (Margulis. 1981).

While all cyanobacteria show photosynthesis, only some bacteria do. Indeed, bacteria show just about every sort of nutrition known and there is little doubt that they should be divided into at least two kingdoms, each as distinct from the other at the biochemical level as animals are from plants or fungi. There is still a great deal that remains to be known about the biochemistry and ecology of the photosynthetic bacteria, but the number of different ways in which they solve the problem of obtaining energy are fascinating. They differ fundamentally from other photoautotrophs in being anaerobic. Nor do they carry out the photolysis of water with the synthesis of NADPH and release of oxygen (Hamilton, 1988). Instead of water being the source of electrons used along the electron transport chain in photosynthesis, various inorganic or organic compounds act as electron donors.

The particular electron donors used are characteristic of different groups of bacteria. In green sulphur bacteria, sulphide is the electron donor. The purple sulphur bacteria can use sulphide and sulphur itself. The non-sulphur purple bacteria are even more complicated. Under anaerobic conditions they are photoautotrophs using either hydrogen or sulphide as their electron donor. Under aerobic conditions, though, they will, in the presence of light, use a simple organic compound such as lactate ($CH_3CH(OH)COO^-$) as their source both of electrons and of carbon. In the dark they can survive and grow as heterotrophs.

In all the bacteria so far discussed, sunlight is trapped by bacteriochlorophyll which absorbs light at a characteristically high wavelength, 870 nm, compared to the 680–700 nm of chlorophyll *a* which is found in eukaryotes. There is, however, an extraordinary bacterium called *Halobacterium halobium*. This species is *unable* to grow at salt concentrations less than 3 M, that is five times as salty as sea water! Accordingly, the bacterium lives in drying saltpans

and in salted fish. *Halobacterium* is photosynthetic, but lacks bacterial chlorophyll. Instead its membrane contains the protein bacteriorhodopsin. At the centre of this protein is a molecule of the visual pigment retinaldehyde. Just as in the rods of the human eye, this substance is bleached by light. The bacterium uses the energy associated with this process to eject protons from its cell. These protons are then allowed to flow back into the cell via an ATPase, thus synthesising ATP just as chloroplasts do.

Photosynthetic bacteria have various ways of obtaining their carbon. In the green sulphur bacteria, the only route of carbon dioxide fixation is by the reductive carboxylic acid cycle. This involves a *reversal* of the tricarboxylic acid (Krebs) cycle together with the reactions acetyl-CoA$\rightarrow$ pyruvate$\rightarrow$ phosphoenol pyruvate$\rightarrow$ oxaloacetate. The purple sulphur bacteria use both the Calvin cycle and this reductive carboxylic acid cycle.

11.2.2 Chemoautotrophs

Different authors use the term **chemoautotroph** in different ways. The easiest is to use it to refer to organisms in which energy is obtained *solely* from the oxidation of inorganic electron donors without the use of light. Organisms covered by this definition are clearly autotrophic yet non-photosynthetic. However, it is *not* the case that non-photosynthetic organisms which can obtain energy from the oxidation of inorganic electron donors are necessarily chemoautotrophic (Hamilton, 1988). For instance, some iron-oxidising bacteria can oxidise Fe^{2+} to Fe^{3+} but are heterotrophic, obtaining their energy principally from the breakdown of organic compounds.

Unlike photosynthetic organisms, chemoautotrophs, or **chemosynthetic organisms** as they are often called, are restricted to just one kingdom – the Prokaryotae. From an ecological perspective, some of the most important chemosynthetic organisms are involved in the nitrogen cycle. These include *Nitrosomonas* and *Nitrococcus*. The nitrogen cycle is covered in Section 13.4. Here we will look at the recently discovered communities which cluster around hot vents deep down on the oceanic floor. These communities rely entirely on chemosynthetic bacteria.

In 1977 American scientists using a deep-sea research submarine were investigating underwater volcanoes erupting from a ridge south of the Galapagos Islands (Attenborough, 1984; Hessler *et al.*, 1988). At a depth of 2500 m the scientists found great concentrations of chemosynthetic bacteria feeding on the hydrogen sulphide (H_2S) thrown out by the hot vents. Across most of the ocean floor animals are dispersed very thinly. Of course, no light can penetrate to such depths, so most animals rely on a slow 'rain' of organic matter from above. Around these volcanic vents, however, are found great concentrations of animals. There are unique tube worms 1.5 m in length and remarkable clams 30 cm long, both of which feed on the bacteria (Figure 11.1). Because the vents are hot, they set up convection currents which draw in organic fragments from the surrounding area. Hitherto unknown fish and blind white crabs act as scavengers which congregate at the vents and benefit from these convection currents.

Similar communities have been found elsewhere, including the Mariana Trough in the western Pacific at a depth of 3600 m (Hessler *et al.*, 1988; Southward, 1989). Here the remarkable tube worms found near the Galapagos Islands are either very rare or absent. Instead there are large numbers of a unique and very primitive sessile barnacle. Particularly interesting are the large snails found crowding around the opening of the vents. These snails have enormous gills which fill about 40% of their body. Transmission electron micrographs show that specially modified gill cells contain chemosynthetic bacteria. Biochemical tests show the presence of enzymes characteristic of sulphur metabolism including ATP-sulphurylase.

The Mariana Trough snail has a well-developed radula in addition to these bacteria. This probably means that it does not depend solely on the bacteria for its nutrition. On the other hand, the tube worm (*Riftia pachyptila*) and the clam (*Calyptogena magnifica*) at the Galapagos Islands vents probably rely on

their chemosynthetic bacteria for all of their nutrition. The tube worms lack a gut and the clam has a gut which is so small that it may be useless.

There is now a tremendous interest in the ecology and evolution of these chemoautotrophic communities (Little *et al.*, 1997). It has been argued that life on Earth evolved some 4200 million years ago in such an environment (Russell & Hall, 1997).

11.3 Decomposers

After the producers, the **decomposers** are the next most important organisms in a community. Without producers, there could be no decomposers, no herbivores and no carnivores. Without decomposers, the remains of the other organisms in a community would simply accumulate. Eventually the world would run out of carbon dioxide or nitrate or phosphate or other inorganic substances essential for life. The decomposers break down the organic waste products and dead remains of organisms into the inorganic substances needed by the producers.

The simplest self-perpetuating community is therefore one consisting just of producers (whether photo- or chemoautotrophic) and decomposers. Such a community is illustrated in Figure 11.2 which shows a community based on photoautotrophs and decomposers. The photoautotrophs obtain their energy from sunlight and their minerals and water from a reservoir such as soil or a lake. Photosynthetic organisms also require carbon dioxide for photosynthesis and oxygen for respiration; these are obtained from the atmosphere, though they may be obtained from the atmosphere via the reservoir, as when submerged aquatics obtain their oxygen and carbon dioxide from the water in which they live. Photoautotrophs produce carbon dioxide in respiration, oxygen in photosynthesis and organic compounds in **anabolism**. Photoautotrophs cannot take up organic compounds. These organic compounds must therefore either accumulate in a sink or be utilised by the decomposers. The decomposers in turn usually require oxygen, unless they are **obligate anaerobes**, and produce carbon dioxide and minerals such as ammonium (NH_4^+) which they return to the atmosphere and reservoir. If the producers are chemoautotrophs, the box labelled 'sun' and the arrow labelled 'sunlight' need to be deleted from Figure 11.2.

Studies of natural communities often pay little attention to decomposers. Maybe this is because decomposers are often microorganisms. Whatever the reasons, decomposers play a vital role in nature and certainly deserve careful investigation.

Decomposers are, of course, heterotrophs. In the language of Chapter 2, they are saprotrophs, feeding

Figure 11.1 Tube worms (*Riftia pachyptila*) and clams (*Calyptogena magnifica*) clustered round an ocean floor volcanic vent at a depth of 2500 m.

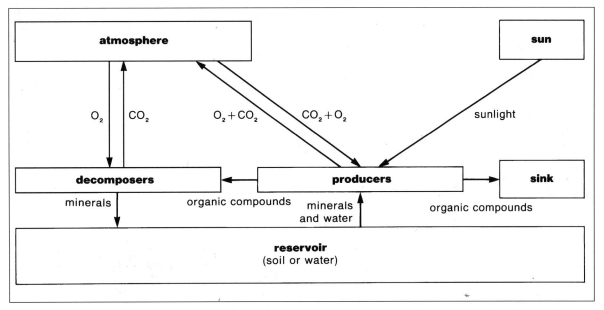

Figure 11.2 The simplest self-sustaining community needs both producers and decomposers. This diagram shows a community based on photoautotrophs and decomposers.

on dead organic matter which they either absorb in solution or ingest as very small pieces. Organisms that ingest small particles of organic matter derived from the dead remains of plants or animals (**detritus**) are known as **detritivores**. Four of the five kingdoms contain decomposers – Prokaryotae, Fungi, Protoctista and Animalia. Accurate generalisations are difficult, but most decomposition is probably carried out by saprotrophic (as opposed to parasitic) fungi, by bacteria and by invertebrates (Mason, 1977; Barnes & Hughes, 1988; Moss, 1988). The best way to see the significance of decomposers is to look at an example where their role has been carefully investigated. As more is known about terrestrial than aquatic decomposers, the example given below is a terrestrial one.

11.3.1 Decomposition on the forest floor

In some communities dead organic matter accumulates, but in most forest environments the rate at which **litter** is added to the soil equals the rate at which it is removed by decomposition. The amount of dead organic matter measured in forests at different latitudes is greatest towards the poles and least at the equator (Table 11.1). As the total biomass per unit area in a forest is *greatest* at the equator, this means that almost all of the organic matter in a tropical rainforest is tied up in the living organisms. This is also the case with the minerals. Tropical rainforests therefore rely on rapid nutrient cycling for their maintenance (Section 15.2.2).

The breakdown of leaf litter is due to physical and biological factors. Breakdown in the tropics is fast for

Table 11.1 The amount of accumulated dead organic matter in different forest types is greatest in forests near the poles and decreases towards the equator. (From Kurihara & Kikkawa, 1986.)

Forest type	Mass of organic matter in the forest layer
Tropical rainforest	2 tons/ha
Subtropical forest	10 tons/ha
Broad-leaved forest	15–30 tons/ha
Taiga	30–45 tons/ha
Shrub tundra	85 tons/ha

a variety of reasons. These forests are hot and humid. The lowland climate is probably less seasonal and the weather less erratic than at higher latitudes. These physical factors in themselves speed up organic breakdown. Additionally, they provide an ideal environment for many decomposers. The relative importance of different organisms in the breakdown of leaf litter has been investigated by a number of authors using litter bags with a variety of mesh sizes (Mason, 1977; Kurihara & Kikkawa, 1986). Mesh sizes of around 5 mm in diameter allow all the decomposers to enter; those of around 1 mm exclude most earthworms (e.g. the larger *Lumbricus* individuals); mesh diameters of around 0.005 mm exclude all invertebrates and allow only microorganisms, including fine fungal hyphae, to enter. Generalisations are difficult, but it seems that organisms of a variety of sizes may be necessary for rapid and complete breakdown. Certainly, exclusion of the larger soil invertebrates (soil **macrofauna**) such as earthworms, millipedes

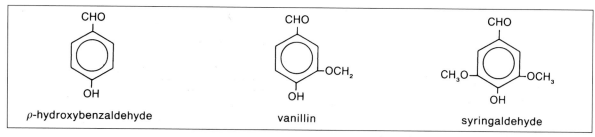

Figure 11.3 The three main aryl aldehydes in the polymers in lignin. (From Lynch, 1988.)

and woodlice tends to slow the rate of decomposition substantially.

11.3.2 Decomposition of dead plant matter

When biological materials decay they undergo **humification. Humus** is the organic matter resulting from the partial decomposition of organic material in the soil. It is typically dark in colour. Functionally, it is of great importance: it acts as a reservoir, holding minerals and water until they are absorbed by plants. Structurally, humus is extremely complicated. Apart from certain inorganic deposits, such as those found in invertebrate shells and in vertebrate teeth and bones, the compounds in humus most resistant to decomposition are the lignins of vascular plants. Lignins, along with cellulose, hemicelluloses and pectins, are found in plant cell walls. Chemically they are diverse, and their biochemistry is still not fully understood. They are composed mainly of phenolic polymers which eventually give rise to humic acid (Lynch, 1988).

The main aryl aldehydes in the polymers in lignin are shown in Figure 11.3. These substances are extremely resistant to microbial breakdown which is why lignin is so resistant to decomposition. Indeed, some fractions of soil organic matter have been radio-carbon dated to over 1000 years old (Perrin *et al.*, 1964). In soils, certain groups of fungi and bacteria are the main organisms responsible for the breakdown of lignin. The reactions produce the dark pigment melanin which accounts for the characteristic colour of humus (Lynch, 1988).

Careful examination of the process of decay of dead plant tissue reveals characteristic patterns in the order in which different decomposers colonise the material. The succession of microbial communities on hay (cut grass) is reviewed by Lacey (1988). Once the grass has been cut and stored, there is a rapid decrease in the microorganisms found on the leaves when the plant was still alive. Simultaneously, there is a build-up of species normally found on stored material. The build-up of these species depends on several factors, especially the water content of the hay. As every farmer and gardener with a compost

---

Box 11.1
Why food rots

We all know that foods smell unpleasant once decomposition has set in to a significant degree. But why? What is the *function* of food rotting? Janzen (1979) has looked at decomposition from the point of view of the decomposers and argues that from their aspect, it is adaptive for food to go rotten.

Janzen argues that **putrefaction**, the rotting of foods by microbes, is the result of explicit efforts by the microbes to render food unattractive to higher animals. The term 'explicit efforts' is a shorthand. Janzen's point is that over the course of evolution, microbes which render food unpalatable to competitors such as ourselves produce more copies of themselves than microbes that do not. This is because, by making food unattractive, they effectively scare off the opposition. Janzen argues, though it has to be admitted that the evidence is largely anecdotal, that vertebrates are more likely to take perfect fruit than fruit colonised by microorganisms which are causing the fruit to rot.

Some of the fungi that colonise grain produce toxins. Ergot (*Claviceps purpurea*) attacks grain and produces a hallucinogenic toxin that may be fatal to livestock and ourselves. **Aflatoxins** are compounds made by the fungus *Aspergillus* and other grain-inhabiting fungi; they are among the most carcinogenic compounds known. It is not surprising that many farm animals will starve to death before eating mouldy feed.

Once microorganisms produced significant amounts of such poisons, natural selection would have started to work on competitors, so as to reduce the effects of these poisons. Of the few species of birds that have been tested, the highly granivorous quail are the least susceptible to aflatoxins. Presumably this is because natural selection for resistance to aflatoxins has been particularly strong in quail. On the other hand, it could be argued that natural selection should be particularly strong on the fungi to evolve toxins poisonous to such a specialised granivore!

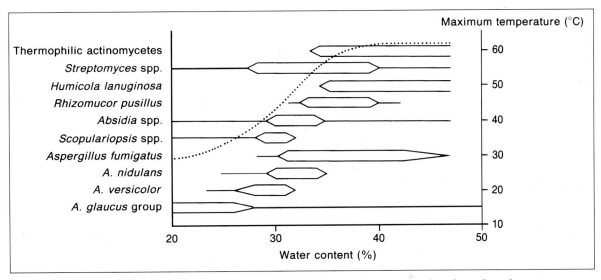

Figure 11.4 The species of microorganism that colonise and grow on decomposing hay depend on the water content of the hay. The diagram shows the range of species likely to be found for any particular water content between 20% and 50%. The greater the initial water content, the higher the temperature of decomposition reached – as shown by the dotted line. (From Lacey, 1988.)

heap knows, growth of microorganisms in damp hay is accompanied by the release of heat. This heat is due to respiration by the microorganisms. Providing the hay is not too dry, temperatures of 60 °C may commonly be reached. Under certain conditions, the temperature may even be such as to lead to spontaneous combustion!

The typical order in which microorganisms colonise hay is shown in Figure 11.4. With a water content of 40%, it usually takes about five days for a temperature of around 60 °C to be reached. The temperature then gradually falls over the next three weeks to that of the surroundings.

11.4 Herbivores and carnivores

Although the distinction between herbivores and carnivores is clear, it has its limitations. Some ecologists consider both plant and animal eaters as predators or parasites. Herbivores can be labelled as predators if they are larger than their prey and consume most or all of a plant in a relatively short amount of time; they are parasites if they live in or on their hosts for long periods of time. According to this definition, most mammalian herbivores are predators; most insect herbivores are parasites. However, locusts are predatory whereas deer browsing on the lower leaves of a tree are parasitic. It might be thought that arguing over terms like this serves little purpose. However, it can help an ecologist to see trophic interactions from a different perspective.

There are some important differences between most plants and most animals from the perspective of the organisms that feed on them. Generally speaking,

plants are immobile. Because they cannot run away from animals, they use a variety of strategies to escape from their herbivores. They may have defence mechanisms, producing spines or tough leaves or manufacturing toxins. Oak trees, for example, produce unpalatable tannins in their leaves; the relationship between the oak leaf roller moth and its host, the oak, is considered in Section 14.2.2. Plants may try to hide from herbivores. Many short-lived plants exist as seeds in the seed bank (Section 4.3) and produce a sudden abundance of growing individuals when environmental conditions are appropriate. This flood of seedlings may be more than the herbivores can eat and increases a plant's chances of reproducing before it is eaten. Such seed dormancy is an example of a plant hiding both in *space* (buried in the soil or leaf litter) and in *time*. Instances of the **evolutionary arms race** that occurs between plants and herbivores are considered in Section 19.2.1.

In some respects carnivores face even greater problems than herbivores in the acquisition of food. In particular, their food is often very mobile. On the other hand, although herbivores may not need to be very mobile to obtain their food, they may need great mobility to escape from carnivores! An advantage of being a carnivore is that food is much less likely to be toxic. It is also easier to digest and tends to contain more protein.

Ecologists have tried to make generalisations about the ecology of herbivores and carnivores, but such generalisations have not proved very successful. For instance, about 25–50 years ago, herbivore populations were often thought to be controlled by the numbers of their predators. Then the theory changed to suggest that predators never regulated the numbers of their prey. Now, the emphasis has switched again and predators are one of a number of factors believed to regulate population sizes (see Chapter 5 and Section 10.9).

11.5 Omnivores

An **omnivore** is an organism that eats both plants and animals. (Many of the technical terms in ecology were coined when organisms were, by and large, classified into two groups – animals and plants. Nowadays, most ecologists would classify a species that fed on fungi and animals, for instance, as an omnivore.) Omnivores are, by definition, **generalists**. That is, they are not **specialists**, but rather obtain their food from a great variety of sources. Humans are omnivores par excellence. Most humans eat both animals and plants, while fungi are also popular with many. Further, the range of animals and plants we eat is huge.

At first it might be thought that omnivores have the best of both worlds. When there are lots of animals available, they can feed on the animals, and if animals are scarce they can switch to plants. However, nothing in life is free. If omnivory has benefits, it must also have costs. The major cost is that omnivores, being generalists, are less efficient at using any single food source than are specialists. For instance, humans lack the gut bacteria capable of digesting cellulose that are found in many specialist herbivores (Section 14.2.4). So we need plant matter with more protein and starch in it than do ruminants and equids. We are also less efficient at getting the last scrap of nutrition from an animal carcass than are dogs (specialist carnivores), for instance. Because we are omnivorous, we need an alimentary canal suited for both animal and plant food. We need teeth suitable for tearing meat and for grinding plant matter. We need enzymes for attacking animal food and those for attacking some of the chemicals made only by plants – starch, for instance.

11.6 Food chains

Food chains are attempts to demonstrate the fate of individual organisms in a particular habitat. Consider a clover plant in a mixed deciduous oak wood in Europe. Suppose it is eaten by a snail, and the snail is killed and eaten by a thrush, which in turn is caught by a sparrowhawk. Figure 11.5 shows this food chain. Traditionally, food chains are drawn with the producer at the bottom. Successive trophic levels are connected by arrows which lead from the organism being eaten to the organism eating it. Sometimes the Sun, as the ultimate source of almost all food chains, is placed at the bottom. Although decomposers play an important role in natural communities, they are usually not included. A generalised food chain is shown in Figure 11.6. Herbivores

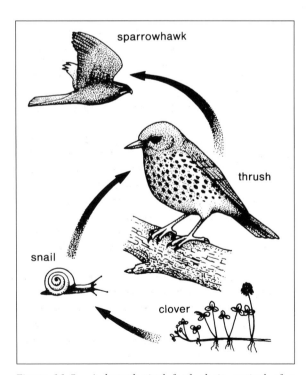

Figure 11.5 A hypothetical food chain typical of a European mixed deciduous oak wood.

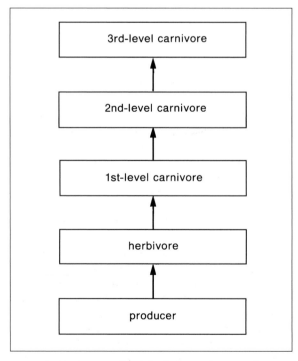

Figure 11.6 A generalised food chain showing the passage of food from producers to third-level carnivores.

are also called **primary consumers**; first-level carnivores **secondary consumers**; second-level carnivores **tertiary consumers**; third-level carnivores **quaternary consumers**; and so on. It is rare for a food chain to have more than five levels.

So why do food chains usually have fewer than six levels? Why do we not find tenth-level carnivores? The answer given by most authors is that there simply would not be enough food to support such a carnivore. As we will consider in Chapter 12, energy is lost as one ascends a food chain. A tenth-level carnivore would have to have a huge **home range area** in order to obtain enough to eat. Pimm (1982) points out that this explanation makes the implicit prediction that sites with high primary productivities should have more trophic levels than those with low productivities. However, the available data do not support this prediction. For instance, tundra communities typically have **annual primary productivities** which average out at about 100–400 mg C/m^2/day, while the most productive site studied as a result of the International Biological Programme of the 1970s was a Czech fish pond, the margins of which had a productivity of at least 4000 mg C/m^2/day. Whatever their productivities, sites usually had four trophic levels. Pimm (1982) also showed that food chains tend to be no longer when they are dominated by ecologically efficient ectotherms (poikilotherms) than by less efficient endotherms (homoiotherms).

A quite different hypothesis to explain why food chains are relatively short has been put forward by Pimm and Lawton (1977). They argue, on theoretical grounds, that the longer a food chain is, the less stable it is in the face of environmental fluctuations. After all, our hypothetical tenth-level carnivore may become extinct if *any* one of the trophic levels beneath it is greatly reduced in number, for whatever reason. At the moment it is difficult to distinguish between Pimm and Lawton's explanation and the energy limitation one. This is partly because communities with considerable environmental fluctuations often also have lower levels of primary productivity and partly because the sorts of experimental interventions needed to test between the explanations have not been carried out (Estes, 1995).

It has to be admitted that we do not have a great deal of data on the length of food chains. It is easy to think up food chains of considerable length. For instance, a plant might support a herbivorous insect which might be eaten by a carnivorous insect. The carnivorous insect might be eaten by a small bird which might support a colony of ectoparasites, and so on. Once one starts to include parasites and hyperparasites (parasites that parasitise parasites) in food chains one can make them almost as long as one likes.

Box 11.2
Do trophic levels exist?

In recent years many ecologists have decided that the concept of the trophic level is more of a hindrance than a help (Cousins, 1987).

The best-known problem with the idea of trophic levels is deciding what to do with organisms that feed at more than one trophic level. This problem is particularly acute with carnivores as it is often the case that carnivores cannot be slotted into discrete trophic levels. Some authors have tried to cope with this problem by allocating species to non-integer trophic positions, so that a carnivore might be described as a level 3.2 carnivore, for instance.

The second problem concerns dead organisms, urine and faeces. Often these are omitted from food webs. At best, they are lumped together. There are two main problems with lumping them together. First, a dead producer is obviously at a different trophic level from a dead herbivore or carnivore. If they are combined into one category, all living heterotrophs might as well be lumped together into another category. A second problem is that some components of the urine and faeces have been assimilated and some have not. The urea in the urine of a herbivore has been made by that herbivore, unlike the roughage in its dung. Accordingly, it could be argued that the saprotrophs feeding on the dung of the herbivore are at a lower trophic level than the microorganisms feeding on the urea in the urine. From a quantitative viewpoint, these objections are not trivial. The majority of the primary productivity in the world probably goes directly to decomposers. In terms of energy flow (Chapter 12) and nutrient cycling (Chapter 13) decomposers are the most important organisms in natural communities.

Various other difficulties with the trophic level concept can be found. The position of symbiotic bacteria in ruminant guts, for instance, is unclear. Are they herbivores or saprotrophs? And whatever they are, what of the hosts that digest them? Are cows really carnivores specialising on herbivorous bacteria? Another difficulty is that it has never been clear whether trophic levels refer to the flow of nutrients or energy. Early ecologists (e.g. Elton, 1927) talked of **food-cycles** and if the fate of individual atoms is considered this is correct. A nitrogen atom in a plant might pass to a herbivore and then to a first-level carnivore and then to a decomposer and then be released as ammonia. It might then be converted to nitrite by one species of bacterium and then to nitrate by another (see Chapter 15) before being taken up by a plant without having passed into the atmosphere at any intervening stage.

Perhaps the most useful way to think of a food chain is as a history of the fate of particular atoms. After all, a snail is very unlikely to eat an *entire* clover plant, though it might nibble a few leaves. As a story, however, of what happens to individual atoms, a food chain is a valid scientific record. On the other hand, a single food chain obviously cannot tell us much about what is happening to an entire community. For example, a number of herbivore species will eat clover, and the sparrowhawk will also take other birds, some small mammals and even beetles. Consequently, a different way of representing such relationships, known as a food web, is often drawn.

11.7 Food webs

A **food web** is really a collection or matrix of food chains. It shows the pattern of energy or nutrient flow throughout a community. Figure 11.7 shows an early food web determined on Bear Island in the Arctic Circle while Figure 11.8 shows a food web for the Antarctic oceans. Food webs do have the advan-tage that they are more realistic than food chains; for instance omnivores can be shown on a food web but not on a food chain. A disadvantage with food webs is that unless you are already familiar with most of the organisms in it, it is difficult to learn much simply from looking at one. Another disadvantage of food webs is that they usually fail to give any indica-tion of the relative importances of the different food chains – one line may represent 90% of an animal's diet, another 10%, yet this is rarely made clear. A final disadvantage with them is that because they usually contain so many species, in their final pub-lished form they may tell us as much about the inter-ests of the people who determined them as about the feeding relationships of the organisms themselves. Figure 11.8, for instance, identifies most of the whales and seals to species level, but lumps together many other organisms into categories such as 'pelagic fish' and 'bacteria and protozoa'. Figure 11.8 undoubtedly reflects the authors' interest in marine mammals!

Various techniques are used to determine food

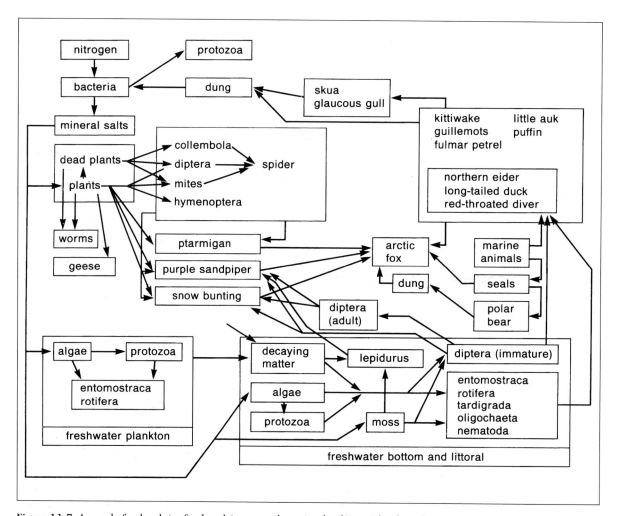

Figure 11.7 An early food web (or food cycle) among the animals of Bear Island, in the Arctic Circle. (From Elton, 1927.)

webs (and food chains). Direct observation may allow many of the feeding relationships to be identified. Analyses of faecal pellets, stomach contents, regurgitated owl pellets and so on may add more information. Radioactive isotopes such as ^{32}P have also been used to trace the fate of atoms in a community.

11.8 Pyramids of numbers

Food webs are useful descriptions of the feeding relationships of the organisms in a community, but they

are non-quantitative. The first attempt to provide a quantitative law concerning the trophic levels in a community was given by Elton (1927) and is worth quoting at some length:

If you are studying the fauna of an oak wood in summer, you will find vast numbers of small herbivorous insects like aphids, a large number of spiders and carnivorous ground beetles, a fair number of small warblers, and only one or two hawks. Similarly in a small pond, the numbers of protozoa may run into millions, those of Daphnia *and* Cyclops *into hundreds*

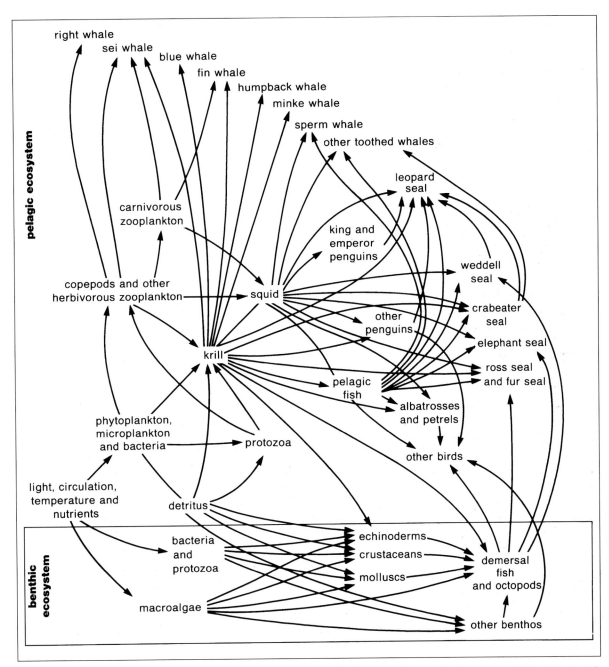

Figure 11.8 A food web for the Antarctic oceans. At one time whales were the principal consumers of krill, but with the hunting of most whale species close to extinction, birds and seals are now the main consumers. (After Krebs. 1988.)

of thousands, while there will be far fewer beetle larvae, and only a very few small fish. To put the matter more definitely, the animals at the base of a food-chain are relatively abundant, while those at the end are relatively few in numbers, and there is a progressive decrease in between the two extremes. The reason for this fact is simple enough. The small herbivorous animals which form the key-industries in the community are able to increase at a very high rate (chiefly by virtue of their small size), and are therefore able to provide a large margin of numbers over and above that which would be necessary to maintain their population in the absence of enemies. This margin supports a set of carnivores, which are larger in size and fewer in numbers. These carnivores in turn can only provide a still smaller margin, owing to their large size which makes them increase more slowly, and to their smaller numbers. Finally, a point is reached at which we find a carnivore (e.g. the lynx or the peregrine falcon) whose numbers are so small that it cannot support any further stage in the food-chain.

(Elton, 1927, p. 69)

It is worth noting that Elton mentions neither producers nor decomposers. He coined the term **pyramid of numbers** to refer to his observation (Figure 11.9). Elton noted that it is hard to obtain the actual numbers of individuals in the different stages of a food chain and the situation has not changed much since then!

Although Elton thought that 'The general existence of this pyramid of numbers hardly requires proving, since it is a matter of common observation in the field' (p. 70), the existence of parasites frequently leads to **inverted pyramids of numbers** (Figure 11.10a). If producers are included, inverted pyramids of numbers may again result as when, for instance, a

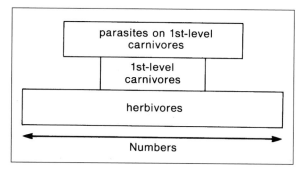

Figure 11.10a Hypothetical inverted pyramid of numbers showing more first-level carnivore parasites than first-level carnivores.

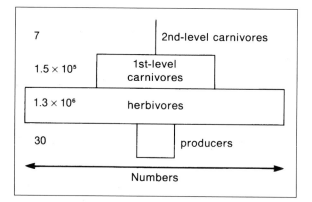

Figure 11.10b Inverted pyramid of numbers in Wytham oak wood, Oxford. The horizontal scale is logarithmic. The numbers at the right-hand side refer to the numbers of the organisms at each trophic level per hectare. Only oaks were counted as producers. (Data from Varley, 1970.)

single tree supports many much smaller organisms. Figure 11.10b shows the inverted pyramid of numbers in the oak wood at Wytham near Oxford.

11.9 Pyramids of biomass

Because some pyramids of numbers are inverted due to the very large size of some producers or the very small size of parasites, ecologists started to determine **pyramids of biomass**. As the term suggests, pyramids of biomass are calculated by determining, for a given unit area, the biomass of the producers, the biomass of the herbivores, the biomass of the first-level carnivores, and so on. It is true that very few pyramids of biomass have ever been determined, as they are even more time-consuming to determine than are pyramids of numbers, but it does seem likely that most are indeed pyramidal in shape and not inverted. The great game communities on the savannahs of Africa appear so abundant compared to the vegetation that it looks as if a pyramid of biomass would be inverted, with the herbivores weighing more than the grass

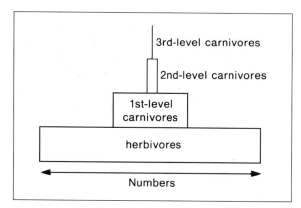

Figure 11.9 Typical pyramid of numbers. Producers have been omitted, following Elton (1927). The width of the boxes indicates the relative numbers of the organisms at each of the trophic levels. The most convenient horizontal scale is probably a logarithmic one. The highest carnivore(s), in this case the third-level carnivore, is/are sometimes referred to as the 'top carnivore(s)'.

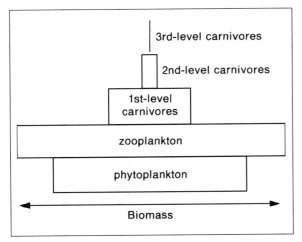

Figure 11.11 Inverted pyramid of biomass of zooplankton feeding on phytoplankton.

and other vegetation. However, even a sparsely covered savannah has a great many roots and the human eye perhaps overestimates the density of the large herbivores.

There is no theoretical reason why pyramids of biomass should never be inverted. Indeed, at least one instance is known where they are. In the oceans a given mass of phytoplankton at certain times in the year supports a larger mass of zooplankton (Figure 11.11). The reason this is possible is largely because the rate of reproduction of the phytoplankton is much greater due to their much smaller size. Consequently, the phytoplankton always produce enough for the zooplankton to eat (Whittaker, 1975). Of course, this means that the turnover of the phytoplankton is much more rapid than the turnover of the zooplankton. A close analogy would be that the mass of food in most people's homes is much less than the mass of the people that live there. Again, this is because the turnover of the food is far quicker than the turnover of the people.

Ecology textbooks sometimes dismiss pyramids of numbers and pyramids of biomass before going on to consider pyramids of energy. We will examine pyramids of energy in the next chapter. Here, though, it is worth emphasising that pyramids of energy are not 'better' than pyramids of numbers or biomass. Each gives an ecologist a different sort of information about the relationships between organisms in a community. Which one prefers will depend on what data one wants to collect and why.

Summary

(1) A convenient way to analyse the feeding relationships of a community is to analyse its trophic levels.

(2) Producers are the basis of natural communities. Most are photoautotrophs, though some are chemoautotrophs.

(3) Decomposers prevent communities rapidly running out of the various inorganic substances necessary for life.

(4) Despite their great biomass, tropical rainforests have little of their nutrients tied up in litter.

(5) Rapid decomposition requires the actions of a variety of organisms from bacteria to termites and earthworms.

(6) Humus results from the partial decomposition of organic matter in the soil.

(7) Lignins are highly resistant to decay and their breakdown is an important stage in humification.

(8) Plants have various strategies to avoid being destroyed by herbivores; in turn, herbivores have various strategies to avoid being eaten by carnivores.

(9) Omnivores are generalist feeders.

(10) Food chains can be used to demonstrate the biological fate of individual atoms.

(11) Food chains rarely have more than five levels. Ecologists disagree as to why this is so.

(12) A food web is the sum of the food chains in a community.

(13) Pyramids of numbers may quite often be inverted; pyramids of biomass only rarely.

(14) The unpleasant smells that accompany decomposition may be an adaptive response by decomposers to interspecific competition.

(15) In recent years, many ecologists have abandoned the notion of trophic levels.

TWELVE

Energy transfer

12.1 Energy and disorder

Living organisms are highly organised. In order to survive and maintain this internal order organisms need supplies of the relevant nutrients, a source of energy and the ability to create a large amount of disorder outside themselves. This last requirement may sound rather odd, but the second law of thermodynamics states that the amount of disorder in a closed system, such as the universe, increases with time. For organisms to create order, as they do when they make new cells, they need energy and/or the ability to create disorder outside themselves. Respiration both releases energy and creates disorder as relatively large molecules, such as glucose, are broken down to smaller and therefore less ordered molecules, i.e. carbon dioxide and water. The trophic levels at which species feed have been considered in Chapter 11, and Chapter 13 will look at how organisms obtain the nutrients they need. This chapter aims to look quantitatively at how organisms obtain their energy and how they pass energy up a food chain.

Because the vast majority of **primary production** in the world is the result of photosynthesis rather than chemosynthesis, we will first examine the environmental factors that determine the amount of photosynthesis in different communities. We will then see whether there are any ecological rules governing the transfer, or movement, of this energy up a food web through the trophic levels.

12.2 Primary production in terrestrial communities

When autotrophs photosynthesise, they turn carbon dioxide and water into larger structural molecules which allow the plant to increase in size. **Gross primary productivity** is a measure of the total amount of dry matter made by a plant in photosynthesis. It is measured in units of dry weight per unit area per unit time. All organisms, including autotrophs, respire. During respiration some of the matter from the gross primary production is converted back into carbon dioxide and water, and dry weight is therefore lost. The overall gain of dry weight by the plant, after respiration has occurred, is called **net primary productivity**. It has the same units as gross primary productivity.

Gross primary productivity, net primary productivity and respiration are therefore related to one another by the equation:

Net primary productivity =
gross primary productivity – respiration

Measurements of the amount of respiration by primary producers in terrestrial communities show that gross primary productivity is approximately equal to 2.7 times net primary productivity. In the oceans, gross primary productivity is only about 1.5 times net primary productivity (Whittaker, 1975).

Different community types differ greatly in their net primary productivities. Table 12.1 lists the approximate mean net primary productivities of the major different sorts of terrestrial communities around the world. Table 12.1 also gives the approximate areas of these different communities and the total amount of net primary productivity they contribute worldwide.

What accounts for this great variation in primary productivity? Many factors can be suggested including nutrient availability, water availability, length of the growing season, temperature, light levels and so on. Major (1963) pointed out that the **evapotranspiration** in a terrestrial environment, that is the amount of water vaporised into the air from evaporation and transpiration, is related to the amount of plant growth. Evapotranspiration may be thought of as the reverse of rain (Rosenzweig, 1968). Its value is given by the equation:

Evapotranspiration =
precipitation – runoff – percolation

In other words, evapotranspiration is the amount of water entering the atmosphere by evaporation from the soil and transpiration through plant leaves

Table 12.1 Net primary productivities (in terms of accumulation of dry organic matter per m² per year) in the major terrestrial community types. The table also gives the global area of these different communities and the total amount of net photosynthate they fix on a worldwide basis. (From Whittaker, 1975.)

Community type	Net primary productivity (dry g/m²/yr)		Area (10⁶ km²)	World net primary production (10⁹ dry t/yr)
	Mean	Range		
Tropical forest	2000	1000–3500	25	50
Temperate forest	1250	600–2500	12	15
Savannah	900	200–2000	15	14
Boreal forest	800	400–2000	12	10
Woodland & shrubland	700	250–1200	9	6
Cultivated land	650	100–3500	14	9
Temperate grassland	600	200–1500	9	5
Tundra & alpine	140	10–400	8	1
Semi-desert & desert	40	0–250	42	2

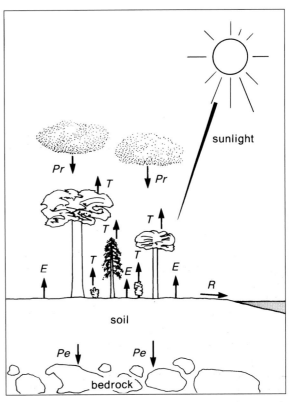

Figure 12.1 The relationship between the evaporation of water from soil, E, transpiration of water through plants, T, precipitation, Pr, runoff, R, and percolation, Pe. These quantities are related by the equation: $E + T = Pr - R - Pe$.

(Figure 12.1). Rosenzweig (1968) extended Major (1963) by analysing data on evapotranspiration and net primary production for 24 different communities. As can be seen from Figure 12.2, there is a good correlation between the evapotranspiration of an area and the net primary production of the vegetation growing there.

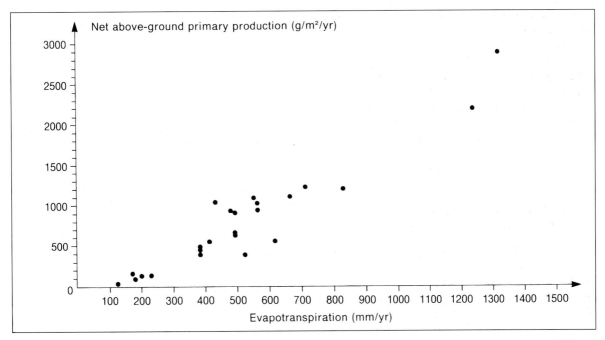

Figure 12.2 Net above-ground primary productivity as a function of actual evapotranspiration. Data on 24 different communities, mainly in the USA. (Data from Rosenzweig, 1968.)

It is not difficult to see why there is a correlation between evapotranspiration and primary production. Evapotranspiration is increased by more rainfall, more sunlight and increased temperature. All of these are factors which we might expect to be limiting primary production. However, there is a problem with Rosenzweig's analysis. It could be objected that evapotranspiration is bound to be greater the more plant leaf area there is in a community, as a larger leaf area increases the chances of water being transpired, or evaporating, rather than percolating through the soil or ending in runoff. Similarly, primary production will obviously be greater in communities with high leaf areas. In other words, evapotranspiration and primary production are not independent variables: they are bound to be correlated. This objection to Rosenzweig's analysis would be decisive except for one fact. Rosenzweig never used data for actual evapotranspiration! There simply had not been enough measurements made. Rather, Rosenzweig used an equation which enabled an estimate of actual evapotranspiration to be calculated from a knowledge of the latitude of a place, its mean monthly temperatures and its mean monthly precipitation.

In a terrestrial plant community which is well supplied with water and sunlight, plants respond by producing many leaves. These leaves trap sunlight and carbon dioxide. In a drier environment, plants cannot afford to produce as many leaves as they would risk transpiring so much water that they would wilt. In a shady environment, there is no point in plants producing great numbers of leaves as there will not be enough light for the leaves to reach their **compensation point**. The compensation point refers to the amount of light necessary for a plant to have a rate of photosynthesis equal to its rate of respiration. If the compensation point is not reached, the leaf respires more than it photosynthesises and so loses weight.

Is there any way we can predict how many leaves a plant should produce? Woodward (1987) attempted to predict the **leaf area index** of different plant communities. Leaf area index is a measure of how many leaves a typical vertical ray of sunlight passes through before hitting the soil (Figure 12.3). What Woodward did was to use an equation which related plant growth to a number of factors including incoming solar radiation, leaf area index, air temperature, water availability and the probability that a leaf will absorb the energy from a photon of light that strikes it. He then calculated the values of leaf area index that would maximise plant growth for the different values of incoming solar radiation, air temperature and water availability found across the globe. These predicted values of leaf area index are shown in Figure 12.4a. Figure 12.4b shows actual values of

Figure 12.3 Leaf area index is a measure of the number of leaves a typical vertical ray of sunlight has to pass through before hitting the soil. The single ray of sunlight shown here has to pass through seven leaves before striking the soil.

leaf area index around the globe. It can be seen that in some parts of the globe there is quite a good agreement between the actual values of leaf area index and those predicted. This is encouraging because it means that we have a good quantitative understanding of why some plant communities, such as tropical rainforests, are lush (have a high leaf area index) and why others, such as savannah and desert, have low leaf area indices.

By comparing Figures 12.4a and 12.4b it can be seen that near the poles Woodward's equation overestimates the leaf area index. These are the places where tundra is found. One reason why tundra may have a lower leaf area index than predicted is that the growing season is so short that trees cannot establish. Another reason is that the piercing winter winds kill any plants that poke above the surface of the snow.

Figure 12.4a Predicted values of leaf area index (LAI) in different plant communities throughout the world. Leaf area indices are calculated from climatological data on the assumption that plant communities have those leaf area indices which maximise their productivity. (From Woodward, 1987.)

Figure 12.4b Observed values of leaf area index (LAI) in different plant communities throughout the world. Compare with the predictions in Figure 12.4a. Unfortunately leaf area indices have not been measured in all parts of the globe. (From Woodward, 1987.)

(a)

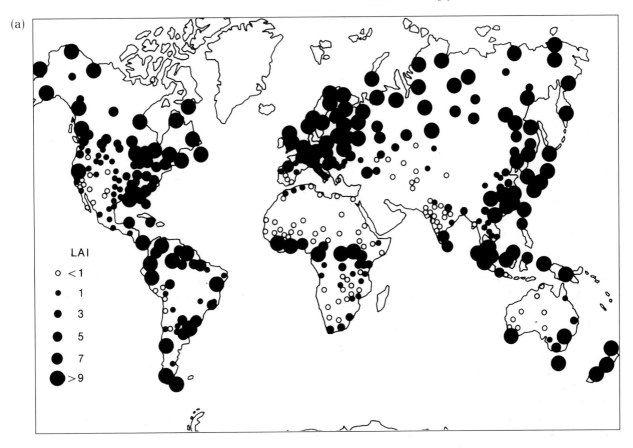

(b)

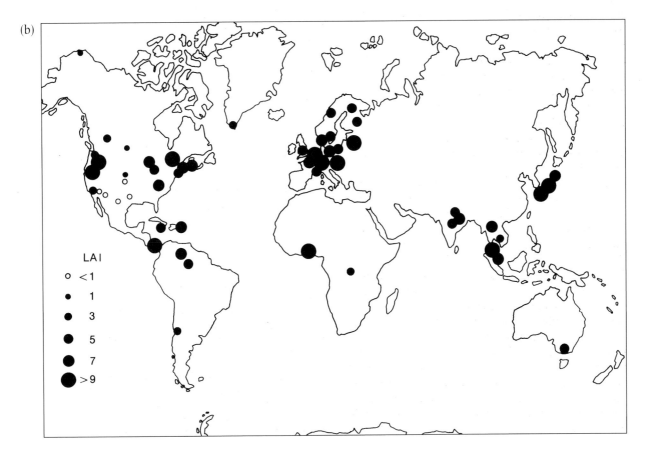

12.3 Primary production in aquatic communities

Seventy-two per cent of the world is covered in water. Yet, this 72% of the globe is only responsible for about 35% of world primary productivity (Table 12.2). Although it is likely that aquatic primary production is underestimated, as a significant amount of it is processed by planktonic bacteria which are difficult to sample (Giorgio *et al.*, 1997), there still seems to be an imbalance between primary productivity on land and primary productivity in water.

Why is this? We have already discussed how, in many terrestrial communities, water availability is what limits plant productivity. Clearly though, water availability cannot be limiting plant and algal growth in aquatic communities!

It is clear from Table 12.2 that certain aquatic habitats are actually very productive. In fact some algal beds, coral reefs and estuaries are *more* productive than the most productive tropical rainforests or cultivated land (Whittaker, 1975; Hatcher, 1988). The community type most responsible for the low overall productivity of the aquatic environment is open ocean. Why is this the case? Colinvaux (1980) poses the same question but in a different form. He asks 'Why is the sea blue?'. The answer a physicist might give is that blue light, with its short wavelength, is less likely to be absorbed by water than red or green light. Consequently, light reflected from water back to our eyes is most likely to be blue. However, a biologist's answer, Colinvaux argues, is that the sea is blue because it is not green with plants. So why, then, are terrestrial environments usually green, unless they are dominated by desert or ice, whereas aquatic environments are usually blue?

Table 12.2 Net primary productivities (in terms of accumulation of dry organic matter per m² per year) in the major aquatic community types. The table also gives the global area of these different communities and the total amount of net photosynthate they fix on a worldwide basis. (From Whittaker, 1975.)

Community type	Net primary productivity (dry g/m²/yr)		Area (10⁶ km²)	World net primary production (10⁹ dry t/yr)
	Mean	Range		
Swamp & marsh	2000	800–3500	2	4
Algal beds, reefs & estuaries	1800	500–4000	2	4
Continental shelf	360	200–1000	27	10
Lake & stream	250	100–1500	2	1
Open ocean	125	2–400	332	42

The answer seems to be that most aquatic communities are starved of nutrients. Indeed, it is the exceptions that prove the rule. Most of the great fishing areas of the world are situated in places where the water is not a beautiful clear blue, but rather a murky green. This green is due to the presence of huge numbers of **phytoplankton**. These phytoplankton are the start of the aquatic food web. They occur in their greatest densities in places where currents bring in large supplies of phosphate, nitrate, potassium and the other minerals usually in short supply. In the sea the most nutrient-rich waters are close to land, where terrestrial input is high, and in areas of upwelling, often at the margins of continental shelves, where nutrient-rich bottom waters rise to the surface. Out in the open ocean, though, such nutrients as there are are gradually lost as they descend to depths at which it is too dark for photosynthesis to take place, and from which negligible nutrient return occurs. So, aquatic communities are usually less productive than terrestrial ones, because they have less effective ways of circulating nutrients in short supply. We will examine the importance of nutrient transfer between organisms in more detail in Chapter 13.

12.4 The capture of light by plants

Even in the most productive communities, plants trap only about 1–3% of the energy which they receive in sunlight. The first estimate of the **efficiency of photosynthesis** was provided by Transeau (1926). Transeau calculated the efficiency with which a field of maize (*Zea mays*) converted the energy from sunlight into energy stored in glucose. To understand the principles of what he did, it helps to study the following equations:

Efficiency of photosynthesis =

$$\frac{\text{total energy fixed in photosynthesis}}{\text{total energy falling on the field}}$$

Total energy fixed in photosynthesis = net primary productivity + respiration

Net primary productivity = dry mass of harvested maize plants

So, by knowing the mass of the harvested maize, the amount of energy it had respired and the total amount of sunlight falling on the field, the efficiency of photosynthesis can be calculated. Of these three pieces of information, the one that was the most difficult to determine was the amount of energy respired. Plenty of data were available on the yields of crops and on the amount of **solar radiation**. Transeau measured the energy lost in respiration by keeping maize plants in the dark and seeing how much carbon dioxide they respired. The assumption was then made that the rate of respiration in the light would be the same as the rate of respiration in the dark.

Finally, Transeau ensured that all his measurements were converted to the same units – kcal (1 kcal = 4.19 kJ). He used bomb calorimetry to convert plant yield (measured in kg) to the amount of energy stored (measured in kcal). The figure Transeau arrived at was that over a typical growing season of 100 days, the photosynthetic efficiency of good farm land in Illinois was 1.6%.

This figure at first seems rather low. It means that over 98% of the energy striking the area was not used in photosynthesis. Subsequent measurements on a wide variety of communities have shown that in the field, photosynthetic efficiencies rarely if ever exceed 3.5% (Phillipson, 1966; Odum, 1971; Colinvaux, 1986).

So, why are plants apparently rather inefficient at photosynthesis? A number of reasons can be given. It has been determined by plant biochemists that a minimum of 12 photons of light are required for one molecule of carbon dioxide to be fixed (Salisbury & Ross, 1985). Twelve moles of photons of a wavelength of 550 nm have an energy of 2.6×10^6 J. One mole of fixed carbon in carbohydrate has an energy of about 4.8×10^5 J. The maximum biochemical efficiency of photosynthesis is therefore equal to $4.8 \times 10^5/(2.6 \times 10^6) = 18\%$.

Cultures of the unicellular alga *Chlorella* have indeed given photosynthetic efficiencies of close to 20% (Wassink, 1959). However, such studies also show that these efficiencies are only reached at very *low* light intensities. At the light intensities typical of full sunlight, photosynthetic efficiency fell to less than 8%. The reason for this is that at these higher light intensities, the rate of photosynthesis is limited not by the amount of sunlight, but by the amount of carbon dioxide. Indeed, some valuable glasshouse crops are sometimes commercially grown under conditions of artificially enhanced carbon dioxide concentration so as to increase photosynthetic efficiency.

There are further reasons why photosynthetic efficiencies fail even to reach 8%. Perhaps the most important is that much of the sunlight incident on a habitat is never trapped by the chloroplasts. In the first place, the light may never even strike a chloroplast: it may be absorbed by the soil or reflected by the plant. Even if a photon of light does reach a chloroplast, its energy may go to warming up the leaf rather than be used in photosynthesis.

12.5 Efficiencies in ecology

Before we consider, in Section 12.6, energy flow in natural communities, it is worth drawing up a list of the various measures of efficiency used in ecology. We have already considered assimilation efficiency, growth efficiency and reproductive efficiency in Chapter 2 and photosynthetic efficiency in Section 12.4, but for completeness these will be listed here too.

Photosynthetic efficiency
This is the percentage of the available sunlight that is used in photosynthesis, i.e. used to convert carbon dioxide into carbohydrate, whether or not this carbohydrate is subsequently respired.

Exploitation efficiency
This is the percentage of the production of one trophic level that is ingested by the trophic level above it. If some of the production of a trophic level is not ingested by the trophic level above, it either passes directly to the decomposers or accumulates.

Assimilation efficiency
This is the percentage of the energy ingested that is actually absorbed across the wall of the gut rather than egested.

Growth efficiency
This is the percentage of the energy assimilated that is devoted to growth rather than being respired or devoted to reproduction.

Reproductive efficiency
This is the percentage of the energy assimilated that is devoted to reproduction rather than being respired or devoted to growth.

Production efficiency
This is the percentage of the energy assimilated that is devoted to production (i.e. growth or reproduction) rather than respired.

Trophic efficiency
This is the efficiency of energy transfer from a trophic level to the one above. The trophic efficiency of herbivores, for instance, equals the percentage of the net primary production that is converted to herbivore production. Similarly, the trophic production of third-level carnivores equals the percentage of the production of the second-level carnivores that is converted to third-level carnivore production.

We will use these measures of efficiency to look at the flow of energy in the communities described in Section 12.6.

12.6 Energy flow in natural communities

In Section 11.7 we discussed the amount of effort required to determine a food web with any accuracy. Even more time is required to quantify the flow of energy through a community. Here we will look at three attempts to measure the flow of energy through the trophic levels of a community. Then in Section 12.7 we will see whether any general laws can be suggested from the results of these and other studies on energy flow in natural communities.

12.6.1 Odum's (1957) study at Silver Springs, Florida

One of the earliest attempts to quantify energy transfer in a natural community was carried out from 1952–5 at Silver Springs, a five-mile river fed by a spring in central Florida (Odum, 1957). Most primary productivity in this community was due to matted diatoms, filamentous green algae and cyanobacteria, all of which were found covering the strap-shaped leaves of eelgrass (*Sagittaria*). This eelgrass covered the bottom of the river. The main herbivores were large snails, numerous other invertebrates, mullet (fish) and turtles. Fish comprised the main secondary consumers. The tertiary, top consumers were made up of three species of fish and a number of invertebrates parasitic on the secondary consumers.

Odum and his field assistants used a large number of techniques to determine the energy flow at Silver Springs. The most fruitful method proved to involve the use of bell jars as shown in Figure 12.5. Because the great majority of the organisms in the community were attached to the bottom, rather than floating free in the water, reasonable measures of community respiration could be obtained by measuring the

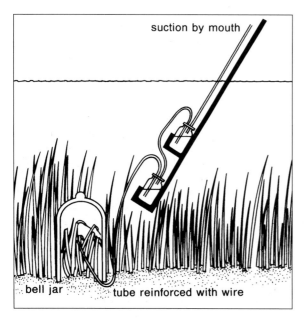

Figure 12.5 Bell jar apparatus used by Odum (1957) to help measure production and respiration at Silver Springs, Florida. Chemical methods were used to determine the changes in the oxygen and carbon dioxide concentrations of the bell jar over time.

changes in the oxygen and carbon dioxide concentrations of the water in the bell jars after a measured time interval. Measurements were taken both during the day and at night. They were also taken both in areas with and in areas without eelgrass. Production by the primary producers could also be determined from measurements made in cage enclosures erected under water to allow the eelgrass and accompanying diatoms and algae to grow in the absence of herbivores. Attempts were also made to determine the growth rates and the rates of respiration of the various animals at Silver Springs.

A summary of the energy flow at Silver Springs as revealed by Odum's study is given in Figure 12.6. Careful examination of this figure reveals a number of interesting findings. As discussed in Section 12.4, the efficiency of the primary production (photosynthesis) equals the amount of energy fixed in photosynthesis divided by the total insolation. We therefore have:

Efficiency of photosynthesis
= $20\,810$ kcal/m^2/yr $\div$ $1\,700\,000$ kcal/m^2/yr
= 1.2%

This figure for the efficiency of photosynthesis in a natural aquatic community dominated by unicellular producers is similar to the one of 1.6% obtained by Transeau (1926) in his study of the efficiency of photosynthesis in a field of maize.

We can also calculate the trophic efficiencies of the herbivores, the first-level carnivores and the second-level carnivores:

Trophic efficiency of herbivores
= 1478 kcal/m^2/yr $\div$ 9319 kcal/m^2/yr
= 15.9%

Trophic efficiency of first-level carnivores
= 67 kcal/m^2/yr $\div$ 1478 kcal/m^2/yr
= 4.5%

Trophic efficiency of second-level carnivores
= 6 kcal/m^2/yr $\div$ 67 kcal/m^2/yr
= 9.0%

12.6.2 Teal's (1962) study at a salt marsh in Georgia

Teal (1962) carried out a detailed study of the energetics of a salt marsh in Georgia, USA. As with Odum's study described in Section 12.6.1, a large number of field measurements were taken and a number of assumptions had to be made. Teal was helped by the fact that one plant dominated the salt marsh, the grass *Spartina alterniflora*. The trophic organisation of the marsh, however, was a little unusual in that most of the animals there could not feed on the *Spartina* directly. Rather, it had first to be broken down by bacteria. Crabs in the genera *Uca*

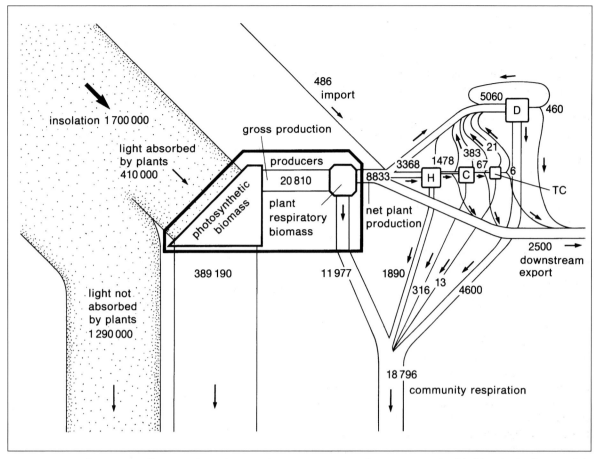

Figure 12.6 The energy flow in the Silver Springs community. The units are kcal/m²/yr (1 kcal= 4.2 kJ). H = herbivores; C = carnivores; TC = top carnivores; D = decomposers. (From Odum, 1957.)

and *Sesarma* were the most conspicuous consumers. Figure 12.7 summarises the energy flow in the marsh. Again we can calculate the efficiency of photosynthesis:

Efficiency of photosynthesis

$$= \frac{\text{total energy fixed in photosynthesis}}{\text{total energy falling on the marsh}}$$

= (net primary productivity of *Spartina* + respiration by *Spartina* + net primary productivity by algae + respiration by algae) ÷ 600 000 kcal/m²/yr

= (6280 + 305 + 27 995 + 1620 + 180) kcal/m²/yr ÷ 600 000 kcal/m²/yr

= 6.1%

It is more difficult to calculate trophic efficiencies than was the case for Odum's study because the organisms cannot easily be categorised into trophic levels: for instance, the crabs and nematodes feed as herbivores on the algae, but they also feed on the bacteria which have fed on the *Spartina*. However, although we cannot really look at the trophic

efficiencies of entire trophic levels, we can look at the trophic efficiencies of groups of species found in the marsh. These are summarised in Table 12.3.

Table 12.3 Trophic efficiencies of groups of species in the salt marsh studied by Teal (1962). The trophic efficiency of the crabs and nematodes, for example, equals their production divided by the production of the organisms on which they feed. That is, the trophic efficiency of the crabs and nematodes equals the amount they pass on to the mud crabs plus the amount they export from the marsh divided by the production of the bacteria and the algae, i.e. (31 + 59) kcal/m²/yr ÷ (2390 + 1620) kcal/m²/yr = 2.2%.

Taxon	Trophic efficiency
Bacteria	38%
Insects	27%
Spiders	6.2%
Crabs & nematodes	2.2%
Mud crabs	6.7%

12.6.3 Varley's (1970) study of Wytham Wood, Oxford

The ecology of Wytham Wood near Oxford has been studied for several decades. We have already looked at the pyramid of numbers in this wood (Figure 11.10b) and the population regulation of its great tits (Section 5.2.4) and in Section 14.2.2 there is a description of the ecology of British oak woodlands of

which Wytham is a good example. Despite the fact that Wytham Wood has been studied for so long, there are so many different species there that it has proved very difficult to provide a detailed quantitative description of the energy flow in the wood. Figure 12.8 gives a simplified food web of the wood with quantitative details of the amount of consumption and production by most of the dominant members of the community.

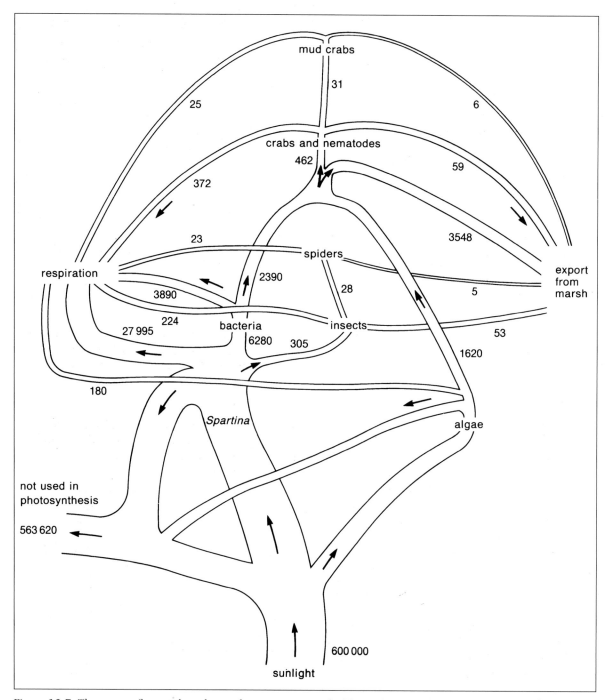

Figure 12.7 The energy flow at the salt marsh community studied by Teal (1962). The units are kcal/m²/yr.

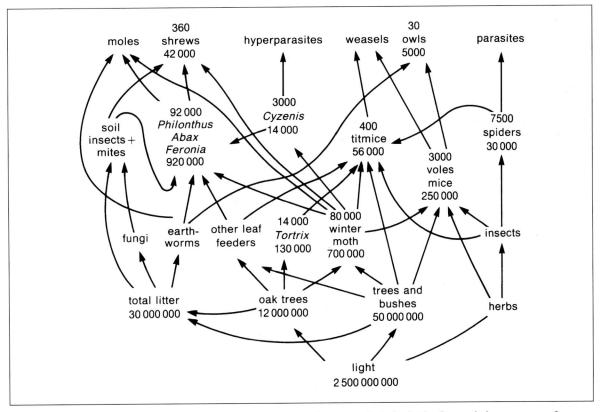

Figure 12.8 A simplified description of the energy flow at Wytham Wood, Oxford. The figures below a group of organisms give their rates of consumption in kcal/ha/yr. The corresponding figures above the organisms give their rates of production in kcal/ha/yr. (After Varley, 1970.)

Only a small proportion of the total photosynthesis in the wood is due to the herbs (Varley, 1970). We can calculate that the efficiency of photosynthesis is approximately given by:

Efficiency of photosynthesis
= 62 000 000 kcal/ha/yr
  ÷ 2 500 000 000 kcal/ha/yr
= 2.5%

It is also apparent from Figure 12.8 that some of the consumers convert only a very small percentage of the energy they consume into production. The percentage of the energy consumed by an organism that it converts into production equals its assimilation efficiency multiplied by its production efficiency (see definitions in Section 12.5). Table 12.4 lists the percentage of the energy consumed that is converted into production for most of the organisms named in Figure 12.8.

From the data in Table 12.4 it is clear that the efficiencies with which different animals convert the food they consume into production vary greatly. It seems as though the poikilotherms (cold-blooded organisms) have efficiencies about an order of magnitude greater than the homoiotherms (birds

and mammals). By comparing Table 12.4 and Figure 12.8 it does *not* look as though the positions of the species within the food web have any obvious connection with these efficiencies.

Table 12.4 The efficiency with which the species in the oak wood at Wytham Wood, Oxford convert the food they consume into growth and reproduction. Data from Figure 12.8.

Group	Percentage of the energy consumed converted into production
Winter moth (*Operophtera*)	11%
Tortrix moth (*Tortrix*)	11%
Voles & mice	1.2%
Spiders	25%
Titmice	0.7%
Tachinid fly parasite (*Cyzenis*)	21%
Predatory beetles (*Philonthus, Abax & Feronia*)	10%
Owls	0.6%
Shrews	0.9%

12.7 The efficiency of energy transfer in ecosystems

There are few laws in ecology. Perhaps one of the best known is **Lindeman's law of trophic efficiency.** Lindeman was an ecologist who worked on the ecology of a small lake in Minnesota. He died in 1942 when he was only 26 years old. However, his last paper (Lindeman, 1942), published after his death, has been very important in the history of quantitative ecology. In it Lindeman produced the first measures of trophic efficiency, that is (as defined in Section 12.5) the percentage of the production of one trophic level that is converted to production by the trophic level above.

Table 12.5 gives the trophic efficiencies at the various trophic levels in two lakes: Lake Mendota, Minnesota studied by Lindeman himself and Cedar Bog Lake studied by Juday (1940). (The trophic efficiencies of the producers equals their photosynthetic efficiencies.)

Table 12.5 The trophic efficiencies of the various trophic levels found in two lakes. Taken from Lindeman (1942).

	Efficiencies	
Trophic level	Lake Mendota	Cedar Bog Lake
Producers	0.4%	0.1%
Primary consumers	8.7%	13.3%
Secondary consumers	5.5%	22.3%
Tertiary consumers	13.0%	—

Lindeman's law of trophic efficiency states that the efficiency of energy transfer from one trophic level to the next is about 10%. In other words, about 10% of the net primary productivity of producers ends up as herbivores, about 10% of the net productivity of herbivores ends up as first-level carnivores, and so on. Although Lindeman was the first to calculate trophic efficiencies, he was too cautious to suggest that his results could be extrapolated to other communities. Indeed the only law of trophic efficiency that Lindeman himself proposed was that trophic efficiencies *increase* as one ascends a food chain: 'These progressively increasing efficiencies may well represent a fundamental trophic principle, namely, the consumers at progressively higher levels in the food cycle are progressively more efficient in the use of their food supply' (Lindeman, 1942, p. 407). You can see this increasing efficiency in the data for both lakes in Table 12.5.

However, more recent data support neither Lindeman's suggestion that trophic efficiencies necessarily increase as one ascends a food chain, nor the generalisation that the efficiency of energy transfer from one trophic level to the one above is about 10%. Rather, it is now realised that trophic efficiencies can vary very greatly in ways which depend more on the behaviour and physiology of the organisms concerned than on their positions in any food web.

As May (1979) has pointed out, to determine trophic efficiencies two questions need to be answered. First, what fraction of the net productivity at one trophic level is actually assimilated by creatures at the next level? Second, how do these organisms apportion their assimilated energy between net productivity (growth and reproduction) and respiration?

The first of these two questions is harder to answer with any precision than the second. The fraction of the net productivity at one trophic level that is actually assimilated by organisms at the next level equals the percentage of the trophic level's production they consume (their exploitation efficiency) multiplied by their assimilation efficiency. For example, if the first-level carnivores in a community ingest 50% of the production of the herbivores, and then assimilate 80% of the energy they ingest, they will assimilate 50% x 80% = 40% of the production of the herbivores. Some data on the exploitation efficiencies of herbivores are summarised in Table 12.6. From this table we can see that considerable variation exists in the efficiency with which organisms exploit their food sources.

Table 12.6 The percentage of plant production consumed by herbivores in various communities. (Data from Pimental *et al.*, 1975; Whittaker, 1975; Krebs, 1978; and Ricklefs, 1980.)

Community	Nature of primary producers	Exploitation efficiencies
30-year-old Michigan field	Perennial herbs & grasses	1.1%
Mature deciduous forest	Trees	1.5–2.5%
Desert shrub	Annual & perennial herbs & shrubs	5.5%
Georgia salt marsh	Herbaceous perennials	8%
7-year-old South Carolina fields	Herbaceous annuals	12%
African grasslands	Perennials	28–60%
Oceans	Phytoplankton	40–99%

Data on assimilation efficiencies were reviewed in Section 2.6.1. Assimilation efficiencies may exceed 90% (e.g. carnivores feeding on vertebrates). On the other hand, they may be much lower than this. Some herbivores have assimilation efficiencies of only 20%. Detritivores may have assimilation efficiencies even lower than this.

The second piece of information May pointed out that is needed to determine trophic efficiencies is the efficiency with which organisms apportion the energy they have assimilated to production. The most careful review of this was carried out by Humphreys (1979) and his results are summarised in Table 12.7. The most notable feature of this table is that production efficiencies vary greatly. It is also clear that mammals and birds have production efficiencies at least an order of magnitude less than poikilotherms. It might be thought that this is simply because they have to spend so much of the energy which they assimilate on keeping warm. However, although a

Table 12.7 Mean production efficiencies (= production/[production + respiration]) for various groups of animals. (Data from Humphreys, 1979.)

Organisms	Mean production efficiency	Standard error	Sample size
Mammalian insectivores	0.9%	0.1%	6
Birds	1.3%	0.03%	9
Small mammal communities	1.5%	0.1%	8
Other mammals	3.1%	0.3%	56
Fish & social insects	9.8%	0.9%	22
Non-insect invertebrates:			
Herbivores	21%	1.4%	15
Carnivores	28%	5.1%	11
Detritivores	36%	4.8%	23
Non-social insects:			
Herbivores	39%	1.9%	49
Detritivores	47%	1.6%	6
Carnivores	56%	0.6%	5

homoiotherm has much greater energetic requirements than a poikilotherm of the same size, it also assimilates food at a much faster rate. On average a homoiotherm assimilates energy about 22 times faster than a poikilotherm of the same size. But its metabolic rate is almost 30 times greater than the poikilotherm's. The fact that homoiotherms have lower production efficiencies than poikilotherms rests on the observation that 22 is less than 30! Theoretically it is quite possible for a homoiotherm to have a greater production efficiency than a poikilotherm as long as the rate at which it assimilates energy more than compensates for the extra rate at which it expends energy on everything except growth and reproduction (Reiss, 1989).

In general, trophic efficiency is related to exploitation efficiency, assimilation efficiency and production efficiency by the following equation:

Trophic efficiency = exploitation efficiency
  × assimilation efficiency
  × production efficiency

This relationship is shown diagrammatically in Figure 12.9. This shows the flow of energy in a generalised food web. For simplification it is assumed that organisms can be allocated to trophic levels. It is also assumed that there is no accumulation of organic matter, for example as peat, and that the third-level carnivores are the top-level carnivores. The simplest way to understand the figure is probably to begin with the Sun and work up. The total amount of sunlight shining on the community equals A + B. Only a small amount of this, B, lands on the producers. Much of the energy that strikes the producers is reflected or lost as heat, C. However, an amount of sunlight equal to D + E is actually used in photosynthesis (gross primary productivity). Of this, D is respired and E ends up as net primary production. Some of this net primary productivity is consumed by herbivores, F, and some goes to the decomposers, G. Of the energy ingested by the herbivores, F, some is never assimilated but passes directly to the decomposers in the faeces, H. The rest is assimilated and is either lost in respiration, I, or goes to production, J, and so on up the diagram.

From this rather daunting figure, one can draw up a table of the various ecological efficiencies of the various trophic levels (Table 12.8).

Table 12.8 Ecological efficiencies for a generalised food web. See Figure 12.9. Definitions of the various efficiencies are given in Section 12.5. You should be able to work out the exploitation efficiency, assimilation efficiency, production efficiency and trophic efficiency of the third-level carnivores in terms of U, V, X and Z.

Efficiency	General formula with reference to Figure 12.9
Photosynthetic efficiency	$100(D + E)/(A + B)\%$
Exploitation efficiency of herbivores	$100F/E\%$
Assimilation efficiency of herbivores	$100(F–H)/F\%$
Production efficiency of herbivores	$100J/(F–H)\%$
Trophic efficiency of herbivores	$100J/E\%$
Exploitation efficiency of 1st level carnivores	$100K/J\%$
Assimilation efficiency of 1st level carnivores	$100(K–M)/K\%$
Production efficiency of 1st level carnivores	$100P(K–M)\%$
Trophic efficiency of 1st level carnivores	$100P/J\%$
Exploitation efficiency of 2nd level carnivores	$100Q/P\%$
Assimilation efficiency of 2nd level carnivores	$100(Q–S)/Q\%$
Production efficiency of 2nd level carnivores	$100U/(Q–S)\%$
Trophic efficiency of 2nd level carnivores	$100U/P\%$

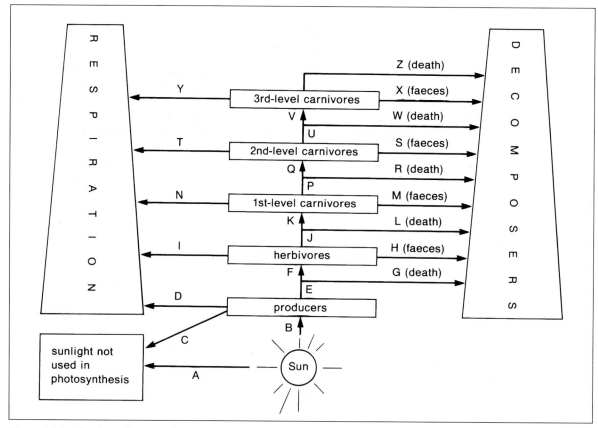

Figure 12.9 The flow of energy through a generalised food web. The meanings of the letters A, B, C, etc. are given in the text.

If one combines the data in Tables 12.6 and 12.7 and remembers the variation that exists in assimilation efficiencies, then it is apparent that rather than trophic efficiencies being fairly constant at around 10% as the law of trophic efficiency states, they actually vary over more than two orders of magnitude. For instance, a community of herbivorous small mammals in a grassland might have a trophic efficiency of less than 0.1%; herbivorous zooplankton feeding on phytoplankton might have a trophic efficiency of over 20%.

12.8 Pyramids of energy

In Chapter 11 we looked at pyramids of numbers and pyramids of biomass. There is a third sort of ecological pyramid used to describe the trophic relationships within a community. This is the **pyramid of energy**. It is important to realise that a pyramid of energy shows the flow of energy from one trophic level of a community to the next. The units of pyramids of energy are therefore energy/area/time, e.g. kJ/ha/yr. In other words, pyramids of energy show the *rate* at which energy flows up a food web: they are measured over a stated period of time. This is different from pyramids of numbers or pyramids of biomass.

Pyramids of numbers have the units of individuals/area; pyramids of biomass have the units of mass/area. One *cannot* convert pyramids of biomass into pyramids of energy by using a conversion factor to convert the mass of each trophic level into its energy content.

Pyramids of energy can never be inverted. This follows directly from the first law of thermodynamics, the **law of conservation of energy**. This states that energy can neither be created nor destroyed, but is always conserved. Look at Figure 12.9 and consider the fate of the net primary production, E. From the law of conservation of energy we have:

$$E = F + G$$

and

$$F = H + I + J$$

From these two equations we have that:

$$J \leq E$$

so that the productivity of the herbivores cannot exceed the net primary productivity. Indeed, the only way the productivity of the herbivores could equal the net primary productivity would be for the values of G, H and I all to equal zero. This is impossible: at the very least, I (herbivore respiration) cannot be zero.

Box 12
The pyramid of energy in a dinosaur community

A remarkable use of the principle that pyramids of energy cannot be inverted is provided by Farlow (1976). Farlow was interested in the question of whether or not dinosaurs were cold-blooded. Traditionally dinosaurs had been assumed to be poikilotherms, like modern-day reptiles, but in the 1970s several workers began to gather evidence which suggested that dinosaurs were actually warm-blooded.

Farlow worked on a collection of fossils from the Oldman Formation. During the Late Cretaceous (about 85 million years ago) the Oldman environment of western North America resembled a warm temperate marsh. The Oldman large herbivores were ornithischian dinosaurs: ankylosaurs, ceratopsians and hadrosaurs. The adults of these species probably had a mass of between 3 and 6 tonnes, about the mass of a large African elephant. Farlow calculated that these three dinosaur groups comprised 97.9% of the biomass of the Oldman large-dinosaur fauna. The other 2.1% were made up by carnivorous tyrannosaurs each with an adult mass of about 2 tonnes.

On the assumption that the dinosaurs were homoiothermic, Farlow then used modern-day data on the food consumption rates of mammals to obtain crude estimates of the ingestion rates and productivities of the dinosaurs. He calculated that the productivity of the herbivorous dinosaurs would have been about 3×10^6 kcal/km^2/yr. However, the food requirements of the carnivorous tyrannosaurs were calculated to have been about 4×10^6 kcal/km^2/yr. When, however, Farlow assumed that the dinosaurs were poikilothermic and used modern-day data on the food consumption rates of lizards, he calculated that the productivity of the herbivorous dinosaurs would have been about 2×10^7 kcal/km^2/yr while the food requirements of the carnivorous tyrannosaurs would have been about 5×10^6 kcal/km^2/yr. In other words, the assumption that the dinosaurs were warm-blooded leads to an inverted pyramid of energy. There were simply too many tyrannosaurs in the fossil assemblage for them to have been warm-blooded.

Farlow recognised that so many assumptions had to be made in his calculations that the figures can be adjusted to suggest that the dinosaurs were warm-blooded. Nevertheless, his study stands as a fascinating attempt to use the principle of pyramids of energy to solve a major question about dinosaur physiology.

In the same way the productivity of the first-level carnivores must be less than the productivity of the herbivores, the productivity of the second-level carnivores must be less than the productivity of the first-level carnivores, and so on. Thus when these productivities, E, J, P and U, are arranged in a pyramid, the pyramid can never be inverted.

An example of a pyramid of energy is given in Figure 12.10. This has been drawn using the data for the food web based on the oak trees at Wytham Wood given in Figure 12.8. It is at once apparent that this pyramid does appear to be inverted. This raises our suspicions that some important data are missing. Indeed, on reading Varley's (1970) paper it is clear that many of the first-level carnivores in the wood are feeding on herbivores whose productivity Varley was unable to measure. To some extent this is shown in the food web of the wood given in Figure 12.8. First-level carnivores such as *Philonthus*, *Abax* and *Feronia* feed on earthworms, soil insects, mites and other species whose productivity is unknown.

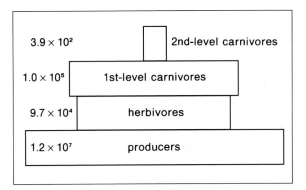

Figure 12.10 Pyramid of energy based on oak trees, *Quercus robur*, at Wytham Wood, Oxford. Energy flows are given in kcal/ha/yr. The production of the first-level carnivores appears to exceed the production of the herbivores because of the difficulties in measuring herbivore production. (Data from Varley, 1970.)

Summary

(1) Net primary productivity is the amount of dry weight accumulated by autotrophs per unit area per unit time.

(2) Gross primary productivity equals net primary productivity plus respiration.

(3) On a worldwide basis, the amount of available water is probably the single factor that most limits plant production in low latitude terrestrial environments. In polar latitudes the low temperatures and short growing season are more important.

(4) The leaf area indices of some plant communities can be predicted with some accuracy solely on the basis of climatological data.

(5) Lack of nutrients limits primary production in most aquatic environments.

(6) Aquatic communities where the nutrient supply is good, as is the case for some algal beds, coral reefs and estuaries, have primary productivities greater than any found on land.

(7) The efficiency of photosynthesis can reach almost 20% in laboratory cultures of green algae, but in nature rarely, if ever, exceeds 3.5%.

(8) A tremendous amount of work is needed to determine the energy flow in a natural community.

(9) Exploitation efficiencies vary in nature from about 1% to about 99%.

(10) Production efficiencies vary from about 1–3% for endotherms to over 50% for some ectotherms.

(11) Trophic efficiencies measure the percentage of the energy transferred from one level of a food web to the one above.

(12) Trophic efficiencies were once thought to be always about 10%. It is now known that they vary from less than 0.1% to over 20%.

(13) Pyramids of energy show the flow of energy up a food web.

(14) Pyramids of energy can never be inverted.

THIRTEEN

Nutrient cycling and pollution

13.1 The pattern of nutrient transfer and its connection with pollution

As we discussed in Chapter 12, organisms need a supply of energy in order to survive, grow and reproduce. They also, of course, need to obtain the elements of which they are composed (Figure 13.1). Energy transfer can be represented as a pyramid where each trophic level obtains less energy than the trophic level beneath it (Section 12.8). This is because at each trophic level of a food web most of the energy is lost from the community as the heat of respiration. This heat ends up in the physical envi-

ronment and is rarely of direct use to organisms. This means that energy transfer is linear. Energy passes from the Sun to primary producers and so forth up the food web.

The fundamental difference between energy transfer and nutrient transfer is that the pattern of nutrient transfer is basically circular or cyclical. The elements which form the molecules of which organisms are made are unalterable in natural conditions on Earth, so they remain in circulation when molecules pass from one trophic level to another. They can be recycled over and over again. As Shakespeare put it 'Imperious Caesar, dead and turn'd to clay, might stop a hole to keep the wind away' (*Hamlet*, Act 5,

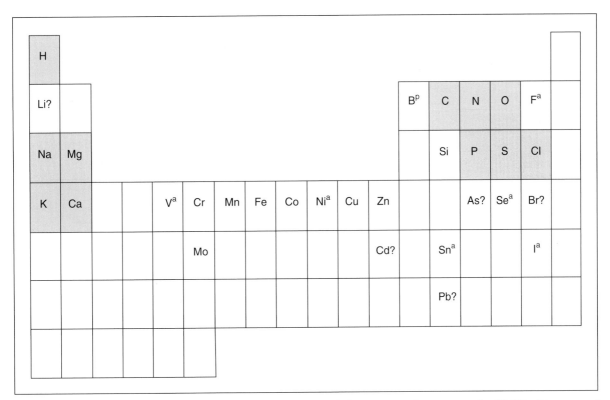

Figure 13.1 The elements required by organisms. The 11 shaded elements together account for 99.9% of the mass of most organisms. Elements marked 'p' are required only by plants; those marked 'a' only by animals. It is still unclear whether or not elements marked '?' are essential.

Scene 1). Usually, decomposers play a vital role in this **nutrient cycling** in returning small molecules and ions back from higher trophic levels to the soil, water or air, where they are available for reuse by the primary producers.

In this chapter we will look at nutrient cycling, paying particular attention to how organisms obtain their carbon, nitrogen and phosphorus. We will look at the roles played by the atmosphere, the terrestrial environment and the aquatic environment, and will compare nutrient cycling in different communities. Throughout the chapter, pollution will be considered. This is because most sorts of pollution occur when certain substances are either too abundant or not plentiful enough as a result of human activity. A knowledge of nutrient cycling therefore helps an understanding of pollution.

13.2 The carbon cycle

The carbon cycle is unusual among nutrient cycles in that it does not necessarily have to involve decomposers. This is because photoautotrophs take in their carbon from the atmosphere in the same form, carbon dioxide, as carbon is excreted by all organisms in

respiration. Figure 13.2 shows a generalised carbon cycle, but it is worth noticing that even if the decomposers were removed, carbon would still be able to cycle round the community. In practice, though, carbon cycles do include decomposers.

From Figure 13.2 we can see that fossil fuels represent a **sink** for carbon. Almost the only way the carbon in fossil fuels can be returned to the carbon cycle is by burning them. Figure 13.2 also shows the importance of the physical environment in the carbon cycle. Carbon dioxide (CO_2) is found in the atmosphere and the ions hydrogencarbonate (HCO_3^-) and carbonate (CO_3^{2-}) occur in water. Sedimentation gives rise to carbonate rocks such as limestones and dolomites. Volcanos can return carbon dioxide to air and water by their eruptions.

An attempt to quantify the carbon cycle on a world-wide basis is given in Figure 13.3. This shows the amount of carbon at any one time in the various compartments of the carbon cycle – the atmosphere, oceans, living organisms, etc. It also shows the annual rate at which carbon moves between these compartments. The units are 10^{15} g for the sizes of the compartments and 10^{15} g/yr for the rate of movement between the compartments. This means, for instance, that the total mass of carbon currently in

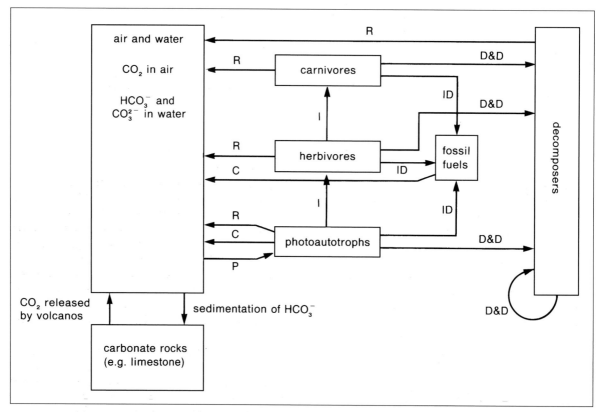

Figure 13.2 The carbon cycle. I = ingestion of organic compounds containing carbon; P = uptake of CO_2 in photosynthesis; R = release of CO_2 in respiration; C = release of CO_2 in combustion; D & D = passage of organic compounds containing carbon to decomposers via death and decay; ID = conversion of organic compounds into fossil fuels via incomplete decay.

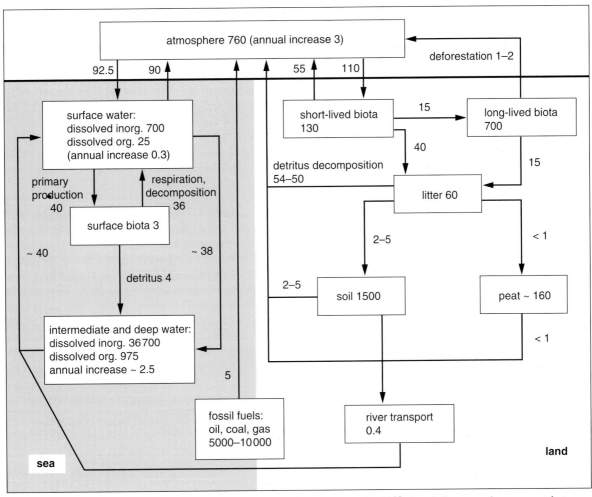

Figure 13.3 The global carbon cycle showing the sizes of the reservoirs (in 10^{15} g) and the annual movement between these reservoirs (in 10^{15} g/yr). (From Bolin, 1983; Holland, 1995.)

the atmosphere is about 7×10^{17} g $= 7 \times 10^{11}$ tonnes $= 700\,000$ million tonnes.

What Figure 13.3 does not show is that there is also a huge mass of carbon in the rocks of the earth. However, the rate at which this is released to organisms is very, very slow. On average, the **turnover time** of a carbon atom in rock is of the order of 100 million years (Kempe, 1979). This means that it takes about 100 million years for a carbon atom in rock to have a 50 : 50 chance of ending up in the oceans, in the atmosphere or in a living organism as opposed to remaining in the rock. Once a carbon atom does get into one of the compartments of Figure 13.3, its average turnover time depends on which compartment it is in, but is at the most a thousand years.

13.3 The greenhouse effect

If the Earth had no atmosphere, its average surface temperature would be about −18 °C, the same as that of the Moon. In fact, the average temperature of the

Earth at ground level is about 15 °C. The reason why the Earth is warmer than the Moon is because it has an atmosphere. The Moon's gravitational field, though, is too weak to hold one. Much of the incoming radiation from the Sun is absorbed by the atmosphere and then re-radiated back to the Earth's surface. On the Moon there is no atmosphere to trap the radiation so any heating up of the ground is lost as the heat is re-radiated out into space (Gribbin & Gribbin, 1996). The atmosphere acts like a greenhouse, trapping the heat. Humans can only survive on the Earth because of this **greenhouse effect**, but as we will see below, it can lead to quite rapid global warming.

The natural gases in the atmosphere most responsible for keeping the surface of the Earth relatively warm are carbon dioxide and water vapour. However, as is well known, each year humans are contributing to the greenhouse effect by releasing into the atmosphere huge amounts of carbon dioxide, methane, chlorofluorocarbons (CFCs), nitrous oxides and other gases. The most important of these gases from the point of view of the greenhouse effect is carbon dioxide. Figure 13.3 shows how the amount of carbon in the atmosphere is increasing at an annual rate of about 3×10^{15} g = 3000 million tonnes, almost one tonne for each person in the world. This increase comes mainly from the burning of fossil fuels and trees, particularly those in the remaining tropical rainforests.

The total amount of carbon dioxide fixed each year in photosynthesis by the world's trees is probably not now decreasing. This is because, although it is difficult to be certain, the loss of natural rainforests and other forest ecosystems seems to be more or less balanced out by increases in the area given over to plantations of trees. However, forest fires burning out of control in drought years can upset this balance.

Although carbon dioxide is the most important greenhouse gas, other gases are important too (Figure 13.4). Indeed, as Figure 13.4 indicates, there is a certain degree of interaction between the greenhouse effect, the depletion of stratospheric ozone (the 'hole in the ozone layer'), acid rain and photochemical smog. Actually, a single molecule of carbon dioxide contributes less to the greenhouse effect than a single molecule of the other greenhouse gases (Table 13.1). For example, a single molecule of the most common CFC has a greenhouse effect equivalent to

Table 13.1 The contributions of the various greenhouse gases to potential global warming relative to the same number of molecules of carbon dioxide. (Data from Trenberth, 1992; Adger & Brown, 1993.)

Gas	Atmospheric concentration (p.p.m. in 1996)	Residency (years)	Potential for global warming (over 20 years)	(over 100 years)
CO_2	360	120	1	1
CH_4	2	10	60	20
N_2O	0.3	130	270	290
CFCs	0.0003	50-120	4500-7000	3500-7000

7000 carbon dioxide molecules. It's just that the extra quantity of carbon dioxide produced as a result of human activities far exceeds the extra quantity of these other gases. The potential for global warming depends on the timescale. For example, methane lasts a much shorter time in the atmosphere than carbon dioxide, nitrous oxide or CFCs. This means that if methane emissions were stopped now, the contribution of methane to global warming would trail off more quickly than if we stopped emissions of the other greenhouse gases. In fact, the atmospheric concentrations of some of these other gases is actually increasing. Methane is present in the atmosphere at a concentration of almost 2 p.p.m., twice what is was

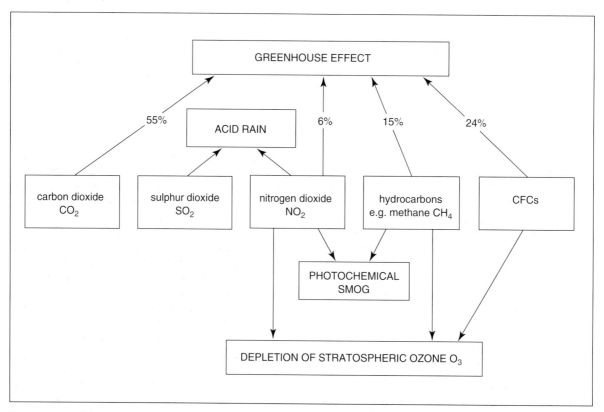

Figure 13.4 The contribution of various atmospheric pollutants to the greenhouse effect, acid rain, photochemical smog and the depletion of stratospheric ozone.

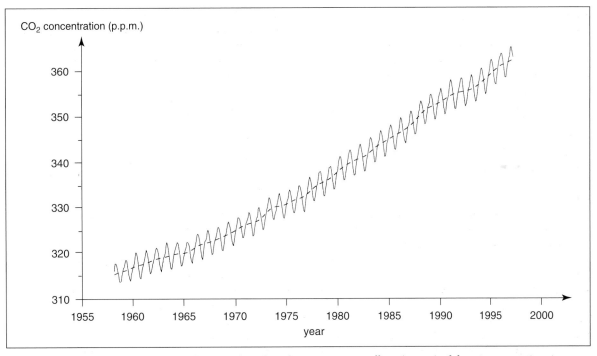

Figure 13.5 Concentration of atmospheric carbon dioxide in parts per million (p.p.m.) of dry air versus time in years at Mauna Loa Observatory, Hawaii. The dashed line indicates the annual trend.

in 1900 (Glantz & Krenz, 1992). For a discussion of CFC levels see Section 13.8.7.

The best data on the rise in the atmospheric concentration of carbon dioxide come from Mauna Loa Observatory in Hawaii and from the South Pole where continuous carbon dioxide measurements have been made since 1958 (Keeling *et al.*, 1986). Figure 13.5 gives the atmospheric carbon dioxide concentrations measured at Mauna Loa Observatory. Each summer they fall due to the effect of plant photosynthesis, but each winter they rise more than they fell the previous summer. In 1988 the average concentration reached 350 p.p.m.; in 1996 it reached 360 p.p.m. (Pearce, 1996a).

Data on atmospheric carbon dioxide concentrations before 1958 are more difficult to come by. Some measurements done in the Smithsonian Solar Constant Program are given in Table 13.2. These show an annual increase in carbon dioxide concentrations from 1935 to 1945 of about 0.6 ± 0.3 p.p.m. An ingenious technique to determine atmospheric concentrations of carbon dioxide before this century has been to measure the carbon dioxide concentration of air bubbles trapped in the deep ice of Greenland and Antarctica hundreds or thousands of years ago! Various methods are used to date these samples. The most valid data come from samples dated at between 500 BC and AD 1880. During this period there is no evidence for any changes in carbon dioxide concentrations which were measured at about 260–280 p.p.m. (Barnola *et al.*,

Table 13.2 Atmospheric concentrations of carbon dioxide from 1935 to 1945 measured as part of the Smithsonian Solar Constant Program. (Data from Stokes & Barnard, 1986.)

Year	Atmospheric concentration of carbon dioxide (p.p.m.) (mean ± standard error)
1935	297.7 ± 5.6
1941	298.4 ± 6.3
1945	308.3 ± 4.6

1983; Oeschger & Stauffer, 1986). It seems clear that the 30% rise in the atmospheric concentration of carbon dioxide over the last hundred years is due to the effects of humans and especially the industrialisation of the northern hemisphere.

What are the chances that the greenhouse effect may lead to significant changes in our climate? Could the greenhouse effect raise world temperatures sufficiently to melt the polar icecaps and flood low-lying cities and much of the coastal land? In 1972 the distinguished biologist Kenneth Mellanby wrote of this suggestion that he was 'personally not acutely worried on this account, as the most competent authorities seem to consider that the effects of man's efforts on the world's climate are likely to be much less than the fluctuations which occur without his intervention' (Mellanby, 1972, p.5). Today, many people are less confident than Mellanby and his competent authorities were.

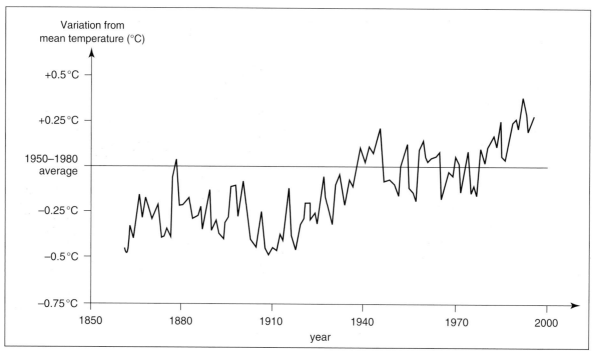

Figure 13.6 Annual variations from the 1950–1980 mean for global temperatures from 1861 to 1994.

Figure 13.6 shows that world temperatures have *already* risen by approximately 0.5 °C over the past 100 years. Although most scientists who work on climates agree that this is due to the greenhouse effect, there are some who disagree. It has been argued that recent readings are affected by the heat generated by cities and so overestimate true global averages. In Section 21.4.1 we shall examine what the future holds as far as the greenhouse effect is concerned. How likely is it that global warming will take place? Just how much warmer is the world likely to become? What are the consequences of global warming likely to be? And, perhaps most importantly, should we be doing something about the greenhouse effect and, if so, what?

13.4 The nitrogen cycle

Water contains only hydrogen and oxygen. Carbohydrates and fats contain only carbon, hydrogen and oxygen. All other molecules of biological importance contain nitrogen. For instance, proteins, nucleic acids and chlorophyll all contain nitrogen. Nitrogen makes up almost 13% of our dry weight. Only carbon (48%) and oxygen (24%) are more abundant (Rankama & Sahama, 1950).

Despite the importance of nitrogen to living systems, it still is not possible to present a detailed quantitative account of the nitrogen cycle. Figure 13.7 shows the global nitrogen cycle with estimates of the annual rates of flow between the various compart-

ments. It is worth emphasising that the annual movement of nitrogen around the cycle is dwarfed by the amount of nitrogen present in the atmosphere and in rocks. The atmosphere contains approximately 4×10^{21} g and rocks approximately 2×10^{23} g. By comparison, as Figure 13.7 shows, annual rates of nitrogen fixation by terrestrial bacteria are about 1.1×10^{14} g. In terrestrial systems, the great majority of the nitrogen (97%) is found in soil organic matter, litter and soil inorganic nitrogen; only 3% is found in plants and animals (Rosswall, 1983; Jefferies & Maron, 1997).

From an ecological perspective it is most convenient to break the nitrogen cycle down into a number of stages. For the sake of brevity, we will concentrate on the terrestrial environment.

Ammonification
When organisms excrete nitrogenous waste or die, their nitrogen is converted to ammonium ions by the action of saprotrophic fungi and bacteria. This process is known as **ammonification**. The saprotroph microbes use the ammonia to synthesise their own proteins and other nitrogen-containing organic compounds. Inevitably, though, some of the ammonia leaks into the surrounding soil and so becomes available to other bacteria and to plants.

Nitrification
In warm, moist soils with a pH near to 7, ammonium ions (NH_4^+) are oxidised within a few days of their formation or their addition as a fertiliser (Salisbury & Ross, 1985). This oxidation benefits the bacteria

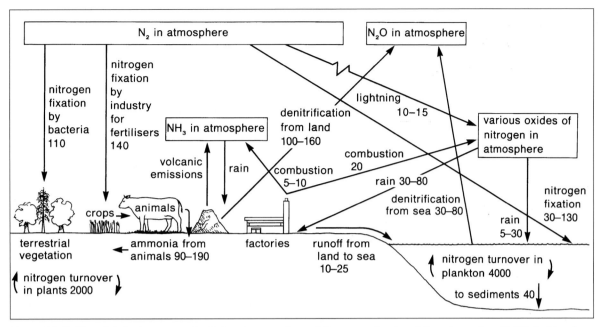

Figure 13.7 The global nitrogen cycle. Estimates are given of the annual rates of flow (in units of 10^{12} g). (Based on Rosswall, 1983; Jefferies & Maron, 1997; Pearce, 1997a.)

performing the reactions by releasing energy which the bacteria can use for the synthesis of ATP. **Nitrification** takes place in two stages. First, ammonium (NH_4^+) is converted to nitrite (NO_2^-), then nitrite is converted to nitrate (NO_3^-). These reactions can only be carried out by certain **nitrifying bacteria**.

In the first stage of nitrification, ammonia is oxidised to nitrite by bacteria of the genera *Nitrosomonas*, *Nitrosospira*, *Nitrosococcus* and *Nitrosolobus* (Hamilton, 1988). The first reaction involves the addition of oxygen to ammonia, giving rise to hydroxylamine (NH_2OH). This is then further oxidised to nitrite.

In the second stage of nitrification, nitrite is oxidised to nitrate by bacteria of the genera *Nitrobacter*, *Nitrospira* and *Nitrococcus*. The reaction proceeds by the addition of water followed by the removal of hydrogen (Hamilton, 1988). The association between *Nitrosomonas* and *Nitrobacter* has been described as one of **commensalism** (Gooday, 1988). The two species exist in the same soil environment without much mutual influence except that *Nitrobacter* depends on *Nitrosomonas* for its nitrite.

Uptake of nitrogen by plants

Most plants absorb the majority of their nitrogen as nitrate. However, many plants also absorb ammonium. In acidic soils, conversion of ammonium to nitrate is slow and in forests on acidic soils, many trees absorb most of their nitrogen as ammonium (Salisbury & Ross, 1985). Indeed, some conifers are ammonium specialists, and have only a limited capacity to absorb nitrates (Kronzucker *et al.*, 1997).

The economic consequences of this can be considerable. After the disturbance caused by the logging of planted conifers, soil pH generally rises and a new microbial environment arises which mostly converts ammonium to nitrate. As a result, when such sites are replanted with conifers, deciduous species (such as aspen) which can take up nitrate frequently invade. In British Columbia alone, more than 1.5 million hectares of productive forest lands have been classed as failed replantings of this sort.

Nitrogen fixation

Nitrogen fixation is the reduction of atmospheric nitrogen (N_2) to the ammonium ion (NH_4^+). It is of great importance to organisms. Together with lightning (Figure 13.7), it is the natural way in which organisms gain access to the huge reserves of nitrogen in the atmosphere. Nitrogen fixation can only be carried out by certain species of bacteria and cyanobacteria (Postgate, 1988). Some of these species are free-living, occurring in soil or in water. Others exist in symbiotic relationships with higher plants.

The most well-known of the nitrogen-fixing bacteria are in the genus *Rhizobium*. These bacteria form symbiotic associations in the **root nodules** of many plants in the family Leguminosae, which includes such important crops as peas, groundnuts, beans and clovers. Until recently it was thought that no members of the grass family, Poaceae, could perform nitrogen fixation. However, new techniques introduced in the 1970s showed that many grasses have nitrogen-fixing bacteria associated with their roots in a region called the **rhizosphere**. This is a transition zone between the root and the soil. How important such nitrogen fixation is to grasses, including the world's cereals, is still under dispute (Salisbury & Ross, 1985). However, it is clear that fixation rates in grasses are certainly less than for legumes (Fabaceae) and other species which possess root nodules.

The formation of root nodules like those shown in Figure 13.8 has been extensively investigated in the Fabaceae. Species in the genus *Rhizobium* persist saprotrophically in the soil until they react to flavenoid molecules produced by the legume (Mestel, 1997). The flavenoids enter the bacterium and cause a gene regulator to turn on certain genes in the bacterium called *nod* (for 'nodulation') genes. The enzymes encoded by the nod genes make a chemical signal, the 'Nod factor', which diffuses through the soil to the plant and tells it to make a nodule. The chemical structures of a number of these Nod factors have now been worked out. The structure of pea Nod factor is shown in Figure 13.9.

Of particular interest is the observation that root nodules contain **leghaemoglobin**. Leghaemoglobin functions in much the same way as our own haemoglobin: it transports oxygen. It is thought that leghaemoglobin allows the rate at which oxygen reaches the nitrogen-fixing bacteria to be controlled. Too much oxygen inactivates the enzyme that catalyses nitrogen fixation, possibly because nitrogen and oxygen resemble each other in size and shape. However, some oxygen is required for the bacteria to respire, as they are aerobic.

The enzyme responsible for nitrogen fixation is called **nitrogenase**. Although its chemical structure is still not completely known, it consists of two distinct proteins, often called components I and II.

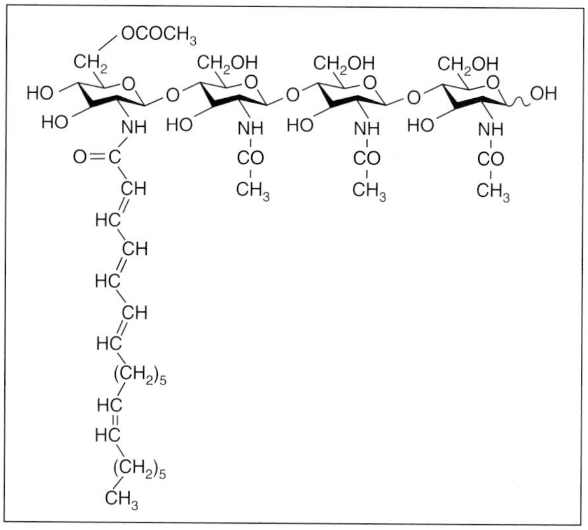

Figure 13.9 The structure of pea Nod factor, the chemical that tells pea plants to make root nodules.

Component I is a molybdenum–iron (Mo–Fe) protein; component II is an iron (Fe) protein (Yates, 1980; Moffat, 1992). More recently, a second form of nitrogenase has been found which contains vanadium instead of molybdenum (Eady *et al.*, 1987). This vanadium nitrogenase closely resembles molybdenum nitrogenase.

So why do so few organisms make direct use of the nitrogen in the atmosphere, as nitrogen is often in short supply? The nitrogen molecule (N_2) is very stable and it is thought that 16 molecules of ATP are needed for each molecule of nitrogen that is fixed! As well as being energetically very expensive, the process of fixing nitrogen also requires a number of enzymes. In the soil micro-organism *Klebsiella pneumoniae* a total of 17 genes, called *nif* genes, are known to be responsible for making the polypeptides involved in nitrogen fixation.

Despite the complexities of nitrogen fixation, many botanists hope that advances in biotechnology will allow these nitrogenase genes to be inserted into crops. Imagine wheat that needed no nitrogen fertiliser because it could fix its own dimolecular nitrogen. Not only would less money have to be spent on fertilisers; these new wheat varieties would reduce the risks associated with high levels of nitrate in fresh water (see Section 13.8.2). At the moment, most high-yielding artificially bred crops are fertilised with ammonium nitrate, anhydrous ammonia, liquid ammonia or ammonium sulphate. These fertilisers are made in the **Haber–Bosch process**. This involves the reaction of nitrogen and hydrogen at high temperatures and pressures in the presence of industrial catalysts. The Haber–Bosch process is expensive and crops typically only take up about 50–80% of the applied nitrogen. The rest is lost due to leaching and denitrification.

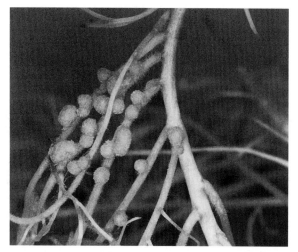

Figure 13.8 Root nodules on a cultivated variety of the garden pea (*Pisum sativum*). Each nodule is between 1 and 3 mm in diameter and contains large numbers of the bacterium *Rhizobium leguminosarum*.

Denitrification

Denitrification involves the reduction of the nitrate ion (NO_3^-) to nitrogen dioxide (NO_2), dinitrogen oxide N_2O), nitrogen monoxide (NO) or nitrogen (N_2) by certain anaerobic bacteria which have the ability to use the nitrate ion as an electron acceptor in respiration. Aerobic organisms, of course, use oxygen as their electron acceptor in respiration, making water. Under anaerobic conditions, however, oxygen is unavailable. Some species of bacteria in a few genera, including *Pseudomonas* and *Clostridium*, can make nitrite from nitrate under waterlogged conditions, when oxygen tends to be in short supply. A few species, including *Thiobacillus denitrificans*, go further than this and reduce nitrate (NO_3^-) to dimolecular nitrogen (N_2) (Hamilton, 1988). It is also now known that plants lose small amounts of nitrogen to the atmosphere as gaseous ammonia, dinitrogen oxide, nitrogen dioxide and nitrogen monoxide, especially when well fertilised with nitrogen (Wetselaar & Farquhar, 1980).

13.5 The phosphorus cycle

One can go on drawing nutrient cycles almost for ever. Usually, they are drawn only for certain elements, e.g. nitrogen, carbon, phosphorus and sulphur. However, cycles can also be drawn for compounds such as water. We have already looked at the carbon and nitrogen cycles, which are of particular ecological importance. We will look at one other cycle – the phosphorus cycle. This is interesting because the phosphorus cycle, unlike those of nitrogen and carbon, lacks an atmospheric component. The phosphorus cycle is also interesting because of the role certain specialised fungi play in it.

The phosphorus cycle does not really warrant a diagram. Plants obtain their phosphorus from the soil either as dihydrogenphosphate ($H_2PO_4^-$) (below pH 7) or more slowly as hydrogenphosphate (HPO_4^{2-}) (above pH 7). Once in an organism, though, phosphorus does not undergo reduction; it remains as phosphate (PO_4^{3-}). In this form it is found in a number of compounds, including nucleic acids, phosphorylated carbohydrates and fats. Animals obtain their phosphorus from plants, if they are herbivores, or from other animals, if they are carnivores. Decomposers return phosphorus to the soil as the phosphate ion. In most soils and waters, phosphorus is in short supply. A phosphorus atom, unless deep in the earth, is therefore likely to spend much of its time in organisms.

Ultimately, organisms obtain their phosphorus from rocks. Such release is very slow. Partly because of this, and partly because phosphorus is very immobile in the soil, phosphorus, along with nitrogen, is

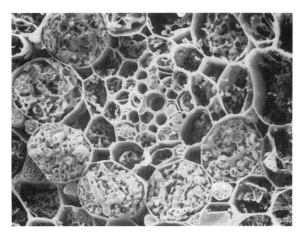

Figure 13.10 A scanning electron micrograph of a root of a plantain (*Plantago* sp.) showing mycorrhizae. Fungal hyphae can be seen in many of the plant cells.

the mineral that most often limits plant growth. This is particularly true for aquatic plants. The absorption of phosphate, along with ammonium, nitrate and the potassium ion (K^+), is, though, greatly aided by the presence of **mycorrhizae**.

A mycorrhiza (fungus-root) is a **mutualistic** (beneficial to both parties) association between a non-pathogenic or weakly pathogenic fungus and living plant root cells (Salisbury & Ross, 1985). In nature the great majority of terrestrial plant species have mycorrhizae (Figure 13.10). Experiments with radioactively labelled elements show that the plant benefits from an increase in the rate at which it can take up limiting nutrients, particularly phosphate, or water; the fungus in return receives organic compounds from the plant. Mycorrhizae therefore play an important role in soil nutrient cycling.

13.6 Interactions between the nutrient cycles

It is all too easy to consider the various nutrient cycles in isolation from one another. In fact, they are interconnected and depend on one another to a great extent. For example, the burning of fossil fuels not only puts large amounts of carbon into the atmosphere, it also increases the amount of atmospheric nitrogen, phosphorus and sulphur. Melillo and Gosz (1983) calculate that the amount of nitrogen released into the atmosphere by the combustion of fossil fuels may significantly increase primary productivity, which is often limited by lack of available nitrogen, and so increase the uptake by plants of carbon dioxide from the atmosphere. However, the inexorable rise in atmospheric carbon dioxide concentrations (Section 13.3) suggests that this effect has so far been, at most, small. In fact, the burning of fossil fuels has released aerosols of sulphates and soot which may have reduced the consequences of global warming. In 1995 the Intergovernmental Panel on Climate Change concluded that without this effect global warming this century would be of the order of 0.9 °C rather than the observed 0.5 °C (Pearce, 1995).

The interdependence of the nutrient cycles is obvious when one considers nutrient cycling through organisms. When a herbivore eats a plant, or a carnivore an animal, it ingests at one go not just carbon, nitrogen and phosphorus, but oxygen, calcium, potassium, chlorine and all the other elements which are found in organisms. Furthermore, the extent to which two elements are linked into a shared nutrient cycle varies considerably. Phosphorus, for instance, is always found covalently bound to oxygen.

There are other ways in which the various nutrient cycles can be linked to one another. From an analysis of nutrient cycling in grassland ecosystems, Stewart *et al.* (1983) concluded that management practices that accelerate the loss of carbon from ecosystems result in losses of nitrogen, sulphur and phosphorus. This is because organic forms of nitrogen, sulphur and phosphorus are more likely to be converted to inorganic forms. Inorganic forms of nitrogen and sulphur are subject to leaching and loss to the atmosphere. Inorganic phosphorus may become tied up in unavailable soil minerals.

13.7 The importance of nutrient availability

Organisms require many nutrients for healthy growth and reproduction. Some, such as carbon, oxygen, hydrogen and nitrogen, are required in large quantities; others, such as iron, magnesium and potassium, in smaller amounts. Nutrient availability varies from habitat to habitat, so that in some places a particular nutrient may be scarce, while in other places it is abundant. A good example of this is common salt, sodium chloride. In some parts of Africa, animals go to considerable lengths to get to salt deposits. One famous elephant troupe, now sadly badly depleted by hunting, goes deep into a dark cave to mine salt, using their tusks, from the cave wall. It is even thought that thousands of years of elephant activity have actually created the caves in the first place. However, at the sea's edge, salt is extremely abundant, to the point where many organisms find it difficult to survive unless they possess special mechanisms that protect them from the high concentrations of sodium and chloride ions.

13.7.1 The response of organisms to nutrient availability

When one looks at the response of organisms to the availability of various essential nutrients, a pattern emerges. If the nutrient is present at too low a concentration for the organism, then the growth and reproduction of the organism are affected. At these concentrations the nutrient is said to be **limiting** and an increase in nutrient availability will cause an increase in the organism's growth and subsequent reproductive success. Figure 13.11 shows the general pattern of organism growth or population size in response to changes in nutrient availability. At very low concentrations the organism cannot survive. As nutrient availability increases, organism growth or population size rises to a point at which the particular nutrient is no longer limiting. However, if the nutrient abundance continues to increase, there comes a point when the nutrient becomes toxic. At this point growth and reproduction are adversely affected and the organism may even be killed.

No figures have been placed on the axes of Figure 13.11, as these depend on the species and nutrient in question. However, even some substances which we are used to thinking of as poisonous turn out to be essential in small quantities. For example, arsenic has long been known to be toxic to humans. However, observations on hospital patients kept on drip feeds for several years show that they need minute amounts of arsenic to remain healthy. Perhaps unsurprisingly, the original formulations for drip feeds failed to include arsenic. It was then found that after many months patients receiving these arsenic-free drip feeds developed symptoms that only disappeared when minute amounts of arsenic were added. Of course, there are substances that are not required, even in trace amounts. These would not conform to the pattern shown in Figure 13.11, as an organism would be able to grow and reproduce per-

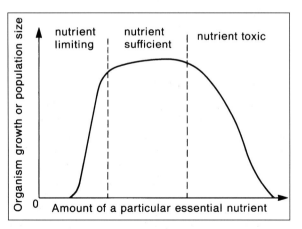

Figure 13.11 The effect that the amount of an essential nutrient has on the growth of an organism or the population size of a species. Fitness is reduced when the level of the nutrient falls outside a particular range. The shape of this curve depends both on the nutrient and on the species.

Box 13

Why are the oceans salty?

Oceans are salty but rivers are not. Why is this? One possibility could be that rivers are weakly salty. The oceans, then, might essentially be concentrated rivers formed by the evaporation of pure water from the weak solutions fed into them by the rivers. However, as Table 13.3 shows, the chemical composition of the oceans differs substantially from that of rivers.

The sodium in the oceans is thought to come mainly from

Table 13.3 The concentration (in parts per million, or mg/l) of the 11 most abundant elements dissolved in sea-water compared with their concentrations in river water. It is clear that oceans are not simply concentrated rivers. (Data from Colinvaux, 1986 after Garrells *et al.*, 1975.)

	Concentrations (p.p.m.)	
Element	Oceans	Rivers
Chlorine	19 000	7.8
Sodium	10 500	6.3
Magnesium	1 300	4.1
Sulphur	904	5.6
Calcium	400	15.0
Potassium	380	2.3
Bromine	65	0.02
Strontium	8	0.07
Boron	4.5	0.01
Silicon	2.9	6.1
Fluorine	1.3	0.1

rivers. However, the chlorine in the oceans is thought to come partly from terrestrial volcanos and partly from the submarine volcanos of the mid-ocean ridges. Both sorts of volcano produce hydrochloric acid (HCl). This reacts with hydrogen-carbonate ions (HCO_3^-) according to the following reaction:

$$HCl(aq) + HCO_3^-(aq) \longrightarrow H_2O + CO_2(g) + Cl^-(aq)$$

The carbon dioxide returns to the atmosphere leaving the chlorine ion to take the place of the hydrogencarbonate ion, balancing the sodium ion (Colinvaux, 1986).

This helps to explain why chlorine is so abundant in the oceans. The reasons why potassium and calcium are relatively rare in the oceans are also understood, at least in part. Potassium is lost from oceans as the potassium-containing clay illite is formed. This is denser than water and so settles to the ocean floor. Calcium, on the other hand, is deposited as calcium carbonate trapped in the shells and skeletons of protoctists, plants and animals. Limestone and chalk rocks represent past events when calcium was trapped in this way.

As the oceans collect minerals from rivers, it might be thought that they are continually getting saltier. However, measurements on rocks formed from marine sediments over the last 500 million years strongly suggest that the oceans have not changed substantially in their composition during this time (Garrells & Mackenzie, 1971). It seems that winds blowing across the oceans pick up ocean spray and subsequently deposit the minerals on land as rain. Ocean salts are also lost when they form sediment on the ocean floor (Colinvaux, 1993).

---

fectly satisfactorily in the complete absence of the substance, though it might be harmed by the presence of large concentrations of the substance.

In the next two sections we will look at two very different instances where nutrient availability limits plant growth.

13.7.2 China clay waste tips

In about 1770 mining for china clay started in Cornwall, England. At first the china clay was used for making chinaware, but the clays are now also used in the manufacture of high-quality paper. The deposits are worked in open pits. Once the valuable mineral kaolinite has been extracted, the huge amounts of waste are left as tall conical tips of white sand (Figure 13.12). About 35 square miles of Cornwall and Devon have been affected in this way (Bradshaw & Chadwick, 1980). The interesting thing about these tips is that they are nutritionally very poor: there are considerable deficiencies in nitrogen, phosphorus, potassium, magnesium and calcium. In addition, the tips are very low in silt and clay,

consisting almost entirely of sand particles. Consequently, the water-holding capacity of the tips is very low and they are quite acidic (about pH 4). These features are responsible for the fact that the natural colonisation of these tips by vegetation is extremely slow (Bradshaw & Chadwick, 1980). It

Figure 13.12 A characteristic feature of much of the Cornish landscape: a china clay waste tip near Treviscoe, Cornwall.

takes at least 20 years for even the base of a tip to develop a natural plant community. As one might predict, such a community contains large numbers of leguminous shrubs as these are able to fix nitrogen and so provide their own nutrients to some extent (Section 13.4). Eventually the vegetation develops to form acid oak woodland.

Because natural colonisation is so slow, nowadays the tips are treated to ensure that they soon become covered with vegetation. A variety of techniques is used. The most common is to contour the tips so that the slopes are not too steep, and then to cover the tips with topsoil removed from parts of Cornwall where there is an active road-building programme. The topsoil is then seeded and trees are planted. The net result is that the tips soon become quite densely vegetated.

13.7.3 *Nutrient cycling in tropical forests*

There are several different types of tropical forest (Spurr & Barnes, 1980), so that simply to generalise about 'tropical forests' is to run the risk of oversimplification. Nevertheless, it is true to say that tropical forests are found on soils which are nutrient poor (Jordan, 1985). This fact may seem surprising as we are used to thinking of tropical forests being lush. Indeed, tropical forests are very productive (Table 12.1). However, high productivity does not require soils to contain large nutrient reserves. What seems to happen in undisturbed tropical rainforests is that any organic matter that falls to the ground is rapidly decomposed. The nutrients thus released into the soil are then rapidly taken up by the surface roots of trees and other plants or, occasionally, leached from the soil.

Tropical trees have a number of features which help them to obtain nutrients (Jordan, 1985). They produce a large root biomass which is concentrated near the surface. These roots seem to be very effective in absorbing nutrients. For example, when Stark and Jordan (1978) added radioactive calcium and phosphorus to the soil surface of an Amazonian rainforest, they found that 99.9% of the radioactivity ended up in roots; only 0.1% was lost by leaching. Another feature of tropical trees which favours nutrient retention is that their leaves are often long lived, tough and resistant to insect attack. These features help to conserve nutrients within the tree but seem to carry with them the disadvantage of reducing the rate of photosynthesis (Medina & Klinge, 1983).

The nutrient-poor status of the soils found under tropical rainforest means that they are usually unsuitable for the cultivation of agricultural crops (see Section 15.2.2). When such cultivation is attempted, the decline in yield can be dramatic. In Zaire, the yield of peanuts in the second year of cultivation on forest soil is only about 15% of the yield in the first year of cultivation. For rice, the figure is about 25%; for cassava, 65% (Norman, 1979). It seems particularly sad, therefore, that so much of the world's tropical rainforests continue to be cut down so that cultivation of crops like these can take place.

13.8 Pollution

Pollution occurs when substances are released into the environment in harmful amounts as a direct result of human activity. Most pollution is due to the presence of excessively high concentrations of substances, though the example of the nutrient-poor china clay waste tips considered in Section 13.7.2 might be considered an instance of pollution. By definition, pollution is the result of human activity, though the huge build-ups of bird droppings (guano) at some seabird colonies come close to being examples of 'natural pollution' as the concentrations of nutrients in them is so high that they prevent almost any living organism from colonising them. Examples of such 'natural pollution' are rare because natural selection nearly always leads to the evolution of organisms that can utilise the metabolic products of other organisms.

13.8.1 *Different forms of pollution*

Most pollution is due to the excessive release of chemicals into the environment by humans, though **thermal pollution** involves the release, not of a harmful chemical, but of excessive amounts of heated water. These are discharged by many industries into rivers and seas. The main problem is that warm water holds less oxygen than cold water, so that the release of large quantities of warm water may kill fish and aquatic invertebrates through oxygen starvation.

Pollutants can be classified into three types. First, there are those substances that occur in nature, but, as a result of human activity, are found in unusually large concentrations. An example is carbon dioxide. We have already seen that the raised atmospheric concentration of this gas is having significant effects on our climate (Section 13.3). A second type of pollution occurs when, as a result of human activity, toxic substances are produced that are not found in nature. An example is the use of pesticides. Such unnatural substances often remain intact in the environment for considerable lengths of time before being broken down or dispersed. A third type of pollution occurs when substances which are not themselves toxic are released into the environment as a result of human activity, but which then go on to have unfortunate consequences. An example is the effect of certain substances on the ozone layer.

With this classification in mind, we will now consider some particular examples of pollution in more detail.

13.8.2 *Eutrophication*

Eutrophication is the name given to the release of large amounts of phosphate and nitrate or organic matter into water resulting in a lowering of oxygen levels and change in the fauna of the water. In their natural state most waters, whether freshwater or marine, contain only low levels of nitrate and phosphate, although natural examples of eutrophic lakes are known. If substantial quantities of these minerals enter water, they allow large numbers of algae, cyanobacteria and aerobic bacteria to build up. These organisms can require so much oxygen that they lower the amount of free oxygen in the water to the point at which aerobic bacteria are unable to decompose organic matter in the water.

One standard way of determining the oxygen requirements of an area of water is to measure the rate at which the oxygen level of a sealed sample of the water falls when kept in the dark for five days at 20 °C. (Keeping the sample in the dark stops photosynthesis which would otherwise lead to oxygen being evolved by any photoautotrophs in the sample.) The **biological oxygen demand (BOD)** of unpolluted river water is typically less than 5 mg $O_2/dm^3/5$ day. Crude sewage has a BOD of around 600 mg $O_2/dm^3/5$ day. It is not surprising that pollution due to sewage can lead to permanent changes in the organisms found in the affected waters. However, good sewage treatment reduces the BOD of discharged effluent to less than 30 mg $O_2/dm^3/5$ day (Moss, 1988).

Biological indices can be used to measure the extent of organic pollution in water systems. In England and Wales, a survey of rivers in 1980 (National Water Council, 1981) classified rivers into five types on the basis of their dissolved oxygen concentrations, ammonia concentrations and ability to support fish. At the same time the invertebrates were sampled according to the system given in Table 13.4. In this system, points are allocated according to whether or not certain **key groups** are present. One advantage of this method is that invertebrates need only to be identified to family level. Oligochaetes and chironomids are found even in highly polluted streams and rivers with very low levels of dissolved oxygen (<10% saturation). At the other extreme, most families of mayflies and stoneflies are found only in waters with high levels of dissolved oxygen (>80% saturation). One major advantage, to an ecologist, of using a biological index such as this one is that it provides a measure of the organic pollution in the water over a reasonably long period of time. A

Table 13.4 The National Water Council system for monitoring organic pollution in rivers. The river is given a score equal to the highest scored family found in the river. For example, if the river contains families which score 1, 2, 3 and 5, then the river scores 5. The higher the score, the less polluted the river. (From Moss, 1988.)

Families	Score
(a) Siphlonuridae, Heptageniidae, Leptophlebiidae, Ephemerellidae, Potamanthidae, Ephemeridae (mayflies)	
(b) Taeniopterygidae, Leuctridae, Capniidae, Perlodidae, Perlidae, Chloroperlidae (stoneflies)	
(c) Aphelocheiridae (beetles)	10
(d) Phryganeidae, Molannidae, Beraeidae, Odontoceridae, Leptoceridae, Goeridae, Lepidostomatidae, Brachycentridae, Sericostomatidae (caddis-flies)	
(a) Astacidae (crayfish)	
(b) Lestidae, Agriidae, Gomphidae, Cordulegasteridae, Aeshnidae, Corduliidae Libellulidae (dragonflies)	8
(c) Psychomyiidae, Philopotamidae (net-spinning caddis-flies)	
(a) Caenidae (mayflies)	
(b) Nemouridae (stoneflies)	7
(c) Rhyacophilidae, Polycentropodidae, Limnephilidae (net-spinning caddis-flies)	
(a) Neritidae, Viviparidae, Ancylidae (snails)	
(b) Hydroptilidae (caddis-flies)	
(c) Unionidae (bivalve molluscs)	6
(d) Corophiidae, Gammaridae (Crustacea)	
(e) Platycnemididae, Coenagriidae (dragonflies)	
(a) Mesovelidae, Hydrometridae, Gerridae, Nepidae, Naucoridae, Notonectidae, Pleidae, Corixidae (bugs)	
(b) Haliplidae, Hygrobiidae, Dytiscidae, Gyrinidae, Hydrophilidae, Clambidae, Helodidae, Dryopidae, Elminthidae, Crysomelidae, Curculionidae (beetles)	5
(c) Hydropsychidae (caddis-flies)	
(d) Tipulidae, Simuliidae (dipteran flies)	
(e) Planariidae, Dendrocoelidae (triclads)	
(a) Baetidae (mayflies)	
(b) Sialidae (alderfly)	4
(c) Piscicolidae (leeches)	
(a) Valvatidae, Hydrobiidae, Lymnaeidae, Physidae, Planorbidae, Sphaeriidae (snails, bivalves)	
(b) Glossiphoniidae, Hirudidae, Erpobdellidae (leeches)	3
(c) Asellidae (Crustacea)	
(a) Chironomidae (Diptera)	2
(a) Oligochaeta (whole class) (worms)	1

single release of fertiliser, for example, into a river may pass undetected by chemical analyses a few weeks later. However, that one incident may have killed key indicator families which can still be absent months later.

There are several causes of eutrophication. The

main ones are sewage input (including that from farm animals and from fish farms) and the runoff from fertilisers applied to crops. Occasionally, the release of concentrated sugars or other organic substances may be a problem. The ecological consequences of eutrophication can be extensive. We have already mentioned likely changes in the invertebrate fauna. Fish numbers may be lower and some fish species may be eliminated altogether. Dense algal mats may out-compete marginal aquatic plants. These changes may result in fewer fish-eating and plant-eating birds. The water will become cloudier and look less attractive. Off Australia, eutrophication has led to significant increases in phytoplankton concentrations since 1930 (Bell & Elmetri, 1995). Eutrophication threatens many reefs in the Great Barrier Reef through algal overgrowth and possibly also by promoting the crown of thorns starfish (*Acanthaster planci*) which has devastated large regions of the Great Barrier Reef in recent times (see Figure 17.20).

Eutrophication can be removed, but only if the source of the pollution is removed. In some countries, particularly in North America and on the mainland of Europe, governments have recently passed quite strict anti-eutrophication laws. It is probably easier to control the amount of phosphorus entering aquatic communities than the amount of nitrogen. This is because nitrate enters from a variety of sources, but phosphorus, in significant amounts, comes only from sewage (Moss, 1988).

13.8.3 Heavy metal toxicity

We saw in Section 13.7.2 how mining for china clay produces waste tips which are low in essential nutrients. Mining for other substances, however, can produce spoil that is poisonous to many organisms. Heavy metal toxicity occurs when elements such as mercury, zinc and selenium are present in superabundance. In California, Nevada and Utah, for instance, selenium toxicity has caused horrific deformities in pelicans and other birds (Figure 13.13). It looks as though agricultural irrigation washes out **trace elements** from the soil which then enter the food chain and get concentrated at higher trophic levels, accumulating in coots, ducks, pelicans and other water birds (Anderson, 1987). (Trace elements are elements such as boron, molybdenum, zinc and copper which are required by living organisms, but only in very small amounts.)

In Spain, the country's 1.3 million hunters add some 3000 tonnes of lead to the environment each year, much of it in wetlands. A two-year study commissioned by the Spanish government concluded that stray lead shot is responsible for the deaths of between 25 000 and 30 000 birds each year (Luke,

Figure 13.13 Brown pelican (*Pelecanus occidentalis*) showing the effects of selenium toxicity.

1995). Some countries have banned the use of lead shot in such circumstances but the Spanish Hunting Federation has reacted unenthusiastically to the suggestion on the grounds of cost.

The fundamental problem with heavy metals is that although some of them are needed by organisms in trace amounts, when present in excess they cause enzymes to denature. In a classic study Bradshaw (1952) found that plants of the grass *Agrostis tenuis* that grew on the waste by old lead mines in Wales had evolved the ability to tolerate the high levels of lead in the waste. Since then, such **tolerant ecotypes** have been found in a number of other species, mostly grasses. Tolerance to several heavy metals, including zinc, copper, lead and nickel has been demonstrated (Gemmell, 1977). Such tolerance is usually specific to a particular metal, so that a population tolerant to lead will usually not be tolerant to any other metal.

Plants use a variety of methods to reduce their susceptibility to metal toxicity. In some species uptake of the poisonous metals is greatly reduced. In other species the elements are taken into the plant but mainly remain in the roots (Salisbury & Ross, 1985). In some plants the metals reach other parts of the plant but then accumulate in the cell walls. Here the metal is bound as a stable complex with either a pectin-like or a proteinaceous substance (Bradshaw & McNeilly, 1981). Plants have also been found which actively transport a large proportion of their absorbed zinc across the tonoplast into the cell vacuole. This prevents the zinc from interfering with the activity of cytoplasmic enzymes (Bradshaw & McNeilly, 1981). There seems to be a biochemical cost to being tolerant to heavy metal pollution. Such plants are often poor competitors and are seldom found growing away from the area where the heavy metals are concentrated.

The evolution of resistance to metal toxicity has

also been demonstrated in aquatic systems. The hulls of ships are often painted with a copper-based antifouling paint. This is designed to prevent the growth of marine organisms such as barnacles. The green alga *Ectocarpus siliculosus* collected from ships has been found to be able to survive ten times the concentrations of copper that can be tolerated by a population growing on an unpolluted rocky shore (Russell & Morris, 1970).

It should not be assumed that heavy metal pollution is only a recent phenomenon. For example, records from peat bogs and even Greenland ice reveal significantly raised levels of arsenic, antimony, copper, lead, mercury and zinc going back to Roman and Greek times (Hong *et al.*, 1996; Shotyk *et al.*, 1996).

The dangers of heavy metal pollution for human health have been known for some time. For example, in the 1950s many people suffered mercury pollution as a result of methyl mercury waste being dumped into the Minimata Bay, Japan. The mercury was consumed by fish and subsequently by women who then gave birth to deformed babies. More recently, it has been claimed that heavy metals at even very low concentrations disrupt neurological control mechanisms and so make a major contribution to violent crime and antisocial behaviour (Motluk, 1997). Roger Masters, a political scientist from the USA, has analysed crime figures from the FBI (Federal Bureau of Investigation) and information on industrial discharges of lead and manganese. After controlling for such variables as income and population density, he has found that environmental pollution seems to have a major effect on the rate of violent crimes – defined as homicide, aggravated assault, sexual assault and robbery. Counties in the USA with the highest levels of lead and manganese pollution typically have crime rates three times the national average.

13.8.4 Alkaline wastes

Many manufacturing processes produce large amounts of alkaline waste. Alkaline waste tips from the Leblanc soda ash process are initially very alkaline (with a pH of about 11) because of the presence of calcium hydroxide (Bradshaw & Chadwick, 1980). In time, however, the tips weather, and when the pH falls to about 8, some of them are colonised by large numbers of very interesting plants which are characteristic of alkaline soils. Huge numbers of orchids are found on some alkaline waste tips in north-west England. Table 13.5 lists some of the rare plants found at the alkaline waste disposal site in the Croal–Irwell Valley near Bolton 80 years after it was last used (Gemmell, 1977).

In the light of this, it is worth thinking about the possibility that many of today's pollution blackspots

Table 13.5 Some of the plant species found on an alkaline waste disposal site near Bolton, Lancashire that are rare or absent elsewhere in the area. (From Gemmell, 1977.)

Carline thistle (*Carlina vulgaris*)
Common centaury (*Centaurium erythraea*)
Common spotted orchid (*Dactylorhiza fuchsii*)
Early marsh orchid (*D. incarnata*)
Northern marsh orchid (*B. purpurella*)
Fragrant orchid (*Gymnadenia conopsea*)
Purging flax (*Linum catharticum*)
Common broomrape (*Orobanche minor*)

may be tomorrow's nature reserves! It is certainly the case that much pollution leads to imbalances in nutrient availability and thus leads to diversification in the physical environment. Now this does not mean that we should be complacent about pollution. Rather, we should not always assume that *all* pollutants need to be completely removed. This often simply is not possible. It may be more fruitful sometimes to try to imagine how an industrial site can be landscaped so that after a few years it will provide a variety of habitats not previously available in that locality. Of course, if a locality is a particularly rare one, or one of outstanding natural beauty, then it may be best to minimise pollution there in the first place. It is worth remembering that in industrial countries many of the features of the countryside which people nowadays find most attractive are the result of human activity. In Britain, for example, the Broads in East Anglia are the result of mediaeval peat diggings; chalk grassland is largely a consequence of sheep grazing; and most hedges are due to human activity, whether Anglo-Saxon, mediaeval or a product of the eighteenth and nineteenth century Enclosure Acts (Rackham, 1986; Taylor, 1988).

13.8.5 Acid rain

Acid rain is the collective name given to a number of processes which all involve the deposition of acidic gases from the atmosphere (Lee, 1988). Unpolluted rain has a pH of about 5.6 due to the presence of dissolved carbon dioxide. However, in Britain and many other industrialised countries, rain water often has a pH of between 4 and 4.5 (Lee, 1988; Swinbanks, 1989a). The difference is due to various oxides of nitrogen and sulphur, often collectively described as NO_x and SO_x. These are the result of the combustion of fossil fuels, whether petrol in vehicles, or coal, oil or gas in power stations.

The ecological importance of acid rain is still under debate. Many environmentalists blame acid rain for the damage to trees seen in European and North American forests. Especially at high altitudes and at the edges of forests, large areas of land are covered by damaged or dying trees (Figure 13.14). In Britain good evidence exists from the distribution of lichens to show that sulphur dioxide (SO₂) pollution has been responsible for the elimination of many species of lichens from large areas of the country. In a classic study Hawksworth and Rose (1970) classified different species of lichens into ten groups, which they called zones. Zone 10 contained lichens most susceptible to sulphur dioxide. Lichens in zone 9 could tolerate mean levels of winter sulphur dioxide deposition of up to 30 μg/m³. Lichens in zone 8 could tolerate mean winter sulphur dioxide deposition levels of between 30 and 35 μg/m³, and so on. The only lichens that could tolerate mean winter sulphur dioxide deposition levels of between about 150 and 170 μg/m³ were *Lecanora conizaeoides* and, less often, *L. expallens*. These were placed in zone 2. Zone 1 contained only the green alga *Pleurococcus viridis*. Zone 0 contained neither lichens nor algae. Figure 13.15 shows the distribution of these lichen zones in England and Wales. The higher zones, as one would expect, are found only in the most rural areas of south-west England and west Wales.

More recent experimental studies have involved fumigating conifers with varying concentrations of sulphur dioxide and then seeing which lichens

colonise. Such work has confirmed that lichen species differ in the extent to which they can tolerate sulphur dioxide pollution, much as expected from lichen distributions (Bates *et al.*, 1996). Interestingly, *Lecanora conizaeoides* (in zone 2 of Figure 13.15) turns out not only to be tolerant of sulphur dioxide but actually to require it. *Lecanora* was so rare 200 years ago that it had not even been discovered. It may well only have lived around sulphur springs; pollution has given it the chance to become widespread.

Acid rain also has an important effect on fresh water. During the 1960s and 1970s many Scandinavian lakes lost large numbers of their fish. These deaths correlated with increasing acidification of the lakes. The problem was most acute in lakes based on granite rock, where the soils were incapable of buffering the acid (Lee, 1988). Increased concentration of hydrogen ions (H⁺) is probably not directly to blame for such effects, but at low pH more aluminium can exist in solution. Aluminium is known to be toxic to many species, particularly some plants and fish. Acid rain has also been implicated in failures of a number of birds to breed normally in the Netherlands (Drent & Woldendorp, 1989).

Recent decades have shown a significant reduction in sulphur dioxide emissions in a number of industrialised countries. Figure 13.16 shows how mean winter sulphur dioxide concentrations have fallen in southern England (Bates, 1993). This reduction is continuing. For example, National Power, one of the major electricity generating businesses in the UK, reduced its sulphur dioxide emissions by 46% from 1990 to 1996 (National Power, 1996). This was achieved through a combination of Drax flue gas desulphurisation (which removes sulphur dioxide before it is pumped into the atmosphere) and the building of combined cycle gas turbine power stations (which produce much less sulphur dioxide than power stations based on coal). The Drax flue gas desulphurisation project alone cost £620 million.

It can take a long time, however, to undo the effects of acid rain. The passage in 1970 of the Clean Air Act in the USA led to major falls in SO₂ emissions. However, freshwater lakes have not shown a commensurate rise in pH. Long-term data from the Hubbard Brook Experimental Forest, New Hampshire suggest that this may be because acid rain led to the leaching of large quantities of calcium and magnesium from the soil. As a result, the recovery of soil and freshwater chemistry will be delayed considerably. In the Hubbard Brook Experimental Forest, average streamwater pH only rose from a pH of 4.85 in 1963 to one of 5.01 in 1993 (Likens *et al.*, 1996). In Sweden, some models suggest that recovery of aquatic ecosystems from acidification may take over a century (Bishop & Hultberg, 1995). At present, Sweden spends over US$25 million a year on the

Figure 13.14 Conifer tree in Germany showing the effects of acid rain.

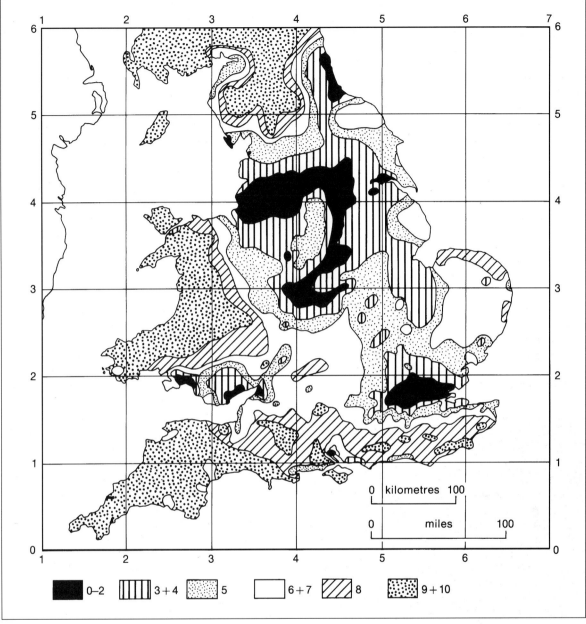

Figure 13.15 Distribution of lichen zones in England and Wales. Zone 10 has the most species of lichens; zone 0 none. (From Hawksworth & Rose. 1976.)

liming of lakes and rivers.

As a final note, it is now thought, somewhat bizarrely, that a lack of sulphur may be responsible for reduced yields of oilseed rape, wheat and barley across Europe (MacKenzie, 1995). The sulphur deposited on European fields fell from about 50 kg per ha in the late 1970s to under 10 kg per ha in 1995 as countries cleaned up their SO_2 emissions. Insufficient sulphur stunts the growth of plants and makes them more susceptible to fungal diseases. Fortunately, the remedy is straightforward for farmers that apply fertilisers: use ammonium sulphate fertilisers in preference to ammonium nitrate ones.

13.8.6 Pesticides

The instances of pollution which we have so far discussed in this chapter have in common that they reflect imbalances in nutrient supply. Essential nutrients may be present in abnormally small amounts, as in the tips associated with china clay extraction. More usually, though, pollution occurs because certain substances, although natural, are found in excess. Examples considered in this chapter have included heavy metals, hydroxide ions and oxides of nitrogen and sulphur. Pesticide pollution is rather different. **Pesticides** are chemicals synthesised expressly

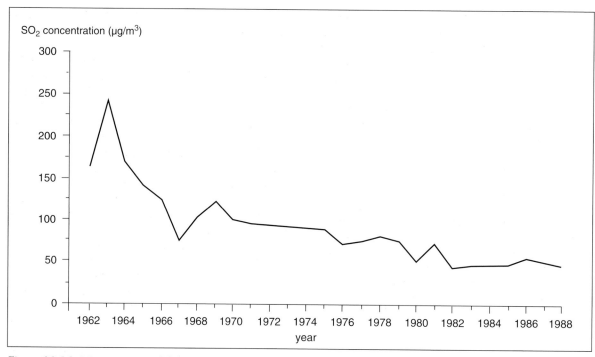

SO₂ concentration (µg/m³)

Figure 13.16 Mean winter sulphur dioxide atmospheric concentrations in southern England from 1962 to 1988.

for the purpose of killing unwanted species. Pesticides may be divided into various categories, such as **herbicides** (chemicals synthesised expressly for the purpose of killing unwanted plants), **insecticides** (which kill insects), **fungicides** (which kill fungi), and so on. Pesticides differ from the examples we have so far considered in that most are unnatural.

Before considering the problems associated with pesticide use, it is important to emphasise that they have proved very valuable. They are used for two main purposes: first, to kill species which carry diseases harmful or fatal to humans; second, to kill species which compete with us for food. There is little doubt that the immediate elimination of all pesticides would lead to a great increase in human misery through disease and starvation. Nevertheless, pesticides have their problems. The first problem is that the pests they are designed to kill soon evolve resistance to the pesticides. This means either that larger and larger doses have to be used or that new pesticides have to be synthesised. The second major problem is that many pesticides are not **biodegradable**. Precisely because they are unnatural substances, they may be resistant to decay. This means that they accumulate in the environment. In particular, they may be concentrated as they pass up a food web, so that although the primary producers and herbivores may only have low levels of contamination, the top carnivores may contain substantial concentrations of these chemicals (e.g. Wania & Mackay, 1993).

In Britain a classic instance of the unintended and

harmful effects of pesticides was the decline in the peregrine falcon (*Falco peregrinus*) (Sheail, 1985). The botanist Derek Ratcliffe devoted much of his spare time to studying peregrine falcons. Ratcliffe established that a decrease in peregrine numbers was associated with the introduction and use of organochlorine pesticides, such as DDT and Dieldrin. By the early 1960s, peregrine numbers had fallen nationally to only about 40% of what they had been 25 years earlier (Ratcliffe, 1963). Careful research failed to implicate changes in the weather or food supply. Rather, the accumulation of organochlorine residues led to breeding failure. One way of quantifying the effect of this pesticide accumulation was to measure the thickness of the peregrines' eggshells. Higher levels of pesticide accumulation led to thinner shells as the pesticides interfered with thyroid and calcium metabolism.

The recent history of the peregrine in Britain is a more encouraging one. Thanks to intensive lobbying by the Nature Conservancy Council and various wildlife organisations including the Royal Society for the Protection of Birds, successive governments introduced progressive restrictions on the use of organochlorine pesticides. By 1975 organochlorine pesticides could no longer legally be used as seed dressings. Over the last 25 years, peregrine shells have got thicker and numbers have increased substantially. By 1980 peregrine numbers had recovered to almost 80% of their 1930–39 levels (Sheail, 1985).

The advent of genetic engineering has led some

people to hope that pesticide use may decrease as crops are genetically engineered for pest resistance. Others are more sceptical, suspecting that pests will evolve the ability to attack such crops in much the same way that they evolve resistance to pesticides. It is probably too early to be certain which of these views is more correct, though there are some early encouraging signs that genetic engineering can reduce pesticide use (Reiss & Straughan, 1996).

13.8.7 CFCs and the ozone layer

Although it is only present in concentrations of a few parts per million or less, ozone (O_3) is concentrated in the Earth's stratosphere 15–50 km above the surface of the Earth (Farman, 1987). During the mid-1980s scientists suddenly realised that above Antarctica ozone levels were falling dramatically. The substances most responsible for this sudden collapse were identified as **chlorofluorocarbons** or **CFCs** for short (Figure 13.17).

What seems to happen is that the chemical $CFCl_3$ may be broken down by a quantum of high-energy light as follows:

$$CFCl_3 + light \longrightarrow CFCl_2\bullet + Cl\bullet$$

The chlorine radical ($Cl\bullet$) thus formed can then cause the conversion of ozone to dimolecular oxygen without itself being used up in the process:

$$Cl\bullet + O_3 \longrightarrow ClO\bullet + O_2$$

and

$$ClO\bullet + O\bullet \longrightarrow Cl\bullet + O_2$$

In this way a single molecule of $CFCl_3$ can remove hundreds of thousands of ozone molecules (Farman, 1987).

CFCs used to be found in every refrigerator and the majority of aerosols. They have also been used in making fast-food containers, cavity insulation and many pieces of firefighting equipment. Intense lobbying by scientists and environmentalists bore fruit in the late 1980s when a number of measures were agreed internationally to reduce the use of CFCs. The signing in 1987 of the **Montreal Protocol** by over 30 countries was particularly significant as it laid down targets to ensure that fewer CFCs were released into the atmosphere. It was followed by increasingly stringent amendments in London in 1990 and in Copenhagen in 1992. The decade after the signing of the Montreal Protocol saw a reduction in CFC use in many countries. However, alternatives to CFCs are expensive and some countries are increasing their use of CFCs. In addition, other long-lived ozone-depleting substances are still being produced (Makhijani & Gurney, 1995).

The data from Antarctica show that ozone levels have continued to decline (Jones & Shanklin, 1995). However, theoretical models suggest that stratospheric ozone levels should begin to recover from the year 2000, though they may not return to their pre-ozone-hole levels until around 2050 (Montzka *et al.*, 1996; Hofmann, 1996).

The environmental consequences of stratospheric ozone depletion are not entirely clear. Stratospheric

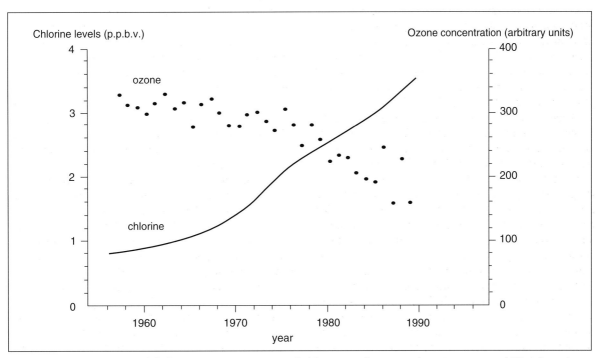

Figure 13.17 Changes in global average concentration of chlorine in the troposphere in parts per billion by volume (p.p.b.v.) held by halocarbons including CFCs (line) and in the total ozone over Antarctica in October (dots).

ozone helps prevent penetration of solar ultraviolet-B (290–315 nm) radiation. UV-B radiation is the most energetic component of sunlight that reaches the Earth's surface. In humans, exposure to UV-B leads to an increase in cataracts and skin cancers and to a decreased immune response to infections. It is probable that for each chronic (i.e. long-term) 1% decrease in stratospheric ozone, there is a 2% increase in non-melanoma skin cancer and a 0.7% increase in cataracts (van der Leun *et al.*, 1995).

The effects on other species are also likely to be significant. Marine phytoplankton and terrestrial plants are harmed by raised UV-B levels (Table 13.6) as are a range of animals including fish, shrimps, crabs, insects and amphibians (Bothwell *et al.*, 1994; Kleiner, 1994). The precise consequences of all this for ecosystems cannot as yet be predicted. Field observations suggest that some amphibian populations may be suffering but, despite the catalogue of woes listed in Table 13.6, there is little indication, from either field observations or experimental studies, that terrestrial producers are, as yet, very significantly affected. However, some effects have been found – for example, evergreen shrubs seem to be more severely affected by UV-B than are deciduous ones (Johanson *et al.*, 1995) – and it should be borne in mind that none of these studies has yet run for more than a few years.

Table 13.6 The damage caused to plants by ultraviolet-B radiation (290–315nm). (From Rozema *et al.*, 1997.)

Absorption of UV-B photons leading to DNA damage
Formation of free radicals leading to DNA damage
Alterations to membrane structure
Disturbance to membrane permeability
Damage to Photosystem II (a pigment system in photosynthesis)
Disruption of thylakoid membranes
Reduction of chlorophyll content
Damage to RuBP carboxylase (the enzyme that binds CO_2)
Reduced levels of indolyl acetic acid (the plant growth substance)

What are also, as yet, unclear are the likely effects of interactions between the effects of ozone depletion, global warming and other consequences of pollution. Long-term studies of lakes in Canada suggest that climate change, ozone depletion and acid rain may have more severe effects when acting together than when acting separately (Gorham, 1996). All three forms of pollution end up leading to deeper penetration by UV-B. Ozone depletion does this simply by increasing the amount of UV-B. Climate changes have included a 25% fall in precipitation over 20 years. This leads to less organic material being carried by streams into the lakes. This organic material helps protect the deeper parts of the lake from UV-B. Acid

rain also reduces the amount of organic material dissolved in the lake, in part by causing it to settle out after it aggregates with aluminium ions released by the acidification.

Finally, while ozone in the stratosphere is a good thing, raised levels of ozone at ground level are not. Here ozone is a pollutant. In humans raised levels of ozone can cause coughing, breathlessness, pain and a progressive loss of lung function (Kelly, 1996).

13.8.8 Radioactivity

Although some ecologists have studied the ecological consequences of radioactivity since the 1960s, it was only really at 01.23 h on 26 April 1986 that the research really got underway. At that time reactor number 4 of the Chernobyl nuclear power plant exploded in northern Ukraine in the then USSR. In an attempt to limit the medical damage to people, approximately 115 000 people living within a 30 km zone were evacuated over the next ten days. By November 1986 the whole of the reactor had been encased in a concrete sarcophagus, though within a few years cracks appeared leading to fears about the further leakage of radioactivity.

It is difficult to be certain, but around 80 kg of radioactive substances were released into the atmosphere, producing somewhere between 2×10^8 and 5×10^8 becquerels (Bq) of radioactivity – one becquerel corresponding to one disintegration of any radionuclide per second.

Understandably, most research has focused on the medical, agricultural, social and economic implications of Chernobyl (Figure 13.18). Rates of cancer and birth malformations in the vicinity have approximately doubled. Children born in the aftermath of the disaster on average have twice as many mutations as those born before (Edwards, 1996). All in all, an estimated 10 000 people will be killed by the radioactivity; countless more will suffer either medically or economically. The economic losses of neighbouring Belarus, the most affected country, are comparable to those of the Second World War.

Considerable work has also been carried out on the wider ecological implications (Savchenko, 1995). Conifers are some ten times more sensitive to radioactivity than are deciduous trees. Over an area of some 4400 ha around the reactor, all the mature pine (*Pinus*) trees died. These are expected to be replaced by birch (*Betula*) and other deciduous trees. Vertebrates are more affected by radioactivity than invertebrates, though it is unclear what the precise consequences of the Chernobyl disaster will be for the fauna (animals) of the area. It is known that the mutation rates for mitochondrial cytochrome *b* in two vole species (*Microtus arvalis* and *M. rossiae-meridionalis*) are literally hundreds of times greater at

Figure 13.18 This deserted village lies close to Chernobyl. It was abandoned soon after the nuclear explosion at Chernobyl in 1986. No one lives here any more.

the Chernobyl site than at unpolluted sites (Baker *et al.*, 1996). Despite this, vole populations seem to thrive in the radioactive regions around the reactor.

Almost every country in the northern hemisphere has been affected by Chernobyl. For example, restrictions were placed on the sale of sheep in certain parts of Britain because the grass they were eating became contaminated by radioactive caesium-137. It was thought that these restrictions would be required for approximately three weeks, but while restrictions on 50 Cumbrian farms were lifted in 1995, nine years after the explosion, restrictions remained in place at a further 60 farms. In some countries, infant leukaemia rates more than doubled as a result of Chernobyl. For instance, some parts of Greece received ten times the radioactivity that other parts of the country did. Children born between 1 July 1986 and 31 December 1987 in these polluted areas were 2.6 times as likely to develop leukaemia as children born between 1 January 1980 and 31 December 1985 or between 1 January 1988 and 31 December 1990 (Petridou *et al.*, 1996).

Summary

(1) Energy is always lost from ecosystems. Nutrients, though, often circulate within ecosystems.

(2) Due to the burning of fossil fuels and trees, atmospheric levels of carbon dioxide have increased by 30% over the last hundred years.

(3) It is possible that the increase in atmospheric carbon dioxide concentrations may account for the observation that the world's climate has got warmer by about 0.5 °C over the last hundred years.

(4) The nitrogen cycle is a complex process involving the activities of several genera of bacteria.

(5) Mycorrhizae help plants to obtain phosphate from soil.

(6) China clay waste tips are very low in several important soil nutrients.

(7) Tropical forests usually grow on soils which are nutrient poor. If the trees are removed and cultivation of agricultural crops attempted, yields rapidly decline as available soil nutrients are exhausted.

(8) Pollution occurs when substances are released into the environment in harmful amounts as a direct result of human activity.

(9) Eutrophication occurs when the levels of phosphate or nitrate rise substantially in fresh water or marine communities.

(10) The biological oxygen demand of water gives a good measure of the organic pollution to which the water has recently been subjected.

(11) The presence or absence of key invertebrate families indicates the amount of organic pollution to which an area of water has been subjected in the longer term.

(12) Many metal ions are poisonous to plants when present in more than trace amounts.

(13) Acid rain is caused by the burning of fossil fuels. It damages trees and lakes.

(14) Long-term sulphur dioxide pollution can be monitored by studying the distribution of sensitive species of lichens.

(15) Pesticides have many uses but may accumulate in food chains and damage wildlife.

(16) Chlorofluorocarbons are destroying the ozone layer. Reversing the damage already caused calls for strict international agreements.

(17) The long-term ecological consequences of radioactivity are still imperfectly understood.

(18) An understanding of nutrient cycling helps to explain why the oceans are salty.

FOURTEEN

Communities

14.1 The community concept

14.1.1 Definitions

The use of a collective term to describe organisms of several species found in association together was first applied to plants. Many definitions of **community** have been botanical ones. Hence a community has been defined as 'an aggregate of living plants having mutual relations among themselves and to the environment' (Oosting, 1956), or more recently, as 'a collection of plant populations found in one habitat type in one area, and integrated to a degree by competition, complementarity and dependence' (Grubb, 1987b). The key points about communities are that they are collections of species which occur together in some common environment or habitat and that the organisms making up the community are somehow integrated or interact as a society.

Of course, communities are not constructed only of plants. Some communities are mainly animals, like the association of fish and invertebrates which make up coral reef communities. Most communities are composed of a mixture of plants, animals, fungi, prokaryotes and protoctists.

14.1.2 Recognition of communities

Communities are usually recognised in two ways. One way a community can be identified is from the form of the environment or habitat in which it occurs. Some communities take their names from physical features; this is the case for rock pools, lakes and sand dunes. Other communities are recognised by the dominant species in the association. These are usually the largest or most abundant plant species present, hence deciduous oak wood, cypress swamp, grassland and sphagnum bog communities. There is no fixed size for a community. They can range from the very small and constrained, like the association of microorganisms found in the mammalian gut, through to huge and variable expanses of grassland or forest.

But why are the same associations of species found again and again in similar habitats (Wilson *et al.*, 1996)? This could be explained if the habitat is a collection of niches. If the same niches occur in similar habitats, then they will be filled with the same species. These could include niches for particular plants, grazing animals, parasites, predators, decomposers and so on. So the community is built up of species with other dependent species in recognisable combinations. The communities found in a particular habitat type will not be identical. Minor differences in the environment, or the effects of chance or history of the site, may mean that some of the species usually found together are missing and other more unusual species are present. The species present in a community will vary depending on where in the world that community is to be found. Hence species growing in pools on rocky shores in Australia will be very different from the species in similar pools in Britain. Even on one coastline the species may be different at either end if there is a cline in environmental conditions like sea temperature or changes in rock type. However, wherever the rock pools are to be found, they will contain similar relationships between species occupying the same collection of niches. If you want to think further about why communities may be predictable like this, and whether this is always the case, read Box 14.

14.2 The structure of communities

14.2.1 The investigation of communities

Communities are almost like living organisms. Within any community is a whole series of complex interactions between individuals of different species. The whole collection of populations then fit together into a working unit. Often a glance at a community does not reveal much about the complexity of the system. More time and observation is needed to study communities.

To understand this complex mesh of interactions, an ecologist has to unravel the associations and events which are important in the community.

Questions like 'Why is the community structured like that?' or 'How does that species fit in?' require an understanding of a whole series of reactions and processes. The easiest way to start an investigation of a community is not to ask 'why?' or 'how?' but to ask 'what?: What is the community structure?'.

First, the abiotic features where the community lives can be noted. For example, the water situation: whether the physical environment is marine, freshwater, marsh, drained soil or desert will be important. So will the mineral nutrients, geology and topography. Initially, this maybe a brief study, but as knowledge of a community develops, the role of the abiotic factors may become more important and require more detailed investigation.

Second, the overall form of the community can be described. In terrestrial communities this is usually determined by the vegetation. Plants can have various growth forms, including:

Trees which create forest or woodland communities;
Bushes which dominate scrub;
Herbs and grasses which are the major types in grasslands like savannah or prairie; and
Wet mosses which build up bogs.

In aquatic habitats, this physical structure is not so obvious except where calcareous reefs occur. Aquatic organisms can often be described by their mode of locomotion, like free swimmers, planktonic drifters, bottom dwellers and sessile adults, or by their position in the food web (see Section 11.7).

After these abiotic and structural features of the community have been recorded, the remaining questions which can be asked are concerned with the species in the community. 'What species are present?' 'How many species live in the community?' 'Is it diverse or species poor?' Further investigation will show which species are dominant. A species may be dominant because of the size of its individuals, such as trees, or because it is abundant. Dominant species may have a strong biotic effect on the environment of all the other species.

Once this mainly descriptive study of the community has been made, a more functional approach can be taken. The trophic food webs within the community can be investigated to assess the importance of primary producers, herbivores, predators, and decomposers. This will provide data on energy and nutrient cycles. At this point, the relevance of the physical environment of soil structure and nutrient availability, mineral input from erosion, toxins, and so on may require re-investigation. The way organisms regenerate and how this affects population dynamics and the stability of the community will also need investigating.

The overall structure of the community will be determined by a combination of several features such as the physical environment, community size and longevity of species present. The community may be stable or unstable, with high or low primary productivity, and may change seasonally or even daily.

The rest of Section 14.2 consists of more detailed accounts of three very different communities to give you some indication of the variety of associations which can occur between organisms in different environments. The three community types are:

Deciduous oak woodlands – high primary productivity and a stable structure. The European and especially British, community is described;
Marine rock pools – low primary productivity in a highly variable rapidly changing environment. The species given are British examples;
Mammalian gut – a very severe and restricted but relatively constant environment. Several examples of this mobile community are given.

14.2.2 Oak woodland communities

Deciduous oak woodlands occur in temperate climates in the northern hemisphere. They experience a high degree of seasonality from cold, subzero winters with frost and snow, to warm mild summers which are often wet, but which may include drought for several weeks at a time. Trees are the dominant life form and the most abundant species are the oaks.

In central and western Europe there are two native species of oak which dominate the deciduous mixed woodland. The pedunculate oak (*Quercus robur*) is deep rooted and most common on heavy wet soils. It is a lowland tree and dominates in southern England below 300 m altitude. The other oak, the sessile oak (*Quercus petraea*) is more abundant where soils are lighter, drier and more acid in southern England and in the higher, shallow soils of the northern hills.

The vegetation structure of an oak woodland takes the form of a canopy of tall trees, about 12–20 m high, with a lower layer or understorey of bushes and a herb ground flora. Associated with this is a diversity of woodland birds and small mammals and a variety of insects. The community has a fairly high diversity (but see Section 14.4.1).

Oak woodland is the major natural vegetation type throughout most of England and is common in Wales, western Ireland and southern Scotland. The community has probably existed in Britain for about 7500 years. At the present time very little of this natural vegetation still exists as much of Britain has been cleared for extensive agriculture or grazing land. Seven thousand years ago most of Britain and north-western Europe was covered with mixed deciduous woodland. Different tree species dominated in different areas including oak, limes (*Tilia* spp.) and beech (*Fagus sylvatica*) but most woodland consisted of a mixture of many tree and shrub species. Most of this mixed, broad-leaved deciduous forest has been destroyed by a succession of human cultures from Neolithic times onwards.

Although oak trees predominate in oak woodland communities in Britain, they occur with a number of other tree species. These include ash (*Fraxinus excelsior*), maple (*Acer campestre*), hornbeam (*Carpinus betulus*) and elms (*Ulmus* spp.), together with some lime (*Tilia cordata*) which is most common where oak woodland grades into the other main southern English woodland community of deciduous lime woodland. There are also a number of shorter trees, which do not usually grow tall enough to reach the full canopy, including holly (*Ilex aquefolium*) shown in Figure 6.3, and the rowan (*Sorbus aucuparia*). The shrub layer under this tree canopy includes hazel (*Corylus avellana*), roses (*Rosa* spp.) and the woodland or midland hawthorn (*Crataegus laevigata*). Figure 14.1 shows the tree structure of the woodland with tall trees in the canopy and a series of shorter trees and bushes underneath them.

The ground layer is very rich in species and consists of a flora which changes in dominance throughout the spring and summer. It includes bluebells (*Scilla non-scripta*), wood anemones (*Anemone nemerosa*), dog's mercury (*Mercurialis perennis*) shown in Figure 9.6, brambles (*Rubus fruticosa* agg.) and bracken (*Pteridium aquilinum*) shown in Figure 3.1. In the more northern woods, other species tend to be included in the ground flora and many southern ones are not found. Bryophytes are also a common element of the ground flora. Detailed lists of plants found in oak woods are given by Tansley (1949). The complex layered structure of the vegetation can be seen clearly in Figure 14.1. With such a large mass of autotrophs, the community is able to fix a large amount of energy during spring, summer and early autumn; so the oak community has a high primary productivity (see Section 12.2).

In association with the plants are a large number of species of small woodland birds such as tits (Paridae), warblers (Sylviidae) and finches (Fringillidae) and other larger woodland birds such as woodpeckers (Picidae) and the jay (Corviidae, *Garrulus glandarius*). Small mammals in such woods include bank vole (*Clethryonomys glareolus*), woodmice

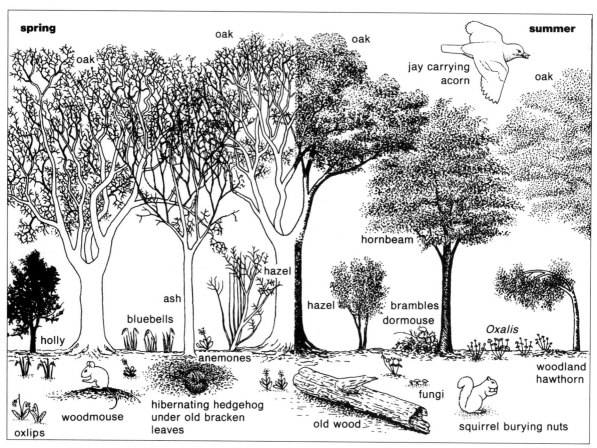

Figure 14.1 The structure of oak woodland in spring, before the leaves have grown on the trees, and later in the year, in summer, when the woodland is in full leaf. The diagram shows the change in ground flora from the 'spring window' to the more shade-tolerant summer plants, and shows some of the associated animals which live in the community.

(*Apodemus sylvaticus*), shrews (*Sorex* spp.) and hedgehogs (*Erinaceus europaeus*). All these are ground-dwelling mammals which live under the cover of the ground flora. Bank voles are almost completely herbivorous, eating a variety of leaves, fungi, moss, roots, fruits and seeds. Shrews and hedgehogs eat mainly invertebrates, woodmice tend to switch from arthropods, buds and seedlings in spring to nuts and fruits in autumn and winter. Other mammals make use of the canopy of the woodland. The dormouse (*Muscardinus avellanarius*) clambers round the understorey of scrub collecting hazelnuts, fruits and berries. The most recent member of the community is the grey squirrel (*Sciurus carolinensis*) which was introduced from America in the nineteenth century. It is very agile and leaps around in the trees where it eats seeds, flowers, buds and leaves, especially from oaks and hazels (Flowerdew, 1977; Corbet, 1977; Tittensor, 1977b). Large numbers of insects are also associated with this community. They include pollinators, leaf-eating caterpillars and gall-producing wasps.

The simplest way to investigate the functioning of the community is to follow the changes which occur in the wood throughout the year. Figure 14.1 may also help you to visualise the structure and changes in the community.

In late autumn all the leaves fall off the trees and shrubs, except for a few evergreen species like holly. The ground flora dies down and survives under the ground as bulbs, roots, rhizomes or tubers. Many of the mammals hibernate through this cold period. Some of the birds, including the warblers, migrate south in winter and the insects often overwinter as a dormant stage such as eggs, larvae or pupae. Because the trees have no leaves, considerably more light can reach the floor of the woodland at this time of year. This extra light is of no use in winter, because it is too cold for active growth. In spring, however, the temperature rises before the trees come into leaf round about April. During this 'spring window' many ground flora species grow and flower in profusion. They include dog's mercury, anemones, bluebells, violets (*Viola* spp., see Figure 6.4), primroses (*Primula vulgaris*) and, in southern woods, the oxlip (*P. elatior*, see Figure 9.6). In many of these species the leaves start to grow in March, or even earlier, and start to die back in June after the trees start to develop leaves.

During the spring and summer the birds and mammals nest and rear their young. The plants flower and begin to set seed. The large nuts of the hazel and oak trees and the berries of rowan, hawthorn, rose and bramble provide food for many of the woodland animals. The dormouse concentrates on collecting hazelnuts and blackberries, while the grey squirrel strips the young hazelnuts from the

bushes and may be responsible for the failure of this species to regenerate in many British oak woods (Rackham, 1986). Both grey squirrels and jays collect acorns for their winter stores to which they return for food during the winter and early spring. The grey squirrel often chews the embryo out of the nut before burying it, so that even if it misses a nut store, the acorns are unlikely to germinate and grow into seedlings. However, jays hide acorns undamaged, so that some acorns have the chance to grow if left unrecovered. The transport of acorns by jays may be very important in the regeneration of oak trees. Acorns can germinate in the dark, but oak seedlings require a light environment to grow past the seedling stage. A mature tree produces deeper shade than its seedlings can regenerate in, so new oak trees cannot grow under the canopy of a mature oak wood. Acorns are heavy, so they tend to fall straight down from the tree. Thus any remaining on the ground after mice, squirrels, voles, dormice and jays have got them will not reach adulthood unless a large gap in the vegetation opens up due to a tree fall close by. A jay carrying an acorn into a clearing may be the best chance an oak has of leaving progeny.

To maintain the population of adult oaks does not require many acorns to grow into trees. Many oaks live for 200 to 400 years. Only one of the thousands of acorns a tree produces during its lifetime has to survive to maturity for the population to be stable. By producing so many acorns, oaks provide a food source for many of the community animals, including jays, squirrels, voles and even some of the insects such as knopper gall wasps (Cynipidae), a recent invader from the continent, and weevils (Curculioniidae) whose larvae develop inside the growing acorn.

Oaks have to suffer the attacks of many insect species, including several other cynipids which create galls on twigs and leaves or in buds. The other most notable insect pest is the oak leaf roller moth (*Tortrix viridana*). The adult moth lays eggs on the twigs of oaks. In spring the eggs hatch and tiny caterpillars creep up to the oak buds where they crawl between the bud scales as these open up. They are then in an ideal position to begin eating the new protein-rich leaves, which are low in protective tannins. In some years the moths can totally defoliate an oak tree. This does not kill the tree, but it does mean that the tree has to grow a new set of leaves which requires an extra input of energy. Extensive defoliation like this means the tree gets a late start to its productive photosynthesis. This probably results in a lower production of acorns that year, but bearing in mind the number of acorns which are eaten anyway, this may not seriously damage the tree's chances of leaving progeny.

The best chance an oak has of avoiding defoliation

is having its buds still tightly closed when the caterpillars hatch. If the caterpillars cannot get between the bud scales into the edible bud, they will either die of starvation or be forced to migrate to another tree. The caterpillars are very small and by spinning a silken thread, they can drop off the oak twigs and float to other trees and bushes. The oak does open its buds later than most of the woodland species, and this may be an evolutionary response to such caterpillar activity. However, the later the buds open, the more photosynthetic opportunity is lost. Trees which put out their leaves earlier than the oak will thus have more time to develop their leaves fully and invade the oak's leaf space. The oak has to balance the costs and benefits of bud opening times to maximise its opportunities in the community.

The oak woodland communities in the south-east of England are tiny remnants of the original post-glacial woods that once grew in the area and even the surviving relics have long been influenced by human activity. Research has shown that such areas of wood have been managed for hundreds of years. Many tree and shrub species including oak, ash and hazel are not killed if they are felled. Instead they sprout up from their stumps in clusters of thin straight shoots. These can then be cut off every few years and a new crop will grow. These regrowing cut stumps are called **coppice** and the coppice poles obtained by this method can be used for a variety of purposes such as fence-making, hurdles and firewood. Old wooden timber houses were often built with coppice twigs woven into walls (wattle) and plastered with various mixtures which would harden to form a weatherproof coating (daub).

The art of coppice and hurdle making was prac-tised on the Somerset levels as long as 5500 years ago (Coles & Coles, 1986, see Figure 15.14). There is much evidence that whole woodlands have been managed for coppice from at least Anglo-Saxon times (pre-AD 1066). The good tall straight trees, often oaks, were left uncut as tall standards which could later be felled when needed for timber. The huge trunks of these standards were used for the timber frames of houses or ships. Coppicing opens up the woodland allowing much more light on to the ground for a year or two until the coppice poles grow up. This has a startling effect in the second spring after coppicing. The whole ground in the light area becomes covered in plants such as oxlips, wood anemones and violets which flower in profusion. Much of the evidence for the history of these ancient managed woods in England was discovered by Oliver Rackham (1980). His popular book on the British countryside (1986) is well worth reading.

14.2.3 Marine rock pools

Marine rock pools can occur on any rocky coastline and are found in both temperate and tropical climates. The shoreline is an extremely variable place as it is the area between extreme high and low tides. The tides advance and retreat in a cycle lasting on average 12 hours 26 minutes, so high and low tide slowly advance around the clock. This means that not only is the degree of exposure changing on the shore as the tides move in and out, but the time at which the exposure occurs also changes. So sometimes areas will be exposed at night, when it is wet and cold, and sometimes during the heat of a midday sun. The kind of rocky shore where rock pools are found can be seen in Figure 14.2. In this picture the tide is

Figure 14.2 Rock pools on a rocky coastline at Wembury, Devon.

at its lowest and most of the shore is exposed to the air, except where the rock pools hold some water. You can see what a bare and desolate place it looks, but it is a fascinating habitat with hundreds of species, each with its own pattern of distribution on the shore.

When the tide goes out, the sea level falls and the shore becomes exposed. Seawater is trapped and held in the rock pools until the next incoming tide covers the shore. If the pools are exposed in the daytime, the water will be warmed up by the sun. Water will evaporate making the pool more saline (salty). If it is raining heavily, the pool may become less saline, although usually the less dense rainwater floats in a layer on top of the salty pool water, so that only the top of the pool is affected.

Vascular plants (pteridophytes, gymnosperms and angiosperms) and mammals are totally absent. The primary fixers of solar energy are photosynthetic multicellular algae or seaweeds. Most of the animals are invertebrates, with the exception of a few fish and the occasional visit from a predatory bird such as the oystercatcher (*Haematopus ostralegus*) or purple sandpiper (*Calidris maritima*). No organism on the rocky shore is large and obvious like the oak tree. This is because the energy of the waves crashing on the rocks would probably break up a large rigid organism and wash it away. As no organism dominates in size and few are very abundant in the pools, the distinctive nature of the habitat is more obvious than the organisms in it. Thus the rock pool community is named after the environment rather than any of the organisms present.

Rock pools differ greatly from one another. Some of the variation is caused by their position on the shore. Those high up the shore will be exposed to the air for long periods while the tide is out, while some very low down on the beach will only be exposed at the lowest tides of the year and then for only a short time. The size and depth of a pool is also important. If it is small and shallow it will be affected to a greater extent by daytime temperatures and rainfall than a large deep pool. If a pool contains a lot of seaweed, then this will provide shade and shelter for other organisms. A pool with a large amount of seaweed will differ in other respects too. The seaweed photosynthesises during the day taking up carbon dioxide and releasing oxygen into the water. This high oxygen level will benefit some pool dwellers. In turn, many of the animals release nitrogen-rich waste products which are used by the algae. Hence the pool community cycles oxygen and mineral nutrients which benefits both the animals and the algae.

During the night respiration of the whole community tends to lower the oxygen level, sometimes to the point where it is in short supply. The seaweeds cannot photosynthesise in the dark, so they do not take in carbon dioxide. Because they are not using up the carbon dioxide it increases in concentration and the pool water becomes acidic. This switch in pH, from alkaline during the day to acidic at night, is another changeable aspect of pool life which is much less variable in the open seas. Most organisms of the open seas cannot survive the changes which occur in pools every day; the pool-dwelling organism is a very tolerant one.

In winter, violent storms may occur which can destroy some of the algae and sweep organisms out of the pool. In summer new algae may establish themselves. The pools have a very variable fauna: which species occur in any one pool, and how abundant each one is, seems to be largely a matter of chance. Summer is also a time when predators, such as crabs and prawns, invade the higher shore with its pools. A cross-section through a typical pool, showing the mixed community present, is given in diagrammatic form in Figure 14.3.

The species of algae which occur vary depending on where on the shore the rock pool is situated. Seaweeds tend to survive a little higher up the shore when growing in a pool than they do when attached directly to exposed rocks. For example, in Britain, the brown algae, the serrated wrack (*Fucus serratus*) occurs in some mid-shore pools, although the species is normally found only on the lower shore. Other species, like the bladderwrack (*Fucus vesiculosus*) do not seem able to occupy pools at all unless herbivores and other algae are absent. Still others, like the channel wrack (*Pelvetia canaliculata*) are absent because they cannot withstand being permanently submerged, although it is not known why this is (Norton, 1985). Other algal colonisers of rock pools include the red algae *Corallina officinalis* and *Chondrus crispus* and the green alga *Enteromorpha intestinalis*.

Many of the animal species found in rock pools are also members of the less sheltered, open rocky shore community. These include gastropod molluscs, like limpets, and shore snails (winkles) in the genus *Littorina*. Arthropods such as the green shore crab (*Carcinus maenas*) also roam the whole shore including the pools. Species which are normally only found in the sheltered rock-pool community include a wide variety of animals. Among them are gastropod molluscs, starfish (Asteroidea) and brittle stars (Ophiuroidea), sea anemones (Anthozoa) and some solitary corals, bristle worms (Polychaeta), shore-lice (Isopoda), sandhoppers (Amphipoda), shrimps (Decapoda) and some fish. The typical pool community consists of a variety of species of these groups, with no species dominant. Each pool has a selection of species which must to some extent depend on pool size, depth, position on the shore, the proximity of other similar pools and chance.

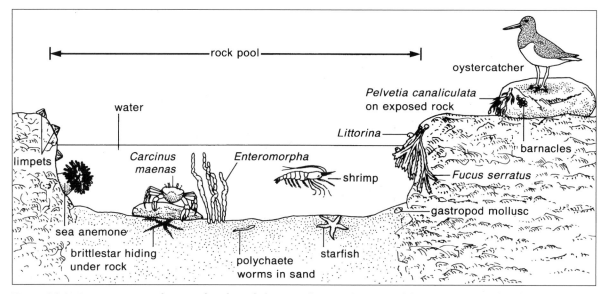

Figure 14.3 A cross-section of a typical rock pool showing the mixed nature of the community and some of the other organisms which inhabit the more open, rocky shore.

Reproduction in such a variable and turbulent environment, with the sea crashing in and retreating twice a day, is obviously going to be a hazardous procedure. So is the establishment of the young so produced. As rock pools are distributed across the whole of the rocky shore, it might be expected that pool organisms would reproduce by releasing large numbers of young into the coastal waters so that a few might find their way into some new pools. This is the strategy which many weedy plant species use when distributing numerous small seeds on land (see Box 5, p. 49). Many marine organisms, such as corals and sea urchins on reefs, and jellyfish (Scyphozoa), use the sea to disperse thousands of eggs and sperm. Fertilisation takes place in the open sea and millions of larval young are carried off suspended in the ocean currents.

In fact this widespread release of young does not seem to occur very often in rock pool communities. One factor may be that the species are rather small. This may limit the amount of energy they can put into reproduction, so that only a small number of eggs are produced. It may also be that the organisms expend energy adapting to the difficult environment they are in and have less energy for reproduction. If larval numbers are low, then releasing them into the dangers of the open sea will substantially reduce the chance of any surviving to colonise a new rock pool. Many species, therefore, care for their young. This is wise as predatory fish, prawns and polychaetes abound in these communities. Although young algae seem to be well dispersed and are quick to colonise pools, establishment of new animals into pool communities seems to be a rare event (Emson, 1985).

If dispersal from the parental pool is unusual and

larvae only travel short distances to new pools, then many of the new recruits to a population in a pool may be related. This will result in a high degree of inbreeding (see Section 6.3.3). The amount of inbreeding in rock pool species is unknown and requires further study. However, inbreeding could be a benefit in this situation. It has been mentioned that pools differ from one another in many ways, including temperature, salinity and pH range. An organism surviving well in one of these pools will be suited to that environment, but may not be so well able to survive in a different pool elsewhere on the shore. Producing a small number of well cared for, well adapted offspring which will not travel far beyond their own pool may be the best strategy in this situation.

14.2.4 Mammalian gut communities

The mammalian gut is a long tube which stretches from the mouth of a mammal, where food is ingested, to the anus where the waste products of digestion are egested. In between the mouth and anus, the gut can be divided into various regions where food is broken down by enzyme activity and other regions where food is absorbed by the organism. These regions are summarised in the diagram in Figure 14.4.

The gut environment is obviously an unusual one; constant in temperature at the blood heat of the mammal, but severe in terms of the anaerobic conditions, varying pH and the presence of lysozymes. The gut communities are also unusual in that there are no autotrophs or holozoic feeders. All the organisms present either break down food eaten by the mammal, or absorb the breakdown products

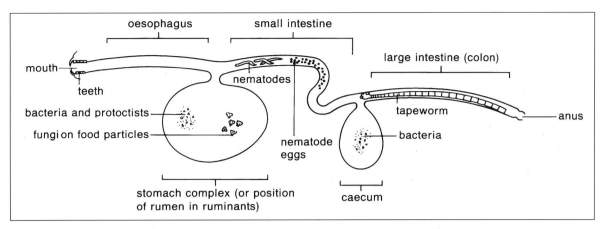

Figure 14.4 The general structure of the mammalian gut. The different sections of the gut, with their functions, are labelled, and organisms from the communities present are shown in diagrammatic form (but not to scale).

of the mammal's own digestive system. The main variability in the environment is the supply and type of food being eaten by the mammal and, in some systems, the problem of the host organism hibernating and therefore not eating for weeks on end. Little is known about what happens to gut communities under such circumstances.

The organisms present in the mammalian gut communities fall into two main categories: symbionts and parasites. The symbionts include bacteria and protoctists. These are present in the guts of all mammals, but are especially important in mammals which are herbivores and eat a large amount of cellulose in their diet. In **ruminants**, such as cattle, sheep and deer, huge populations of bacteria and protoctists can be found in an enlarged pouch in the foregut called the **rumen** (see Figure 14.4). An adult ruminant has many billions of bacteria and millions of protoctists in its rumen. Over 100 protoctist species and about 1000 strains of bacteria are known to occur in rumen communities, but any one community will have only a selection of this variety.

The symbiotic bacteria and protoctists are important to the mammal because they produce the enzyme cellulase which breaks down cellulose. Ruminants cannot produce their own cellulase so without the symbionts they are unable to digest the cellulose plant cell walls in their food. For leaf eaters, cellulose is such a high proportion of the plant matter they ingest that without cellulase they cannot obtain enough energy from their food to support themselves. A ruminant which is fed material that harms the organisms in the rumen so that it cannot break down its food quickly loses its appetite, because its rumen remains full of cellulose, and eventually dies.

The rumen organisms are provided with an anaerobic, slightly acidic environment, which they require, and a constant supply of food. The bacteria also seem able to use nitrogen which has been secreted or ingested into the rumen in the form of urea. They use

this to synthesise proteins. The relationship is, perhaps, more complex than the term symbionts suggests as many members of the community are swept, with the digested food, into the small intestine to be digested and absorbed by the mammal, which obtains a source of essential amino acids in this way (Moen, 1973; Orpin & Anderson, 1988).

It is only recently that another group of organisms, fungi, have been recognised in the symbiotic gut community. The uniflagellate spores of these fungi had been noted but were mistaken for protoctists. These mobile spores move around the rumen until they come into contact with a piece of swallowed vegetable matter. They then grow long projections, called rhizoids, into the food. The rhizoids produce enzymes which digest the cell walls of the food and it disintegrates into pulp. Eventually a sporangium develops and more flagellate spores are released. Spore production appears to be triggered (in as little as 18 minutes!) when a fresh meal of food is eaten by the host mammal. Such fungi appear to be widespread and have now been found in red deer, reindeer, sheep, cattle, kangaroos and elephants (Heath, 1988).

Most leaf eaters have an enlarged area of the gut, like the rumen in ruminants, where a community of bacteria can aid them in cellulose breakdown. Many leaf-eating monkeys have enlarged stomach areas which function like the rumen. Tree sloths also have huge complex stomachs as do some marsupials, including the tree kangaroos. In other species the stomach seems to be mainly a store for the food, and the communities of symbiotic bacteria inhabit a huge elongated hind-gut (Bauchop, 1978). In the rabbit (*Oryctolagus cuniculus*), for example, the caecum is greatly enlarged to about half the total gut. As the main gut community is situated after the major digestive and absorptive areas, the rabbit cannot readily utilise the bodies of gut organisms as they are swept along the hind-gut. The rabbit has overcome

Box 14
The environment and the concept of community

The word 'community' has been used by ecologists in many different ways. The Committee on Nomenclature of the Ecological Society of America defined the word in 1933 as 'a general term to designate sociological units of every degree from the simplest (as an unrooted mat of algae) to the most complex biocoenosis (as a multistoried rain forest)'. More recently, it has been described as 'a particular limited area of vegetation which seems homogenous . . . there is no marked, progressive change within it towards a different kind of vegetation' (Whittaker, 1967, p. 211).

Ecologists differ in their opinions of how these definitions can be interpreted when confronted with real communities. Some ecologists, described as phytosociologists, think that plant communities can be identified and classified as discreet units just like species and genera. Braun-Blanquet, one of the major early exponents of **phytosociology**, saw the community as a social unit. He stated that 'every natural aggregation of plants is the product of definite conditions, present and past, and can exist only when these conditions are filled' (1932, p. vii). He went on to describe that the phytosociologist's aim is to 'catalogue and describe the plant communities of the earth, to discover their causal explanation, to study their development and geographic distribution, and to arrange them according to a natural system of classification'. The groupings of the communities in this classification (species into associations, and associations into alliances, orders and classes) are given Latin names, usually based on the generic names of some of the more obvious constituent species.

Other ecologists see the vegetation, not as a patchwork quilt of distinct classifiable communities, but as a continuously variable continuum. Vegetational study cannot, therefore, be by identification of a community type and its boundaries, but by mapping the alteration in vegetation across a wide area of environmental changes. This approach is called **gradient analysis**. Whittaker (1967) states 'it may be fair to say that gradient analysis has changed the conception of vegetation as much as research on the genetic basis of variation and evolution has changed the concept of plant species' (p. 207). Whittaker distinguishes two forms: direct gradient analysis where the changes in vegetation along a known environmental gradient are studied; and indirect gradient analysis where the differences in species composition at various sites are compared and arranged without prior knowledge of environmental differences.

But which approach is right – phytosociology or gradient analysis?

The answer may lie in a consideration of the environment. Obviously in some places the natural environment can change very suddenly. For example, abiotic change could occur where a cliff plunges into the sea, or along the edge of a lake or river; on one side of the line will lie a terrestrial community of grassland or wood, on the other an aquatic environment

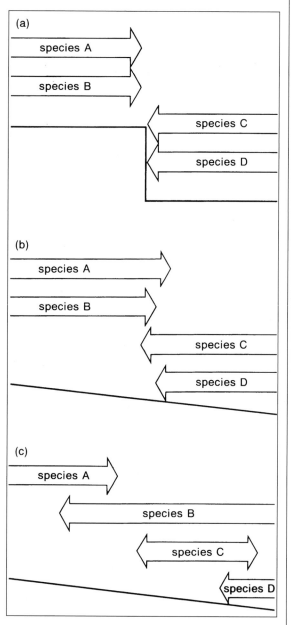

Figure 14.5 The distribution of species across abrupt and gradual changes in environment: (a) shows an abrupt environmental change such as a cliff edge where there is no overlap in species between the two distinct communities; (b) and (c) show cases where the environment changes gradually such as up a mountainside. In (b) the distribution of several species changes at about the same point on the gradient giving the impression of distinct communities; in (c) species distribution is not so similar and a continuum of variation is found.

will support marine, lacustrine or riverine communities. Figure 14.5a shows a diagrammatic representation of this: the environment changes suddenly, shown as a step, with species A and B occurring up to the edge of the step but not beyond it (they could represent terrestrial grassland species). Species C and D only occur in the alternative environment (they could represent aquatic species). Rapid changes in the biotic environment could also occur at the edge of a forest with the rapid drop in light intensity at ground level and increased litter build-up underneath the trees. So some boundaries between communities may be clear cut because of abrupt and considerable changes in environment.

But this may be the exception rather than the rule. In most cases the environment will change spatially only gradually with respect to parameters such as altitude, depth of soil, amount of waterlogging, length of snow cover in winter and so on. On these gradients of more gradual change, a point will occur where a particular species becomes less common, and, as one moves further along the gradient, the species will cease to occur. The sharpness of the changeover from the species occurring as part of the community to being absent will depend on how the species interacts with the environment. If several species have the same response to this particular gradient change in the environment, then several species will become less common or absent at the same place along the gradient and a distinct line between two obvious communities will be observed (as in Figure 14.5b). If the species in the area have differing responses along the environmental gradient, then a continuum of variation in the structure of the vegetation will be found (Figure 14.5c).

So both the spacial rate of change of the environment and the response of individuals in a species to the change will affect species distribution and therefore patterns of communities.

---

this problem by producing soft faeces which it eats. After passing for a second time through the gut, the faeces are discarded as small hard droppings.

The second major category of organisms in the mammalian gut communities are the parasites. Symbionts are often concentrated in the fore-gut where the largest quantities of undigested matter are found. Parasites, though, tend to be concentrated in the hind-gut in the large intestine and caecum. By the time the food reaches these areas it has been at least partially digested. It is therefore in a form which is relatively easy for the parasite to utilise. Gut parasites include various protoctists including amoeboids and flagellates; but although the parasites have to be small to fit into the gut, many are multicellular. These include a variety of small invertebrates such as nematodes, and rather larger tapeworms (cestodes). Many other parasites which do not normally live in the gut community pass through at least part of it on their way to other organs of the host. Such parasites often enter through the mouth but burrow through the gut wall to reach the bloodstream or other organs of the body.

Gut-dwelling parasites live in a severe but fairly constant environment. They do not have to hunt for food, or hide from predators. In fact the two main things the parasite has to do in the mammalian gut are to tolerate the environment and reproduce. The main structures and behaviour of the parasites are suited for just these two problems.

The environment a gut parasite faces is harsh because of the function of the gut as a place where organic matter is broken down into absorbable molecules. Any organism in the gut runs the risk of being digested, unless it can protect itself from the gut enzymes. Gut parasites also have to contend with the immune reactions of the host to parasite infection. Within mammal populations, some hosts seem to be genetically susceptible to parasite infection. Other hosts develop immune responses which may eliminate the parasites present in the gut. In many cases, parasite numbers remain low in the gut community, below the level of infection which would trigger immune responses from the host. This seems to be the case, for example, in mice (Wakelin, 1987).

Many of the smaller multicellular parasites are very mobile in the gut. The larger ones, such as tapeworms, are at risk of being ejected by gut and food movements. They tend to have special hooks and suckers which they use to attach themselves to the host gut wall.

The size of the parasite also influences its mode of reproduction. The whipworm (*Ascaris lumbricoides*) is a small nematode which parasitises humans. There are separate males and females and sexual reproduction takes place. The female lays large numbers of tough, knobbly-walled eggs which leave the gut community in the faeces. In the soil, a larva develops within each egg case. It can survive there for many years, until eaten with food. The larval stage temporarily rejoins the fore-gut community, where the eggshell dissolves. It then burrows through the small intestine wall and travels to the lungs. After further development it breaks through into the lung cavity, moves up the trachea (the air passage to the lungs) and back into the fore-gut down the oesophagus. Once in the hind-gut it matures into an adult and the cycle is complete.

Larger parasites, like the human tapeworm (*Taenia*), have a problem in that their population size in the gut is usually very small. Often only one tapeworm is present there. To rely on outcrossing of separate sexes is risky as individuals might end up alone in the gut or only with members of the same sex! The adult tapeworms are **hermaphrodite**. Each mature segment, or proglottid, of the tapeworm contains both male and female sex organs. Self-fertilisation occurs and the ripe proglottids, full of eggs, pass out of the intestines in the faeces. Infection of the secondary host, the pig, then occurs.

Many gut parasites have extremely complex life cycles, infecting one or more other species. With the human tapeworm, both host species are mammals, but in other parasites, hosts such as insects, gastropods and birds may be involved. Our understanding of how these complex methods of reproduction and dispersal evolved is poor as internal parasites have almost no chance of being fossilised. Some of the earliest *in situ* gut communities of which we know are only about 2–3000 years old. Corpses of Iron Age humans in Denmark (Tollund and Grauballe man) and England (Lindow man) have been found as subfossils in peat with their internal organs preserved. All three had parasitic nematode eggs in their guts. Lindow man had two genera of nematode eggs, including the whipworm *Ascaris* in his small intestine (Jones, 1986).

14.3 Global distribution of terrestrial communities

The importance of the environment in influencing the structure of communities has already been mentioned in Section 14.2. Climate is an environmental factor which produces global patterns in features such as temperature, rainfall and light. These patterns are linked to seasonality. The causes for the variation in seasonality from tropics to poles are explained in Section 9.2.4. The changes which occur in the climate with increasing latitude produce different conditions in communities depending on where they are situated in the world. These different climatic conditions favour different kinds of plants.

The hot, wet climate of the tropical lowlands supports a very diverse community with tall trees, shorter understorey trees, a large number of **climbers** and **epiphytes** and a scattered ground flora able to exist in the extreme shade cast by the taller vegetation. (If you look at Figure 14.6 you will see the tropical forest portrayed on the left-hand side; you can follow the changes seen in vegetation with latitude described in the next few paragraphs by moving across the diagram to the right.) If there is adequate rainfall throughout the year, the tropical climate has very little seasonality. This means plants can grow throughout the year, consequently the tropical lowland forests are evergreen, with huge leaves to take advantage of the light.

Seasonality increases with increasing latitude. The most obvious change is in the minimum temperatures, which drop from above 10 °C in lowland tropics down to below freezing point (0 °C) in temperate

regions. Low winter temperatures coincide with short days and low light intensities. Below about 4 °C, plants find it difficult to take in water, so plants in temperate communities may experience water shortages even though the soil is quite wet. Trees in leaf in temperate forests (such as the oak woodlands in Section 14.2.2) lose a lot of water by transpiration. This means that keeping the leaves during winter would result in too much transpiration leading to damaging water loss. So the leaves are dropped: lowland, temperate forests are deciduous. Because these leaves are dropped when the cold winter approaches, they never suffer from heavy frosts, so the leaves do not have to survive freezing temperatures. This means that expensive biochemical mechanisms to prevent frost damage (see Section 9.2.5) are not needed in the leaves which can be cheap and 'throw-away'.

At still higher latitudes, the winter is longer and colder. The forest communities in these areas, the **boreal forests**, are composed mainly of evergreen conifers (gymnosperms). These trees keep their leaves through the winter and usually for several years: they drop only a portion of their leaves at a time. There are some deciduous angiosperm trees in the boreal forests such as birch (*Betula*) and poplars (*Populus*) and a few deciduous conifers: larches (*Larix*).

The leaves of evergreens have to be frost tolerant and need to reduce their water loss in winter. To cut down evapotranspiration the leaves are needle-like and have thick cuticles; usually the stomata are in grooves too. Because of the antifreeze mechanisms and thick cuticles, they are expensive to make. So the plant does not throw them away the following spring, but keeps the needles as long as possible. They therefore have to be tough so as to last many years without disintegrating under heavy snowfall or in high winds. This also reduces damage by insects or fungi. Having such resistant leaves has problems: when eventually they are dropped, the needles do not break down easily. The boreal soils are very cold during the winter which also reduces the rate at which leaf litter is broken down. So the needle leaves tend to lie around on the surface of the soil where they lock up nutrients for years, making the soil underneath nutrient poor (see Section 15.2.2).

The fact that most of the dominant trees in the boreal forest are evergreen gymnosperms suggests that they have an advantage over the deciduous angiosperms in the harsh climate. There is no really adequate explanation for why this is so. If some angiosperm trees, such as poplars and birches, can survive in the boreal forests, why are so many temperate tree genera absent? Also the larches are deciduous gymnosperms, and they form the most northerly forests in Russia. Broad-leaved deciduous forests

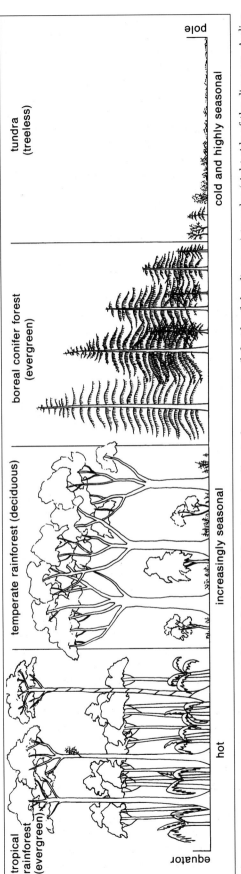

Figure 14.6 The vegetation types associated with the latitudinal variation in climate from equator (left side of the diagram) to poles (right side of the diagram. A slice through the four major wet lowland vegetation types associated with the tropical, temperate, boreal and tundra zones shows the general changes which occur in community structure with increasing latitude.

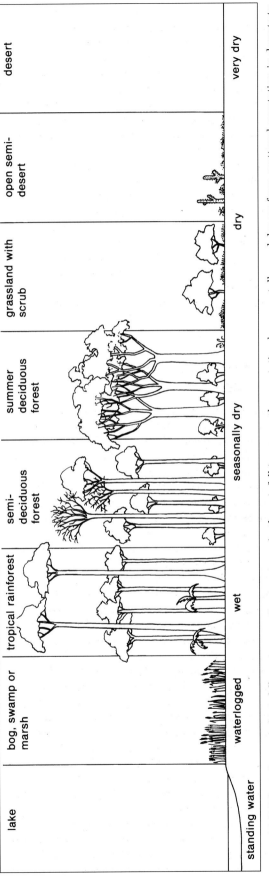

Figure 14.7 The influence of rainfall on community structure. As the rainfall increases the vegetation becomes taller and denser from scattered vegetation in deserts to tall forest. If rainfall causes waterlogged conditions specialised wetland vegetation occurs.

grew at much higher latitudes in the past (Spicer & Chapman, 1990).

As the poles are approached (the right of Figure 14.6), the growing season gets shorter and shorter. The height of the boreal trees declines, and communities eventually become treeless. Because most of the land masses are in the northern hemisphere this trend is best seen there. This treeless vegetation is called **tundra** and it consists of low growing bushes such as willow (*Salix*) and juniper (*Juniperus*), tiny flowering plants, mosses and lichens. Tundra is so cold that much of the soil is permanently frozen (**permafrost**) and only the top metre or so thaws during the brief summer. The shortness of the growing season, permafrost and the extremely harsh winters seem to prevent colonisation by trees. The vegetation is dwarfed so that it is covered with snow in winter. The snow protects the plants like a blanket during the very cold winter months.

Similar vegetation types to the latitudinal zones of evergreen, deciduous and boreal forests and tundra can be seen in a much smaller space from the base to the top of mountains. Mountains have similar zones of vegetation with forests on the lower slopes gradually dwindling to a **tree line** and **alpine** vegetation above. In latitudinal zones the seasonality changes affect the vegetation; in altitudinal zones the variation is in temperature alone. If the mountain is on the equator there will be no seasonality in day length; if it is at higher latitudes then there will be greater seasonality. Isolated mountains often have unusual communities because of coloniser effects and niche filling (see Section 18.3.5).

Another important parameter in the climate is rainfall. We can start again with the lowland, tropical, evergreen rainforest, but this time in Figure 14.7, we can follow the effects of rainfall in communities. We know that the tropical forest has high rainfall, so although evaporation and transpiration are high because of the temperature, there is plenty of water to sustain the vegetation.

Some areas in the tropics have more seasonality than this (moving to the right in Figure 14.7). Rainfall is concentrated into a wet season and much less falls at other times of the year. The wet season tends to be cooler, because there is cloud cover. Conversely, the dry season is very hot, because there are no clouds and the sun beats down. During the dry season the air is dry and any available water evaporates very quickly. As the leaves transpire the soil gets drier and drier. Even if the soil is quite wet, near a lake for example, the drying power of the air may cause the trees to lose water through their leaves faster than it can be supplied by the roots. In these circumstances the trees are in danger of drying out. The stress of water shortage in the leaves induces abscisic acid production which causes the

leaves to drop: so the trees are dry-season deciduous.

If the total yearly rainfall declines, then the large trees with big leaves seen in the tropical rainforest, and even large dry-season deciduous trees, will not be able to survive. The trees get smaller and more scattered until, in dry climates, the vegetation opens up into grassland communities with bushes. If the hot dry summers cause frequent fires then bushes also fail to grow (see Section 5.2.8). Even drier conditions tend to support more specialised plants such as cacti and succulents which maintain a store of water in their leaves or stems. These specialised, drought-tolerant species occur with low bushes (the vegetation is not dense enough for frequent fire) and grass in **semi-desert** communities. Succulents do not survive subzero temperatures very well, so in temperate latitudes, cold semi-desert vegetation is dominated by scattered bushes. If the climate is any drier it cannot support vegetation at all, except in the valleys where water collects, or where groundwater reaches the surface at oases. In such dry areas true **deserts** are found where plants survive as seeds for many years until freak rainfalls provide conditions for a burst of growth and flowering. Some deserts never even get these cloudbursts, and so lack even the seeds of plants.

As temperatures decline towards the poles, the amount of water needed for evapotranspiration decreases. The effects of rainfall on community structure become less important than the effects of the harsh winters, so the effects of variation in amounts of rain are most evident in low latitudes.

In places where rainwater collects, or where drainage is impeded by underlying rock or clay, the soil may become waterlogged (move to the left from the tropical forest vegetation in Figure 14.7). Many plants are unable to grow in such conditions, among them most trees and shrubs. Such waterlogged habitats include fens, marshes, swamps and bogs. If waterlogging increases and the topography allows standing water to accumulate, pools or lakes form. These are dominated by aquatic communities (see Section 15.3).

Other major environmental factors which influence the distribution of communities include **topography** and geology. Topography includes the effects of altitude which have already been discussed, and factors such as aspect, steepness of slope (which affects drainage) and so on (see Section 9.2.3 and Figure 9.5). The geology in turn affects topography. For example, soft chalk produces rolling countryside while harder rocks such as limestones outcrop as cliffs. Rock types also affect soils which, because of their complex interactions with vegetation, fauna and climate are discussed further in the next chapter (Section 15.2).

14.4 Patterns of diversity

14.4.1 Global diversity

As we have just seen, climate has an important influence on the distribution of species and thus the development of communities and community structure. If a transect of communities is taken from the equator to the North Pole, the **diversity**, or number of species in each community, is found to decrease from low to high latitude. This decline in species diversity is seen in plants and many groups of animals including insects, snakes, lizards, frogs, birds and mammals. Figure 14.8 shows this pattern of decline in species number for mammals. The graph records data running north up America from Mexico to Alaska. The degree of seasonality is plotted as the temperature range experienced from winter to summer against the number of mammals found in similar types of community along the west coast of America. The number of species of mammal declines sharply into the temperate region. The same shaped curve of decreasing diversity also occurs in birds (MacArthur, 1975).

So why do communities differ in species richness from equator to poles? Before considering possible answers to this question we must first look at the factors which are likely to influence the species richness or paucity of a single community. Then we can return in Section 14.4.3 to the cline of diversity of communities across the Earth.

14.4.2 Species richness in a community

For an individual community there are many factors which can affect the number of species present. Some of the influences on a community are external ones:

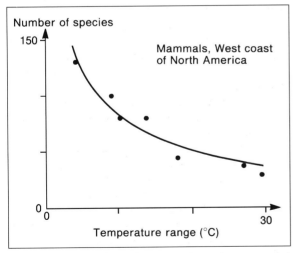

Figure 14.8 The decline in diversity with increasing latitude of mammal species within communities. Diversity increases as the variability of the climate (temperature variation) decreases. (From MacArthur, 1975.)

the **abiotic** effects of the habitat. Other influences will come from within the community: the individuals present will themselves affect the species diversity (Grubb, 1987; Pianka, 1988).

Taking the abiotic characters of the habitat first:

Size of habitat obviously affects the number of species which can live in it. For a rock pool, which is only a few metres square, only a limited number of species can maintain populations there and all must be small species. A huge habitat, like that which supports oak woodland, is large enough to maintain populations of many different species. This much is obvious, but consider the case of a single plant and how much surface area it has to support insects. Somewhat surprisingly, it has been argued that there is, in absolute terms, more habitat surface area available on plants if you are a small insect than if you are a big insect. This is because plant surfaces are fractals (Lawton, 1986). Mandelbrot (1967) introduced the idea of a fractal by asking 'How long is the coast of Britain?'. The answer depends on your measuring instrument: the finer your ruler, the longer the coastline. The mathematics of fractals are quite complicated but the basic idea is straightforward: the smaller you are, the more space there is available for you. Lawton uses this notion to suggest that this may be one reason why small species are so much more common than large ones – they can, as it were, carve out for themselves small niches that simply don't exist for larger species.

Spatial patchiness in a habitat allows organisms with different requirements to live together. A rocky pool with crevices and a sandy floor with small rocks will provide more microhabitats for species than does a completely smooth-walled, rock-bottomed pool. Statistical studies of mammalian species richness in North America show that although no single, all-important factor explains this richness, habitat heterogeneity and potential evapotranspiration are the most important factors (Kerr & Packer, 1997). Potential evapotranspiration is the amount of water that would evaporate from a saturated surface and is a good measure of the amount of primary productivity in a habitat (Section 12.2).

Harshness of habitat may affect the number of species able to survive in the community. The difficult environment of the mammalian gut excludes all plants and most animals from the community. Similarly, rock pools show so much daily change in their environment that animals of the open sea cannot survive there, and a rock pool which is frequently scoured by storms has very few species.

Predictability of change in the habitat. If the changes occurring in an environment are cyclic and therefore predictable, then different species will be able to use the habitat at different times in the cycle. Typical cyclic changes include seasonality in temperature or rainfall. In oak woods, for example, the trees lose their leaves each year because of low winter temperatures. Because of this, there is always a 'spring window' in which light-demanding species of the ground flora can grow and flower. As spring progresses, the woodland floor becomes darker as the tree leaves develop and shade-tolerant species take over the space.

Disturbance in a habitat. Occasional severe or frequent low-level disturbance affects the population size of species in the community. If population numbers are kept low by disturbance, then there will be more 'room' in the habitat for other populations. For example, in the rock pool community a heavy winter storm which removes algae from the rocks will provide space for other species to colonise the bare area. If there is too much disturbance, then the habitat becomes too harsh, and species diversity may decline.

Isolation of a habitat. The absence of similar habitats close by decreases the chance of some species reaching the habitat. The further away an area is from similar habitats, the fewer species are likely to colonise it. This is especially noticeable on islands a large distance from the mainland (see Section 18.3).

The remaining factors are biotic ones:

Age of community type. The length of time the community type has existed in evolutionary time may affect the diversity. The various species normally found in the community will be evolving in association together. One theory is that the longer the community type has existed, the more species will have had the opportunity to join the community. Niche volumes will have decreased as more species enter the community structure (see Section 10.2.1). The way in which community richness can increase with time has been shown by looking at tree species in Britain. John Birks (1980) showed that the longer a tree species had been in the British woodland communities, the more insect species were associated with it. The two oak species, which have been in Britain for about 9000 years, have 284 species of insect living on them. The hawthorns (in Britain for 7000 years) have 149 species of insect, while the hornbeam (*Carpinus betulus*) and maple (*Acer campestre*), both in Britain about 5000 years, have 28 and 26 species of insect respectively. Introduced species have even fewer insects: the sycamore (*Acer pseudoplatanus*), introduced 650 years ago, has 15 species of insect and the horse chestnut (*Aesculus hippocastanum*), in Britain for 400 years, has only 4 species of insect. The overall picture is not as simple as these data might suggest. Some species are better at rejecting grazers and the harmful effects of insects or have a scattered distribution which makes insect colonisation difficult. The rowan (*Sorbus aucuparia*),

see Figure 14.9, seems to have been in Britain about 12 000 years, but only has 28 insect species on it, and tough-leaved trees like holly (see Figure 6.3) and conifers also have fewer insect species.

Age of particular community. The time any one community has existed at a site affects its species richness. A recently developed community may contain only a few species and the number will increase as more species invade and establish. Rock pool communities have animals which are very slow to establish and disperse to other pools. So it may take many years for a full community to develop in a new pool, or one completely devastated by storms. The degree of isolation, mentioned earlier, will also alter colonisation time. The importance of the age of a single community is different from that of the age of the general community type. The diversity in the general community type often depends on the co-evolution of organisms, while the diversity of a particular community is primarily due to the immigration of and colonisation by existing species.

Primary productivity of a community may influence species richness. If primary productivity is high, then food for herbivores will be abundant. Many species of herbivore may be able to utilise the food

Figure 14.9 Rowan trees (*Sorbus aucuparia*) in flower, showing the large insect pollinated flowers.

without competition for a scarce resource. High herbivore numbers and diversity will influence the carnivores and so on along the food chains.

If *community structure* is complex, then the community will contain more niches. For instance the trees, bush and ground flora structure of oak woods provides nest sites for birds, a huge leaf and bark area for insects, support for climbers and ground cover for small mammals. Many of these niches are not available in grassland.

Competition between species in a community may alter species richness. If some species are highly competitive for resources such as space or food, others may be excluded. Sometimes a new species invades a community and becomes very abundant, taking the place of individuals of other species. The populations of these species dwindle and may disappear. The invasion of bracken into open oak woodland can have this effect (see Section 3.2.2). Competition is related to the degree of disturbance in a community. If disturbance prevents a competitive species from dominating a community, then species that might otherwise disappear may be able to survive. The harshness of the environment in some habitats alters the relative competitive ability of species, giving species which are well adapted to such environments the chance to survive there when they would be excluded from an easier one. The effects of competition in structuring communities is still not fully understood and there is much debate as to its importance (see Sections 9.3 and 9.4).

Some or all of these factors affect every community and influence species richness. As you can see, there are many possible causes of species richness or paucity. A single factor can, under different conditions, cause an increase or decrease in diversity. In some situations, for example harshness of environment or disturbance preventing invasion by one vigorous species, could increase the diversity, while in other situations too harsh an environment or too much disturbance could exclude most species. This provides quite a problem when trying to decide what is affecting even a single community.

14.4.3 Stability–diversity relationships

Are diverse communities – ones that have more species – more stable? Common sense may at first suggest this might be the case. After all, consider a community with only a handful of species. The food web will mostly consist of only simple food chains in which each predator has only a single prey species. The chance extinction of a single prey species would then be expected to lead to the extinction of its predator. This, in turn, will lead to the extinction of that predator's predator and so on. By way of contrast, in a community in which there are many species and the food webs are very complicated, we might predict that the chance extinction of a single species would have less extreme consequences. It might be expected to lead to changes in the population sizes of other species, but not to wholesale extinctions.

This common-sense view received considerable support from early theoretical studies carried out in the 1950s, and some data exist in support of it. For example, in grasslands in Minnesota, USA, species-rich plots change less and recover from disturbance more rapidly than species-poor plots (Tilman & Downing, 1994). In other words, the floristically more diverse communities are more stable. However, other field studies and more recent theoretical work have challenged this view (Johnson *et al.*, 1996). Today's view is that it is the nature of the interactions between the species, rather than simply the diversity of species, that determines the stability of a community. One interesting proposal is that, within a community, species can be likened to the rivets that hold an aeroplane together (Ehrlich & Ehrlich, 1981). You can safely lose a certain – and rather difficult to predict – number of rivets from an aeroplane. However, the loss of a few more leads to catastrophe. If this analogy is correct, the relationship between stability and diversity within a community may be something like that shown in Figure 14.10. Others have argued that a better analogy is to think of species as the passengers and crew in an aeroplane. Most of the people on a flight are passengers and you can remove as many of them as you like without endangering the aircraft. However, remove the crew, and the plane crashes – no matter how many passengers there are. In other words, a small number of key species may maintain community stability.

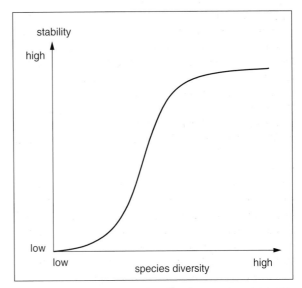

Figure 14.10 One suggestion for the way in which the stability of a community depends on the number of species in it.

Sadly, it is only very recently that experimental studies have begun to investigate the relationship between stability and diversity. For too long ecology has relied on conclusions drawn from natural variations or from theoretical models. What are needed are carefully controlled experiments. However, these are expensive, time consuming and difficult both to control and to replicate.

14.4.4 The global cline

Now that we have looked at the factors which determine species diversity within communities, we can return to the cline in species richness found in communities at different latitudes. Are some of these factors responsible for the global trend in species diversity? The most obvious explanation for the decline in diversity with increasing latitude is the change in community structure. The major community type changes from tall tropical forest, with two or more layers of trees, through shorter temperate woodland, to boreal forest, which has a poorly developed understorey and ground flora because of the deep shade, to tundra with no trees and few bushes. This results in a decline in the total number of niches available for species from equator to pole.

High diversity is associated with high primary productivity. This is so in temperate forests and probably also in tropical lowland forest, where high rainfall and soil fertility correlate with high numbers of tree and climber species (Adams, 1989). A hectare of tropical forest, like that shown in Figure 14.11, can have between 100 and 300 tree species. The argument breaks down, however, when aquatic habitats are considered. Many lakes and estuaries have high productivity, but they also have very low diversity. This is also the case in swamp forests (see Section 15.3.5). A small number of species occur at high densities. In terrestrial environments, primary productivity is probably indirectly linked to community structure. It is interesting that in aquatic environments, community structure is less complex than in terrestrial communities. The fact that the species diversity in such habitats is lower supports the idea that it is the complexity of structure of the community, not its primary productivity, which is important in diversity.

The apparent importance of community structure might explain why there are more climbers, birds or insects in the tropics, but it does not explain why, among the trees themselves, there are many more species in tropical forests than in temperate or boreal ones. There may, however, be an additional factor involved in tree diversity: pollination biology.

All the conifers in the boreal zone are wind pollinated. They rely on breezes to pick up pollen grains from male cones and transport them to little droplets

Figure 14.11 Aerial photograph of the Amazonian rainforest in Ecuador, showing the tremendous diversity of tree species.

on female cones which catch pollen and draw it inside the cone to the eggs where fertilisation occurs. The action of the wind is random: it cannot take pollen precisely to one tree. Instead the pollen blows about and chance dictates where it lands. This means that conifers are limited to living in groups of one or a few species. If many species were mixed up together, then almost all the pollen a female cone received would be from another species. Within temperate woods, many dominant species including the oaks, elms, hazel and ash are wind pollinated. The more scattered species such as rowan (shown in flower in Figure 14.9), holly and hawthorn are insect pollinated.

In the tropics most of the trees are pollinated by insects and some by birds, bats or other animals. Some flower all year round and have a very specific pollination relationship with a single species of pollen vector. Others use more general pollinators, but the whole population flowers all at once, in glorious profusion, so that the many animals which are attracted to the flowers are bound to visit another tree of the same species in flower and pass on the pollen. Because animal pollen vectors can be relied on to visit flowers directly, the randomness of using wind is eliminated. Trees do not have to grow in clumps of one species to ensure pollination, they can be distributed more sparsely in the forest. Hence very many more **entomophilous** (insect-pollinated) species can grow in an area than can **anemophilous** (wind-pollinated) ones: pollination by insects and other animals allows more species of tree to exist. Tree diversity may be dictated by the niche numbers associated with pollination mechanisms, but this does not, of course, explain why insect pollination is most prevalent in the tropics or why the boreal conifers have never co-evolved with insects to be entomophilous.

Two other explanations can be proposed for why the decline in species richness occurs. First, in the tropics more evolution could have taken place, so that more species have evolved by splitting existing niches into smaller and smaller units. A second explanation is that more extinctions could have occurred in high latitudes, cancelling out the effect of any evolution or migration into those regions.

It used to be thought that tropical rainforests were extremely stable and had been there for millions of years, really since angiosperms evolved to dominance, 65 or more million years ago. The vegetation of tropical forests is dominated by angiosperms and ferns. The angiosperms rely on animal pollinators with which they have evolved. This co-evolution was considered the result of the long-term stable environment which allowed a gradual build-up of complexity and diversification. Certainly, this insect–plant co-evolution must be a major reason for diversity in the tropical forest because, as mentioned above, it allows many plant species to live intermixed together, yet still be pollinated. The insects have provided the tool for splitting the reproduction part of the plant niche into many parts and in the process the insect niches have subdivided too.

However, the long-term stability of tropical forest has been called into doubt. First, on a minor scale, habitat disturbance is known to be continuous. Old trees crash down, carrying all their climbers and epiphytes with them, leaving a gap which is quickly colonised. Hurricanes may bring down larger patches of forest. Thus the forest is continually changing and this may help to maintain diversity as small, or larger, gaps appear and are filled by new individuals.

Disturbance on a much longer time-scale may have occurred to tropical forest during the glaciation of the ice ages. It has been suggested that much of the area which is now tropical forest in the Amazon basin and in Africa was grassland 13 000 years ago! The tropical forest may have diminished to a few isolated patches or refugia. After the last glaciation, these patches expanded and merged to form the present forest. The forest must have been dissected about 110 000 years ago at the onset of glaciation, so each of these refugia would have been isolated for about 100 000 years before it merged back into one forest. Populations in each refugia would be cut off from any genetic exchange with populations in other refugia. During the glacial phases, and there are

thought to have been at least 15 of them in the last 2 million years, genetic drift or natural selection could have altered populations in each site. When the refugia merged, populations could have been sufficiently different that they formed different species. Thus the degree of disturbance occurring in the tropics could have caused extra speciation to have occurred and thus increased diversity. Evidence to support the notion of refugia comes from the patchy way species are distributed in the Amazon (see, for example, the different forms of *Heliconius* butterfly in Figure 19.2, each possibly representing a refuge population).

Further evidence to support or disprove this theory is still required (Connor, 1986). Recently, though, a continuous pollen history of more than 40 000 years has been obtained from a lake in lowland Amazonian rainforest (Colinvaux *et al.*, 1996). The species present show that tropical rainforest occupied the area continuously and that grassland was not present. This is the first decent piece of 'hard' evidence against the refugia hypothesis but it only comes from one site, so most refugia supporters are not convinced it proves anything. Interestingly, the presence in this lake of pollen from species now restricted to cooler areas suggests that during the last glaciation this region was some 5 to 6 °C cooler than today. There does also seem to have been some decrease in rainfall, but not enough, Colinvaux argues, to dissect the forest with great swathes of grassland. The hunt is now on for more lake sites to analyse.

These glacial phases would have had much more catastrophic effects in temperate to polar regions. The advance of glaciers as far south as Norfolk in England and the Great Lakes in America wiped out all the communities in the region. The tundra, boreal and temperate zones advanced further south. Here is ample explanation for why high-latitude communties may be species poor compared to tropical ones. In Europe the diversity gradient of woody species at the beginning of the present warm phase, 13 000 years ago, was only half as steep as it is now: the difference in diversity with latitude across Europe increased as more woody species colonised in the south than in the northern part. These communities have only been stable for about 6000 years (Silvertown, 1985). Before this many species were found in different community structures which are not known today (see for example the 'English tundra' discussed in Box 5, p. 49). Also these species had to undergo long-distance migration when many extinctions may have occurred. For example, a number of large mammals went extinct during the glacial phases: although it is difficult to disentangle the effect of the rise to dominance of humans during this period on both the flora and fauna of these areas.

So the same global occurrence of a series of glacial

FIFTEEN

Ecosystems

15.1 The first use of ecosystem

The word **ecosystem** is a relatively recent term. Sir Arthur Tansley invented the word in 1935 to apply to a whole community of organisms and its environment as one unit. Tansley realised that the community could not be separated from the particular environment in which it lived. The physical features of the habitat plus the climatic influences determine which species form the basic structure of the community. For many years, ecologists had realised that a habitat, and the community it contains, are a single working system. They applied many terms to this concept, but apart from ecosystem, only one earlier word, **biocoenose** (Möbius, 1877), is frequently used, especially in Europe and Russia.

The ecosystem, or biocoenose, consists of the community of organisms plus the associated physical environment. The main features of the abiotic environment (as discussed in Section 9.2) are climate, soil and water status (land, freshwater aquatic or marine); other features include geology, topography and depth below sea level, or altitude above it. We have already looked at some of the organisms which occur in ecosystems: both **autecological** studies of species (see Chapter 3) and species in communities (Chapter 14). The study of ecosystems is included in **synecology** as all the organisms and environmental factors are studied as an integrated unit.

Many features of the ecosystem are difficult to separate out and look at in isolation from the communities they support. This is the case, for instance, with the soils of terrestrial ecosystems and the water relationships within ecosystems. Some of the importance of a marine environment has already been described in the rock-pool ecosystem in Section 14.2.3. In this chapter the importance of water in freshwater ecosystems, including bogs and lakes, is described. The relationships of soils to vegetation and climate is also investigated, including soil classification and distribution on a global scale.

15.2 Soils

15.2.1 The structure of soils

All soils are made of the same major components: mineral particles, of various sizes and chemical structures, and organic matter in various stages of decomposition. The different proportions of these components, and the way in which they are distributed in soil, create different soil types. Soils are an integrated and essential part of terrestrial ecosystems. The soil affects which types of plant community grow on it, and the plants, in turn, affect the soil.

The mineral structure of the soil comes from the weathered or eroded rock which is deposited as sand or finer-grained silts and clays (Heimsath *et al.*, 1997). All soils have a mixture of particles of different sizes grading from large sand grains through smaller silts to small clay particles. The simplest classification of soil types is based on the relative proportions of these particles of different size, as summarised in Figure 15.1. The particle size and composition of soil in an area depends on the source material: the rock types available for weathering (see Section 9.2.2), the presence of **glacial tills** left by melting glaciers and wind-deposited particles.

Sand is composed of large quartz grains which are very hard and are often worn into rounded shapes with smooth surfaces. The smaller clay particles have a layered structure, like the pages of a book. Mineral ions can be held around the outer edges of the layers in each clay particle. From the clay particles nutrients enter the water in the soil and diffuse, or are drawn in moving water, towards the root surfaces of plants.

In a developed soil the sand, silt and clay particles are mixed with input from the biotic environment. Most of this is organic plant matter, such as leaves and wood. The plant matter is broken down by fungi and bacteria, or chewed up by insects which eject most of it as faecal pellets (sometimes called **frass**). Various other animal products, including calcareous snail shells and the exoskeletons of arthropods. also

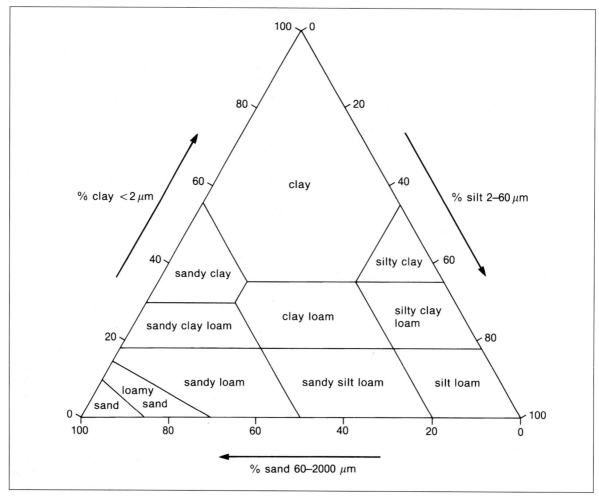

Figure 15.1 The simplest classification of soil texture based on three sizes of particle. Each point of the triangle represents a soil made only from one particle size: within the triangle are soils composed of a mixture of particle sizes.

add to the soil organic matter, but to a lesser extent than plant debris. The amount of organic matter content affects the water retention of the soil. A peaty soil, high in organic matter, holds considerably more water than does a mostly inorganic soil.

Water content is obviously very important, as plants with their roots in the soil have to extract water for requirements such as transpiration, the nutrients dissolved in it, maintenance of cell contents, turgidity and photosynthesis. This means that particle size in soil is important as it influences the drainage properties of the soil. If you take a plant pot tightly packed with sand and pour water on to it, the water will quickly run through the sand and out of the pot. Water poured into a pot full of clay may not drain through at all, but remain on the surface. Sand and clay have very different textures too: sand is soft and crumbly; clay is sticky if wet, or very hard if dry.

Soils which contain a high proportion of sand are free draining and easily cultivated, but water quickly drains from them into the groundwater, leaving the surface dry and prone to drought. Clay, on the other hand, holds water and resists drainage. As clay particles are tiny, the total surface area of clay is large compared to the surface area of sand. The grains in clay are tightly packed so that the **pore spaces** between the grains are small and can fill completely with water. Clay particles hold on to water tightly, so that drainage from a clay soil is almost negligible. This makes wet soils with high clay contents heavy, waterlogged and sometimes anaerobic.

It is very difficult to measure how easy it is for plants to obtain water from a particular soil. The total water content may not indicate just how tightly a soil is holding the water or how efficient each individual plant is at extracting it. Clay particles, as discussed above, hold water tightly on their surfaces. Thus much of the water in a clay soil is unavailable to plants because they cannot get it to move from the clay surfaces into their roots. As clay dries out, the particles begin to pack closer together as they lose their coating of water; this causes the clay to shrink

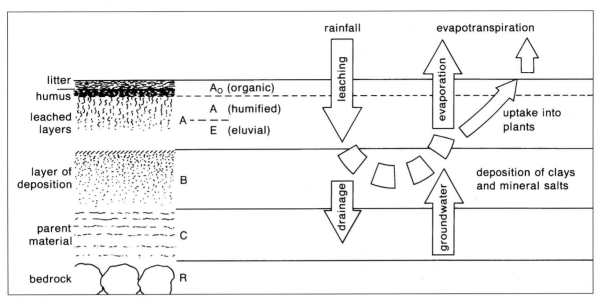

Figure 15.2 A generalised soil profile showing the classification of horizons within the soil and the major water movements which occur in soils.

and crack at the soil surface. When it rains again, the water runs down the cracks, and eventually enough seeps back into the soil to re-wet it.

Often the particles in a soil are clustered together into clumps called **peds**. In a very sandy soil, the sand may remain as individual particles, but as these are large and often roughly shaped, there is quite a lot of air in the soil. The formation of peds creates larger pore spaces between the lumps. This produces a fairly well-drained soil structure, unlike pure clays where the pores are too small and do not link up to provide drainage channels for water. The presence of a ped structure helps to create a free-draining, well-aerated soil. Factors which damage the ped structure, like heavy trampling from large grazing animals, alter these properties making it more difficult for vegetation to grow. The soil particles are mixed up by soil organisms such as earthworms which can help to improve soil structure (see Section 15.2.3).

When rainfall lands on the surface of the soil and drains through it (see Figure 15.2), the water picks up small particles of clay and dissolves minerals which are transported to lower levels. This is called **leaching**. Heavy rains can leach a lot of nutrients out of the upper part of the soil; these may then be redeposited further down in the ground. If a lot of soil water is lost from the upper surfaces by evaporation or passes into plant roots as a result of transpiration, then water may move upwards in the soil from deeper, more moist parts. This results in upward movement of minerals (for example calcium carbonate, see Section 15.2.2).

If you dig a deep hole through soil, you will see that it is not uniform in texture or colour. It is made up of different layers called horizons. The left side of

Figure 15.2 shows a section through a soil and the system used to name the different horizons. Usually the top horizons are dark in colour, almost black, because this is where the organic matter, or **humus**, is concentrated. Beneath this is often a dark layer of sandy soil stained with the decay products of the organic matter. Then there is a light coloured sandy layer from which the clay and minerals have been leached. Beneath this is a darker, clay-rich horizon, often reddish or dark brown because of deposition of mineral salts and products of the humus. Underneath this is raw weathered material from the underlying rock or from sedimentary deposits. Usually some of the texture of the soil has been derived from these underlying layers. This is often called **parent material**. This series of horizons forms the **soil profile**.

The different horizons of the general soil profile in Figure 15.2 have been labelled with letters. There are three basic sections to the soil (A, B and C). These sit on unaltered bedrock (R). The A section of the soil profile consists of horizons where leaching occurs; B horizons are where the mineral salts, humic products and clays are deposited; and C is the shattered bedrock which is unaltered by leaching or deposition. Usually the A section is subdivided to identify a litter and humus layer (known as the A_O layer – O for organic) and a lower paler horizon which is often called the E horizon (E stands for **eluvial**. which is really another term for leached).

There is considerable variety in the structure of soil profiles in different parts of the world. Because of this, soils have been classified by soil taxonomists in the way that species or communities can be classified. Most of the world's soil types can be placed into a few major categories called great soil groups. Some of the names of the great soil groups are Russian, because some of the first attempts at soil classification were made in Russia. These are listed in the first column of Table 15.1 with a very general description of their character in column two. They tend to be associated with particular vegetation types (given in the third column in Table 15.1).

Table 15.1 The UNESCO scheme of classification for the soils of the world and the vegetation types most commonly associated with them.

Great soil group	Soil form	Vegetation type
Laterite (latosol)	Weathered red	Wet tropical forest
Margalite (tropical black)	Base rich black	Seasonal tropical forest
Desert soils	Hot dry	Desert
Chernozem	Dry black (with calcite)	Grasslands (steppe)
Chestnut soils	Dry organic black or red-brown	Grasslands (prairie)
Grey-brown podzol (brown earth)	Grey-brown mull	Deciduous forest
Podzol	Grey banded mor	Boreal conifer forest
Peats	Organic peat	Bog
Tundra soil or podzol (if well drained)	Organic on mineral base (with permafrost)	Tundra

The type of soil, the climate and the vegetation are all associated and inter-related (more about this follows). There are several classification systems in use around the world which have a whole proliferation of unpronounceable names for soil types (Fitzpatrick, 1986). For example, just one soil type, formed on poorly drained, continental plains, can be called a planosol, pseudogley, stagnogley, aqualf, xeralf, ustalf, aquult, podbel or a solod; all of which sound more like a set of dwarves in J. R. R. Tolkien's *The Hobbit* than a soil type!

At the global scale of the great soil groups, the climate seems to be the most important determinant of soil structure. The important features of the climate are rainfall and yearly temperature cycles. Rainfall produces downward movement of water in the soil profile (see the right-hand side of Figure 15.2). If there is a lot of rain it drains right through the soil. As it goes it leaches out the minerals and humic products from the top part of the profile (the A horizons) and carries them further down the profile. It also washes down small soil particles to a lower level in the soil. If the soil is in a hot dry climate, then evaporation of water from the soil surface will be important. If evaporation is high, then water is drawn up, by **capillary action**, from lower down in the soil profile. If this happens, the rising groundwater may carry dissolved minerals which are

deposited in the soil when the water evaporates. This movement of water downwards, causing leaching, and then up again during evaporation, concentrates the minerals at a particular level in the soil in the B horizons (see Figure 15.2). Deposition of substances in the B horizon is most obvious in seasonally wet and dry climates where each year the movement of water down (leaching) in the winter, then up again (concentrating minerals) in the dry summer produces considerable deposits.

Apart from water movements in the soil, climate also has an effect on the organic contents. A humid, hot climate promotes the action of decomposers, which break down and recycle the organic matter. Cold climates slow down this activity. If the ground is very cold or wet, then considerable amounts of organic matter can build up in the soil. The consequences of climate can be seen in the distribution of the great soil groups of the world.

15.2.2 The great soil groups

The global distribution of these soil groups (listed in Table 15.1) are similar to the major vegetation zones described in Section 14.3. They do not match these zones exactly, but are related to them quite closely. We can look at the different soil types as they change from equator to poles, just as we did for vegetation. Figure 15.3 shows the same basic zonation of vegetation as seen in Figure 14.6, but this time the soil profiles appear under the vegetation.

In the humid tropics, the structure of the soil is often very simple. There is usually a layer of leaf litter only a few centimetres deep at the surface. Beneath this is a red-brown layer where some of the humus from the litter has stained the soil. This layer blends gradually into a strongly coloured red horizon of laterite: iron- and aluminium-enriched clay soils which have been strongly weathered by heavy rainfall in the hot tropical climate. This red horizon may be 25 metres deep! The silicates and other minerals have been leached out of the soils to leave oxides and hydrites.

The main organisms in this type of soil responsible for mixing the layers are termites. Termites build tall mounds of soil above ground level, but much of their activity and a mass of burrows extend below soil level. Most of the other organisms responsible for the decay of organic matter are in the very shallow surface layer of litter. Most tropical soils are extremely deficient in nutrients, a fact which is difficult to appreciate as the vegetation they support looks so lush. However, in the tropical forest ecosystem nearly all the nutrients and organic matter are in the **standing crop**, that is, in the vegetation. Any leaves, twigs or fruit cases which fall to the forest floor quickly decompose in the hot, humid conditions and are immediately cycled back into the vegetation via the surface roots of the trees. Hence, if a tropical forest is cut down and burned, much of the nutrient and organic content of the ecosystem is lost as gases, wind-blown ash and leached out of the soils by the heavy rains. The soil beneath may be quickly exhausted and susceptible to rapid erosion so that crops will only flourish for a few years. If carefully tended, with some natural vegetation left to reduce erosion, addition of green plant matter to improve soil fertility and use of complementary mixtures of crops, tropical soils can be farmed successfully. It is difficult to carry out these practices in large-scale clearance, so in some circumstances it may be better to harvest the tropical forest for its natural products, such as rubber and nuts, rather than clear it for other cash crops (see Section 21.3.1).

As we go polewards the soil becomes more complex in structure. Under temperate deciduous forest (Figure 15.3), such as the oak woodland communities of Section 14.2.2, lies a brownish soil type called **brown earth**. These brown earths are very fertile and, throughout the world, the original woodland vegetation has been cleared from this soil type for agricultural land. The soil profile is much more complex

than in the tropical laterites. On top is a layer of leaf litter which, as more litter falls, mainly in autumn, breaks down into humus. The humus is mixed in to the mineral soil beneath by the action of large earthworms. This forms a **mull** soil about 15 cm thick which is dark in colour and easy to cultivate. This dark layer grades into a slightly leached layer, and below this lies the B horizon of reddish mineral clays. This grades gradually downwards into the parent material. Such mull soils are close to neutral in pH and rich in soil organisms. They develop where the climate is cool, so that organic recycling is not as rapid as in the tropics.

In still higher latitudes the true podzols are found. The main feature of the podzolic soil profile is the clarity of the different horizons. Podzols develop under boreal forest, so the litter layer on the soil surface is a deep pile of tough, undecomposed conifer needles. This grades down into a thin organic layer of humus called **mor**. The mor, unlike the crumbly fertile mull of the deciduous forest soil, is sticky and structureless. The dark humus stains the layer beneath brown. Below this organic-rich sequence is the band of grey sandy soil. This well-developed E horizon can be turned up to the surface by plough-

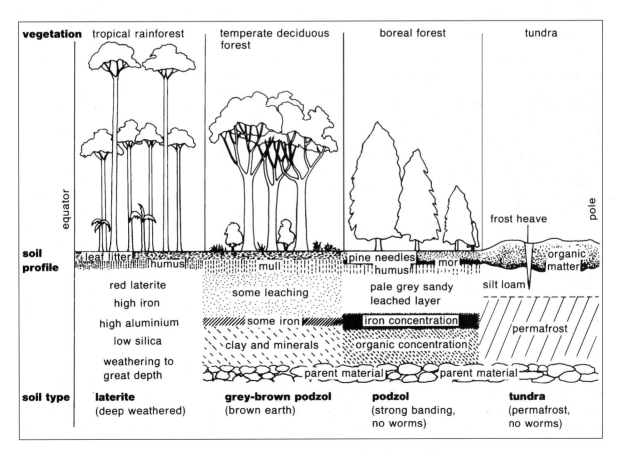

Figure 15.3 Soil profiles for the great soil groups which are found under the major terrestrial vegetation zones shown in Figure 14.6.

ing, where it looks like ashes from a fire, hence the soil name podzol. The word 'podzol' is Russian for ash beneath.

Podzols are freely draining soils, and the grey layer forms because rainwater percolating through the soil leaches out all the minerals and clays, leaving a layer of almost pure sand. The leached nutrients are deposited in the B horizons just beneath the grey or white sands. This deposited layer is often quite bright red-brown with iron oxides and humic deposits. Below this is the parent material, which is often a different colour again, depending on the form of weathered rock in the region. Podzols are products of cold, wet climates where the litter fall is of tough leaves. These mor soils are very acidic, with a pH of about 4.5 or even less, because of the humic acids released by the decaying litter. The acidic water passing through the soil is then responsible for leaching out the nutrients which are redeposited lower in the profile where the pH rises to about 5.5. Here the tree roots take up the nutrients. The striking layered effect of the horizons is due to the absence of earthworms in the profile, which means that there is no mixing of the layers.

At very high latitudes, the very cold winters create permanently frozen ground or permafrost (Figure 15.3). In the summer, the top few centimetres of the soil melt, but the permafrost below impedes the drainage, so that unless the soil is on a slope it gets very wet. These unusual circumstances produce a very different soil type in the tundra. The constant cycles of freeze and thaw churn up the ground, bringing raw mineral soil and many stones up to the surface in a pattern called frost polygons. The ground has many cracks full of ice called ice wedges. Organic matter builds up on top of this disturbed silty loam. The very low temperatures slow the decay of organic matter almost to a standstill. The impeded drainage promotes the growth of mosses, especially the bog moss *Sphagnum*, which builds up on the soil surface (see Section 15.3.6 and Figure 15.11).

So far in Section 15.2.2 we have looked at the relationship between soil structure and the latitudinal distribution of climate and vegetation. Rainfall is important in leaching nutrients from the soil, but temperature and vegetation influence the effects of rainfall by determining the degree of evapotranspiration. Temperature is also important as it regulates the rate of decomposition of the organic litter and thus the speed of nutrient recycling.

In Figure 14.7 the relationship between vegetation and variation in rainfall within the same latitudinal zone was illustrated. The same kind of relationship can be seen in the distribution of soil types and is shown in Figure 15.4.

We can start again in the tropics with the now familiar red, weathered lateritic soil associated with the tropical lowland rainforest shown in Figure 15.4. In drier areas of the tropics deep, uniformly black soils develop. They always seem to be associated with base rich parent material, like volcanic ash or lava flows. An alternative name for these soils is margalitic soils. Unlike the weathered laterites, they are fertile, and tend to have quite a high pH of 8 or above. They form in seasonally dry climates such as in central India. Unlike most soils, tropical black soils do not seem to have normal identifiable horizons.

At higher latitudes, where rainfall is low, grassland vegetation grows. The soils in these regions, in the Russian steppe for example, are very fertile, deep and black. This is how they get their name **chernozem**, which means black earth. The low and seasonal rainfall results in a considerable amount of movement of water up and down in the soil profile. This does not leach many nutrients out of the soil, hence its richness. It does, however, cause calcium carbonate to be precipitated as thin filaments or nodules lower down in the B horizons of the profile. The area of steppe where chernozems are produced is composed largely of wind-blown deposits called **loess**, which originated as **glacial tills** after the ice age. They are often deep yellow in colour (they give the name to the Yellow River which is coloured by eroded loess). The one- or two-metre thick black soils of the chernozem grade fairly quickly into the yellow horizons beneath. This results in a very distinctive profile. These soils are very rich in animal life. They are riddled with burrows of earthworms. In Europe the large burrows of mole rats (*Spalax* spp.) are obvious in the profile. The mole rats transport soil while digging so that the tunnels in the yellow loess are often filled with black soil and those in the upper horizons are full of paler soil. These rich soils are now heavily cultivated, and the mole rats seem on the decline because of ploughing practices. In America, the prairie has similar soil, but the carbonate layer is not so obvious. Here too there is often a clay layer between the organic horizons and the parent material and burrowing mammals: the prairie dog (*Cynomys* spp.).

Where total rainfall is about 10 cm a year, the calcium carbonate layer appears closer to the surface. This is a measure of how much upward water movement there is. The organic surface layer is very thin as there is not much primary productivity to produce litter. These are **desert soils** where productivity is low, because of a lack of available water. High evaporation rates can often lead to certain salts accumulating in the surface layers and crystallising on the soil surface.

To complete Figure 15.4 so that it corresponds to Figure 14.7, two wet soils have been added on the left-hand side. These profiles are of peat bogs where massive amounts of organic matter build up. These

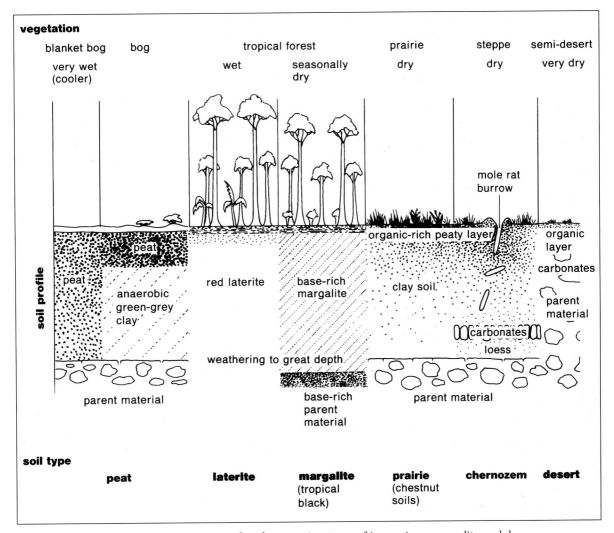

vegetation

| blanket bog | bog | tropical forest | | prairie | steppe | semi-desert |
| very wet (cooler) | | wet | seasonally dry | dry | dry | very dry |

mole rat burrow

organic-rich peaty layer

organic layer

carbonates

peat

peat

anaerobic green-grey clay

red laterite

base-rich margalite

clay soil

parent material

carbonates

loess

weathering to great depth

parent material

base-rich parent material

parent material

soil type

| peat | laterite | margalite (tropical black) | prairie (chestnut soils) | chernozem | desert |

Figure 15.4 The great soil groups associated with vegetation types of increasing seasonality and dryness.

soils link this section with the next, Section 15.3 on wetlands. They are considered in much more detail in Section 15.3.6.

So soils are greatly affected by climate, especially the temperature, which can alter the rate of decomposition of organic matter, and the rainfall. Rainfall and evapotranspiration between them affect the severity of leaching and the movement of water in the soil profile. It is also obvious that, because the organic layers are produced directly from the litter of the vegetation above, the vegetation type also affects soil structure. The soil organisms too have an effect by mixing the organic matter into the mineral layers beneath: earthworms seem to be especially important in soil development.

If the soil is very shallow, the effects of the C and R levels in the profile are much greater. This can override the climatic effects just described. Shallow soils, less than half a metre thick, often develop over limestones and chalk whatever the climate. This is because chalk and limestone weather by dissolving in

groundwater. The two rock types are composed almost entirely of calcium carbonate, so that when they dissolve, nothing is left to make the soil bulk. Normally parent rock contains a lot of sand or silt which ends up in the soil profile, but this is not so with carbonates. The B horizon is very poorly developed or absent altogether. Because the soil is so shallow, there are often chunks of parent rock in the soil. Such soils are given the name **rendzina**, which means 'to tremble' in Polish: the rocky consistency of the soil causes ploughs to shake as they strike the stones. Rendzinas are usually very fertile and, because of the influence of the bedrock, their pH is usually about 7. However, because they are so shallow and rocky, they can dry out considerably in hot summers.

All the great soil groups described in Section 15.2.2 are large-scale categories for soil structures, just like the large vegetation groups or zones described in Section 14.3. To read more about the soils of the world, see Fitzpatrick (1986) or Colinvaux

(1993). On a much smaller scale, every individual patch of soil is different, just as every community is different, although all can be grouped into larger categories.

15.2.3 The effect of vegetation on soil – two case studies

So far we have looked in detail at the effects of climate and particle composition on the structure of soils. The effects of plants have been indicated briefly as the producers of the organic matter in the soil. Particular vegetation types seem to be associated with some soil types, for example, brown earths with deciduous broad-leaved woodland and podzols with boreal conifer forest. This supports the theory that, at a large scale, vegetation is important in soil formation. Such soil groups could be climatically determined: the importance of rainfall, water movement and temperature in affecting mineral movement and humic decomposition have already been shown. The correlation of soils with vegetation could therefore simply be because they are both dependent on the climate.

To investigate the effect of vegetation on soil requires a much smaller-scale study in an area small enough for the climate to be uniform. At this level, the level of the single ecosystem, the soil–vegetation interactions will be much clearer. Two such studies have been chosen here: one in an ecosystem created by the planting of trees, the other in a rather more natural environment of heathland.

Case study 1: Soil developments under lime and beech trees at Holt Down, Hampshire, England (Pigott, 1989)

In 1935 the Forestry Commission set up an experiment on tree growth by planting a plot of small-leaved lime trees (*Tilia cordata*), 15 m wide by 130 m long, next to a young stand of beech (*Fagus sylvatica*), which had been planted in 1931. The site had originally been arable land and rabbit-grazed grassland and the soil type was a brown loam, overlying a one-metre layer of clay with flints over chalk.

A walk through this plantation in summer revealed the remarkable influence of the tree species on the ground flora. Underneath the rectangle of limes, and remarkably well correlated within the outline of the trees, was a similar rectangle of bright green vegetation. This consisted of great areas of dog's mercury (*Mercurialis perennis*, see Figure 9.6) and smaller patches of moschatel (*Adoxa moschatellina*), as well as a grass, wood false-brome (*Brachypodium sylvaticum*), sweet woodruff (*Galium odoratum*) and other, less abundant species of flowers, ferns and mosses. These species were virtually absent under the beech trees. The flora under the limes included 24 low flowering species, 2 species of under-storey bushes, seedlings of 2 tree species and 9 species of ferns and bryophytes. The beech ground flora consisted of only 8 flowering species, one species of understorey tree, seedlings of 2 tree species and 3 fern and bryophyte species.

Various pits were dug in the two areas described

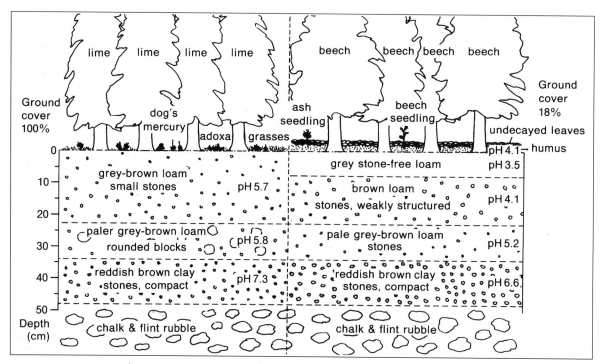

Figure 15.5 Separate soil profiles developed under small-leaved lime and beech at Holt Down, Hampshire. The soil profiles are drawn to scale but the plants are not. (From data in Pigott, 1989.)

above to investigate the soil structure. A devastating storm which swept across southern England on the 16 October 1987 blew over several trees which allowed more soil profiles to be observed. The results of these observations are summarised in Figure 15.5. In the diagram the general soil profiles are drawn to scale as found in the pits under the two tree species so that they can be compared horizon by horizon. The woodland structure above is drawn to a different, much smaller, scale than the soil to give an idea of the difference in ground flora between the two areas.

Both soil profiles, as revealed in the pits, are shallow brown earths (grey-brown podzols) as would be expected under temperate deciduous forest, but there are quite obvious differences between the two in the top few horizons. The loams in the top 30 cm both have 20% clay and 50% fine sand and are well drained. The most obvious difference between them is the build-up of a litter layer 4 cm deep under the beech trees and the absence of such a litter layer under the limes. Analysis also showed that the pH was 5.1 ± 0.29 under the limes and 3.8 ± 0.17 under the beeches. The other main difference was in the concentration of mineral ions which were obtained from the soil. The soil under the limes had very high levels of soluble mineral ions (calcium 70–80 μmol/g, magnesium 2–7 μmol/g, sodium 3–7 μmol/g) while under the beech the mineral ions were rarer (calcium 7–45 μmol/g, magnesium 2–4 μmol/g, sodium 1–3 μmol/g) indicating a less fertile soil under the beech.

The leaching under the beech, and the beginnings of the formation of a deep litter layer and grey leached E horizon, suggests that the soil was developing into a podzol. This soil had only a few small pigmented worms at the bottom of the litter layer and soil mixing was reduced. The soil under the limes, however, was full of large earthworms (*Lumbricus terrestris* and *Allelobophora* spp.). The earthworms seem to be encouraged by the production of palatable leaf litter under the limes. Once there in numbers they mix the upper layers in the soil profile. Unlike the soft lime leaves, beech leaves are dry, papery and high in polyphenols; the worms are deterred and no mixing occurs causing the strongly stratified soil to develop, leading to podzolisation.

The differences produced in the soil, especially the higher pH and more aerated loam in the top 10–20 cm in the soil profile under limes, are probably responsible for the diverse ground cover. The acidic, compacted, nutrient-poor soil under beech has an impoverished ground flora with only about 18% cover.

Case study 2: Soil variation in chalk heathland at Lullington Heath, Sussex, England (Grubb *et al.*, 1969)

At Lullington Heath National Nature Reserve there is a small area of chalk heathland. Heathland like this is one of the most endangered ecosystems in Britain: many sites of heathland have been destroyed during the last hundred years. The heathland ecosystem consists of a mixture of grasses, especially the red fescue (*Festuca rubra*) and low bushes of heather (*Calluna vulgaris*), heath (*Erica cinerea*) and gorse (*Ulex europaeus*), with many other flowering plant species, including dropwort (*Filipendula vulgaris*), salad burnet (*Sanguisorba minor*) and sorrel (*Rumex acetosa*). The vegetation on the heathland has grown taller since the number of rabbits on the heath was greatly reduced because of myxomatosis in 1954. During this time the grasses have probably increased in dominance. The heather, heath and gorse bushes have grown up since then too.

The community grows on a shallow, free-draining dark brown loam; below this is a layer of flints in loam, then a narrow band of chalk rubble and the bedrock of chalk below. This soil profile can be seen in Figure 15.6, to the left-hand side of the diagram. Earthworms are fairly rare but ant hills of the yellow ant (*Lasius flavus*) are common and move soil around in the profile. In places the soil thins to a rendzina-type structure. Some areas of the site had been cultivated extensively from Iron Age times onwards, a period of over 2000 years. The soils seem to have developed partly from the underlying chalk, and partly from the loess deposited from the glaciers of the last cold phase.

The species which occur on the chalk heaths today are an unusual mixture. Some, like the dropwort and some of the grasses, are species which are associated with soils rich in calcium and of high pH. They are **calcicoles**. Other species, like the heather, heath, gorse and sorrel, do not thrive in calcareous soils and are called **calcifuges**. So how do both calcicoles and calcifuges come to be growing together?

It seems that in the chalk heath situation both the calcicoles and the calcifuges can grow and regenerate in a pH of about 5–6. Some of the calcifuges only had their roots in the top few centimetres of soil where the pH was lowest. However, the roots of some of the calcifuges were found to be penetrating the chalk where the pH must be much higher, that is around pH 7–8. The yellow ants, which live in the heath, disturb the lower horizons of soil and carry it up to the surface when constructing their ant hills. These mounds of soil have a much higher pH than the rest of the soil surface and provide valuable sites for the germination of the more extreme calcicoles.

When the pH of the soil under the calcifuge bushes was investigated, it was found to be much more acidic than the soil elsewhere. The larger the bush, the lower the pH found under it; some large bushes had pH readings of 4 or lower. The acidification of the soil under such bushes is shown on the right-hand side of the soil profile diagram in Figure 15.6. These

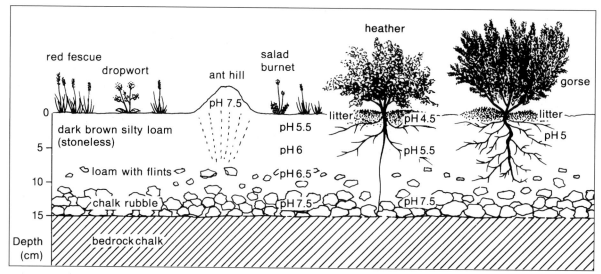

Figure 15.6 Soil profile found at Lullington Heath, Sussex. The right-hand side of the diagram shows how the profile and pH change under bushes of heather and gorse. (From data in Grubb *et al.*, 1969.)

bushes have only grown up since the disappearance of rabbits in 1954. So the degree of acidification has happened rather rapidly. This acidification is due to the build-up of a litter layer 1–5 cm deep underneath the bushes. The deeper the litter, the lower the soil pH. The longer the gorse and heather bushes remain on the chalk heath, the more acid the surface soil will become. Within only a few tens of years, certain species in the community have been able to change the character of the soil. No podzolisation was detected under the gorse or heather scrub, but it appears that the bushes are changing the whole character of the soil from chalk grassland to acid heath.

15.3 Wetland and aquatic ecosystems

15.3.1 *Water – the important factor*

We have just looked in detail at a whole series of terrestrial ecosystems where soils were a very important part of the inter-relationships between climate and communities. With wetlands and aquatic ecosystems the climate is a less important environmental factor than it is for terrestrial ecosystems. The effect of water is predominant and all-important in determining the type of wetland or aquatic ecosystem.

Aquatic habitats are, perhaps, easier to identify than wetlands. They have free-standing water covering the land surface which can either be fast moving, as in streams and rivers, or relatively still, as in lakes and ponds. If we consider all aquatic environments, then we must also include areas of marine influence such as brackish estuaries, the tidal zone (see Section 14.2.3) and the open sea. Both ponds and lakes tend

to be shallow at their edges and deeper at the centre. There are many ways in which lakes can form by build-up of standing water, including in depressions caused by earth movements and in valleys carved by glaciers and dammed by the debris deposited when the ice melts. Large lakes, such as the Norfolk Broads in England, have even been produced by human excavation. The Broads were created, from Roman times onwards, when peat was excavated from the site and the depressions were subsequently flooded.

Linear aquatic habitats are formed by moving water: streams form as rainwater runs off high ground and flows down valley bottoms from mountainous areas. Streams join up to form larger rivers by the time they reach the coastal lowlands. Streams are usually fast moving with little volume of water. When the water volume increases due to high rainfall or snow melt, the water has a lot of energy which is capable of eroding and carrying quite large particles away with it to leave a rocky or stony stream bed. Rivers usually flow more slowly than streams; they are large and carry enormous loads of debris from higher up the water flow. This is often deposited in the lower reaches of the river, so that it has a muddy floor. Such rivers often meander across large flat plains which flood periodically when the river bursts its banks (see Section 15.3.4).

Wetlands, as their name suggests, are wet ground, rather than standing water. Unlike aquatic ecosystems, they develop an organic soil profile, but the saturation with water alters the community structure. The waterlogging of the soil tends to cause a problem for plants as anaerobic conditions are produced with little or no oxygen available. Wetlands have been described as 'a halfway world between terrestrial and aquatic ecosystems' (Smith, 1980, p. 225). They are a transitional link between two extremes, and range from very nearly aquatic to almost dry. Indeed several wetland habitats dry out during dry seasons in the year and are then very difficult to identify as wetland at all!

15.3.2 Types of wetlands

There is an enormous variety of wetland ecosystems throughout the world. This has led to a large number of descriptive names for wetland types. Unfortunately, the same names often apply to different wetland types in different countries and many local terms are used for wetlands common in the region. Because water is more important than climate, very similar wetlands occur in many regions. The species in the community are often different, but the basic structure of the ecosystem is similar.

This variation, and the wide range of environments in which similar ecosystems occur, makes it quite difficult to classify wetlands into some wider system. We cannot use climate or global position as was possible for terrestrial ecosystems and soil types. It is perhaps easiest to divide wetlands according to the source and nature of the water which maintains the system. There are four possible sources of water, each of which have different properties:

The sea – sea water is highly saline and tidal;
Streams and rivers – sediment rich and seasonally flooded;
Drainage from surface runoff or groundwater – usually nutrient rich and basic; and
Directly from rain or snow – rain is nutrient poor and acidic.

The distribution of these water sources depends on the occurrence of topographical features. For example, sea water will obviously only influence coastal regions, rivers tend to flood most in lowlands and rainfall is often highest in mountain regions close to oceans. For these reasons, the source of water is used here as the basis of the classification of wetland ecosystems.

15.3.3 Marine wetland ecosystems

Mangrove swamps

The coastline in the tropical and subtropical regions is fringed with a strip of swampland which is inundated every high tide with marine or brackish waters. Wherever the wave action is not too strong to prevent regeneration, these coastal wetlands are densely vegetated with thickets of mangrove trees. There are about 70 species of mangrove around the world, the most important genera being *Rhizophora* and *Avicennia*. Mangroves are capable of growing in fresh water, but it is thought that the saline conditions of the intertidal zone give these poor competi-

tors an advantage over other species. They are well adapted to the salty conditions as they can prevent high concentrations of salt entering the roots and can secrete excess salt from their leaves. They may also jettison extra salt when they drop their leaves. Just like the ecosystems on the rocky shore (Section 14.2.3), mangrove swamps are influenced strongly by the tides. Incoming tides import nutrients to the system, and tides are also responsible for dispersing seeds. Mangrove seeds often germinate while still held on the tree. They grow a thick spear-like **hypocotyl** so when the seed falls off the tree, the hypocotyl is embedded in the mud like a spear and then grows anchoring roots (Mitsch & Grosselink, 1986).

A variety of animals live within the mangrove including fiddler crabs (*Uca* spp.) and mudskippers (*Periophthalmus* spp.). The food web is based on the presence of many detritivores, but ascends to major carnivores like alligators, crocodiles and big cats such as tigers (*Panthera tigris*).

The most unusual feature of mangrove swamps is the structure of the trees. Figure 15.7 is a photograph of a typical swamp at low tide. You can see that each tree has a whole series of aerial roots arising quite high up on the trunk and plunging into the mud beneath. These stilt roots help to support the

Figure 15.7 Mangrove trees in Cuba, showing pneumatophores and seedlings in the foreground.

trees and lessen the wave action of incoming tides. Sometimes considerable amounts of organic mud build up under the mangrove forest. Figure 15.7 shows many stick-like structures protruding from the mud: these are called **pneumatophores.** Both these and the supporting aerial roots have little pores on them that can take in oxygen from the air. One of the problems of living in wetlands is the lack of oxygen for roots: mangroves have overcome this problem with these pneumatophores.

Salt marshes

In higher latitudes the mangrove ecosystems disappear because the trees cannot tolerate even minor frosts. They are replaced on the coastal strip by much lower growing ecosystems: **salt marshes**. Like the mangrove swamps, the salt marshes are within the tidal influence of the sea and are inundated with salt water every high tide. As you can see from Figure 15.8 they consist of a patchwork of low vegetation separated by tidal creeks. Salt marshes tend to develop in sheltered intertidal regions where wave action is not too strong.

Salt marsh vegetation is dominated by grasses such as *Spartina* and rushes (*Juncus* spp.). The lower lying pools, where the salt concentration is higher, have different vegetation including glasswort (*Salicornia europaea*), a very succulent plant which is eaten as a (rather salty) vegetable in Europe. Salt concentration can vary considerably depending on several factors including the structure of the marsh, how often it is flooded by tides and the amount of rainfall. If rainfall is high, areas of the marsh are washed free of some of the salt and are colonised by other plant species such as sea lavender (*Limonium* spp.) and arrow grass (*Triglochin* spp.).

Salt marsh soils are high in phosphates but low in nitrogen, which may limit plant growth. They can also be toxic to plants if sulphur from sea water collects in marsh soils as iron pyrites and hydrogen sulphide. When the soils dry out, sulphuric acid may form which lowers soil pH. In these marshes up to 70% of the net primary productivity is due to chemosynthetic bacteria which utilise the sulphur (Howarth *et al.*, 1983). Mats of cyanobacteria on the mud are important as fixers of light energy. Salt marshes have high primary productivity, but much of the organic matter produced gets washed out of the marsh with the retreating tides.

15.3.4 *Floodland ecosystems*

The second group of wetland ecosystems are those which obtain their water from rivers. These wetlands are often extremely seasonal. They flood deeply with water when the river overflows its banks. This may happen during a rainy season, or when winter snows melt in the catchment area. Floodlands tend to occur in lowland, flat-bottomed valleys through which a large river meanders. Such areas can be very large on the flood plains of great rivers. These large lowland rivers are associated with a number of features illustrated in Figure 15.9. Normally, the mature river will meander across a wide, flat plain which was created by sediment deposited in the past by the river. The channel in which the river flows builds up at the side into sandy banks called **levées**. It is only when the water reaches the top of these banks and overflows that the valley is flooded. The flood plain often has small permanent wetland areas which form when the river takes a new path on the valley floor, leaving behind an **oxbow lake**. These lakes slowly fill in with silt to form boggy areas. Often a flood plain will have several oxbow lakes in various stages of infill.

When the river floods the valley, it carries with it clays, silt and mineral nutrients which are deposited on the flood plain. This is because as the water slows down it loses its capacity to carry particles. The largest particles settle out of the water to form the sandy levées. The smaller silt and clay particles are carried further by the floods and settle out only when the water stops flowing. This frequent addition of new material makes flooded valley soils nutrient rich. During the period of flooding, anaerobic conditions are created in the soil. This temporary condition seems to be the limiting factor which determines which communities grow on the site.

The topography of the valley floor in relation to

Figure 15.8 A salt marsh in Udale Bay, Cromarty Firth, Black Isle, Scotland.

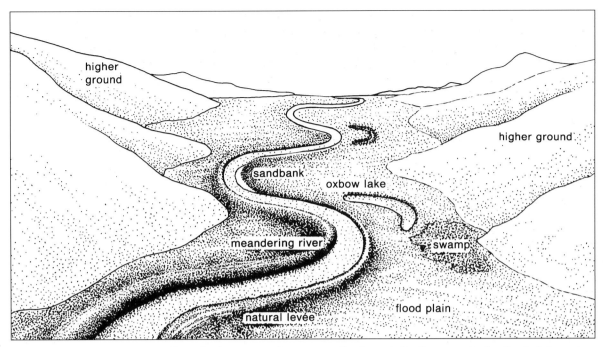

Figure 15.9 A mature river valley. The large meandering river floods periodically to create the flood plain and as meanders are cut off from the main river flow they form oxbow lakes.

the river affects the number and duration of floods. Areas closest to the river tend to flood most often and those further away and on higher ground may only flood in very wet years. The species composition in the ecosystem changes with distance from the river.

The natural vegetation in temperate flood plains is mixed deciduous forest. In temperate North America, for example, five associations of forest species can be identified. The most extreme is in areas flooded all year round. These have a small number of specialised swamp trees like the swamp cypress (*Taxodium distichum*) and tupelo (*Nyssa aquatica*). In areas flooded for more than half the year, the swamp species are joined by specialist oaks (*Quercus* spp.) and ashes (*Fraxinus* spp.). With less than 50% flooding the swamp species disappear and a larger variety of oaks, elms (*Ulmus* spp.) and many other genera grow in their place (Larson *et al.*, 1981). These mixed woodlands have a rich fauna of birds, amphibians, reptiles and mammals.

In many parts of the world floodlands have been cleared of their natural vegetation. The frequent flooding constantly enriches the soils and the proximity of the river makes it ideal for irrigation. These areas are very valuable for cultivation and many such areas, like the Nile valley in Egypt, have been farmed for thousands of years. In temperate zones, flood plains make excellent water meadows of rich grazing land, although the wet environment can harbour diseases and parasites which may infect grazing animals.

15.3.5 *Swamp and marsh ecosystems*

The ecosystems in this category are found in areas of impeded drainage, where water runs off the surrounding land and collects, or where groundwater lies close to the surface. In some cases rivers and streams may also feed into the area. The land is flooded all year round, except in very dry years, unlike most of the floodlands described in Section 15.3.4. Swamp and marsh ecosystems are very variable in size and form, depth of soil and plant community structure. In general they can be divided into two major types: **swamps**, in which trees are the dominant vegetation; and **marshes**, which have large open areas of grasses and reeds.

Swamp wetlands are dominated by trees similar to those found in the very wettest areas of flooded river valleys. Some of the most famous swamp areas occur in Florida (USA), where the very low-lying land is often flooded. Dominant tree species here include the swamp cypress and water tupelo. Figure 15.10 shows a close-up of swamp in South Carolina. You can see that the floor of the forest is under water. The tall stately trees form a very impressive structure; those in the photograph are swamp cypresses with fanned out butresses at the bases of their trunks. The knobs you can see projecting out of the water are not rotted off tree stumps, but knee roots. Several species in the swamp have such knee roots. They project above the surface of the water and are usually one or more metres high. These pneumatophores may act just like the ones in the mangrove swamps in Figure

Figure 15.10 Close-up of the bases of swamp cypress trees (*Taxodium* spp.) in South Carolina, USA showing knee roots.

15.7, as aerators for the roots which are permanently under water. Carbon dioxide has been shown to leave these knee roots, though it has not been proved that oxygen is taken in (Mitsch & Grosselink, 1986). Unlike the mangrove trees, which produce seedlings able to survive and become established in the inundated muds, swamp trees produce seeds which only germinate and grow in dry conditions. This means that regeneration of the swamp trees can only occur after the swamp has temporarily dried out. This is the key feature which determines the existence of a swamp ecosystem or lake ecosystem. If the swamp never dries out but remains with standing water, then the trees will never regenerate and the swamp will become a lake. However, swamp trees are very long lived, so drying out need only happen every hundred years or so for the swamp forest to regenerate.

Marshes are wetland ecosystems which are dominated by grasses (Poaceae), sedges (Cyperaceae) and reeds (Juncaceae) rather than trees. The movement of groundwater, runoff and the addition of stream water usually make marshes nutrient rich with a fairly high pH. Marshes are common in temperate zones. In the northern hemisphere, the same genera are usually dominant in both North America and Eurasia. These include *Typha*, *Scirpus*, *Phragmites*, *Cladium* and *Juncus*. Each genus tends to dominate in a particular habitat where the marsh conditions suit it best. The species which live in marshes often have tough or very sharp-edged leaves (*Cladium* and *Phragmites* were used to roof houses because they were long lasting). These tough leaves may have evolved to deter large herbivores, such as swamp rhinoceros (*Teleoceras* and *Elasmotherium* spp.) and straight tusked elephant (*Loxodonta antiqua*) in the Pliocene and Pleistocene (Tomkins, pers. com.). Since these animals have become extinct, their absence has left the tough grass species relatively untouched (see also Section 21.3.4).

Most leaf matter falls to the wet marsh surface and is attacked by decomposers which form the base of the rest of the food web. The leaf litter often builds up into a peaty layer. This is especially so in fenland, although the thickness of fen peat is never as great as seen in bogs (Section 15.3.6). There is more on the inter-relationship between lakes, marshes and bogs in Sections 16.3.2 and 16.4.1.

15.3.6 Bog ecosystems

The fourth category of wetlands is true bogland. It receives water only from rainfall, not from streams, rivers or groundwater. This sole source of water greatly affects the nutrient content of boglands. Rainwater has very little nutrient content, unlike river- and groundwater, and as it drains through the soil profile it tends to leach out any remaining nutrients. The dominant species in bogs are mosses in the genus *Sphagnum* (Figure 15.11). Many different *Sphagnum* species are found in bogs, each with slightly different water requirements. The mosses

Figure 15.11 The surface of a *Sphagnum* bog. The tops of the *Sphagnum* can be seen in the extreme foreground; the insectivorous sundews are *Drosera anglica*.

grow upwards and their lowermost leaves decay and join the peat building up beneath.

Wherever rainfall is high and temperatures cool, so that water loss by evapotranspiration is less than water gain from rain, *Sphagnum* bog tends to develop. The main areas where bogs occur are in the temperate and boreal regions. The bog may slowly grow up to form a huge dome of peat called a **raised bog**. Raised bogs are usually confined by some feature of topography such as a hollow formed by an infilled lake (see Section 16.4.1). If rainfall is high, above about 1000 mm a year, **blanket bog** can develop even on shallow slopes. The peat in blanket bog can build up to several metres thick. This peat is vulnerable to erosion as it is easily washed away once the surface vegetation has been damaged by trampling, grazing or pollution. The peat is washed away leaving deep gullies which gradually widen as more peat washes into them. Figure 15.12 shows eroded blanket peat in the Pennines, England. The high rainfall aids this process and often several metres of peat are eroded away to reveal the underlying gritstone.

The high rainfall leaches nutrients from the bog. The pH is also very low, often about 3–4, caused by humic acids and sulphuric acid formed when organic sulphur is oxidised. Few plant species other than *Sphagnum* mosses can grow in these conditions. Those which do well have often evolved methods of increasing their nutrient intake. Some are carnivorous plants which lure insects into traps, or have sticky leaves that catch flies. Such plants include sundews (*Drosera* spp.) and butterworts (*Pinguicula* spp.) with sticky leaves, and pitcher plants (*Sarracenia* spp.) which have a modified leaf shaped like a tube with slippery sides and digestive juices in the bottom. There are sticky-leaved *Drosera* plants growing on the *Sphagnum* in Figure 15.11. Other plants are more orthodox and gain extra nutrients from nitrogen-fixing bacteria in root nodules. An example is the bog myrtle (*Myrica gale*).

Figure 15.12 Eroded blanket bog in the Yorkshire Pennines.

The whole bog community is slow growing and short. Primary productivity is low, with only small populations of herbivores such as insects, hares and bog lemmings and a few predators such as spiders and owls. The larger herbivores and predators like deer, caribou and bears roam over a much wider area, although they may enter bogs from time to time (Mitsch & Grosselink, 1986).

15.3.7 Aquatic ecosystems

Aquatic ecosystems are found in marine habitats, brackish estuaries, rivers, streams, lakes and ponds. Lake and pond ecosystems are considered here as they are the most likely habitats to occur in association with the wetlands already described.

Ponds are really at one end of the size range of standing water bodies. Many of the smaller and shallower ponds are ephemeral: they turn into muddy hollows in dry seasons. The pond life in these ecosystems has to be well adapted for survival of such extreme conditions, or mobile enough to move to another pond. Once a pond reaches a fairly arbitrary size, it can be called a lake. Lakes may be supplied by rivers, or only be fed by rainfall and surface runoff. Lake ecosystems have a wide variety of communities depending on their size, depth, edge structure and water source; but they do have many features in common.

The presence of deep standing water in the ecosystem produces an environment with many unusual properties. First, the amount of light penetrating the lake varies with depth and clarity of water. Really deep or murky lakes have much of their environment below the **photic** zone within which active photosynthesis takes place. Most of the primary producers have floating leaves, or grow underwater only in the shallower parts.

Second, the properties of water when heated can have quite astounding effects on lake ecosystems. In summer, lakes receive heat from the sun. This warms up the surface of the water. This warmed water expands and therefore, because it is less dense than the cooler water beneath, remains as a surface layer. Figure 15.14 shows the temperature gradient found in such a lake. The surface waters may be many degrees warmer than the bottom water. The region of changeover of temperature is called the **metalimnion**. This rapid changeover in temperature between the warmer **epilimnion** and cooler **hypolimnion** is called the **thermocline**. This stratification prevents the mixing of the layers of water, so that although debris from above can sink to the bottom and decay there, there is little transfer of soluble substances such as minerals and dissolved oxygen or carbon dioxide.

If the lake is fertile (**eutrophic**), primary productivity will be high, and considerable amounts of debris

Box 15
Closed ecological systems

If you have ever tried to keep an aquarium or even an indoor pet, you have begun to try to keep a more-or-less closed ecological system. A closed ecological system or ecosystem is one which is self-contained. Actually, it is doubtful if any ecosystems are absolutely closed except, perhaps, for certain very distinctive, deep caves found in Romania and a few other places.

The science of artificial closed ecological systems was pioneered by Clair Folsome who first suggested that continuous life in a closed container was possible (Shaffer, 1993). Folsome's interest began when he was a candidate in the NASA scientist programme in the 1960s. Here he became interested in the difficulties that would have to be overcome if there were ever to be permanent colonies established on the Moon or on Mars. Folsome and other scientists accepted that humans would have to grow their own food and recycle their own waste. Folsome soon found that sealed glass globes half-filled with seawater and containing decomposers and photosynthetic algae could persist for years. A commercial form known as the EcoSphere was devised. This contained snails and small shrimps in addition to algae and microorganisms and came in three sizes: $3\frac{1}{4}$", $5\frac{1}{4}$" and $6\frac{1}{2}$" diameter.

In September 1991 four women and four men entered a somewhat larger closed ecological system called Biosphere II (Figure 15.13). Biosphere II was designed to be totally self-contained except for sunlight. The intention was that for two years the eight scientists would be sealed off from the world outside. They would grow their own food and be part of a carefully designed ecosystem that would provide oxygen and water and recycle carbon dioxide and nutrients (Anderson, 1989).

Biosphere II was constructed in the Arizona desert at a cost of around $150 million, provided by a Texan multimillionaire, Edward Bass. The building has a volume of 200 000 m^3 and contained a total of about 4000 species of animals and plants ranging from three African pygmy goats (to provide milk), about 30 chickens (eggs and meat) and banana trees (fruit) to rufous-tailed hummingbirds (pollinators) and termites (decomposers). Insect pests were biologically controlled by predators such as lacewings and small wasps. All in all, there were a total of 140 different crops, including coffee. There was even a small rainforest, a mini-ocean (in reality a saltwater lake with 15 cm high artificially generated waves), a coral reef and a small desert.

Biosphere II ran into various problems (Veggeberg, 1993). The crops never produced enough food so that the eight 'bionauts' lost weight, and then in January 1993 oxygen levels suddenly began to decline. By the time they had fallen to just 15% — making the atmosphere as thin as it is at an altitude of 4000 metres — the crew were literally gasping for breath and the project managers were forced to inject oxygen into the system, breaching the integrity of the closed system. Perhaps rather unfairly, many scientists were very critical of the project from the start. There was even a spoof movie made — 'Bio-dome' — in which two young men who are motoring through the Arizona desert come across what appears to be a new shopping mall. Once inside they're stuck for a year! Joking apart, scientists can sometimes learn more from a complex experiment that goes wrong than from a simple one that works flawlessly.

Figure 15.13 A view of the inside of Biosphere II.

More recently, Biosphere II has begun to be used to investigate the effects of raised carbon dioxide levels (Macilwain, 1996; Walker, 1996). Its sheer size and ecological complexity means that a considerable amount should be learned.

Mid-way between the $3\frac{1}{4}$" to $6\frac{1}{2}$" EcoSpheres and the 200 000 m³ Biosphere II is the Ecotron at Imperial College London. The Ecotron is a 16-chambered, controlled environment facility (Lawton *et al.*, 1996). Each chamber has a floor area of 2 x 2 m. The first experiment in the Ecotron consisted simply of a study of detritivores (earthworms, snails and springtails) and plants (a clover, a grass and a composite). Despite the simplicity of the communities, some of the results were not only unexpected but could not be explained by any of the processes that standard ecological textbooks say are important in structuring ecological communities! In particular, the earthworms (*Lumbricus terrestris* and *Aporrectodea* spp.) increased root nodulation of the clover (*Trifolium repens*) for reasons which are still unclear. The earthworms also increased the germination of the clover by creating safe sites in worm casts and decreased competition from the grass (*Poa annua*) and the composite (*Senecio vulgaris*) by burying their seeds too deeply for successful germination.

The second Ecotron experiment looked at the effect of decreasing species diversity on ecosystem processes (Naeem *et al.*, 1995). High, medium and low diversity communities had 31, 15 or 9 animal and plant species respectively and were followed for 206 days. Among the findings was that the lower the diversity, the lower the overall productivity of the ecosystem.

The fundamental importance of closed ecological systems such as Biosphere II and the Ecotron is that they give us a chance not only of understanding natural ecosystems but also of investigating the likely consequences of global climate change before it becomes too late for us to do anything about it.

---

may fall to the hypolimnion. As the organic matter decomposes, the oxygen in the water is used up, but cannot be replaced because of the stratification. On the other hand, the nutrients in the epilimnion are taken in by the plants, then carried down to the hypolimnion when they die. This results in a temporary drop in fertility in the upper zones. If the lake is infertile there will be very low primary productivity, so very little debris falls below the thermocline. These infertile (**oligotrophic**) lakes, therefore, do not develop anoxic bottom waters. The stratification in lakes periodically breaks down and the water remixes. In eutrophic lakes the bottom waters are reoxygenated and the upper ones replenished with nutrients. In most lakes partial mixing occurs several times a year, and total mixing at least once a year.

Temperate lakes become very cold in winter. Ice forms on the surface of lakes at 0 °C yet the bottom waters are at about 4 °C (see Figure 15.14). This is because water at 4 °C is denser than colder or warmer water, so it sinks to the bottom. The lake waters are well mixed from the autumn turnover and the temperatures are cool enough to prevent anoxic conditions or mineral depletion. In spring, as the lake warms up, phytoplankton flourish and stratification begins to occur but daily turnover is easily triggered. Later in the summer stratification becomes more severe, nutrients are depleted, the phytoplankton decreases and sinks to the bottom, and the thermocline becomes well developed. At this time of year the temperature difference across the thermocline may be as much as 20 °C. The lake mixes again as it cools in autumn when the surface waters are replenished with nutrients and the bottom waters with oxygen. There may be another bloom of phytoplankton before the winter temperatures fall too low. In the tropics

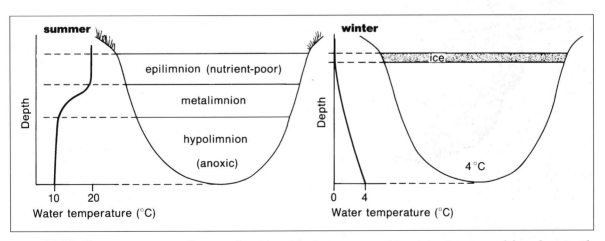

Figure 15.14 The temperature gradients produced in a lake in summer, and in winter. In summer lakes often stratify into a layer of warm water (the epilimnion) which floats on denser cooler water (the hypolimnion). In winter the lake is warmest at the bottom, usually 4 °C, and gradually becomes cooler towards the ice layer at the surface.

the difference between the epilimnion and the hypo-limnion may be only 2 or 3 °C, yet in some lakes the thermocline is very stable so that hardly any mixing occurs (Payne, 1986).

15.4 Inter-relationships of ecosystems

The main wetland and aquatic ecosystems have been discussed here under five different headings based on their water sources. These ecosystems are less discrete than this suggests. Salt marshes are influenced by freshwater influxes at their landward side; freshwater marshes close to the sea may also be influenced by the tide. The wettest parts of flooded river valleys are really often small swamp forests in structure. As the water becomes deeper and more permanent, such swamp forests and marshes blend into the open water of lakes. Fen marshland in wetter areas can develop into acid raised bogs, as can infilled lakes.

Wetlands are often threatened by human activity. This is because wetlands can easily be converted to drier, more agriculturally productive land by draining them. Nutrient-rich marshlands, like the Fens of East

Anglia in England, produce dark, rich soils and are most at risk. Other wetlands may be cultivated in their waterlogged state as paddy fields for rice crops. In general, peat bogs are not threatened for agriculture, but peat has always been cut and dried to use as fuel. The sedge peats of old marshlands are now excavated for peat for use as a gardening material. In Britain, this excavation of the lowland peat of the Somerset Levels has revealed a whole network of ancient trackways (Figure 15.15) made by Neolithic people, presumably to help them walk across the wetlands which were extensive in the area 5000 years ago (Coles & Coles, 1986).

The soil systems described in Section 15.2 are also inter-related. They can be changed by the communities living on, or in, them. The presence of certain species, such as lime trees or heather, can alter the soil and the soil organisms. This in turn alters the whole community structure.

It is relatively easy to put ecosystems into boxes based on soils or water relationships, but in fact the categories merge into each other in a continuum. Often one ecosystem gradually changes into another one as species invade and others die out. Chapter 16 is all about this slow process of change.

Figure 15.15 The Walton Heath Track is about 5000 years old. It is one of the earliest known hurdle tracks using coppiced poles of hazel and alder.

Summary

(1) An ecosystem, or biocoenose, is composed of a community and the physical environment it occupies.

(2) Soil is an important part of terrestrial ecosystems.

(3) Soils are composed of mineral particles, including sand and clay, and organic matter, including plant litter and insect droppings.

(4) Soil usually has a layered structure caused by the build-up of organic matter on the surface and the effects of water movements which leach the soil and deposit nutrients, humus and clay particles within it.

(5) Animals such as worms and termites are important in soil as they mix up the different horizons: in soils where worms are absent the horizons are very pronounced (podzols).

(6) Soils are affected by climate (rainfall and temperature) and, to some extent, by the vegetation growing on them. As a result soils are distributed on a latitudinal pattern corresponding approximately to vegetation zones.

(7) Wetland ecosystems include mangrove swamps, salt marshes, flooded river valleys, swamps, marshes and bogs.

(8) The environment of a wetland habitat depends on the source of its water: sea water is saline, river water is sediment rich, drainage water is nutrient rich and rainfall is nutrient poor.

(9) Mangrove swamps and salt marshes are dominated by the influence of tides and by the high salt content of the soil and water.

(10) Bogs are mainly fed by rainfall and are therefore nutrient poor. Many bog plants augment their nutrient intake by being insectivorous.

(11) Aquatic ecosystems include the open sea, ponds, lakes and rivers.

(12) Ponds are small and some are ephemeral, drying out occasionally in years of drought or regularly every dry season.

(13) Lakes may be fertile (eutrophic) or nutrient poor (oligotrophic).

(14) Lakes may become stratified in summer due to the heating of the surface waters. As debris sinks and decays, the bottom waters, especially in eutrophic lakes, can become anoxic and nutrient rich while the surface waters become nutrient poor.

(15) Closed ecological systems can be used to study ecosystem processes and to investigate the possible effects of climate change.

SIXTEEN

Succession

16.1 Vegetation changes

Vegetation is not static and unchanging: it can be altered in many ways. For example, wetland ecosystems, like those described in Section 15.3, are often drained by humans. When this happens the species in the community which are adapted to living in the waterlogged habitats die out and are replaced by other species which are more characteristic of drier conditions. Drained land can be ploughed and planted with crops or used for other agricultural purposes. If such drained land is abandoned it is quickly invaded by other species and turns into grassland, scrub or, given time, woodland. Often vegetation changes occur which are not caused by human intervention. In Section 15.3.5, we saw how swamp cypress needed a brief phase of drying out to regenerate and how, if the area remained under water for hundreds of years, swamp could turn into shallow lake. The reverse can also occur: a shallow lake which dries out occasionally can be invaded by *Taxodium* and become swampland.

These are examples of vegetation changes, but not all can be called succession. **Succession** is a natural change in the structure and species composition of a community. Changes caused by the direct influence of humans in clearing and replanting land have never been included in the definition of succession. If human influence declines the changes which then take place are usually classified as succession.

Succession is often taken to mean changes in vegetation alone. Other organisms associated with the vegetation types also change, but it is the plants which dominate and most influence the community structure. When applied to vegetation, succession implies a directional change from one community to another. The original concept of vegetation succession, as proposed by Clements (1916), was that succession represented a unidirectional series of changes which could not be reversed.

Clements' ideas on succession and community classification are now thought to be too precise and rigidly organised to reflect the realities of ecosystem change. We have already seen that, under the right conditions, swamps can become lakes, and lakes can become swamps: so reversals in succession may occur by the alteration of an environmental factor, in this case flooding. Because of criticisms of Clements' views, definitions of succession were broadened to include all community changes including regeneration of the same vegetation type and minor fluctuations in community structure (Gleason, 1927). This broader view was perhaps rather too all-encompassing to be helpful for the understanding of vegetation changes. The term 'succession', as used by ecologists today, falls somewhat in between these two earlier views. It is not simply the one-way changes in vegetation in a developing or maturing habitat, nor is it now taken to include the minor fluctuations or changes found in communities due to seasonality or other cyclical changes. Succession is thought of as a series of changes in community structure and species composition.

At its most extreme, this sequence of succession starts on bare, uncolonised ground which has never had any vegetation growing on it before or in newly formed lakes. This is known as **primary succession**. It is uncommon to find the early stages of primary succession today. Bare areas are rare, but they may occur on sand dunes, the lava flows of volcanos, new volcanic islands (see Section 18.3.2), landslips and glacial debris left by melting glacier ice. The remains of glacial tills left by retreating ice sheets and lakes forming in hollows and blocked glacial valleys were very common at the end of the last ice age. At that time, much of Canada and north-eastern Europe would have been exposed and subject to primary succession.

Most bare areas of land today are exposed because existing vegetation has been destroyed by fire or covered with flood silt or volcanic ash (see Figure 5.7). The vegetation recolonising the area will be affected by conditions not available in primary successions. This includes the remnants of soil, organic matter, seeds, and even some plants which have survived the changes. The sequences of vegetation developing on these previously vegetated and disturbed areas are called **secondary successions**. The changes on abandoned cultivated land are also secondary successions.

The sequence of vegetation types which occur in primary successions is often called a **sere**. Hence a **hydrosere** refers to the series of aquatic communities found in a lake, for example, and a **xerosere** is the sequence on dry land. The type of xerosere which develops on rock is often called a **lithosere**. Typical examples of a xerosere and a hydrosere are described later in the chapter (Section 16.3).

16.2 The causes of change

We have seen in Chapters 14 and 15 how a certain community will develop in particular environmental conditions. If successional change occurs in the community this suggests that the environmental conditions have altered. The vegetation change can be caused by effects of the plants themselves, **autogenic succession**, or from external environmental influences, that is by **allogenic succession**. A large number of factors can influence vegetation change. Many of these are summarised in Figure 16.1. Some of these are due to human interference, catastrophic events or the effects of seasonality and are not part of succession, although they may have considerable effects on the communities. Both human influence and catastrophes can change or destroy vegetation so that succession can occur, but the changes they create are not strictly themselves part of the succession. Other autogenic and allogenic factors are important in succession and are listed on the lefthand side of the diagram.

Autogenic succession can be brought about by changes in the soil caused by the organisms growing there. These changes include accumulation of organic matter in the litter or humic layers, alteration of soil nutrient status or change in pH caused by the plants growing there. Such autogenic soil changes were described in Section 15.2.3. The structure of the plants themselves can also alter the community.

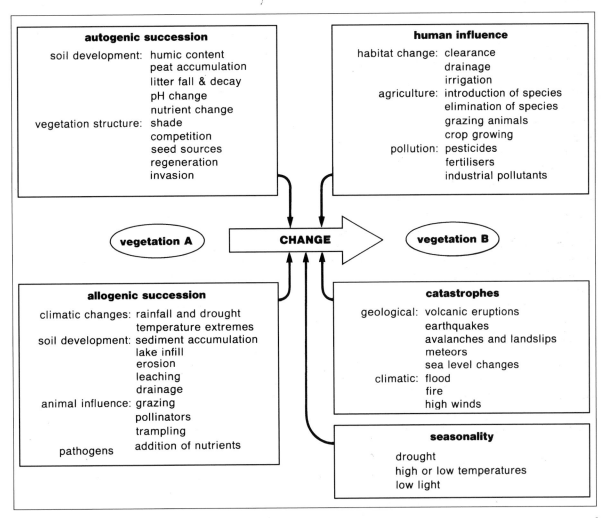

Figure 16.1 The various factors which influence vegetation change. Those on the left are the main promoters of succession. Those on the right tend to produce vegetation change not considered as succession.

For example, as larger species like trees mature, they cast shade on to the developing woodland floor which tends to exclude low-growing light-demanding species. Shade-tolerant species will invade the area instead.

Allogenic changes are caused by external environmental influences, not by the vegetation. For example, soil changes due to erosion, leaching or the deposition of silts and clays can alter the nutrient content and water relationships in the ecosystem. Animals can also be important as pollinators, seed dispersers and herbivores. They can also increase the nutrient content of the soil in certain areas, or shift soil about (as termites, ants and moles do) creating patchiness in the habitat. This may create regeneration sites which favour certain species.

Climatic factors may be very important, but on a much longer time-scale than any others. Changes in temperature and rainfall patterns will promote changes in communities. Evidence for the great effect of climate on vegetation is preserved in the fossil record of the last 500 000 years or so. Fossil pollen, preserved in peat and infilled lakes, show that the huge oscillations in climate which caused the icecaps to advance and retreat also greatly affected the vegetation of most of the world. In cold phases, tropical forest was succeeded by savannah and deciduous forest became treeless tundra. As the climate warmed at the end of each ice age, great successional changes took place. In Britain the tundra vegetation of the south of Britain and bare glacial till deposits in the north underwent succession to mixed deciduous forest. The greenhouse effect, caused by the release of massive amounts of carbon dioxide into the atmos-

phere, is now increasing global temperatures quite rapidly. In the next century we may be able to watch allogenic succession caused by changing climate to a degree never before seen by ecologists. For more on the greenhouse effect see Sections 13.3 and 21.4.1.

16.3 Examples of primary seres

16.3.1 Xeroseres

Where bare ground is colonised by organisms the subsequent primary succession is called a xerosere. About 13 000 years ago, the melting of the ice sheets from the land in the northern hemisphere must have created vast expanses where primary xeroseres could occur. Primary succession can still occur on landslips and where valley glaciers retreat to provide ground for such seres. Our understanding of the primary colonisation and later sequence of succession comes from two sources. The study of the early phases of seres can be made by direct observation of recently exposed sites. The later stages of a sere can also be investigated by analysing fossil records from regions exposed at the end of the last glacial period and by examining recent community structure.

Bare rocky areas are often invaded first by animals such as spiders which can hide in the cracks between boulders or stones. Such animals are carnivores or scavengers and they live by feeding on insects which have been blown in or flown in by mistake. There is as yet no plant matter for herbivores to survive.

The process of succession usually starts when autotrophs begin to live in the area. Both the structure and the nutrient content of exposed ground

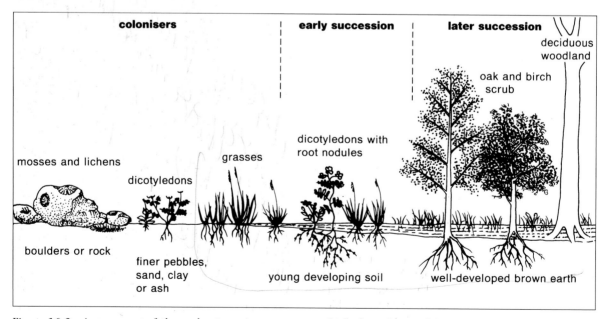

Figure 16.2   A summary of the early stages in a xerosere, which depend on substrate type, and the subsequent successional changes which occur after colonisation.

will affect which species become the first colonisers. Retreating ice sheets can leave a variety of structures from piles of large rounded boulders right through to very fine clays. Figure 16.2 summarises the xerosere on exposed glacial surfaces. If the ground consists of solid rock or very large boulders, there is no purchase for rooting plants and the first colonisers are usually lichens and mosses. These can cling directly to the rock surfaces. Slowly, the presence of moss or lichens promotes the build-up of soil in shallow depressions or cracks in the rocks as wind-blown particles are caught and added to by direct weathering of the rock surface. This creates small pockets where vascular plants can grow, but the whole process is extremely slow and may take hundreds or thousands of years before the pioneers are succeeded by another community.

If finer particles, such as silts or clays, are present in quantity in the substrate, then the initial pattern of colonisation is very different from that on rock. Such areas already have more nutrients and a more balanced water regime than bare rock. The substrate can be penetrated by roots, so that vasular plants can colonise immediately. Such sites tend to be invaded by dicotyledonous herbs, or, if they are quite nutrient rich, by grasses (Grubb, 1986). Even clay and silt substrates are poorly structured and nutrient poor compared with mature soils, so it is not surprising that the second community in these sites is rich in plants with nitrogen-fixing bacteria held in root nodules. Such communities include some horsetails (*Equisetum* spp., see Figure 16.3), legumes like *Astragalus*, *Lupinus* and *Trifolium*, the mountain avens (*Dryas* spp.) and, in areas where trees can grow, alders (*Alnus* spp.).

In some cases the nitrogen-fixing species are the primary colonisers, as is the case on the ash falls of the Mount Saint Helens eruption where *Equisetum* (Figure 16.3) and *Lupinus* were the dominant species. It is difficult to understand why nitrogen fixers are not always the first species to colonise a site. It has been suggested (Grubb, 1986) that such species have relatively large seeds because they need an initial store of nitrogen to produce their first root nodules. The heavy seed is a poor dispersal unit, so that these plants are unlikely to spread far into open areas. Most colonisers have lightweight seeds which are ideal for long-distance dispersal. It may also be that plants with nitrogen-fixing bacteria are only at an advantage where the roots of plants are crowded close together and competition for available nitrogen is great. *Equisetum* is not a seed plant: it is dispersed as spores, so this may explain its abundance on the slopes of Mount Saint Helens. The primary coloniser on volcanic ash on the Mexican volcano El Chicon (erupted March–April 1982) was also a sporebearing plant, in this case a fern, *Pityrogramma calomelanos*

(Spicer *et al.*, 1985).

Later in the xerosere, the grass and herb mixture is invaded by scrub. Early invasion of scrub is slow, but once a few bushes have become established, they form an ideal site for roosting birds. Once birds are common, bush invasion speeds up as the birds deposit many seeds on the site in their droppings which are also rich in nutrients. In northern Europe, most mid-successional species like hawthorn (*Crataegus monogyna*), privet (*Ligustrum vulgare*), blackthorn (*Prunus spinosa*), wayfaring tree (*Viburnum lantana*) and spindle (*Euonymus europaeus*) have berries which are eaten by birds as a means of seed dispersal (Fenner, 1987). Colonising trees such as the wind-dispersed birches (*Betula* spp.) and bird-dispersed oaks (*Quercus* spp.) may also establish at this stage. Birch and oak are both species which require high light intensities for their seedlings to establish. They are pioneer trees. The birches are usually short lived and fail to regenerate once the oaks mature and cast too much shade. Once woodland is established it may be invaded by more shade-tolerant species such as limes (*Tilia* spp.), ash (*Fraxinus excelsior*) and hornbeam (*Carpinus betulus*).

Figure 16.3 A horsetail (*Equisetum* sp.) colonising land covered by volcanic ash from the eruption of Mount Saint Helens.

16.3.2 Hydroseres

Hydroseres are primary succession sequences which develop in aquatic environments such as in lakes. After the glaciers of the last ice age retreated, the new lakes collected inorganic particles such as sand and clay washed in from the bare hillsides around. Once primary xeroseres developed on the bare land, sediment inwash decreased: as the vegetation reduced erosion, so the infilling of the lakes slowed. The early changes in plant succession are allogenic and very slow as a deep lake has to become shallow enough for rooted aquatics to grow. This may take hundreds to many thousands of years depending on lake depth, erosion and drainage patterns in the catchment area.

The open water of lakes may be colonised by free floating aquatic plants like the duckweeds (*Lemna* spp.) or the alga *Chara*. At the edges of the lake where the light penetrates to the floor submerged aquatics, with their roots in the sediment, grow. These often include genera such as *Potamogeton* and *Myriophyllum*. Once plants like this root in a lake, the biomass is increased and productivity increases too. More organic matter falls to the lake floor creating a fine mud.

Once submerged aquatics colonise a lake, the successional changes become more rapid and are mainly autogenic as organic matter accumulates. Inorganic sediment is still entering the lake and this is trapped more quickly by the net of plant roots and rhizomes growing on the lake floor. Areas of the lake soon become sufficiently shallow for floating-leaved aquatics to grow. These plants are rooted in the mud, but some or all of their leaves float on the water surface. These include water lilies (*Nuphar* and *Nymphaea*). The large surface leaves cast deep shade which reduces the submerged aquatic species. The plants decay to form organic muds which, as the lake becomes shallower, are invaded by emergent vegetation, such as *Phragmites* and *Typha*, to form reed swamp. These plants have creeping rhizomes which knit the mud together and produce large quantities of leaf litter as they die back each autumn. This litter is resistant to decomposition and reed peat rapidly builds up, accelerating the autogenic change. If the peat becomes raised above the water surface at dry times of the year, the community changes again to one dominated by sedges (*Carex* spp.). Sedge peat accumulates above the water level and so is no longer totally waterlogged. It can then be invaded by trees: usually alders (*Alnus*) and willows (*Salix*). The development of wet woodland (called fen carr) is often seen as the last stage in the hydrosere before the area dries out completely and typical deciduous oak woodland develops.

Often many of these communities can be seen together at a single lake. Figure 16.4 shows the edge of a lake with emergent rooted aquatics in the foreground; in the centre are emergent reed beds with irises. Bushes are invading this area and the trees in the background are on dry land. The vegetation types in the succession are mainly determined by water depth. In the lake centre it is still too deep for rooted aquatics to grow, but towards the edges the lake is shallower so that rooted aquatics can grow there.

As a lake fills in with sediment, the open area decreases in size and the vegetation types move inwards as the water becomes shallower. Figure 16.5 shows a cross-section of a lake. You can see how the different vegetation types are associated with different water depths and also how the sediments build up and fill the lake. Notice how the different plant types produce different sediments beneath them. Because different vegetation types produce different sediments a record of what has grown in the lake is fixed in its infill deposits. Old lakes, which have completely filled in, can be cored to give a long tube of sediment which goes right down to the bottom of the original lake. Analysis of the sediment shows the successional changes which have occurred in the lake during its history. The right-hand side of Figure 16.5 shows what a core taken from such a lake might show. Such a permanent record of successional changes cannot be obtained from terrestrial seres as the top surface of the soil is changing and being

Figure 16.4 A hydrosere at Thornton Ings, Yorkshire showing successional stages from rooted aquatics through to woodland.

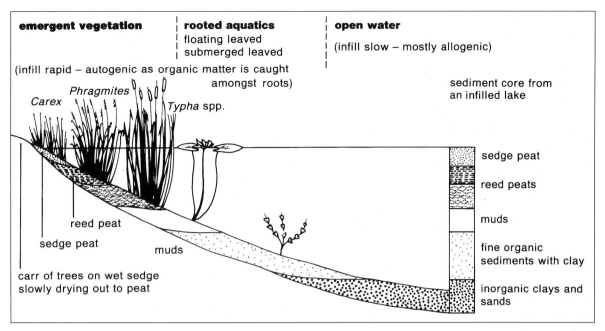

Figure 16.5 The vegetation zones and sediment types usually found in an infilling lake. If an old, infilled lake is cored to take out a column of sediment (illustrated on the right), a record of the lake's history can be discovered.

reworked rather than building up in a sequence. So hydroseres provide us with more accurate information about succession than do xeroseres. The patterns revealed by this record are discussed in Section 16.4.

16.4 Patterns of succession

16.4.1 *Variation in seres*

As mentioned in Section 16.1, early concepts of succession described seres as unidirectional vegetation changes. The presentation of 'typical seres' like those described in Sections 16.3.1 and 16.3.2 still suggests that seres are predictable sequences of change along a linear path from bare rock to forest, or from open water to sedge fen. But is this in fact the case?

Succession is typically far too slow for ecologists to observe, so that direct study of vegetation changes on one spot is impossible. Usually our understanding of succession is built up from study of several sites where more than one stage of the sere occur. This situation occurs in lakes (see Section 16.3.2) or along the valleys of retreating glaciers. Such studies are not completely satisfactory: they show the rough direction of the seral changes, but not the actual history of change. What is needed is a record of vegetation at one locality for thousands of years. This is only possible where vegetation build-up is sequential and records plant types. This is where the infilled lake cores mentioned in Section 16.3.2 are so valuable.

Figure 16.5 shows how the different sediments record the vegetation changes from a core which can be retrieved for analysis. Radio-carbon dating of the organic matter in the core provides evidence of sediment accumulation rates and the length of time any one vegetation type in the succession lasted.

Analysis of many such cores from British wetland sites shows that what actually occurs in hydroseres is a fascinating complexity of succession (Walker, 1970). The pattern of succession found in the lake hydroseres investigated is shown in Figure 16.6. The typical sequence as described in Section 16.3.2, which is often considered as the most likely course of succession, would follow the arrowed lines running round the outside of the diagram as A → B → C → D → F → oak woodland. As you can see from Figure 16.6 this is by no means the only sequence of vegetation change recorded in lake cores. Submerged aquatic vegetation (**A**) is more likely to be succeeded by reed swamp (**C**) than by floating-leaved aquatics (**B**). Although **B** is usually succeeded by **C** it can revert to **A**, presumably due to an increase in water depth in the lake. Similarly reed swamp (**C**) can be followed directly by fen carr (**F**) without the intermediate stage of sedge fen (**D**).

The study shows that seral succession is not a fixed one-way sequence of vegetation change, but a much more complex process where each vegetation type may be invaded by one of a number of different communities. So reed swamp vegetation will not always develop into sedge fen but may be invaded by trees or by *Sphagnum* bog (see Figure 15.11).

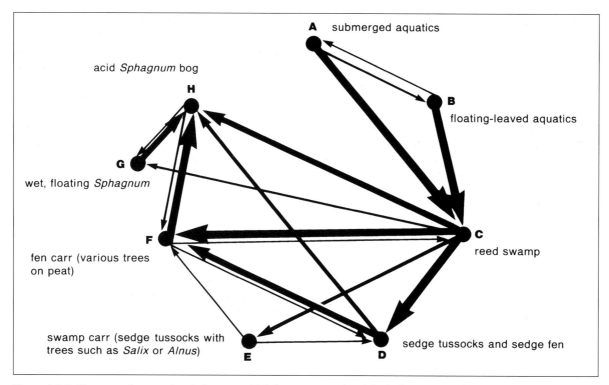

Figure 16.6 Diagram of successional changes which have occurred in lake hydroseres. Evidence comes from sediment types and pollen analysis of cores recovered from British lakes. The thickness of arrow between vegetation types indicates how often the first type was succeeded by the second. (Modified from Walker, 1970.)

In other words, there are several possible outcomes of any successional change; some will be more likely to occur than others. From the data used to produce Figure 16.6 it can be calculated that in lakes in Britain the overall probabilities (*p*) of reed swamp being succeeded by various other vegetation types are as follows:

succession by: floating-leaved aquatics $p = 0$;
   sedge fen $p = 0.26$;
   swamp carr $p = 0.16$;
   fen carr $p = 0.26$;
   wet floating *Sphagnum* $p = 0.1$;
   Sphagnum bog $p = 0.21$.

At any site, the probability of a vegetation type being succeeded by another specified type will depend on the environment at that site.

Another point to notice from Figure 16.6 is that fen carr is not the final stage in the hydrosere succession. It appears that carr usually gives way to *Sphagnum* bog. In fact *Sphagnum* communities can develop directly from several seral stages including reed swamp, sedge fen and carr. *Sphagnum* growth is promoted by several conditions. If the water in the hydrosere becomes very nutrient poor, for example where reeds fringe the lake and act as a nutrient sink by catching the richer runoff waters from the lake edge, the *Sphagnum* will develop inside the bank of reeds towards the centre of the lake. If fen peats

become oxidised and acidification occurs, because there is only infrequent flooding, then *Sphagnum* can invade in these conditions too. In carr, the shade cast by the trees tends to exclude many shade-intolerant hydrosere communities and again *Sphagnum* is a coloniser of such pools. Where the input of rain, which is nutrient poor and acidic because of carbon dioxide dissolved in it as carbonic acid, is high (over about 1000 mm a year) *Sphagnum* can even envelop the slopes around the lake hydrosere to form blanket bog (Figure 15.12).

Whether or not seres actually have a final stable end is discussed in Section 16.4.2.

16.4.2 *The end of the succession*

The crucial question is whether the vegetation changes of seres progress to an end point of a final community type. In other words, does succession cease when a particular vegetation type develops, or does it have no end, progressing so far, then changing back to an earlier community?

In the early days of the concept, the answer was thought to be obvious. Clements believed that succession progressed steadily from communities with small simple plants, like mosses, through to a final stable community with large dominant advanced plants like trees. He called this final community type the **climax vegetation**. The nature of the final community was understood to be well suited to the environment in the area, in particular the climate, and was often called the **climatic climax** vegetation. We have already met the kinds of communities which Clements would have called climatic climax as representatives of the major vegetation zones in Section 14.3. They include evergreen broad-leaved tropical forest, temperate deciduous forest and boreal conifer forest.

A problem with the concept of climatic climax came with the realisation that the climate is not stable. Oscillating climatic changes of glacial cold phases and warm interglacials occur. Even within our apparently uniform interglacial, oxygen isotopes show a gradual cooling since about 3000 BC. Historic records also show shorter periods of climatic change such as the little ice age of the seventeenth century (see Section 9.2.5) and now the greenhouse effect may well be reversing the cooling trend. The realisation that climate varied substantially led many ecologists to reject the concept of the climatic climax.

This rejection does not alter the fact that we can recognise vegetation types which are usually associated with climatic zones, as described in Section 14.3. These are mature communities which seem to develop at the end of primary or secondary seres. The term **late successional vegetation**, as a replacement for the discarded term climatic climax, has been suggested to describe them. In some respects this phrase is no better. If vegetation changes are cyclical, or at least very variable, as Walker discovered for hydroseres (Section 16.4.1), then what is 'late' in the succession when one community can be replaced by another apparently earlier in the sere?

Can we classify vegetation on the basis of its stability? A particular phase in the sere, such as the rock with lichens stage in a xerosere or open water with floating aquatics in a hydrosere, may last for thousands of years. In this sense it is very stable, but in the end allogenic factors alter the environment and change occurs. For a succession to come to an end, the final vegetation type must be able to regenerate on the site. It is not really clear if this is the case for communities such as forest or *Sphagnum* bog. Certainly trees are very long lived, and therefore, within our time-scale, seem to form a highly stable vegetation type. But do they regenerate? Many trees seem unable to establish directly underneath adults of the same species. This creates a shifting patchwork of regeneration. Often major phases of regeneration only occur after some catastrophe has destroyed part of the forest. In such cases, which species regenerate may depend on the type of catastrophe. For example, fire may promote pines (*Pinus* spp.) or *Eucalyptus*, while wind damage favours broad-leaved species rather than evergreens. So regeneration of damaged communities may not always end in the same species structure.

The second requirement for stability is no autogenic or allogenic effects. Is this true for tree communities? We saw in Section 15.2.3 that individual tree species can, in a relatively short time, promote changes in the soil profile and consequently in ground flora and soil organisms. So autogenic change is possible under trees. In other words, examples exist which show that neither of the requirements for a totally stable community are met consistently by what might be recognised as climax or late successional forest vegetation.

Perhaps succession is still continuing in tree communities, but on a time-scale which we have difficulty recognising, and so we see such communities as stable. One problem is the apparent similarity between forest ecosystems of all types. All have canopies, trunk space and a shaded ground flora. Wet tropical forest looks pretty much the same whatever the tree species and ground flora composition. These obvious similarities may simply mask the variation between communities in forest systems. As an analogy, reed and sedge fen are easily distinguished because we can readily distinguish *Phragmites* from *Carex*, but if we were the size of a small beetle crawling through the stems, the succession from one to the other might be harder to identify!

16.4.3 *Diverted seres*

So far we have looked at the network of successions which may (or may not!) lead to a stable final vegetation type such as deciduous forest or *Sphagnum* bog. Sometimes the natural sequence of succession is halted or diverted by some external factor which influences the vegetation structure or community composition. At some stage in the sere the autogenic changes, which would usually occur to promote the development of the natural succession, are prevented. Autogenic change is usually prevented in two ways: by the removal of vegetable matter which would normally build up and alter the substrate; and by the exclusion of invading plants of the next seral stage. Both these environmental constraints are applied by grazing animals and by fire. Quite often diverted seres end in a stable grassland community maintained by heavy grazing pressure or regular fires. If the environmental constraint ceases, then the grassland is rapidly invaded by species of other communities.

The effect of intensive grazing on succession has been dramatically illustrated in Europe and Australia in the last 50 years. In the early twentieth century rabbits (*Oryctolagus cuniculus*) were extremely common in Europe, especially in Britain where they are thought to have been introduced by the Normans in the eleventh or twelfth century. In 1896 a viral disease of rabbits was discovered in South America and called myxomatosis. It was introduced into France in June 1952 and reached Britain in October 1953. In only two years the disease had spread and reduced the British rabbit population to about 1% of its former level (Sumption & Flowerdew, 1985).

The effect on chalk grassland was rapid. Before 1954 rabbits had heavily grazed such areas and were responsible for keeping the grass very short (as in Figure 16.7) and thus maintaining the downland communities (Watt, 1957). Once rabbits had died out in an area, the previously short-cropped grasses grew taller and more diverse as species which had been eaten by the rabbits became established. Many plants flowered in profusion including *Primula* species and orchids. Soon, however, the diversity disappeared again as the more vigorous grasses increased in number and size. Many more seedlings survived,

especially those of woody plants such as heathers (*Calluna*, *Erica*) and gorse (*Ulex*) and scrub composed of hawthorn (*Crataegus*), juniper (*Juniperus*) and oak (*Quercus*) grew up. In other words, the diverted sere of grassland began to change into the typical scrub-woodland succession. The effects on the chalk heath of the invasion of woody species and subsequent autogenic acidification of the soil is described in Section 15.2.3. Such changes to the soil mean that even if the bushes are subsequently cleared, the soil is too acidic for the original chalk heath vegetation to re-establish. Autogenic changes also occur under hawthorn which seems to enrich the nutrient content of the soil. If it is cleared, weeds like nettle (*Urtica dioica*) and goosegrass (*Galium aparine*) grow instead of the grassland flora.

The changes in the vegetation caused by the col-

Box 16
Order and chaos

Much of this book reflects the desire of ecologists to understand the complexities of living organisms and their environment. Many ecologists, especially those who worked in the early years of the subject, felt the desire to order and classify what they studied. Different communities were given Latin sounding names, almost like individual species (see Box 14, p. 176). Similarly concepts such as succession were confidently defined and described (Section 16.1).

Recently, a new concept has entered science, that of **chaos.** The theory of chaos expresses the unpredictability of many processes. For example, weather forecasting has improved considerably this century with the use of recording stations, satellite pictures and computer predictions. There was a general feeling of optimism that ever-improving technology would produce more and more accurate predictions for further and further into the future. It turns out that this is not the case: weather is chaotic (those who live in temperate areas like Britain will say they knew this all the time!). It cannot be predicted more than about 14 days ahead however good the data collected. This is because very small differences in features such as wind speed or temperature will develop into extremely different patterns of weather. This is often called the butterfly effect: in other words, even the microturbulence of a butterfly flapping its wings can alter the weather experienced halfway round the world several days later!

Many ecosystems seem to be potentially chaotic (Berryman & Millstein, 1989). They possess positive feedback processes which cause populations to grow continuously (an example is reproductive rate; the faster the rate, the more offspring there are to have offspring and the more rapidly the population grows). They also possess negative feedback components which have a time lag before they operate. These are density-dependent factors and decrease the population size (for example, overgrazing eventually causes starvation and

reduced population growth, predation increases when there are more prey as more predators can be raised to adulthood, but when the predators increase in number they kill more of the prey species and reduce its numbers).

Slight irregularities in such systems, which are often identified as 'ecological noise', could well be governed by equations which are chaotic (Schaffer & Kot, 1986). The extent to which chaotic fluctuations are important in natural ecosystems remains unclear, though there is increasing evidence that laboratory populations of the flour beetle (*Tribolium castaneum*) sometimes show chaotic changes in population size (Costantino *et al.*, 1997; Rohani & Earn, 1997). If ecosystems are chaotic, then predicting their future becomes impossible. The oscillations in population sizes would be the result of a complex of environmental factors which we will never be able to understand fully because we cannot measure them accurately enough (see Sugihara, Grenfell & May, 1990). The outcome of many other ecological processes, like competition, succession, regeneration and features of genetic flow in populations, could all be chaotic.

Berryman and Millstein (1989) have commented on the fact that if a population is behaving chaotically, its population size will oscillate widely and will frequently fall to very low numbers. At such a time it will be more likely to become extinct. So chaotic population changes may lead to higher extinction rates. They suggest that, because of this, natural selection will tend to produce reproductive rates which keep the population within non-chaotic limits. Hereditary lines with reproductive behaviour which tends to reach chaotic levels will, in the long run, be less likely to survive than lines which remain within non-chaotic limits. They point out, however, that such biological systems have the potential to be chaotic, and may well be pushed into a chaotic regime by human interference which could alter processes such as reproductive rate or population size.

Figure 16.7 Rabbits (*Oryctolagus cuniculus*) near the entrance to their warren. In places the vegetation is so grazed that bare earth has been exposed.

lapse in rabbit numbers affected various animals. Many insect species increased in numbers, but yellow ants (*Lasius flavus*) decreased as their ant hills were overshadowed. Perhaps the most serious effect of myxomatosis on a single species, other than the rabbit, was to the large blue butterfly (*Maculinea arion*). This species requires the presence of the ant *Myrmica sabuleti* to complete its life cycle (see Figure 9.11). The ant only occurs in turf less than 1 cm tall and the large blue declined and finally became extinct in Britain in 1979 (Thomas, 1980a and 1980b). The sudden decrease in rabbit numbers had a catastrophic effect on their predators too. Without this ready supply of food, many carnivores such as foxes, stoats, weasels and birds of prey failed to breed and declined in numbers. This decrease in predators, plus the growth of ground cover, led to an overabundance of small mammals such as voles (*Microtus*) by 1956–7. Weasels, owls and other carnivores soon made use of this increase in small mammals and recovered to their earlier population sizes. More details of the amazingly complex interplay of flora and fauna revealed when a single species is removed from an apparently stable ecosystem can be found in Sumption & Flowerdew (1985).

Grazing by domestic animals such as cattle and sheep can maintain short grassland and prevent the natural succession towards scrub and woodland. Frequent fire also has an environmental effect which can produce diverted seres. Bushes and trees, especially in their young stages, are susceptible to fire and are killed by it. Grasses, on the other hand, can survive fires and regenerate from basal meristems just below ground. Tall grass, which dies back in the dry season, produces a dry, highly flammable layer of surface litter. Such grasslands will catch fire and burn quite regularly and in fact require frequent burns to maintain them and prevent succession to woodland.

Sometimes grazing can have the opposite effect from diverting succession to grassland, and instead promotes scrub growth. In the African savannahs, overgrazing of areas can prevent the fires which would kill young woody species as there is no dry grass to burn. In such conditions dense unpalatable thorny thickets (*Capparis* spp.) can grow up (Lind & Morrison, 1974). This sometimes occurs near rivers where hippopotamuses (*Hippopotamus amphibius*) live (Figure 16.8). Their close cropping of the vegetation prevents fires and may allow fire-susceptible woody plants to establish.

16.5 Human influence on succession

This chapter has concentrated so far on the natural autogenic and allogenic changes which lead to succession in vegetation or to diverted seres. But a large amount of the terrestrial surface of the Earth carries vegetation which is no longer in natural succession. Human interference for agricultural production has halted vegetation change or completely destroyed the natural vegetation.

The most obvious effects of farming practices are found on arable land where the natural vegetation has been cleared and monocultures of crops are grown instead. Such plants are usually only grown for a few months before they are harvested, the soil turned over and a new crop planted. The ground is maintained permanently at the colonisation stage of a secondary sere. The only wild plants able to survive such practices are quick-growing annual weeds which can flower and set seed before the crop is harvested (see Box 5, p. 49). Another major managed vegetation type is grassland. Grazing by sheep, cattle or other domestic animals, or mowing, either for short turf or once a year for hay, all maintain grassland and prevent succession to scrub.

The effects of human land management on preventing or diverting succession is obvious all around us. Such land use has been widespread in Europe for at least 6000 years. Within the last 500 years agricultural expansion on other continents has increased dramatically. This has resulted in the clearance of large areas of natural vegetation. Most controversial, perhaps, is the clearance of tropical rainforest for the creation of grassland for cattle. Such large-scale clearance seriously alters the ecosystems involved in a way which small-scale shifting cultivation does not. There is nutrient loss from the system via export of wood, burning and leaching. This is often followed by

Figure 16.8 Hippopotamuses grazing on the Okavango Delta, Botswana, showing the close-cropped grass that results.

erosion of the topsoil, further damaging the environment. It is uncertain what would happen if a secondary sere were allowed to develop on such land. Even quite small amounts of tropical forest which have been cleared and allowed to regenerate are usually invaded by fast-growing coloniser trees. Some areas in Brazil, known to have been cleared over 800 years ago, are still clearly recognisable as they do not have the native forest growing there (Prance & Schubart, 1978).

Summary

(1) Succession occus when the structure and species composition of a community change because of natural causes. Change due to the direct influence of humans is not considered to be succession.

(2) Primary succession is vegetation change on bare ground (xeroseres), on rock (lithoseres) or in new lakes (hydroseres).

(3) Secondary succession is colonisation and change on areas disturbed by fire, flood or cultivation where some seeds, vegetation, animals or soil structure remain.

(4) Vegetation changes caused by the effects of the plants themselves is called autogenic succession; if an external influence produces change it is allogenic succession.

(5) In xeroseres, early colonisers include sporebearing plants, such as ferns and horsetails, and species with nitrogen-fixing root nodules which can augment the nitrogen in the nutrient-poor soils.

(6) The grass and herb stage of a xerosere is usually invaded by scrub species and later by trees.

(7) In hydroseres, the initial stages of succession in deep lakes are usually slow, until the water becomes shallow enough for rooted aquatics to grow.

(8) Environmental factors, such as fire and heavy grazing pressure, can halt or divert succession.

SEVENTEEN

Biomes

17.1 How many biomes are there?

Many ecologists have tried to classify the principal vegetation types into which the world is divided (Mueller-Dombois, 1984). The world vegetation types have come to be known as **biomes** and aquatic biomes have also been recognised. It is impossible in a single chapter to look at all the biomes in any detail. Instead, we will try to give a flavour of what life must be like for some of the organisms found there. We will also consider why particular biomes are found where they are. Some biomes have already been considered in earlier chapters and we will refer to these again only briefly.

No two ecologists seem to agree on how many biomes there are. This is hardly surprising as a biome is not a natural unit. If individual ecosystems are the 'species' of ecology, then biomes are the phyla or divisions. No two taxonomists seem to agree on a system of classification, so it is hardly surprising that a definitive list of biomes cannot be produced. This does not mean, however, that the concept of the biome has no use. Biomes provide a convenient shorthand for describing the world's flora and fauna. Traditionally biomes have been defined mainly in terms of their vegetation. However, in our descriptions below we will look at the animals in them (and sometimes the fungi, protoctists and prokaryotes too) as well as at the plants.

Whittaker (1975) provides a comprehensive list of the world's biomes and we will follow his approach. Friday and Ingram (1985) and Archibold (1995) provide valuable updatings of Whittaker's treatment.

17.2 The world's terrestrial biomes

The distribution of the world's major terrestrial biomes is shown in Figure 17.2.

17.2.1 Tropical rainforest

Tropical rainforests occur near the equator where rainfall is abundant and occurs throughout the year. They are found in South and Central America, West and Equatorial Africa, South-east Asia, Indonesia and North-east Australia (Whittaker. 1975). This continual combination of warmth and moisture allows continuous plant growth to occur. Even today we still know desperately little about the ecology of these tremendously impressive and productive ecosystems. We do not even know to within a factor of two how many different species they contain. It has been said that if all the soil invertebrates found in a cubic metre of tropical rainforest soil are collected, there will probably be at least one species there which has never been named or described by scientists. With the possible exception of the ocean benthos (Section 17.4.10), there is little doubt that tropical rainforests contain the greatest diversity of life of any of the world's biomes, though the reasons for this remain unclear (Grubb, 1996).

Contrary to popular belief, undisturbed tropical rainforest is not impenetrable (Figure 17.1). This is because so little light is able to get through to the forest floor that relatively few plants can grow there.

Figure 17.1 Tropical rainforest on Monte Verde, Costa Rica.

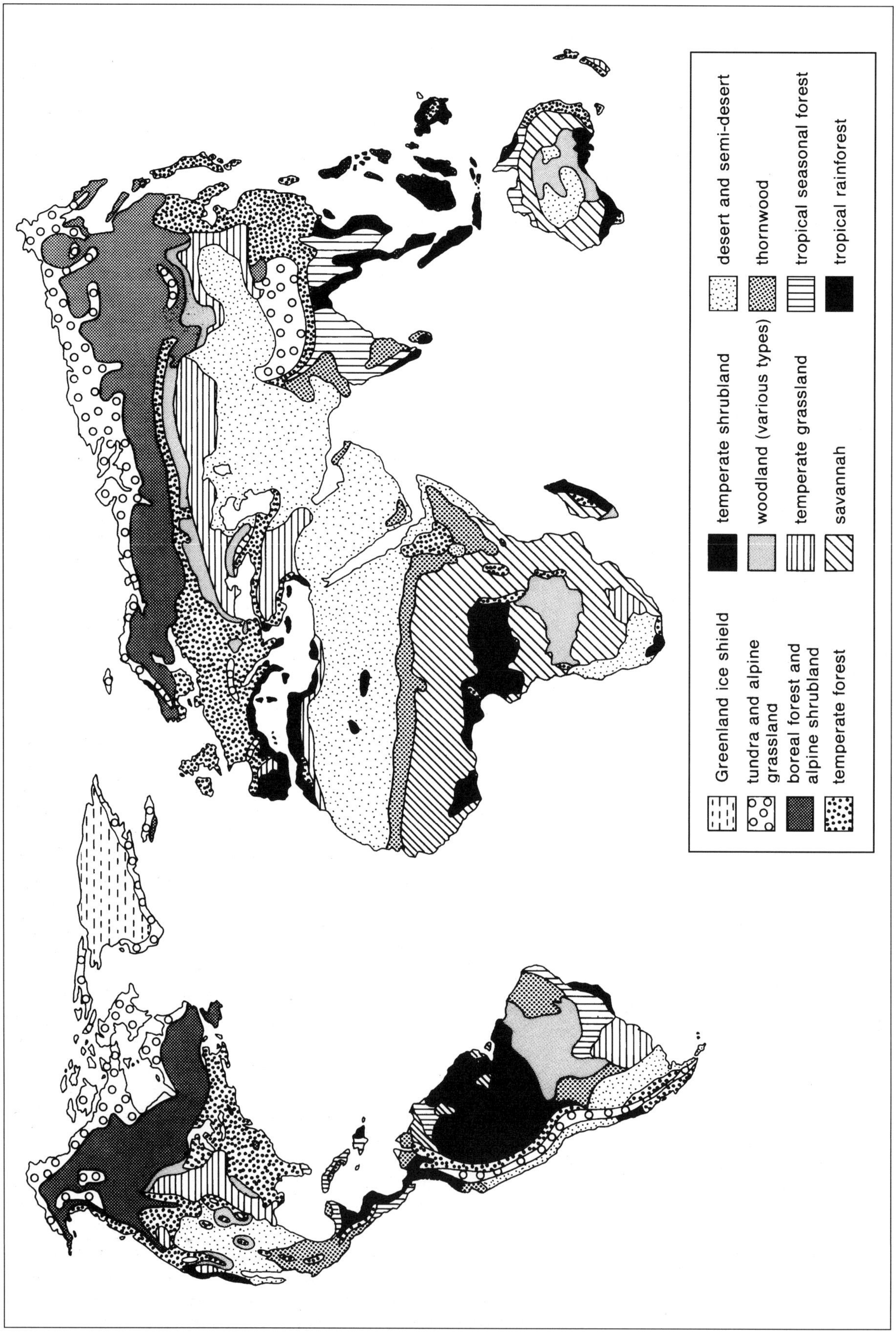

Figure 17.2 The world's major terrestrial vegetation types. (From Friday & Ingram, 1985.)

It is up in the canopy, some 30–50 m high, that most photosynthesis occurs. It is only recently that ecologists have begun to study these canopy trees. One technique is straightforward in principle: climb the trees (Attenborough, 1984). This method is increasingly being used, though it has its dangers: some people have fallen and been killed in the process.

Many trees in the canopy are covered by **epiphytes** – plants growing on other plants. Whole communities are found up here. Indeed, nutrient cycling within the canopy is probably important and many trees produce aerial roots which absorb nutrients just as roots in soil do. Much of the animal life too is confined for most of the time to the canopy. Fruits are found throughout the year and specialised fruit-eaters have evolved among the insects, the birds and the primates.

Other animals concentrate on leaves. All species of sloth (*Choloepus* and *Bradypus*), for instance, have extremely large multi-compartmented stomachs which hold cellulose-digesting bacteria. A full stomach may account for 30% of the body weight of a sloth, and meals may be digested there for more than a month before passing into the relatively short intestine (Dickman, 1985). Some of the adaptations shown by sloths to this indigestible diet are remarkable. Their body temperature is low, about 30–34 °C, and variable, falling at night, during wet weather and when the animals are inactive. This helps conserve energy (Montgomery & Sunquist, 1978). One apparent disadvantage with their slow life style is that they cannot run from predators. Camouflage, provided by two species of cyanobacteria which live in their fur and turn it green, helps reduce predation. At any event, they seem quite successful as sloths are the most abundant large mammals in tropical South American forest (Dickman, 1985).

17.2.2 Elfinwood

To some extent **elfinwoods** resemble tropical rainforests in miniature. They get their name from elves, the mythical little people recorded in many European fairy stories. Elfinwoods occur high up on tropical mountains in Africa, South America and New Guinea. Here it is cold, but the climate is non-seasonal. The cold leads the trees to be small and stunted, with contorted branches and a low canopy of broad evergreen leaves (Whittaker, 1975). The branches are covered with curtains of lichens, mosses and filmy ferns (Walter, 1979). These epiphytes occur because the relative humidity is typically very high and elfinwoods are sometimes called cloud forests.

17.2.3 Tropical seasonal forest

Tropical seasonal forests occur in humid tropical climates with a clear dry season during which trees may lose their leaves. They are found in India, Southeast Asia, West and East Africa, South and Central America, the West Indies and Northern Australia (Whittaker, 1975). It is their seasonality that probably accounts for their being less diverse than the rainforests. Tropical seasonal forests are often found where monsoons occur as these provide seasonal rain.

17.2.4 Tropical broad-leaved woodland

Tropical broad-leaved woodlands contain small trees and replace tropical seasonal forests when the climate gets drier and the soils poorer. The canopy is typically only 3–10 m high and consists of trees or shrubs with twisted branches and thick, fire-adapted bark (Whittaker, 1975). Tropical broad-leaved woodland occurs in northern South America, the West Indies, southern Africa and Burma.

17.2.5 Thornwood

Thornwoods occur in the same parts of the world as tropical broad-leaved woodlands and in other parts of Asia and in Madagascar. The climate, though, is more arid (Whittaker, 1975). The name of the biome comes from the presence of spiny species of trees in the genus *Acacia* and in other genera of the pea family (Fabaceae).

17.2.6 Temperate rainforest

Temperate rainforests occur along the Pacific Coast of North America and in New Zealand, Australia and Chile. Their climate is cool and maritime, lacking great variation in temperature and with abundant summer rain and much cloudiness and fog (Whittaker, 1975). In common with tropical rainforest, they have rain throughout the year, though at some times of the year the 'rain' is condensed fog. This fog comes from moisture brought in by winds from the sea.

The trees in temperate rainforest are the tallest in the world. In Australia, the dominant tree of these forests is the mountain ash (*Eucalyptus regnans*) which can grow to over 90 m in height (Bergamini, 1965). In North America, the dominant tree is the redwood (*Sequoia sempervirens*) which may reach 100 m (Figure 17.3). The New Zealand forests include trees in the genus *Podocarpus*. Unfortunately, the flora of the New Zealand forests is being severely damaged by introduced deer (Friday & Ingram, 1985). Indeed, the slogan 'Save the rainforests' applies even more urgently to temperate than to tropical rainforest. Canada's rainforests are disappearing at an alarming rate, mainly due to large-scale logging. According to the World Resources Institute, two-thirds of British Columbia's temperate rainforest has been degraded

Figure 17.3 Giant redwood trees (*Sequoia sempervirens*) from temperate rainforest in the Muir Woods National Monument, California, USA.

by logging or other development (Greenpeace, 1997a). These rainforests are home to the rare and remarkably beautiful 'white Spirit Bear' – a white form of the grizzly bear (*Ursus arctos*) which is threatened with extinction. In April 1997, the environmental campaigning organisation Greenpeace began a campaign to save these bears with a march of sixty 'bears' (Figure 17.4) to the headquarters of the three major logging companies in Vancouver.

Figure 17.4 Greenpeace activist dressed as a bear protesting against the logging of temperate rainforest in British Columbia, Canada.

17.2.7 Temperate deciduous forest

Temperate deciduous forests grow in continental climates with summer rainfall and severe winters. They are dominated by broad-leaved deciduous trees and are found in Europe, Asia and the Americas (Whittaker, 1975). Oak woodlands are discussed in detail in Section 14.2.2 and so temperate deciduous forest will not be considered here any further.

17.2.8 Temperate evergreen forest

Temperate evergreen forests occur in a variety of climates. Those in California, the Mediterranean and Southern Australia are dominated by trees with tough, evergreen and broad, but relatively small, leaves. North of California in Western USA, conifers predominate, including Douglas fir (*Pseudotsuga menziesii*) and sitka spruce (*Picea sitchensis*). In New Zealand, various trees are found including southern beech (*Nothofagus*), which also occurs in the temperate evergreen forests of Chile.

The New Zealand forests include some remarkable birds (Figure 17.5). Perhaps the most unusual are the flightless kiwis (*Apteryx* spp.). New Zealand has no native mammalian ground predators and so many of the indigenous birds have lost the ability to fly. Although kiwis have poor eyesight and cannot fly, their sense of smell is excellent and they use this to hunt out worms on which they feed (Friday & Ingram, 1985). Also in New Zealand is the kakapo (*Strigops habroptilus*), a very rare bird that is almost unable to fly. It was long thought extinct, but a few are known to survive, feeding on leaves, young shoots, berries and moss. A third unusual bird of these New Zealand forests is the kea (*Nestor notabilis*). This is one of the few New Zealand species to have benefited from introduced species. Previously vegetarian, the kea has learned to attack sheep and

Figure 17.5 The common or brown kiwi (*Apteryx australis*), a flightless bird which forages on the ground using its long beak with sensitive hair-like feathers.

is now the world's only carnivorous parrot.

Across much of Europe a form of temperate ever-green forest is found on mountains where the growing season is too short for deciduous trees. The trees include silver fir (*Abies abies*), Norway spruce (*Picea abies*) and several species of pine (*Pinus*). The forest has the structure of the mountain form of the boreal forest described in Section 17.2.11. Mammals include the herbivorous red squirrel (*Sciurus vulgaris*) and chamois (*Rupicapra rupicapra*), and the carnivorous pine martin (*Martes martes*) and lynx (*Lynx lynx*). Among the birds are the nutcracker (*Nucifraga carypcatactes*) and crested tit (*Parus cristatus*).

17.2.9 Temperate woodland

In Whittaker's (1975) classification, temperate woodland occurs when the climate is too dry for true forest. It is dominated by small trees, but these often provide only an incomplete and open canopy, so that the appearance may be of grassland with occasional scattered trees. Temperate woodland therefore lies on a continuum between forest and shrubland, grassland and semi-desert. An example of temperate woodland is pygmy conifer woodland in the western United States (Whittaker, 1975). Here, open woodland dominated by pines (*Pinus* spp.) and junipers (*Juniperus* spp.) is found with shrubs and grass beneath.

17.2.10 Temperate shrubland

Around the shores of the Mediterranean Sea, the annual rainfall is between about 300 and 800 mm, and during the summer there is usually no rain for about four months. In these months the mean temperature is 20–25 °C. In the coldest months the mean temperature is about 10 °C and frosts are only sporadic (Friday & Ingram, 1985; Walter, 1979). Very similar climates occur in southern California, parts of Chile, the Cape of Good Hope in southern Africa and South-western Australia. The vegetation is very similar in these separated regions, even though the individual plant species differ greatly between the regions.

The shrubs are between 1 and 5 m high, and have small thick drought-resistant leaves. Mediterranean **maquis** and Californian **chaparral** are among the best known examples of this biome. **Heathland** is also included in this classification, although it can only just be classified as shrubland. All of these types of shrubland are greatly affected by fire. These fires burn off the above-ground stems. Regrowth then occurs from root systems that survive the fire and from buried seeds (Gimingham, 1972; Whittaker, 1975).

Californian chaparral contains shrubs of *Quercus*, *Ceanothus*, *Arctostaphylus* and many others forming a low scrub which is highly inflammable. Fires are often started by lightning. If a fire occurs every 12 years or so, the chaparral can perpetuate itself. However, if fires occur too often, those species which rely on regeneration from seed, rather than on regrowth from roots, are eliminated and the composition of the vegetation changes. If fires occur much less often than once every 12 years, fire-intolerant species such as *Prunus ilicifolia* and *Rhamnus crocea* invade (Walter, 1979).

Heathland occurs in Ireland, Scotland, England, Denmark and parts of northern Germany and southern Sweden (Friday & Ingram, 1985). The vegetation is dominated by low shrubs in the heather family (Ericaceae). These heaths generally occur on very poor and shallow sandy podzols. The origins of heathland have been widely debated. One theory is that they are derived from forest by grazing and burning. Another theory is that they have never been forest; rather, they represent a 'sub-climax' community. Evidence from archaeological sites and from pollen analysis now favours the former theory. Further, the disappearance of tree pollen often coincides with the appearance of weed and crop pollen, strongly implicating the activities of early farmers. Much of the original clearance probably occurred in the Bronze Age or Iron Age.

Much of the heathland of northern Britain is still managed for grouse (*Lagopus scoticus*), a slow-flying bird which is easy for people to shoot. Regular burning of the heather at intervals of 10–15 years stimulates new growth on which the birds feed. The older, taller heather affords cover for breeding sites. Throughout much of Europe, heathland is being destroyed for conifer plantations, agricultural reclamation and housing developments. In Dorset, for instance, the Dartford warbler (*Sylvia undata*) is declining as the gorse (*Ulex* spp.) and tall heather in which it lives are destroyed (Figure 17.6) (Simms, 1985).

17.2.11 Boreal forest

Boreal forest is also known as **taiga**. It extends from north-eastern Europe across Russia to the Pacific Ocean, and right across North America from Alaska to Newfoundland (Friday & Ingram, 1985; Archibold, 1995). To the north it merges into tundra; to the south it grades into deciduous forest or grassland. It can be very cold in the taiga: Eastern Siberia is the coldest area in the northern hemisphere with a mean January temperature of −50 to −60 °C. Permafrost is widespread, so that the deepest layers of the soil remain frozen all year round, and for much of the year snow lies on the ground. The soils are podzols (see Section 15.2.2).

The vegetation of the boreal forest is dominated by

Figure 17.6 Dartford warbler (*Sylvia undata*) nesting in gorse on Winfred Heath, Dorset.

Figure 17.7 A pair of crossbills (*Loxia curvirostra*) at their nest in a conifer tree in Scotland.

17.2.12 Savannah

Savannahs are tropical grasslands, often with scattered trees (Whittaker, 1975). They are most extensive in Africa, but are also found in Australia, South America and southern Asia. Savannah is subject to fire. Fires may result from lightning or be started by humans. Much of the African savannah is burnt each year. Savannah trees have thick bark which insulates the living cambium from the heat of the fire (Friday & Ingram, 1985).

The grasses that dominate the savannah are largely perennial. Most of them have coarse leaves that die back to form a protective layer around the meristems which give rise to new growth the following year (Archibold, 1995). Most of the grasses have rhizomes, and new plants usually arise vegetatively. Maximum growth rates are found at around 35 °C, some 10 °C warmer than the optimum for temperate grasses, and a light intensity twice the optimum for temperate grasses. The reason for this is that most tropical grasses have a different photosynthetic mechanism to temperate plants. The initial compound formed when carbon dioxide is fixed has four carbon atoms. Because of this, these plants are known as C_4 plants. In temperate plants, the initial compound formed when carbon dioxide is fixed has three carbon atoms: such plants are called C_3 plants. It is this biochemical difference which allows C_4 plants to make better use of carbon dioxide at high temperatures and light intensities.

The savannah or grassland of East Africa is famed for its abundance of wildlife. Despite the activities of hunters, killing elephants for their ivory, rhinoceroses

coniferous trees. With the exception of the larches (*Larix* spp.), these are evergreen. The important genera are pines (*Pinus* spp.), spruces (*Picea* spp.), firs (*Abies* spp.) and larches. Some broad-leaved species also occur, notably poplars (*Populus* spp.) and birches (*Betula* spp.). The ground flora includes some exquisite and delicate flowering plants such as twin flower (*Linnaea borealis*), one-flowered wintergreen (*Moneses uniflora*) and creeping lady's tresses (*Goodyera repens*).

With its harsh winters, the taiga poses severe problems for the animals found in it. Some of the birds and mammals migrate, some remain active during the long dark winter and some of the mammals hibernate. Most of the invertebrates survive as resting stages, typically eggs or pupae (Friday & Ingram, 1985).

Many of the animals in the taiga have evolved specific adaptations to the conifers. The conifers produce large numbers of seeds, which provide a valuable food source for many birds. One group of finches, the crossbills, are particularly interesting because their mandibles, which cross at the tips, enable them to get at the seeds before the cones open (Figure 17.7). Other seed-eating birds have to wait until the cones open before they can feed. Many ecological studies have been done on the animals in the taiga. Some of the best evidence for interspecific competition comes from studies on the birds there (Oksanen, 1987).

Figure 17.8 Migrating wildebeest (*Connochaetes taurinus*) in the Ngorongoro Crater, Tanzania.

for their horns and cats for their skins, enormous numbers of large herbivores and carnivores still roam East Africa. More is known about the natural behaviour of the animals and their ecology in the Serengeti of northern Tanzania and southern Kenya than about any other region in the world (Sinclair & Norton-Griffiths, 1979).

The Serengeti still has vast herds of wildebeest (*Connochaetes taurinus*). These move around, following the same path that the rain does (McNaughton, 1979), as they seek the best grass on which to feed and calve (Figure 17.8). Other large grazers found in great numbers include zebra (*Equus burchelli*), buffalo (*Syncerus caffer*) and Thomson's gazelle (*Gazella thomsoni*). Browsers include impala (*Aepyceros melampus*), giraffe (*Giraffa camelopardalis*), eland (*Tautotragus oryx*), black rhinoceros (*Diceros bicornis*) and gerenuk (*Litocranius walleri*). These herbivores support large numbers of mammalian carnivores including spotted hyena (*Crocuta crocuta*), lion (*Panthera leo*), leopard (*P. pardus*), cheetah (*Acinonyx jubatus*), African wild dog (*Lycaon pictus*) and three species of jackal (*Canis* spp.).

Ecologists are only beginning to unravel the interspecific relationships in savannah ecosystems. In Africa, for example, the near extinction of elephants in certain areas may lead to an increase in woodland and accompanying decrease in grassland. This has obvious consequences for the relative numbers of grazers and browsers. The removal of one species may therefore greatly affect the populations and species composition both of the flora and the fauna.

17.2.13 Temperate grassland

Temperate grasslands are found across large areas of eastern Europe and Asia (**steppe**), central North America (**prairie**) and Argentina (**pampas**). The climate is moderately dry (between 200 and 750 mm of rain each year) and continental, so the summers are hot and the winters cold. The flora of these

Figure 17.9 Pampas grass (*Cortaderia*) on the pampas of Uruguay, South America.

temperate grasslands is dominated by perennial grasses (Figure 17.9). Associated with the grasses are broad-leaved perennials. The smaller broad-leaved plants flower early in the growing season, before the grasses reach their maximum height; the larger broad-leaved plants flower towards the end of the growing season, after the grasses have begun to die down (Friday & Ingram, 1985).

The Eurasian steppes lie between the forests to the north and the deserts to the south. The soils are chernozems (see Section 15.2.2) and nowadays most of the steppe has been devoted to the production of wheat. Only in a few reserves and in parts of eastern Asia is natural steppe still found. Once the huge areas of wild grassland supported great herds of Przewalski's horse (*Equus przewalskii*). Przewalski's horse is now possibly extinct in the wild, though still found in a few zoos where it breeds well. It is a possible ancestor of the domestic horse which was first domesticated on the steppe about 6000 years ago. Przewalski's horse, with the wild ass (*E. hemionus*), saiga antelope (*Saiga tatarica*) and the now extinct tarpan (*E. gmelini*) are the ecological equivalents of the grazing antelopes and zebra found on the African savannah.

There are not very many species of birds on the steppe, probably because of the uniformity of the vegetation structure. Even some of the eagles (*Aquila* spp.) are forced to nest on the ground, given the absence of trees or cliffs. Nor is the steppe ideal for amphibians or reptiles. The few amphibians found are burrowers, while the short growing season gives little time for eggs to hatch and develop to adulthood.

What reptiles there are tend to be **ovoviviparous** – they produce eggs which hatch within the mother, so that the young are born live (without eggs) and can start to look for food straight away (Friday & Ingram, 1985).

17.2.14 Alpine shrubland

Alpine shrublands are sometimes found above the treeline on mountains. They occur in South America, Africa, the Himalayas and New Zealand. On mountains in Africa a characteristic vegetation is found in these circumstances dominated by large sparsely branched rosette plants which may be 5–8 m in height. The genera *Dendrosenecio* and *Lobelia* are most prominent (Friday & Ingram, 1985). These Afro-alpine biotas are considered in more detail in Section 18.3.5.

17.2.15 Alpine grassland

The climate high up a mountain might be thought to be similar to the climate near the poles. However, day length remains constant up a mountain, while it changes as you approach the poles. Further, mountains often receive more rain and snow than regions near the poles. The thin atmosphere lets through more ultraviolet radiation, and diurnal changes in temperature may be extreme (see Section 18.3.5). The soils may be very thin and large amounts of loose rock are often present. These differences combine to ensure that mountain vegetation differs significantly from the vegetation of the tundra (discussed in Section 17.2.16).

Small alpine plants are often much loved by gardeners. They tend to grow slowly and have beautiful, delicate, brightly coloured flowers. Well known genera include *Pulsatilla*, *Hepatica*, *Saxifraga*, *Primula*, *Androsace*, *Gentiana*, *Campanula*, *Fritillaria* and *Crocus*.

The alpine marmot (*Marmota marmota*) occurs throughout the Alps of central Europe (Figure 17.10) and is well adapted to life in the mountains. Marmots

Figure 17.10 Alpine marmot (*Marmota marmota*) feeding in front of its den.

hibernate at least half the year (MacDonald, 1985). The entire family unit (up to 15 animals) retreats into an excavated den and sleeps huddled together throughout the winter. During hibernation, their heartbeats slow down and their body temperature falls to between 4 and 8 °C. This helps the marmots to conserve energy. During the few months of the year when they are active they seem to spend much of their time standing on their hind legs checking whether there are any predators in the vicinity.

17.2.16 Tundra

Huge areas of northern Asia and northern Canada are covered with low treeless vegetation known as tundra. The growing season is too short, the winters too cold and dry, and the soil too unstable to support trees. Primary production is low, as might be expected (Wielgolaski *et al.*, 1981). The soils are poor and the annual freezing and thawing gradually turns the soil over and sorts the soil particles out by size. On flat ground the result is often a characteristic polygonal pattern which results from the larger stones being forced to the surface (Everett *et al.*, 1981; Friday & Ingram, 1985).

The growing season is so short that few plants in

Figure 17.11 The colourful flowers of the tundra near the Hofsjokull ice cap in Iceland.

the tundra are annuals. Sedges (*Carex* spp.), lichens and dwarf willows (*Salix* spp.) are found. When the short growing season does arrive, plant growth and flowering may be spectacular with enormous areas bathed in colour (Figure 17.11). Tremendous numbers of insects suddenly appear and for a few weeks productivity is high before winter sets in again for most of the year.

17.2.17 Warm semi-desert scrub

Warm semi-desert scrubs occur throughout the world in dry warm-temperate and subtropical climates (Whittaker, 1975). The vegetation is characterised by scattered bushes, while the fauna is often rich in small mammals, birds, lizards and snakes.

17.2.18 Cool semi-desert

Cool semi-deserts are found in parts of North America, central Asia and Iran, Australia, Patagonia and the Andes (Whittaker, 1975). The vegetation often contains scrub and is found in places too dry for grassland.

17.2.19 Arctic–alpine semi-desert

Arctic-alpine semi-deserts are found in arid regions either in the Arctic or on mountains above the tree-line. The vegetation usually consists of low cushion plants.

17.2.20 Desert

Deserts are found throughout the world, mostly in the subtropical zone between 15° and 40° north and south of the equator. Among the great deserts of the world are the Namib and Kalahari deserts in the south of Africa, the Sahara in the north, the Arabian desert, the Gobi to the north of the Himalayas and, perhaps the driest of them all, the coastal Atacama desert of Peru and Chile (Friday & Ingram, 1985).

Some deserts receive up to 200 mm of rain a year, but most get less that 50 mm (Figure 17.12). Not only that, but the rain is unpredictable. For most desert organisms, the key to survival is being able to make use of the occasional heavy rainfalls. There are two extreme strategies open to plants. One is to survive as an **ephemeral**. Some of these ephemerals can complete their life cycle in a few weeks. They can take advantage of sudden rains by germinating, growing, flowering and seeding within the space of just 20–30 days. Other plants survive as perennials. **Geophytes** effectively avoid periods of drought by surviving as underground corms or bulbs. Above-

Figure 17.12 Sand dunes and creosote bushes from Death Valley, California, USA.

Box 17.1
Desertification

Deserts are getting bigger! The spread of deserts into semi-arid lands is known as **desertification**. It is generally agreed that desertification is an important phenomenon but it is very difficult to quantify. Deserts are more difficult to map unambiguously than, say, lakes or forests. In addition, because the rainfall that deserts receive is unpredictable, it is natural for their area to increase in some years and then decrease in others even in the absence of human activity. Table 17.1 shows the changes in the area and southern boundary of the Sahara during the 1980s (Tucker *et al.*, 1991). During this decade the Sahara grew in size by 7.4% and its southern boundary moved south by 132 km. However, the year-on-year changes were dramatic and in either direction: in some years the Sahara grew; in others it shrunk.

Table 17.1 Changes in the area and southern boundary of the Sahara during the 1980s.

Year	Area (km²)	Change from previous year (%)	Distance (km) and direction of movement of southern boundary	
1980	8 630 000			
1981	8 940 000	+3.6	55	South
1982	9 260 000	+3.6	77	South
1983	9 420 000	+1.7	11	South
1984	9 980 000	+5.9	99	South
1985	9 260 000	−7.2	110	North
1986	9 090 000	−1.8	33	North
1987	9 410 000	+3.5	55	South
1988	8 880 000	−5.6	99	North
1989	9 130 000	+2.8	44	South
1990	9 270 000	+1.5	33	South

Most ecologists agree that desertification can be caused by excessive demands made on the environment by people and/or adverse weather conditions. The relative importance of these two factors remains controversial, but the argument that people contribute to desertification goes as follows. An increase in the number of people leads to **overgrazing** by domestic animals such as sheep and goats, or to a decrease in the number of trees as they are cut down for firewood. Overgrazing and/or the loss of trees leads to a loss of vegetation and a sharp increase in **soil erosion**, whether by winds or occasional floods. Soil erosion leads to greater demands being made on the productive land that is left. As a result, the carrying capacity of the environment is grossly exceeded and the fragile soils of semi-arid regions are destroyed. People move to neighbouring semi-arid regions, causing them, in turn, to become desert.

Desertification may also result from a shift to a settled lifestyle by previously nomadic people. Even the provision of water holes by well meaning aid organisations can result in desertification as domestic animals destroy the vegetation around them by their frequent trampling.

Desertification can be reversed. The Kenyan Green Belt Movement has shown how beneficial tree planting in semi-arid regions can be. This project started in the early 1970s and has largely been managed by Professor Wangari Maathai. She has worked with the traditional wood gatherers of Kenya, women who know from bitter experience the problems caused by a lack of trees. In all, some 600 tree nurseries have been established and over 7 million trees planted. Such trees provide wood for fuel and foliage for cattle. In addition, they help prevent soil erosion.

ground parts are produced only after heavy rains. **Succulents**, though, such as the cacti of America and the *Euphorbia* of Africa, have adaptations which enable them to survive above ground throughout the year. Succulents typically have thick cuticles, a very low surface area to volume ratio and sunken stomata which only open at night so as to minimise transpiration losses.

Animals in deserts face a formidable array of problems. Water is obviously scarce, but so is food. It may be very hot during the day, but cold at night. Sand makes locomotion difficult and there may be occasional blinding sandstorms. A wide variety of mechanisms have evolved to deal with these problems (Louw & Seely, 1982: Schmidt-Nielsen, 1983). In Section 8.6.4 we looked at the adaptations of naked mole rats to desert conditions. Here we will concentrate on the camel (*Camelus dromedarius*) (Schmidt-Nielsen *et al.*, 1957; Schmidt-Nielsen, 1964, 1983).

Camels can tolerate a 30% loss of their total water content. Most mammals die if 14% of their water is lost. Even when severely dehydrated, camels continue to eat at almost their normal rate. In most other mammals, food is refused once the animal becomes dehydrated. Recovery from dehydration is spectacular. Camels can drink 20% of their body weight in ten minutes! When a camel becomes short of water, its temperature fluctuates more than is otherwise the case (Figure 17.13). By allowing its daily temperature to vary by as much as 7 °C, the camel saves water which would otherwise have to evaporate to keep it cool. The thick fur of a camel insulates its body from the sun: in one experiment a sheared camel increased its water loss by about 50%. Further adaptations to desert life are seen in the splayed hooves, ideal for walking on sand, and in the hump which stores fat. This fat serves two functions, providing both energy and water on being respired.

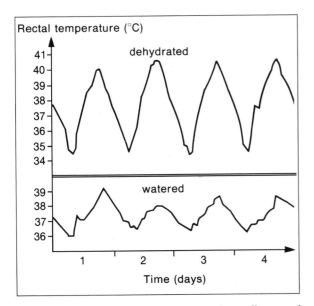

Figure 17.13 The body temperature of a well-watered camel fluctuates on a hot day by only about 2 °C. A dehydrated camel, though, must tolerate fluctuations of its body temperature of up to 7 °C. (From Schmidt-Nielsen, 1983.)

17.2.21 Arctic–alpine desert

Greenland, Antarctica and mountains well above the tree line support almost no plant life. The environment is dominated by rock, ice and snow. Arctic–alpine desert is the home for just a few wonderfully adapted animals. The Arctic is home to the polar bear (*Thalaretos maritimus*), while in Antarctica the emperor penguin (*Aptenodytes forsteri*) breeds in midwinter when it is appallingly cold and there is darkness 24 hours a day (Figure 17.14).

Figure 17.14 Emperor penguins (*Aptenodytes forsteri*).

17.3 Wetland and freshwater biomes

17.3.1 Cool temperate bog

Bogs are wetlands that only receive water from rainfall. They tend to be nutrient poor and dominated by mosses in the genus *Sphagnum*. They are described in Section 15.3.6.

17.3.2 Tropical freshwater swamp forest

As defined in Section 15.3.5, swamps are wetlands in which trees are the dominant vegetation. The best developed tropical freshwater swamp forests are those of the Amazon basin (Friday & Ingram, 1985). Minerals in the waters and accompanying soils tend to be scarce, so that primary productivity is low. Consequently, there are few animals. In the blackwater tributaries of the Amazon, the water is dark brown due to the presence of humic acids which also lower the pH. Here the tamaqui fish (*Colosoma bidens*) has become dependent on the flowers and fruit which fall from the forest trees into the water (Friday & Ingram, 1985).

17.3.3 Temperate freshwater swamp forest

The vegetation of the temperate freshwater swamp forests of the Everglades, Florida is described in Section 15.3.5. Animals are abundant. There are some 250 species of birds, 240 species of fish, 57 species of reptiles, 25 species of mammals and 17 species of amphibians (Moss, 1988). In the 1890s millions of feathers from egrets and other Florida birds were taken for hats and by 1930 over 100 000 alligators were being taken each year for their hides. Stringent laws now protect these animals and the danger to the Everglades' ecosystem, as to so many wetlands, is from interference with the natural drainage.

17.3.4 Lakes and ponds

As discussed in Section 15.3.7, crucial determinants of the ecology of standing freshwater are the extent to which the water is present all the year round (whether or not the pond is ephemeral) and whether nutrients are abundant (when the pond or lake is said to be eutrophic) or scarce (oligotrophic). We will look briefly at the North American Great Lakes (Moss, 1988).

The North American Great Lakes are among the largest in the world. They drain a huge area which until 200 years ago consisted of deciduous forest to the south and conifer forest to the north. The history of the last 200 years has been one of progressive changes in the composition of the fish species.

Fishing for salmon (*Salmo salar*) ceased by 1900. The salmon probably disappeared because of changes in the streams flowing into the lake. These streams were essential for spawning, but in the nineteenth century dams were built to operate sawmills, and large quantities of sawdust polluted the streams. Repeated attempts to re-establish the salmon have failed because the stream flows are now insufficient to maintain the cool, oxygenated waters and gravel bottoms which the salmon require.

By the end of the last century sea lampreys (*Petromyzon marinus*) had entered Lake Ontario. They may have come via the canal built in the early 1880s between the Hudson River and Lake Ontario. By 1946 the lamprey had spread to Lake Erie, Lake Huron, Lake Michigan and Lake Superior. The lamprey is a parasite. It attaches itself to other fish by the sucker which forms its jawless mouth. The advent of the lamprey caused many of the native fish in the Great Lakes to decline as they became parasitised.

The last 200 years has seen progressive eutrophication of the Great Lakes. Summer phosphorus concentrations in the epilimnion are now around $20-30\ \mu g/dm^3$, whereas 200 years ago there is little doubt that the lakes were clear and oligotrophic with phosphorus concentrations of less that $2\ \mu g/dm^3$.

A number of measures are currently being tried to undo some of this ecological damage. There are restrictions on the use of detergents which contain phosphorus; legislation requires effluents to have phosphorus levels of less than $1\ mg/dm^3$; lampricides are being tried; restocking of some fish is being attempted. However, the huge size of the lakes means that it will probably be many years before the benefits, if any, are seen and a more natural balance is restored.

17.3.5 Streams and rivers

Streams and rivers vary greatly. One only has to compare the huge mouth of the Amazon with a steep alpine stream (Figure 17.15). However, as water winds its way along a stream or river from its source to its mouth, certain generalisations usually hold:

The speed at which the water moves (the current) decreases;
The volume of water passing down increases;
Oxygen levels fall;
The bed becomes composed of smaller particles (gravel gives way to silt);
The bed becomes less steep;
The summer temperature of the water often increases;
Eutrophication often occurs;
Invertebrate diversity often decreases;
Human influences increase.

The most obvious factors which distinguish a stream or river from a lake are the unidirectional and relatively rapid flow of the water. These mean that streams and rivers generally have communities quite different from adjacent lakes (Whittaker, 1975) and probably account for the virtual absence of plankton in rivers or streams. The invertebrate families found in British streams and rivers were listed in Table 13.4. The British bird most characteristic of fast streams is the dipper (*Cinclus cinclus*). The dipper can walk along the stream bed and feed on the invertebrates found there.

17.4 Coastal and marine biomes

17.4.1 Marine rocky shore

Often the border between the land and the sea (the **littoral** region) consists of sand or shingle. However, if waves or currents are particularly strong, or if coastal cliffs descend to well below the low-water mark, the **intertidal zone** is rocky (Friday & Ingram, 1985). We have already considered the ecology of marine rock pools in Section 14.2.3. Here we will look at the characteristic **zonation** that occurs on rocky shores.

One of the most obvious biological features of a rocky shore is zonation. Low down on the shore organisms spend almost all their time submerged by water. High up on the shore they may be submerged only at certain times of the month; for the rest of the time they are essentially terrestrial, apart from the influence of sea spray. Because of this unidirectional gradient from the sea to the land, a characteristic zonation of organisms is found in littoral regions.

Figure 17.16 shows the regular zonation of five species of seaweed on the Isle of Cumbrae, Scotland. The ability to tolerate exposure to desiccating conditions increases in species from low to high shore levels. During the spring and summer, prolonged exposure during warm dry weather by neap tides (tides when the position of the Moon relative to the Sun causes the water level to be lower) may result in the death by desiccation of *Ascophyllum nodosum*, *Fucus spiralis* and *Pelvetia canaliculata* at the upper edges of their zones. Experiments have shown that if *Fucus spiralis* is moved up into the *Pelvetia* zone, it grows poorly and eventually dies, even though not crowded out by *Pelvetia*. It is probable that the seaweeds growing high up on the shore are prevented from growing further down the shore by competition from the seaweeds naturally found there. Certainly, *Pelvetia* and *Fucus spiralis* grow well if transplanted downshore (Schonbeck & Norton, 1980).

It is not just the plants that are zoned on a rocky shore. Both sedentary (e.g. barnacles) and mobile

Figure 17.15 (a) An alpine stream near Zermatt, Switzerland. (b) The mouth of the Amazon, the world's largest river.

(e.g. periwinkles) animals show characteristic zonations too. The factors which affect the distribution of the barnacles *Chthamalus montagui* and *Semibalanus balanoides* are considered in Section 9.4.3.

17.4.2 *Marine sandy beach*

Marine sandy beaches occur around the world in the littoral belt. To us they may look like an inviting environment. However, they pose considerable problems for the organisms living there. For a start, sand is abrasive and constantly on the move. That is sufficient to ensure that in the intertidal region no rooted plants can establish themselves. Further, sand is nutrient-poor, being composed almost entirely of silica (SiO_2), though there may be considerable amounts of calcareous matter from seashells. Diatoms and other algae occur in the surface layers of the sand where they can photosynthesise, but in the intertidal region the animals depend largely on organic matter such as plankton and detritus brought in by the waves (Whittaker, 1975).

Microscopic examination of sand grains in the intertidal zone reveals a world in miniature. Tiny creatures live between the grains. The mollusc *Caecum glabrum* and the sea cucumber *Leptosynapta minuta* are each only 2 mm long (Friday & Ingram, 1985). Larger organisms are mainly worm-like or flattened. A large sieve and a long, narrow spade capable of digging down to about 25 cm or more reveals a variety of life on most British beaches. The most obvious are bivalve molluscs (Figure 17.17), including rounded cockles (*Cardium* spp.), flattened tellinids (*Tellina* spp.) and elongated razor shells (*Ensis* spp.). Worms may also be found, including

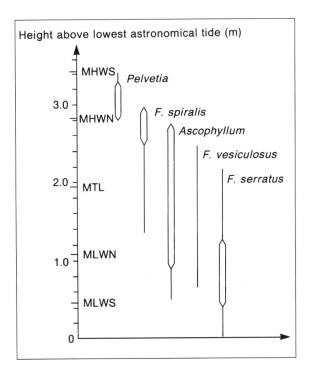

Figure 17.16 Zonation of seaweeds on the Isle of Cumbrae, Scotland. MLWS = mean low water spring tide; MLWN = mean low water neap tide; MTL = mean tide level; MHWN = mean high water neap tide; MHWS = mean high water spring tide. (From Barnes & Hughes, 1988; after Schonbeck & Norton, 1978.)

lugworms (*Arenicola* spp.) which produce characteristic casts (Barrett & Yonge, 1958).

This rich invertebrate life often supports large numbers of wading birds. To some extent, birds with bills of different lengths specialise on different prey, as different invertebrates tend to occur at different depths in the sand. The ringed plover (*Charadrius hiaticula*) and sanderling (*Calidris alba*) have short bills, while the black-tailed godwit (*Limosa limosa*) and curlew (*Numenius arquata*) have much longer bills.

17.4.3 Marine mud flat

Marine mud flats are composed of fine silts. These tend to retain fine organic matter and the relative abundance of organic matter together with the fine particle size of the muds combine to produce an-aerobic conditions only a few millimetres below the surface of the mud (Friday & Ingram, 1985). This can be seen by scraping at the surface of the mud: the grey surface lies immediately above a thick black anoxic mud. Few species are adapted to survive in these conditions, but those that are may reach high densities because of the high productivity of the mud flats. Depending on where in the world the mud flat is situated, the fauna may include clams, crabs, gastropod molluscs, annelid worms and amphipod crustaceans. Marine mud flats often support even larger number of waders than are found on sandy beaches (Figure 17.18).

Many of the mud flats around the British Isles have been colonised by dense stands of the grass *Spartina anglica*. In the early nineteenth century the cord-grass *S. alterniflora* was introduced from America to Southampton Water. There it hybridised with *S. maritima* giving rise to the sterile hybrid *S. × townsendii*. This subsequently doubled its chromosome number, becoming polyploid, and formed the fertile *S. anglica* which vigorously out-competes its parents.

17.4.4 Temperate salt marsh

Salt marsh soil and vegetation is described in Section 15.3.3. Salt marsh primary productivity is often high and the algae and detritus, in particular, support large numbers of insects and other invertebrates. These in turn are fed on by waders such as spotted redshank (*Tringa erythropus*) and oystercatcher (*Haematopus ostralegus*).

17.4.5 Mangrove swamp

Mangrove swamps are described in Section 15.3.3.

17.4.6 Coral reef

In terms of their high productivity, high species diversity, complexities of co-evolution and sheer beauty, coral reefs are the tropical rainforests of the oceans. Corals are coelenterates, relatives of the sea anemones. Most corals are colonial and secrete a protective limestone skeleton from which they extend. At times of danger, the individual polyps can pull themselves down into this skeleton out of harm's way.

Corals are both autotrophic and heterotrophic. The polyps have tiny tentacles which catch food and stuff it into a central mouth. However, the polyps also contain symbiotic unicellular dinoflagellates apparently all belonging to one species, *Symbiodinium microadriaticum* (Barnes & Hughes, 1988). These exist in huge numbers and fix carbon dioxide in photosynthesis. It is the presence of these dino-flagellates which accounts for the high productivity of coral reefs.

Reef-building corals are only found in clear seas

Figure 17.17 Molluscs from sandy beaches are an important food for humans and other animals in many parts of the world.

Figure 17.18 Bar-tailed godwits (*Limosa lapponica*) on the English Solway, West Cumbria.

within about 50 m of the surface and where the temperature remains above 20 °C throughout the year. It was Charles Darwin who recognised the three main categories of coral reef: fringing reef, barrier reef and atoll (Figure 17.19). Darwin knew that a coral reef is made up of the limestone skeletons secreted by innumerable coral polyps, which are left behind, one on top of another. He imagined a coral reef developing around the top of a mountain subsiding into the sea slowly enough for the corals to be able to keep on secreting their skeletons and building a reef up. The end process, argued Darwin in 1837, would be a coral atoll. A critical test of Darwin's theory was provided in 1952 when engineers in the United States Navy sunk a shaft through layer after layer of increasingly ancient coral at Eniwetok atoll. 4222 feet down the drill hit volcanic rock – confirmation of Darwin's brilliant suggestion that beneath an atoll should be the sunken island on which coral polyps had begun building hundreds of thousands or millions of years ago (Engel, 1963).

Many species of corals live together within a single reef. In turn these corals support many other animals. For a start, a number of species simply eat the corals. Many fish, including parrot fish, have teeth and jaws which enable them to attack the coral. Polyps are also eaten by a number of invertebrates of which perhaps the most notorious is the crown-of-thorns starfish (*Acanthaster planci*). These are sometimes found in enormous numbers, when they can cause immense damage to a reef (Figure 17.20). Recovery may take 40 years (Barnes & Hughes, 1988). It is an indication of how little we really know about coral reef ecology that it still is not known for sure whether these population explosions of *Acanthaster* are a natural feature of coral reef life, or are the result of some human activity such as pollution.

Some of the most fascinating instances of coevolution in coral reefs are found in **cleaning symbioses**. In these, one organism, a fish or shrimp, cleans another organism, a fish, of ectoparasites. Numerous species of fish benefit as hosts in this way. About 50 species of fish, and half-a-dozen species of shrimp, are known to be cleaners. Cleaning symbioses appear to have evolved independently many times. The cleaners are almost never eaten by their hosts, despite the fact that they sometimes enter into the gill-chambers and mouth of the host fish to clean. Trivers (1971) argued that cleaners are worth more to hosts alive than dead. If a host eats a cleaner, it may find it difficult to locate another cleaner when it next needs to be cleaned. The importance of cleaners to their hosts has been demonstrated in experiments where divers have removed all the cleaners from an area (Limbaugh, 1961; Feder, 1966). Within a couple of weeks almost all the fish in the area leave. Those that

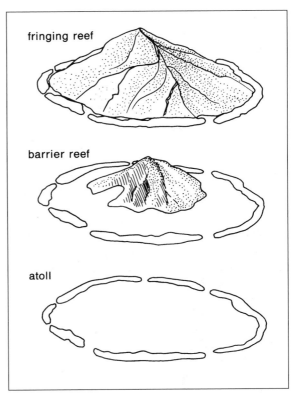

Figure 17.19 Barrier reefs and atolls are formed from fringing reefs either by a rise in sea level or by the subsidence of the land they lie next to. (From Barnes & Hughes, 1988.)

remain are the territorial ones and many of these have white fuzzy infections, ulcerated sores and frayed fins.

17.4.7 Marine surface pelagic

Pelagic communities inhabit the ocean waters, not the sea floor. The marine surface pelagic biome includes ocean waters only down to depths penetrated by light. The maximum depth at which photosynthesis occurs is approximately 250 m, and this is only in full sunlight in the most clear and tranquil open ocean waters (Barnes & Hughes, 1988). Figures of less than 150 m are more typical for open ocean waters, and ones of 30 m for coastal waters. The surface zone where photosynthesis occurs is known as the **photic** zone. It is here that **phytoplankton** occurs. Phytoplankton consists of photosynthetic organisms between 2 and 200 μm in diameter. Cyanobacteria, dinoflagellates, diatoms and green algae are all represented in the phytoplankton.

The phytoplankton of the marine surface pelagic biome supports large numbers of animals (Longhurst & Pauly, 1987). As discussed, though, in Section 12.3, the oceans, despite their huge size, are relatively unproductive. Further, there simply are not that many distinct niches in open water. This is probably

Figure 17.20 Damage to staghorn and plate coral on the Great Barrier Reef, Australia, caused by crown-of-thorns starfish (*Acanthaster planci*).

the reason why 98% of all marine species are benthic (see Section 17.4.9). Animals in the sea may belong to the **plankton**, when they are referred to as **zooplankton**, or to the **nekton**. Phytoplankton and zooplankton are at the mercy of gravity, ocean currents and the wind. The nekton, though, includes larger organisms that can swim. Many animals have young that belong to the plankton before developing into larger more independent individuals.

A variety of animals comprise the nekton of the marine surface pelagic biome, including many species of fish, turtles, penguins and seals, though many of these animals also forage in the marine deep pelagic biome (Section 17.4.8).

17.4.8 Marine deep pelagic

Marine deep pelagic communities are those found beneath the photic zone but still in open water: bottom-dwelling organisms are excluded. As no photosynthesis is possible, the creatures found here chiefly rely on the gravitational shower of food from the communities above. Carnivorous copepods and other crustaceans are frequent, along with Foraminifera and Radiolaria (Whittaker, 1975). Because so little light penetrates from above, the organisms here are either blind or produce their own light. These light-emitting organs serve a variety of functions. They can help in the detection of prey, confuse predators and help individuals to find each other for mating purposes.

The further down one goes in the oceans, the less food there is. Organisms have responded to this lack of food in a variety of ways. Many deep-sea pelagic

fish have hugely distensible stomachs and enormous mouths, so that fish twice as long as themselves may be taken as prey (Friday & Ingram, 1985). Most deepsea pelagic fish minimise their metabolic rate: they move slowly and have poorly developed muscles. With so little food available, population densities are very low and this, coupled with the total darkness, may make finding a mate difficult. A number of species of ceratioid anglerfish have a remarkable way of solving this problem. Males are equipped with specialised teeth with which they grasp hold of a passing female. Attachment is followed by the uniting of the circulatory systems so that the male, whose sole function is to provide sperm, becomes totally dependent on the female for his nutrition (Pietsch, 1975). As a result, the female is far larger than the male, as you can see in Figure 17.21.

17.4.9 Continental shelf benthos

Marine benthic organisms live on the bottom of the ocean. The considerable continental shelves that surround many of the continents lie, on average, 130 m below sea level, so that in their shallower regions benthic algae and plants can photosynthesise. Here are found the impressive kelp forests (Figure 17.22). Kelps are large brown algae, such as *Laminaria*. They attach themselves to the substratum by means of a holdfast and have one or more flat blades or fronds which can grow to more than 50 m in length. Kelps are often found in areas where ocean currents and the action of waves ensure a plentiful supply of nutrients (Friday & Ingram, 1985). Their productivity may be very high, with individual fronds growing 25 cm a day. Surprisingly, few animals feed on them. Most of their production enters the food web in the form of detritus.

The continental shelf benthos supports large numbers of animals from a variety of phyla. Polychaete worms, nemertine worms, molluscs, sea squirts, bryozoans, sponges, sea spiders, crustaceans and echinoderms are all found along with a number of fish.

17.4.10 *Deep ocean benthos*

The deep ocean benthos extends from the edge of the continental shelf to the deepest ocean trenches (Whittaker, 1975). Until recently these communities were assumed to be wholly heterotrophic, but as discussed in Section 11.2.2, it is now known that whole communities depend on the sulphur spewed from deep volcanic vents. The other deep ocean benthic organisms depend on the slow rain of food from above that has escaped the attentions of the deep pelagic organisms. Indeed, there is evidence that decaying whale carcasses serve as 'stepping stones', allowing animals on the deep ocean floor to move from one volcanic vent to another (Putnam, 1994). A surprising variety of organisms is found on the deep floor of the ocean despite the perpetual cold (nearly all the water is always at a steady 4 °C), the continual darkness and the enormous pressures caused by the tremendous weight of water above.

Inevitably, our knowledge of the deep ocean benthos is rather sparse (Edwards, 1988). The limited data available suggest that growth rates may be 50 times less than those typical in shallow waters, so that small molluscs are thought to take 50 years to reach adulthood (Friday & Ingram, 1985).

Recent research has shown that the deep ocean benthos has not remained unchanged for millions of years. The species composition of ostracods (a group of Crustacea) found on the North Atlantic seabed during the Pliocene (some 2.5 million years ago) at a depth of 2400–3400 m varies over a 10^3–10^4 year time scale. Benthic diversity seems to have been coupled with surface productivity, itself affected by the glacial–interglacial cycles at the time (Cronin & Raymo, 1997).

Currently, the deep ocean benthos must be one of the world's few biomes still relatively unaffected by human activity. Whether that continues to be the case remains to be seen. Every few years proposals are put forward for the deep sea mining of the nodules of manganese and other elements found on the ocean floor. Thankfully, such proposals remain uneconomic.

Figure 17.21 Female anglerfish (*Edriolychnus schmidti*) with two parasitic males.

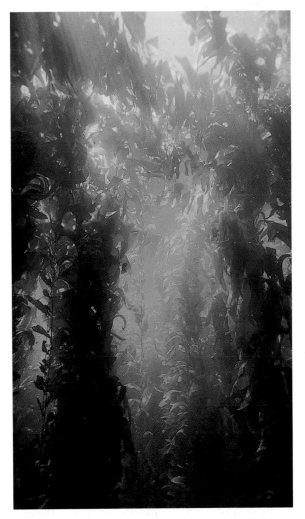

Figure 17.22 Kelp forest of *Macrocystis pyrifera* off the coast of California.

Box 17.2
Life on Mars?

We are beginning to realise that there are a surprising array of microorganisms distributed throughout the Earth's crust (Ghiorse, 1997). Most of our knowledge of this subterranean world comes from drilling and is still very incomplete. These habitats are dark, generally low in organic matter and relatively constant in temperature. Organisms are supported by chemosynthetic primary production. What is, as yet, very uncertain is the extent of these communities. Some ecologists have claimed that the total mass of organisms in subterranean ecosystems exceeds that of all other organisms. It has also been argued that life may have evolved in such circumstances rather than in the 'primaeval soup'.

The evidence that large numbers of microorganisms exist deep in the Earth's crust has encouraged those who hope to find life on the planet Mars. It is known that Mars has some type of tectonic (continental drift) activity and also had a hydrological (water) cycle at some time. At the time of writing no rocks have been retrieved by missions to Mars. However, there are 12 meteorites which are thought to have come from Mars – one 4.5 thousand million years old, the others 180 to 1300 million years old (Grady *et al.*, 1997). It was the oldest

of these that, in 1996, gave rise to great excitement when it was suggested, by reputable scientists, that it might contain the fossilised remains of life. The excitement arose partly because this meteorite contains surprisingly large amounts of carbon compounds, partly because isotope data indicate that these compounds were formed at temperatures between 0 and 80 °C (i.e. at temperatures ideal for water-based life) and partly because of the presence of oval-shaped, elongated structures that resemble, though are significantly smaller than, the remains of fossilised bacteria. By 1997, enthusiasm had cooled somewhat. However, the hunt is now on for more Martian meteorites. Calculations suggest that 100 tonnes of them arrive on Earth each year!

On Mars, it now seems likely that if there is life it will consist of subterranean bacteria-like organisms. This means that the best chance of finding life will be by drilling deep below the surface. So far, those space probes which have landed on Mars have only looked at and analysed surface dust and rocks (Figure 17.23). Exciting as the pictures and data returned to Earth are, just 'scratching the surface' may never provide evidence of life outside our planet.

Figure 17.23 The Sojourner rover investigating a rock (nicknamed Yogi) on the surface of Mars.

Summary

(1) The natural assemblages of organisms in the world can be divided into biomes.

(2) Biomes provide a convenient shorthand for describing the world's flora and fauna even though ecologists disagree about how many biomes there are.

(3) Biomes, especially terrestrial ones, are traditionally defined mainly in terms of their vegetation.

(4) This chapter broadly follows Whittaker's (1975) division of the world's communities into 36 different biomes.

(5) Human activities have affected some biomes much more than others.

EIGHTEEN

Biogeography

18.1 Species distribution – where and why?

So far we have looked at the distribution of organisms in communities and have seen how particular communities are typical of certain environmental conditions. On a larger, global scale a particular species may occur only on one continent or island, only in the tropics or on a single mountain range. Most of us, for example, would recognise Figure 18.2 as African savannah. Often the same ecosystem occurs in several parts of the world, but the species or genera in the community are different in different regions. The study of this geographical distribution of species is called **biogeography**. Biogeographers investigate the location of species and the reasons for the distributions they find. Thus biogeography links ecology with geology, earth history, evolution, climatology and geography.

If you look back at Figure 17.2 you can see the distribution of the major terrestrial biomes of the world. The same biome often occurs on several continents or marine regions. For example, tundra and boreal forest occur in northern America and Europe; temperate grasslands are found in the dry centres of most continental landmasses. If you compare Figure 17.2 with the major **biogeographic regions** of species distribution (Figure 18.1) you will see a very different pattern. In this case the terrestrial regions are associated with landmasses such as major continents, not with climatic zones. Sometimes species have a wide distribution and occur in many biogeographical regions: an example is the bracken fern (*Pteridium aquilinum*) described in Section 3.2. More often, a genus is widespread, but within each region different species occur, for example the oaks: *Quercus robur*, *Q. petraea* and *Q. ilex* in the palaearctic, *Q. borealis*, *Q. coccinea* and many others in North America.

The factors which affect the distribution of individual organisms have consequences for the range of the species and are, therefore, of importance in

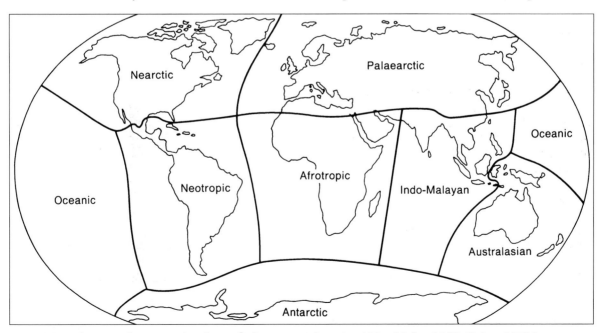

Figure 18.1 The major biogeographical regions for terrestrial species. (After Muller, 1980; Pielou, 1979.)

Figure 18.2 Zebra and wildebeest drinking at a water-hole in the Masai Mara, Kenya.

biogeography. They include abiotic influences of the environment such as climate, soil type and drainage, geological features and topography, all of which have been touched on in earlier chapters. Biotic factors include characteristics such as the mobility of animals and seed or spore dispersal in plants. The other species in communities may also be important if they are a source of food or shelter or have symbiotic or parasitic relationships. A third influence on biogeographical distribution is the history of each species and the geological history of the areas it occupies. Most important in this area is the understanding of the consequences of plate tectonics.

18.2 The historic effects of plate tectonics

18.2.1 Past continental movements

In this section several geological eras are mentioned. Table 18.1 shows the age and duration of these and notes some of the major occurrences recorded in the fossil record.

The theory of **plate tectonics** has only been developing since the early 1960s (Wilson, 1965; McKenzie & Parker, 1967), but its conclusions have had considerable reverberations for biogeography. Many phenomena of global plant and animal distribution can only be understood with the knowledge of plate tectonics which we now possess.

The surface of our planet is made of several large plates about 100–150 km thick which move very slowly across the surface of a more fluid core. Most of these plates have major landmasses or whole continents on them (these are thicker areas in the plate), so that as the plates move around, the continents also move with respect to one another. This movement is called **continental drift**.

Where the plates touch three types of movement are possible. At some edges new plate is created by

Table 18.1 Major events in the geological record

Geological eras		Age (millions of years)	Major events
Quaternary		0	?Greenhouse effect Ice ages
		1.64	
Tertiary	Pliocene		Early hominids
		5.2	
	Miocene		North and South America join Ancestor of honeycreepers arrives on the Hawaiian chain
		23.3	
	Oligocene		
		35.4	Evolution of grasslands and grazing animals India collides with Laurasia
	Eocene		
		56.5	
	Paleocene		Rise of mammals
		65	Extinction of the dinosaurs Rise to dominance of angiosperms Break-up of Gondwanaland
Mesozoic	Cretaceous		Early angiosperms
		145.6	
	Jurassic		Early birds Break-up of Pangea starts
		208	
	Triassic		Early mammals Mass extinction event
		245	
Paleozoic	Permian		
		290	
	Carboniferous		Huge rainforests of tree ferns, club-mosses and early gymnosperms
		362.5	
	Devonian		Early terrestrial vertebrates First seed plants
		408.5	
	Silurian		First land plants
		439	
	Ordovician		
		510	
	Cambrian		Rise of metazoans First fossil exoskeletons
		570	
Pre-Cambrian			Origin of metazoans Origin of life

the upwelling and cooling of molten rock. Such margins are usually under the ocean but if the plates both have land on them, then the land masses move *away* from each other. This type of margin may start down the middle of a continent as a rift valley as shown in Figure 18.3(i). The Great Rift Valley in East Africa is an example of an early phase in the separation of two plates. As the two parts of the continent separate, the sea floods the rift (ii) and forms a new ocean (iii). The two parts of the land may initially have the same flora and fauna but as they separate, extinctions and the evolution of new

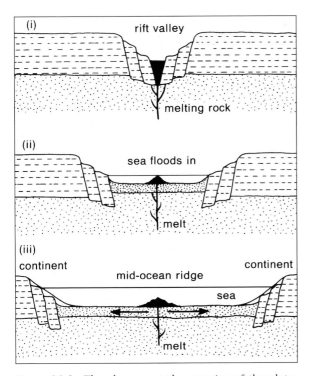

Figure 18.3 The changes at the margins of the plates covering the Earth's surface caused by plate tectonics. Two plates are moving apart. At first the land faults and sinks to form a rift valley (i). This valley floods (ii) and further upwelling of molten rock forms new sea floor. The two parts of the continent move apart at a mid-ocean ridge (iii).

species and genera cause their floras and faunas to become distinct.

The second type of plate movement is where the plates move towards each other. If one or both plate edges are thin ocean crust, then one plate slips under the other as in Figure 18.4(i). As the plates move together the friction between them causes sticking and sudden movement which creates earthquake zones. The ocean crust which sinks melts as it goes deeper, towards the hot core, and the melted crust rises to the surface in volcanic eruptions (ii). If the two plates carry continents, then these will move *towards* each other. Continental crust is less dense than ocean floor so if a continent reaches the edge of a sinking plate it will not slide under the other plate, and therefore stops the plate it is on from sinking at its edge. If both plates have continental masses on them which hit each other, neither will give way. They crush into each other forming long mountain chains (iii). The Himalayas, with the largest mountains in the world, were created when the plate carrying India collided with the rest of Asia 40–45 million years ago. In fact they are still rising, if rather slowly!

The third kind of boundary between plates is where plates may slip sideways against each other. The San Andreas fault in California marks the boundary of a Pacific plate which is slipping north and the North American plate which is moving

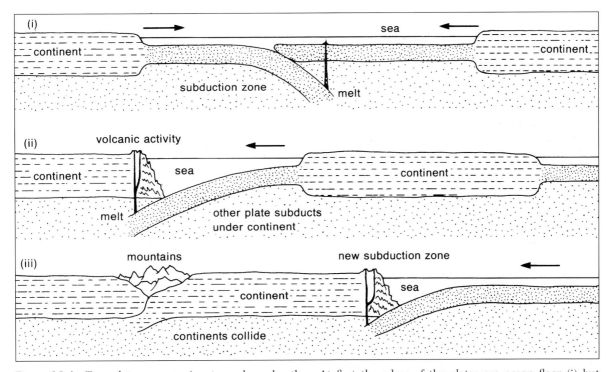

Figure 18.4 Two plates are moving towards each other. At first the edges of the plates are ocean floor (i) but eventually the continent on one plate reaches the edge of the plate. The large volume of less dense continental crust cannot sink beneath the denser ocean floor (ii) so the ocean floor sinks: as it does so it melts and molten rock rises through the continental crust to form volcanos (iii).

south. The slipping is intermittent and sudden as tension builds up across the fault due to friction, then gives suddenly. The earthquake of 1906. which wrecked San Francisco, was caused by the fault blocks slipping 3–7 m past each other. Present movement is about 5 cm a year.

It appears from various types of evidence that at one time all the present landmasses were joined together in one huge supercontinent called Pangea. This began to split in half about 200 million years ago to form Laurasia and Gondwanaland. Figure 18.5 shows how the present continents were arranged in the Triassic. Gondwanaland was a southern hemisphere continent composed of what are now India, Africa, Australia, New Zealand, South America and Antarctica. This supercontinent seems to have begun to break up in the Jurassic about 190 million years ago. Before this, the **palaeobiogeography** (or biogeography of past continents) of Gondwana shows that there was free movement of animals, plant seeds and spores across all the landmasses.

Figure 18.5, apart from showing the present outlines of the continents which made up Gondwanaland, also shows the position of several pre-Jurassic fossils. The Permian vegetation, for example, was dominated by large-leaved gymnosperms (*Glossopteris* spp.). When India moved northwards, it carried its Gondwanan fossil load with it. The present distribution of the fossil-bearing rocks

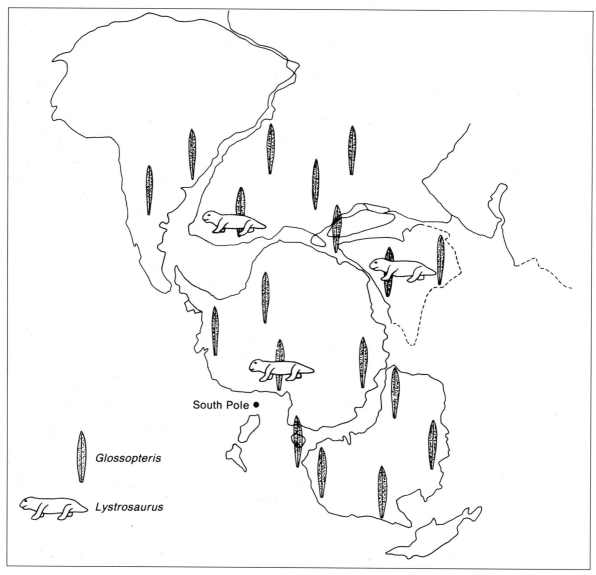

Figure 18.5 Map of Gondwana showing the distribution of Permian fossil leaves of *Glossopteris* and the Triassic reptile *Lystrosaurus*. The outlines of the present continents are shown in their position in the supercontinent of Gondwana in the late Permian–early Triassic. (From Colbert, 1973; Chaloner & Lacey, 1973).

only makes sense when plate tectonics is considered. A similar distribution is found in the fauna. For example, fossil remains of the Triassic mammal-like reptile *Lystrosaurus*, and associated tetrapods, are found in India, southern Africa and Antarctica (Crawford, 1974).

Once the continents began to break up, they became isolated from each other by ever widening oceans. The point in evolutionary time when the fauna and flora of each continent were cut off from contact with other animals and plants, and the subsequent degree of isolation from other areas, is reflected in the present patterns of biogeography.

18.2.2   Present patterns of biogeography

By the early Tertiary, Gondwana and Laurasia had separated into the continents more or less as we know them today. Figure 18.6 shows the distribution of continents during the Tertiary; the continents have moved considerable distances from their positions in Figure 18.5. As you can see, India is moving towards Asia, and South America has yet to touch North America. The map shows the approximate time in millions of years that each continent has been separated from others once close to it. Hence, New Zealand split off from Gondwana about 80 million years ago and Australia separated from Antarctica about 50 million years ago. Some continents, once isolated, have collided with each other and Figure 18.6 also shows when the collisions occurred. Australia became linked to Asia by island

chains about 15 million years ago, and the Americas joined together by a land bridge only 6 million years ago.

Perhaps the best example of a distinct regional biota is found in Australia. Australia started to drift northwards about 90 million years ago and gradually separated off from Gondwanaland until it became completely isolated about 50 million years ago. From then on, like a great ark, it moved northwards, carrying a precious cargo of organisms. Australia is famous for its present-day fauna, especially the **marsupials** and **monotremes**. The monotremes are a unique and obscure group of furry creatures which lay eggs: characters which are not known together in any other group. Two kinds of monotreme are found in Australia. One is the duckbilled platypus (*Ornithorhynchus anatinus*), an aquatic burrowing mammal about the size of a rabbit. with a huge, soft, highly sensitive bill and webbed feet. The platypus feeds on river-dwelling insects and lays eggs in a nest at the end of a long burrow along river banks. The eggs are soft and round and a centimetre or so in diameter. After 10 days the eggs hatch and the young are fed on milk which oozes from glands in the female's skin. There are no teats to suck from and the young simply lick the milk from their mother's fur.

The second kind of monotreme is the short-beaked echidna (*Tachyglossus aculeatus*), a fat spiny creature rather like a large hedgehog, which eats ants and termites. The echidna also lays eggs, but into a pouch on its stomach. The young hatch in the pouch and remain there until their spines make them too

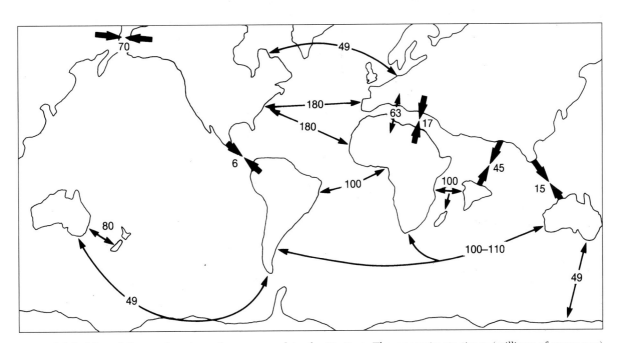

Figure 18.6 Map of the continents as they occurred in the Tertiary. The approximate times (millions of years ago) when continents became isolated from each other are indicated on separating arrows; the times when land masses collided are indicated on converging arrows. (After Raven & Axelrod, 1974.)

uncomfortable for the mother to carry them. They also lick the milk from their mother's fur (Attenborough, 1979).

Although these two animals are very different and highly specialised, they are both obviously from a common and ancient mammal stock. They have incomplete body temperature regulation and their body temperature fluctuates around 30 °C. They seem to combine reptilian and mammalian characteristics. Australia provided the only place where the monotremes survived and the present species are really only just hanging on in the region. The platypus is under threat because it is easily disturbed by rabbits which disrupt its burrowing behaviour. The echidnas seem to have done a little better. They spread north when Australia came into contact with the islands there and a separate species, the long-beaked echidna (*Zaglossus bruijni*) now lives in the mountains of New Guinea.

The diverse Australian marsupials seem to have evolved from an early mammal line in the Cretaceous. The present distinctive features of marsupials are that they give birth to young in an immature state almost like a small embryo. The tiny young then climb through their mother's fur to find a teat where they cling on and are fed with milk. Most marsupials have a pouch over the teats which protects and holds the developing young. The pouch is called a marsupium, and gives its name to the group.

Fossil marsupials can be identified by skeletal characteristics, especially their teeth. The first marsupials seem to have been arboreal opossum-like creatures. They may well have evolved in South America, but had a widespread distribution as similar fossils have also been found in North America and Europe. They seem to have reached Australia via Antarctica. Australia then became isolated from further invasions, before placental mammals reached it. The marsupial group then evolved into many forms to fill available niches in the continent, first as browsers of leaves and later, as the drier grasslands developed, as grazers (Romer, 1966). This evolution of a diversity of species to fill a variety of niches is called **adaptive radiation** (see also Section 18.3.4).

The degree of parallel evolution between marsupials in Australia and placental mammals elsewhere is quite startling. Marsupial shrews, moles, anteaters, squirrels and flying squirrels all occurred in Australia; there was even a marsupial wolf (Figure 20.5(d)). Some animals are different in shape from placental mammals, but have similar feeding styles. For example, the largest grazing marsupials are kangaroos (Figure 18.7) which fill a niche similar to those occupied on most other continents by deer and gazelles. Their foreparts look quite like those of deer, but their mode of locomotion is quite different. They hop on their hind legs using their large, powerful tail to maintain balance. The famous koala bear is also distinctive: it may fill a niche occupied in other countries by leaf-cutting primates.

As the tectonic plate carrying Australia moved northwards, it carried the continent closer to Asia. As it neared New Guinea and the surrounding islands, some placental mammals invaded. These were bats, which could easily fly across the intervening sea, and one ground-dwelling group: placental rats (Murinae) which then radiated and now number about 50 species in Australia. About 3 500 years ago, the wild dog or dingo seems to have been introduced into Australia by humans. It has now become a pest to the native wildlife. Australia has suffered considerably from introduced placental mammals, plants and birds (see Section 19.5); so has New Zealand which has no native mammals but many ground-dwelling birds which are vulnerable to alien predators.

The flora of Australia has also evolved independently of other regions. Angiosperms of several families were once widespread across Gondwana, including several rainforest families. In Australia, these families seemed to have evolved further to produce a slow growing, tough-leaved flora which lives on the old, weathered, nutrient-poor soils which are found over most of Australia. About 45% of genera of these tough-leaved floras are endemic to Australia.

As Australia moved towards the equator, its climate was affected by the cooling and icing up of Antarctica. From the mid Oligocene, 35 million years ago, Australia became drier, so that grassland developed. These grasses (in the family Restionaceae) are

Figure 18.7 A group of Eastern grey kangaroos (*Macropus giganteus*).

another example of parallel evolution between southern and northern hemisphere grasses (in the Poaceae). The development of extensive grasslands in Australia led to the evolution of the grazing marsupials, such as the kangaroos in Figure 18.7 (Cox & Moore, 1985).

As can be seen, the unusual species now found in Australia are the result of the species trapped there when it became isolated, and of its changing position and altered climate. Its unique ecosystems have evolved and been preserved by isolation from other species. The reverse of this situation is found in South America. South America was also once part of Gondwanaland and must have had a similar flora and fauna to Australia. About six million years ago, South America became attached to North America by the Panama Isthmus. Some southern species, notably the marsupial opossums, re-invaded North America, a continent where they had once lived before they went extinct there many millions of years earlier. As they spread through North America the tropical species tended to be stopped by the cold, dry Mexican Plateau. However, species moving into South America from the north had less of a problem. About half of the genera of mammals now found in South America came from the northern continent. This includes llamas and tapirs which then went extinct in their native north because of the effects of the ice ages (Cox & Moore, 1985).

18.3 Island biogeography

18.3.1 The fascination of islands

Islands have always interested biologists. They represent small, isolated microcosms with discrete boundaries. This makes the study of island populations easy as immigration and emigration are greatly reduced. Islands vary greatly in size, age, isolation, shoreline structure, topography and geology. All these features affect the ecology of the communities which live on and around them.

Island biotas are often noted for the absence of whole groups of organisms. Often, as a result of this, they have a small and unusual complement of species which have evolved together into unique ecosystems. Many isolated islands have high percentages of endemic species. Because islands are so numerous and varied they are very valuable to biogeographers, more so in some respects than the continents as these are huge, often have very complex histories, and differ from each other in many ways.

Islands have provided inspiration for evolutionary theories. Both Charles Darwin and Alfred Russell Wallace had the opportunity to study islands and certainly the variation Charles Darwin found in the

Galapagos Islands triggered his thoughts on the origin of species. The effects of evolution are easier to observe on islands, partly because fewer species are involved, so that patterns of evolution are clearer, and partly because of the variation found between islands which occur in chains or clusters, highlighting the individuality of each evolved form.

Islands can be formed in two basic ways: by the disappearance of a connecting land bridge from the mainland; or by land rising from the sea. An island can be cut off from the mainland if the sea rises and floods an area in between, or if the joining land is eroded or sinks. In either case the island will be formed with a flora and fauna already in place. This is not the case for islands which rise out of the sea. Such islands usually form by volcanic activity on the ocean floor gradually building up igneous material until it reaches above sea level: the process is shown in Figure 18.8. Many island chains are formed in this way as the volcanic activity is caused by a hot spot under the Earth's crust. As the crustal plate moves over the hot spot the volcanic activity on one island ceases, and a new island begins to form.

The most famous island chains formed in this way are the Hawaiian Islands and the Galapagos Islands. A feature of such island chains is that they are graded in age. A new island is always being added to one end of the chain. Figure 18.8 also gives a map of the Hawaiian Islands, showing the various island ages. As newly formed volcanic islands and underlying crust cool, the rock contracts and they slowly sink. To the west of the Hawaiian chain are the stumps of even older islands which, because volcanic activity ceased on them a long time ago, have been eroded by the sea. If a cooling island sinks very slowly, coral reefs can build up around it which can remain as an atoll after the volcanic rock is no longer visible (see Figure 17.19).

New volcanic islands are formed from hot, sterile volcanic lava or ash. They form the basis for primary xeroseres, but the colonisers and later successional species all have to reach the island across several or perhaps many hundreds of miles of ocean.

18.3.2 Colonisation of isolated islands

There are several ways in which an organism can reach an isolated island. Active methods involve flying or swimming. More passive means include floating with ocean currents either directly in the sea, or attached to driftwood or other flotsam. Air currents may carry light seeds and spores. Seeds can also be carried on birds' feet or feathers or in their digestive system. Colonisers reaching a freshly formed island may have problems establishing on the hard, unweathered, base rich, hot lava. The order of colonisation will be mostly the result of chance: some

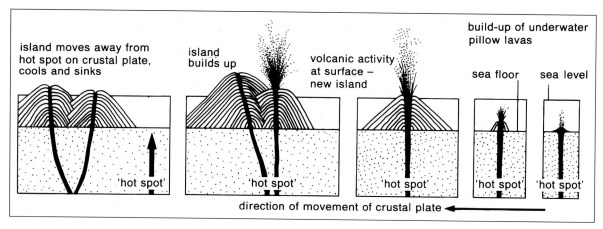

Figure 18.8a The build-up of an island chain due to volcanic activity at a 'hot spot'. As the continental crust moves away from the 'hot spot' the volcanic island becomes dormant, cools and sinks while a new volcanic island builds up above the 'hot spot'.

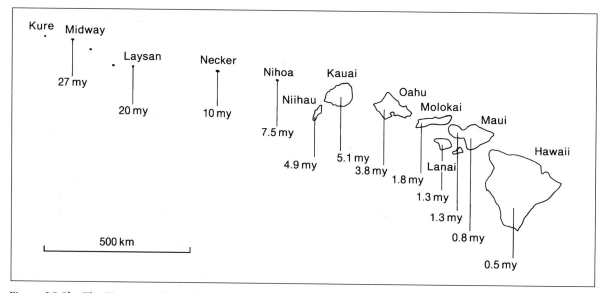

Figure 18.8b The Hawaiian Islands have developed in this way. The ages of the islands (in millions of years (my)) show how the Pacific Ocean plate is moving slowly over the 'hot spot' to create new islands.

species may arrive before there is a suitable habitat for them and die out only to recolonise later when the situation is more favourable.

Usually our understanding of island colonisation comes from studying well developed communities on relatively old islands. Even if an island is only a few thousand years old we will have missed the early stages of succession. Most islands are older than this; their present species reflect not only the species which first colonised them, but also subsequent evolutions and extinctions. Just occasionally, biogeographers are given a tremendous opportunity to study the initial stages of colonisation. On 14 November 1963, a disturbance on the sea floor 30 km south of Iceland was noticed and by 15 November an island, formed of volcanic ash, had risen above sea level. The new island was named Surtsey and it provided

ecologists with the opportunity to study the very first events in the colonisation of new land. The record of that colonisation is given in Table 18.2.

The species which reach an isolated island like Surtsey (or the many other oceanic islands like the Hawaiian and Galapagos Islands, Madeira, the Canaries, St Helena and Tristan da Cunha) must have all undergone long-distance dispersal. Certain conditions must be met for a species to colonise an island in this way. First, the organism must be capable of founding a population. A single male or female, or the seed of a self-incompatible plant, will fail. Organisms which are most likely to succeed are self-fertile plants or a single pregnant female animal. Other possibilities include a single influx of several individuals of a species, or frequent immigration events of solitary organisms. There are around

Table 18.2 The events occurring off the coast of Iceland which resulted in the formation and early colonisation of the island of Surtsey. Data assembled from the Surtsey Research Progress Reports.

Date	Events
14 November 1963	Eruptions on floor of ocean 120 m deep first noticed.
15 November 1963	Land first appeared above sea level.
1 December 1963	Lull in volcanic eruptions. *First life on island:* a flock of unidentified seagulls landed.
End January 1964	Island 1100 m long x 800 m wide x 160 m high made of hot pumice and cinders, easily eroded by storm waves.
April 1964	Explosive eruptions ceased.
Spring–summer 1964	Kittiwakes and gulls seen roosting on island.
14 May 1964	More birds: 7 oystercatchers and 2 dunlin seen.
21 May 1964	Plant fragments found washed up on shore: 5 species of leaves and 1 seed probably all washed there from Vestmann Islands 5–18 km away. Algal fragments also found: thalli of 3 brown seaweeds.
7 June 1964	More birds: 14 red-necked phalarope.
4 October 1964	First non-marine bird: 1 turtle dove.
15 October 1964	Plant fragments and seeds of 9 species, fragments of 5 seaweeds.
31 January 1965	More land birds: 2 ravens.
7 May 1965	Lava flows ceased.
3 June 1965	*First land plants growing:* about 30 seedlings of *Cakile edentula* growing on decayed seaweed, killed by new ash fall two weeks later.
20–29 June 1965	First insects found: 6 species of Diptera, 2 Lepidoptera, also 1 mite carried there on a dipteran.

Date	Events
1965	*First breeding insects:* a fly, *Heleomyza borealis,* breeding on dead fish and bird carcasses.
June 1966	More insects: 22 species including 1 beetle, and arachnoids: 4 species.
22 June 1966	*First algae growing:* a green alga.
2 July 1966	More plant seedlings: 4 *Elymus arenarius,* 1 *Cakile edentula,* probably washed away by surf.
19 August 1966	Re-eruption of lava flows.
6 September 1966	First raptoral bird: 1 merlin.
5 June 1967	Lava flows ceased.
26 June 1967	*First flowers:* of *Cakile edentula.*
Summer 1967	*First mosses found:* probably introduced by scientists; also the plant *Honkenia peploides,* not found on nearest islands so probably from Heimaey, 18 km away.
1970	*First breeding insects in island habitat:* a chironomid midge breeding in the shallow pools.
1970	*First nesting birds:* a pair of fulmar and a pair of black guillemot.
Summer 1975	1449 plants growing: 15 plant species.
Spring 1976	502 plants growing: probably all winter survivors.
Summer 1976	1132 plants growing: 11 species.
Spring 1977	489 plants growing.
Summer 1977	962 plants growing: 11 species.
Spring 1978	698 plants growing.
Summer 1978	3428 plants growing: 11 species. After this *Honkenia peploides* increased rapidly to 50 000 plus plants.

800–1000 species of fruit flies in the family Drosophilidae in Hawaii, and it is thought that they probably all evolved from a single (or, at most, two) fertilised female flies blown there by chance a few million years ago (Kaneshiro, 1995).

The ways in which species have arrived on particular islands can be suggested by studying the characteristics of the present biota. On the Hawaiian Islands, for example, study of the flora suggests several methods of dispersal which may have brought species to the islands (Carlquist, 1974). These different categories are shown in Table 18.3. Fern spores were probably mostly blown in, as they are so light. However, very few seed-bearing plants are likely to have colonised in this way as the Hawaiian Island chain is a considerable distance from other landmasses (1600 km from the nearest island and 4000 km from the mainland of America). Wind dispersal may be much more important on islands closer to shore. Quite a large proportion of the Hawaiian flora (23%) seems to have arrived by ocean drift. Most arrive as seeds which are buoyant and float until

Table 18.3 From the seed morphology of the present flora of the Hawaiian Islands, the most likely method by which the original colonisers reached the islands can be determined. (Data from Carlquist, 1974.) The number of successful colonisation events from which extant flora developed is thought to be 256.

Arrival method	Frequencies
In birds' digestive system	38.9%
Rafting on floating debris	14.3%
Barbed seeds attached to birds' feathers	12.8%
In mud on birds' feet	12.8%
Viscid seed stuck to birds' feathers	10.3%
Drifting by sea	8.5%
Air	1.4%

washed ashore. Some species may even travel as vegetative fragments which are tolerant of salt water and root easily on landing. A Hawaiian example of such a species is *Portulaca lutea* in the Portulaceae.

Some species seem to have been carried by birds,

mostly stuck to their feathers by means of barbs (e.g. species in the Asteraceae), or some sticky substance. Many shore and marsh birds have large feet which get coated in mud. Such mud sometimes contains seeds that drop off when the bird reaches a new island. The Hawaiian flora is thought to have obtained about 13% of its species this way. Birds can also transport seed internally, by eating it. Seeds may be damaged by this process, but many pass out in the droppings in a viable state. Many Hawaiian species have fleshy fruits apparently attractive to birds. It appears that about 39% of the flora may have reached the islands inside birds!

Sea-going birds may visit islands frequently. The species whose seeds they pick up on their feathers or feet are thus widely dispersed to many islands. Constant addition of new seed will increase the genetic variation of the island population and maintain links with the mainland and other island populations. On the other hand, land birds do not travel far and may only reach islands infrequently, when some strong wind blows them out to sea. They too may carry seed to islands, but such events are rare and the species they carry will become isolated populations. As inland birds are likely to pick up or eat different species from coastal or marine birds, this pattern of dispersal and isolation will be different for various plant species. This is indeed the case. Sticky-seeded coastal plants are widespread on the Pacific islands. This difference in dispersal and frequency of immigration of new genetic material can affect the variation and degree of evolution which occurs. This is discussed further in Section 18.3.4.

Many species seem almost incapable of long-distance travel. As a result many islands lack certain groups. The Hawaiian Islands, for example, have no amphibians, no terrestrial reptiles, no gymnosperms and only one living mammal, a recently arrived bat species (Zimmerman, 1970). Even major groups which are abundant on the islands are often represented by only a few taxa. The birds of Hawaii have representatives of only 7% of the world's passerine families although those which did make it to the islands have subsequently undergone much adaptive radiation. Fossil records show, however, that the bird fauna of the islands has become poorer: several honeycreepers, raven-like corvids, a flightless ibis, flightless geese, rails, a hawk and an eagle have all become extinct, possibly due to the pressure of developing human culture (Freed *et al.*, 1987).

The endemic flora of the Hawaiian Islands contains about 2000 species, but it is thought to have arisen from only about 260 immigrants, and lacks many mainland tree species such as oaks, elms, figs, maples, mangroves and willows. A similar story appears for insects, where 6500 species are thought to have descended from about 250 immigrant groups. These data suggest that, because of the extreme distance of the Hawaiian chain from shore, a new species became established on the islands only every 25 000 to 100 000 years! (Zimmerman, 1970). This is a remarkably slow rate of colonisation.

18.3.3 The equilibrium theory

The **equilibrium theory of island biogeography** was proposed by MacArthur and Wilson in 1963 (expanded 1967). To understand the theory, we must consider an island, like Surtsey, risen from the sea and with no species having yet colonised it. At first, almost every organism arriving on the island will be a new species to the island, so the rate of immigration for the island will be high. As the number of species on the island increases, many of the organisms arriving on the island will be of species already present. So the rate of immigration will decrease and eventually fall almost to zero as the number of species rises. This relationship is shown by the graph Figure 18.9(a).

If the island is close to the mainland, it will be easy for species to reach it. Species will arrive quickly, so the immigration rate will be very high. If the island is a long way from the shore, like the Hawaiian Islands, then fewer species will reach it and the rate of immigration will be much lower, even at the beginning of colonisation. These extremes are represented by the two different curves for immigration rate shown in Figure 18.9(a).

Species which have colonised the new island run the risk of extinction. If there are only a few species on the island, then each will probably have a large population size and little competition from the other species, so that the rate of extinction will be low. As the number of species increases, the population size of some of the species will become lower and the probability of some species becoming extinct will therefore be higher. This relationship is illustrated in Figure 18.9(b). The rate of extinction rises as the number of species on the island rises. If the island is large, there will be more room for many viable populations of species than if the island is very small. So the rate of extinction of species on a small island will be higher than the rate of extinction for the same number of species on a larger island.

If we put the immigration and extinction curves together for a single island, we get the situation illustrated in Figure 18.9(c). If the number of species on the island is small, the immigration rate will exceed the extinction rate and the number of species on the island will rise. As it rises, the immigration rate will fall, but the extinction rate will increase. Eventually a point will be reached where the immigration rate equals the extinction rate. At this point the number of species on the island reaches an equilibrium point.

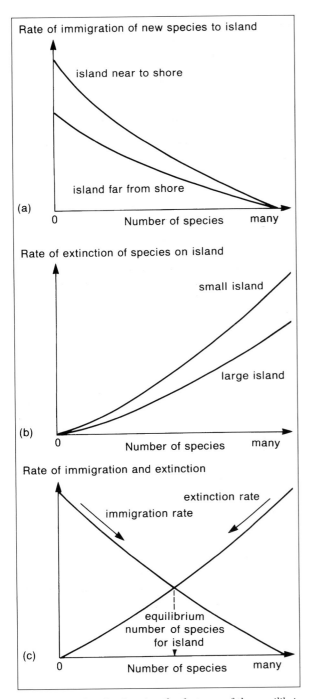

(a)

Rate of immigration of new species to island

island near to shore

island far from shore

0   Number of species   many

(b)

Rate of extinction of species on island

small island

large island

0   Number of species   many

(c)

Rate of immigration and extinction

extinction rate

immigration rate

equilibrium
number of species
for island

0   Number of species   many

Figure 18.9 Graphs showing the features of the equilibrium theory of island biogeography: (a) rate of immigration against number of species on the island for islands close to shore and a long distance from other land; (b) rate of extinction against number of species on the island for large and small islands; (c) immigration and extinction curves combined to show the point of equilibrium.

Addition of new species balances the loss of species by extinction. This is the equilibrium which gives its name to the theory. It predicts that any island will have a balanced number of species which depends on its size and distance from land. The number of species

will be lower on isolated or small islands than on close or large ones.

The equilibrium theory of island biogeography is most elegant in its simplicity. However, since the theory was proposed, there has been a considerable amount of debate about its validity (Pielou, 1979; Gilbert, 1980; Vitousek *et al.*, 1995). One problem is that the theory does not include the addition of species *in situ* by evolution. This is probably very important, especially on islands a long way from the source of new species (see Section 18.3.4). We saw from the Hawaiian Islands that the immigration rate was incredibly low and that the great variety of species found on the islands is mainly the result of evolution from a relatively small number of species. Evolution can increase the number of species on the island without the requirement of immigration. This may make it harder for newly arrived species to colonise successfully as the available niches have already been filled by species which have evolved to be a 'perfect fit' in each niche.

A factor which is probably extremely important in determining species number is habitat diversity. Such diversity may correlate to some extent with size of island, so that island size is really only a secondary measure of the variation of environment on the island.

The attraction of the equilibrium theory also resulted in its application to objects other than real islands. It has been used in the design of nature reserves for conservation (Section 20.5.2), and applied to single species through time in considering how they are 'colonised' by insects. The theory has even been used for individual plants with their load of insects and spiders.

18.3.4 Evolution on islands

Many isolated islands contain endemic species (Figure 18.10). The endemics are often from particular groups: the land birds and plants for example. In the Hawaiian Islands, 70.8% of the ferns and 94.4% of the angiosperms are endemic (Carlquist, 1974, after Fosberg, 1948). The presence of so many endemics could be explained two different ways: they could have evolved from mainland forms which had arrived on the islands, or they could be mainland species which have survived on the islands while the mainland populations went extinct. So island communities could represent totally new or very old relict species.

Are islands covered with new species or are they museums full of relics? To answer this we have to look at the evidence we can obtain from such islands and at the conditions island immigrants experience.

The frequency of immigration of individual members of a species on to an island will affect the

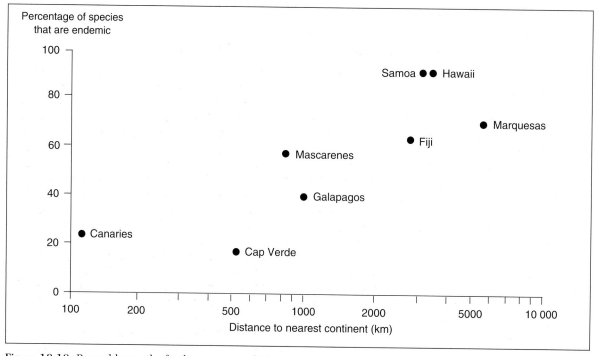

Figure 18.10 By and large, the further a group of islands is from the nearest mainland, the greater is the proportion of species on those islands that is found nowhere else.

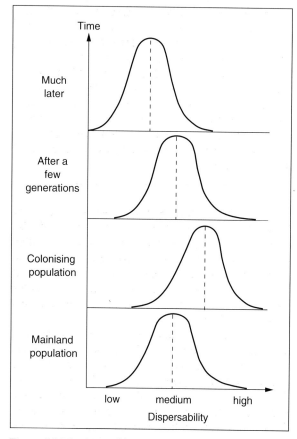

Figure 18.11 A possible evolutionary scenario showing how island organisms often lose much of their ability to disperse with the passage of time.

genetic variation in the island population (Section 6.4.3). This means that species which are easily dispersed, and which are constantly in genetic contact with populations on other islands and the mainland, are species which are widespread and show little variation between populations (Carlquist, 1974). These include marine birds such as frigate birds (Fregatidae) and coastal plant species. Other species, such as land birds and inland plants, may only be introduced once on to an island as a small number of individuals. This new population is then isolated from its parent population and is in a new locality where environmental conditions are different. Such a population is likely to undergo natural selection and become adapted to the new environment: it evolves into a new species.

One generalisation that often holds true is that species, whether animals or plants, on islands are often less good at dispersing than their mainland counterparts. However, in Section 6.4.2 we saw how Cody and Overton (1996) found that individuals of the plant *Lactuca muralis* which were new colonisers to islands were *better* at dispersing. Only with the passage of time did they lose this ability. Cody and Overton propose the evolutionary scenario summarised in Figure 18.11. Founders are more likely to be good dispersers. However, directional selection then selects against good dispersal ability simply because good dispersers are likely to become lost at sea. Eventually, stabilising selection results in old populations with limited dispersal ability.

The species–area curve (Figure 18.12) shows how the number of species in an area depends on the size of that area. It seems obvious that a larger area would contain more species than a small one and that habitat size thus affects species richness (see Section 14.4.2). Traditionally, there are two explanations for this. One is that the larger an area, the more habitat variation there will be in that area so that more species will find suitable niches to occupy. This is called the habitat heterogeneity hypothesis. The alternative argument came from MacArthur and Wilson: that species richness increases with area due to decreasing extinction rates. This is the equilibrium theory of island biogeography illustrated in Figure 18.9.

Another commonly observed pattern of species distribution is that species which have a wide distribution (like the starling and bracken described in Chapter 3) also tend to be locally abundant where they occur, while species with restricted distributions tend to be rare. Again, there have been two

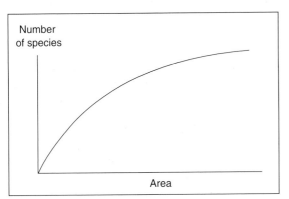

Figure 18.12 The typical relationship between the size of an area and the number of species within the area.

explanations suggested for this. The first is that widespread species are generalists, able to occupy a relatively wide range of conditions and therefore have a wide niche: the niche breadth hypothesis (Brown, 1984). The second explanation sounds rather like the equilibrium theory: it argues that common species have low extinction rates of individual populations and that migrants from large populations 'top up' small populations as well as found new ones, creating a metapopulation (see Section 4.1). The distribution (restricted to wide) and abundance (rare to common) of species form a distribution–abundance curve which is roughly similar to the species–area curve.

Recently Hanski and Gyllenberg (1997) have produced a mathematical model which combines the species–area and species distribution patterns. The model considers a series of islands of different sizes (and differing distances from a mainland) and the incidence of species of different abundances on them. Looking at the number of species on individual islands models species–area relationships, while looking at the distribution of a species on several islands models species distribution and abundance relationships.

The model is the first to produce predictions about both species–area and distribution–abundance patterns. It has also shown that the distribution–abundance curve is important in generating a realistic species–area curve for the model, indicating that species distribution patterns are an important factor in species richness. The model seems to suggest that species distribution and abundance are linked with colonisation and extinction rates. Thus it appears that the alternative explanations for the relationships between area, species number and species distribution can be combined to enhance our understanding of observed species abundance and richness.

Sometimes a founding population arrives on one of a group of islands and, as it colonises each one, each population is changed to suit conditions on that particular island. This results in a group of closely related species on the different islands. This process of speciation on a group of islands is a form of **adaptive radiation**. The striking effects of adaptive radiation were what Darwin noted in the finches of the Galapagos Islands. It was this variation from island to island in a closely related group of finches which seemed to have a common ancestor, which drew his attention to the possibilities of a selection process which could produce new species.

Another example of adaptive radiation in birds is found in the honeycreepers of the Hawaiian chain (see Figure 18.13). These are a diverse group of passerines which seem to have evolved from one colonisation event. Nucleotide substitution rates sug-

gest that these honeycreepers diverged from the basic finch line 15–20 million years ago (Sibley & Ahlquist, 1982). This suggests that the adaptive radiation began long before many of the present Hawaiian Islands existed (see Figure 18.8). Presumably the honeycreepers island-hopped to newer islands as the older ones in the chain decreased in size and sank beneath the Pacific.

Insects too show adaptive radiation to a remarkable extent. The Hawaiian Islands have 50 species of bee, all in one genus, *Hylaeus*, which seem to have radiated from a single bee invasion. Some have very different life styles from the normal pollen and nectar collectors and are parasitic on other bees in the genus. They lay their eggs into the nests of other bees, like cuckoo bees, and leave them to raise the young (Zimmerman, 1970).

The other major evolutionary feature of islands is

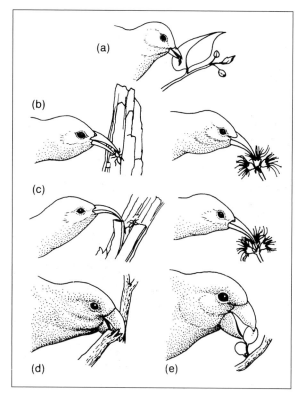

Figure 18.13 Some of the adaptive radiation found in Hawaiian honeycreepers. The shape of the bill reflects the feeding habits of each species. Species illustrated are (a) *Loxops coccinea*, (b) *Loxops virens*, (c) *Hemignathus lucidus*, (d) *Pseudonestor xanthophrys* and (e) *Psittarostra kona*. (From Carlquist, 1974.)

the way in which some species take on very unusual forms or structures which are not typical of the group on the mainland. The lack of predators often seems to result in the evolution of ground-foraging birds which have lost the power of flight because it is not required for escape and because, with no ground-dwelling mammals, there is plenty of food on the ground. This is seen in New Zealand, where the kiwi, and flightless parrots such as the kakapo, take the niches of foraging mammals in the undergrowth. Such birds are very vulnerable if predators are subsequently introduced into the island. In the Hawaiian Islands, flightless birds were once common: 17 fossil species have been found, including ibis, geese and rails (Freed *et al.*, 1987) but all are now extinct.

The apparent rate of evolution on islands is sometimes startlingly rapid. For example, there are 45 species adapted totally to cave life on the Hawaiian Islands. Each of the five main islands has its own endemic species. The caves occur in the lava flows and are extremely inhospitable places. A cave studied on Hawaii itself and known to be about 500 years old had 12 species of obligate cave dwellers. Many of these are blind including a planthopper, a white millipede, a springtail, a terrestrial water treader, a large

earwig and three species of spider (Howarth, 1987). This does not mean that the species have all evolved in 500 years; cave species probably reach new caves by crawling through the porous lava flows. However, on Hawaii, such habitats are probably only about half a million years old and it is difficult to see how such cave dwellers could have island-hopped like the birds.

So far we have investigated the evolution of new species on islands and have seen that, on islands like the Hawaiian and Galapagos groups, there are large numbers of endemic species, many of which may be only a million or so years old. Is this the case for all islands and is there any evidence for relic species on some islands?

St Helena is a volcanic island in the South Atlantic Ocean. It is at least 14.5 million years old and almost 2000 km away from the mainland of Africa. It is quite small with an area of only 122 km² but has an unusual flora (Figure 18.14). If the flora is compared with world floras, an interesting pattern emerges. The island, like most isolated islands, contains many endemic species, and several endemic genera. Endemic species of non-endemic genera on St Helena are often most closely related to species in mainland South Africa or the tropics closest to the island. Endemic genera, however, are often more closely related to genera in Australia, South America, India or Madagascar than to genera in mainland Africa. This seems very strange as St Helena is much further from South America (3000 km when it was formed and even further now) than it is from Africa. Madagascar is even more isolated from St Helena with the whole of Africa between. What this distribution suggests is that the endemic genera are relics of an earlier period when their ancestors were much more widespread than they are now. Such earlier floras survived on the islands like Madagascar and St Helena, and in South America and Australia, but became extinct in Africa (Cronk, 1987). The species from these old floras must have arrived on St Helena long ago. Since then they have evolved to produce not only new species, but even new genera. In one way they are relics of an earlier world flora, but they are also new species.

Thus islands are both museums and places of recent evolution. Island species are often descendants of ancient groups that were once more widespread. However, time has not stopped on such islands; the descendents of these groups, like the marsupials in Australia and the flora of St Helena, have evolved into modern species. Hence in Australia the opossum-like ancestors of the present fauna radiated into large numbers of arboreal and ground-dwelling herbivores as well as into various carnivores. On St Helena, the endemic woody members of the daisy family, the Asteraceae seen in Figure 18.14, are

Figure 18.14 A relict stand of gumwood trees (*Commidendrum robustum*, Asteraceae) on St Helena. The steep slopes and plains were once covered with this tree which is now rare.

probably descendants of an ancient woody species. The Asteraceae are now known mainly for its non-woody continental species, many of which are weeds.

18.3.5 Mountain islands

Any isolated habitat, totally separated from others of its kind, can be considered as an island. For instance, there are many mountains surrounded by low-lying land. Just as real islands are cut off from one another by water, so mountains are isolated by 'seas' of low-lying land. Mountain species have to arrive on a mountain by some form of long-distance dispersal, just like mainland species arriving on an island.

Some of the most interesting isolated mountains are in tropical latitudes where the high altitude ecosystems are surrounded by tropical rainforest or savannah. Mountains in the tropics have a very distinct climate. Above about 3000 m frosts occur at night and this altitude seems to mark the upper limit of mountain forest. The number of nights with frost and the depth of frost increases with altitude, until eventually every night is freezing cold and a permanent snow line is reached at about 5000–6000 m. As seasonality is suppressed so near the equator, there is no obvious period of the year when organisms can avoid the frost by becoming dormant. All the species up on the high mountain have to be frost resistant. Some mountains have definite dry and rainy seasons,

and the frosts tend to be concentrated in the dry season, when cloudless skies make the days very hot and bright and the nights very cold. The heavy cloud cover of the rainy season prevents the sun beaming down during the day and traps the heat at night. This means the temperatures are more even but the clouds may make the light intensities during the day too low for much plant growth (Sarmiento, 1986).

The mountains of East Africa are rather like a cluster of small islands. They are solitary volcanic mountains on or near the equator surrounded by savannah. Just as islands often have very unusual biotas, so these mountains too have unique ecosystems. The vegetation on such mountains is described as Afro-alpine as it is of a low growing, treeless alpine type, but about 80% of the Afro-alpine species are endemic (Hedberg, 1986). Most of the plants can be categorised into one of five different growth forms which are shown in Figure 18.15. These are tussock grass (a), sclerophyllous bushes (b), cushions (c), stemless rosettes (d) and giant rosette plants (e).

Some of the most interesting and remarkable species are the giant rosette plants which dominate the landscape in an almost alien fashion as you can see from Figure 18.16. The East African mountains have several of these species, most of which are in the genera *Dendrosenecio* (the tree senecios of the Asteraceae), and *Lobelia* in the Lobeliaceae. Members of these unusual groups also occur on real islands.

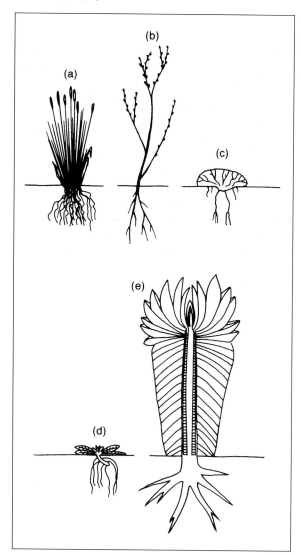

Figure 18.15 The different life forms of plants which grow on mountains in the Afro-alpine ecosystems. (From Hedberg, 1986.)

Giant lobelias also live on the Hawaiian Islands, for example, and woody Asteraceae on St Helena. As on marine islands, a considerable amount of adaptive radiation has taken place in these groups and endemism is frequent. In the African *Dendrosenecio* genus, two are endemic to Mount Kenya, while the most widespread species (*D. johnstonii*) occurs on most of the mountains, but each mountain has a different sub-species. These giant rosette plants are able to survive the rapidly fluctuating daily temperatures; at night the temperature can fall as low as −8 °C. These large plants retain leaf bases all along their stems which act as insulation and provide shelter for many endemic insect and sunbird species. Such plants are slow growing and take a long time to recover if grazed. They contain an anti-mitotic toxin which seems to prevent the hyraxes (*Procavia johnstoni*) which live on the mountains from eating them. The hyrax droppings provide nutrients which promote the growth of the tree senecios and they grow around the entrances of hyrax burrows and hide them (Mabberley, 1986).

In recent years, the tall elegant lobelias and young tree senecios seem to have fared less well. Hyrax, rat and mice populations have greatly increased because of the extra food sources associated with increased tourist activity. When the food is in short supply, the rats and mice attack the endemic species. The lobelias especially have almost vanished on some parts of Mount Kenya in the last 20 years or so (Kokwaro & Beck, 1987).

It is easy to compare the African mountains with the Hawaiian Islands. Both have many endemics which show adaptive radiation, although the Hawaiian Islands show this and speciation to a much greater degree. This is probably because true islands have a wider range of habitats and a less extreme climate than the mountain tops. Both have unusual biotas which only survive because they have evolved in a protected environment and both show the extreme risk to such biotas when the delicate balance of ecosystems is shifted in some way (see Sections 19.5 and 20.4). Both marine islands and mountain tops are important systems which provide valuable information about the processes of evolution and speciation in isolated regions.

Figure 18.16 Tree senecios (*Dendrosenecio keniodendron*) and giant lobelias (*Lobelia telekii*) in the Teleki Valley at about 4100 m on the upper slopes of Mount Kenya. The slender spikes are the lobelias in flower.

Summary

(1) Biogeography is the study of where species occur in the world and the reasons for their patterns of distribution.

(2) Biogeographical regions and biome distribution do not coincide.

(3) The surface of our planet is composed of large moving plates which carry the continents. Continental drift accounts for the distribution of many species which evolved in isolation or migrated to their present locations.

(4) Islands often have an impoverished flora and fauna due to the limited number of colonising species. This is especially so for isolated islands.

(5) Groups which colonise isolated islands often subsequently undergo adaptive radiation.

(6) Adaptive radiation on islands results in large numbers of endemic species, but islands are also places where relict groups can survive.

(7) In their equilibrium theory MacArthur and Wilson predict that island biotas reach an equilibrium where immigrations of new species equal extinctions. Larger islands and islands closer to shore are expected to have more species than smaller and isolated islands.

(8) Many bird species on islands with no predatory mammals have lost the ability to fly.

(9) Isolated mountains can be thought of as islands: they too have many endemic species which are at risk from introduced species.

NINETEEN

Co-evolution

19.1 The different grades of co-evolution

Ehrlich and Raven first introduced the term **co-evolution** in 1964 in a paper on the relationship between butterflies and the host food plants of the larval stage. They were concerned in particular with the 'patterns of interaction between two major groups of organism with a close and evident ecological relationship' (p. 586). Within the paper they also introduced the term **community evolution**: 'evolutionary interactions found among different kinds of organisms where exchange of genetic information among the kinds is assumed to be minimal or absent' (p. 586). Their definition of community evolution is broad, and they seem to consider co-evolution as a smaller subset of community evolution, but their application of the term has been interpreted in a number of ways by ecologists since then.

The most restricted use of the term co-evolution is applied to the evolutionary relationships between two species which are, ecologically, very closely associated. This is often called **pairwise** or **one-on-one co-evolution**. Usually this implies that as one species evolves particular characteristics, the other species also evolves in response to this; the first species then evolves in response to the response of the second, and so on. Examples of pairwise co-evolution include ant species living on acacia trees (see Section 19.2.2), butterfly caterpillars and their host plants, and bees with the plants they pollinate (see Section 19.4.3). Sometimes the term pairwise co-evolution is applied to two species where only one has evolved features to intensify the relationship between it and another species. The distinction of whether one or both species in the association have evolved is difficult to determine as we only see the relationship at one point in time. Such fine distinctions in the use of the term are perhaps not so important.

The broader use of the term, closer to Ehrlich and Raven's community evolution, is called **diffuse co-evolution**. In such cases, more than two species co-evolve together: for example, predators and prey species (see Section 19.3.2) or mammalian grazers and grassland plants (Section 19.3.1). Taken to its extreme, it could be argued that this definition includes all evolution as every species evolves in response to the other species in its community.

A problem with the concept of co-evolution is that it is often extremely difficult to determine whether two species which appear to have co-evolved have in fact done so. For instance, an introduced insect from another geographical area may find and eat a plant species because it has a similar chemical composition to its normal food, or an introduced flower will attract a pollinator which has never seen the species before, as in Figure 19.1 (Janzen, 1980). The fact that such partnerships are successful does not

Figure 19.1 The British bumble-bee *Bombus terrestris* visiting a tomato flower.

necessarily rule out the occurrence of co-evolution as both the plant and insect may have co-evolved, in their native habitat, with similar species to those with which they are now associated.

19.2 Pairwise co-evolution

19.2.1 General aspects of one-on-one relationships

In this restricted sense, one-on-one co-evolution covers close ecological relationships between pairs of species. Nowadays the term **symbiosis** is used to describe *any* close relationship between two species. If both species benefit from the relationship it is said to be one of **mutualism**. Such mutualistic co-evolution can result in extremely complex interconnected life cycles (Section 19.2.2 gives an example of such a relationship between ants and acacias). Under such circumstances, either one or both species are incapable of surviving without the other.

When one of two species has a deleterious effect on the other, such as a herbivore on a plant, a predator on its prey or a parasite on its host, then the co-evolution will be antagonistic. The plant, prey or host often evolves some way of avoiding or deterring the herbivore, predator or parasite. That species may then evolve a counteracting characteristic which once again allows it to eat, catch or parasitise the other species. The result is almost like a battle with the two species evolving in a step by step manner better methods of attack and defence. The characteristics that evolve are often chemical in the form of poisons and antitoxins, but can also be mechanical such as harder shells and stronger claws, or behavioural such as better hiding or hunting methods. This retaliatory relationship is often described as an **arms race** (Dawkins & Krebs, 1979).

There are several possible results to an arms race. One side may win, in which case the other species may go extinct. It may end in a stable equilibrium where neither species has an advantage, or the race may be cyclical, such that if further escalation of the race does not benefit one species, it may 'retreat' and start on another path of evolution. The second species would then have no need of its weapons, lose them, and the race would start again.

As, by definition, pairwise co-evolution must involve this step by step change in the two species, then such species pairs must have been associated in the same community for long enough for evolution to occur. This requires a certain stability of the community. This is uncommon in higher latitudes, where the glacial cycles of the last two million years have frequently mixed communities and produced new associations. Tropical ecosystems have the high-

est species diversity (see Section 14.4) and these are also the richest areas for co-evolved species pairs.

Particular groups seem to be more likely to form co-evolutionary associations than others. Ants feature in many mutualistic relationships; with plants, butterflies, aphids and fungi. Lepidoptera too are often involved, for example, adult butterflies and moths have pollination roles with some plants as do many bee species (see Section 19.4).

Distasteful or dangerous animals are often brightly coloured (wasps and bees, butterflies and some caterpillars) to warn birds not to eat them. The birds learn the patterns of these warnings after trying one or two and avoid them in future. Distasteful and poisonous butterflies and moths often have similar wing patterns in one geographical area. Figure 19.2 shows several different races of two species in the genus *Heliconius*. Because in each geographical area both species look the same, the birds in each area will only have to learn to associate one wing pattern with distastefulness. During the learning process they will kill fewer insects so the insects benefit. The birds too benefit as they catch fewer poisonous insects in learning what to avoid. This convergence of patterning in poisonous insects is called **Müllerian mimicry** (named after Fritz Müller, a nineteenth-century naturalist who investigated convergent mimicry in distasteful species). In the tropics, the effect is so striking that in different patches of forest the same species of butterfly can look completely different if the local warning patterns in the two areas are different.

Insects which are not harmful converge on the warning coloration too. They use it as a disguise, to pretend they are distasteful when in fact they may be quite edible. This is called **Batesian mimicry** (after another nineteenth-century naturalist, H. W. Bates, who studied mimicry in Amazonian butterflies). As long as there is a high enough proportion of really poisonous animals in the area for the birds to learn about, the Batesian mimics are protected. Presumably the evolutionary pressure is on the birds to become better at recognising the fake Batesian mimics so they can eat these. However, it may not be worth the effort as careful observation would be needed: to the human eye at least, the mimics are very accurate. This may be an arms race which has reached an equilibrium point.

The relationships between caterpillars and their host plants provide examples of **antagonistic co-evolution**. Many plant species have toxic substances, especially in their leaves, and many different caterpillars have overcome these poisons. In fact, as was mentioned above, the first use of the term co-evolution was for such relationships (Ehrlich & Raven, 1964). It has been argued recently, however, that chemical compounds in plants are not as important in determining insects' choice of host as was

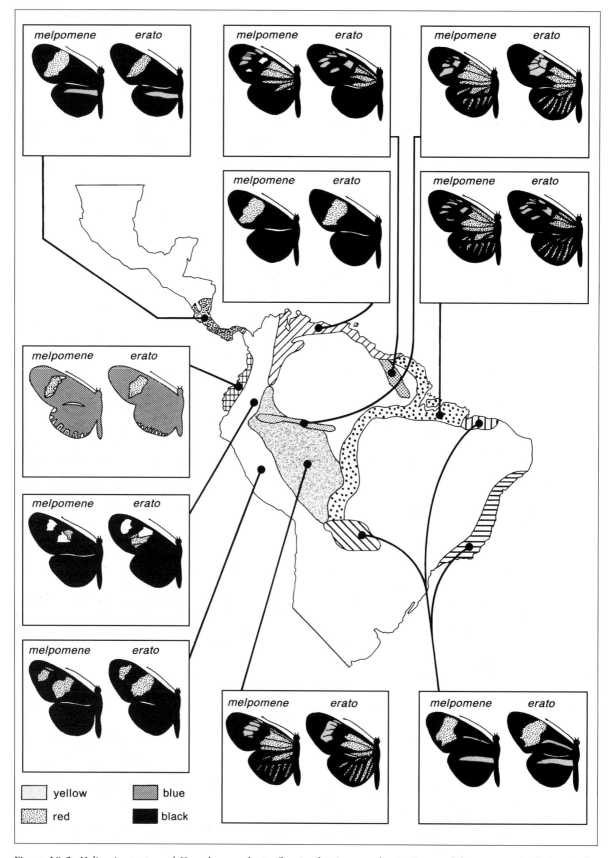

Figure 19.2 *Heliconius erato* and *H. melpomene* butterflies in the Amazon basin. Races of the two species living in the same area have very similar wing patterns. Both species are poisonous, so this is an example of Müllerian mimicry.

previously supposed. In other words, co-evolution in such cases has been overemphasised (Bernays & Graham, 1988). Bernays and Graham suggest that it may be the predators of the insects that cause a specialist insect to have a restricted range of host plants. The insect may specialise on one plant so that it can use the plant toxins to deter predators and parasites, or so it can become camouflaged there. Another possibility is that it may have moved to a plant with toxins because other herbivorous insects were rare there so that it was not worth a predator's time hunting on such plants with few insects. In response to this argument, it has been pointed out that the importance of chemical compounds in plants should not be underemphasised as not enough is known about plant chemical–insect reactions to dismiss them (Schultz, 1988). Many factors may influence which plant species a herbivore chooses to eat including chemical compounds, external plant structure, climate, plant availability, predators and parasites. All these parameters probably interact too. This is an area of ecology in which careful research will tell us a great deal about the relationships between insects and plants in the years to come.

19.2.2 The ant–acacia example

Perhaps some of the most intriguing examples of mutualistic co-evolution are those found between ants (Formicidae) and plants. As Beattie (1985) puts it: 'the ways in which plants manipulate ants, and vice versa, can be so complex and subtle as to severely stretch the credence of the observer' (p. ix). The relationship between ants and bracken was described briefly in Section 3.2.3, but the tropical examples of ant relationships are far more intricate than this.

The first person to unravel the truly complex features of the one-on-one co-evolution between ants and the species of *Acacia* (Mimosoidea, Fabaceae) was Janzen (1966). Acacias are elegant trees with pinnate leaves divided into many rounded leaflets along a central leaf-vein. They have clusters of tiny flowers which are usually bright yellow. Many ant species have mutualistic relationships with a group of acacias known as the swollen thorn acacias for reasons which will become apparent later. Here, just one example, that between the ant *Pseudomyrmex ferruginea* and the tree *Acacia cornigera*, will be described.

The relationship starts when a queen ant finds a young acacia seedling or sucker which is not already occupied by ants. As they grow, acacias produce swollen stipules at the bases of leaves as shown in Figure 19.3. These swollen stipules are sharp and look like the horns of a bull, hence the name of these trees. The queen lands on a swollen stipule and cuts a hole into it: she then clears out the soft filling of

parenchyma to leave a hollow cavity. Here she lays her eggs. She wanders over the plant which provides her with food by secreting nectar from special glands, called foliar nectaries, which are at the base of the leaves (this is rather like the provision of food from nectaries on the bracken fern in Figure 3.4). The queen also gathers small swollen nodules from the ends of the pinnae of the leaves. These modified leaf tips (see Figure 19.3) are called Beltian bodies and contain proteins and lipids. With these the queen feeds her developing larvae and a colony of ants develops. After about a month the colony will have developed to about 1200 worker ants which live in various swollen stipules on the tree. After three years or more, there can be as many as 30 000 ants in the colony.

Once the workers become numerous, they swarm all over their acacia attacking insects and other herbivores they find. They also attack any plants which come into contact with their acacia. They even clear the ground of plants under the tree, leaving a bare circle up to a metre in diameter around the trunk.

If acacias have their ants removed, they are rapidly overrun by leaf-eating insects and climbing vines. They seldom survive more than one or two years in this state. These acacias do not have the normal features of tough new shoots and chemicals which are found in other *Acacia* species to deter herbivores. The ant acacias use their ant colony to protect them. As a result they do not need other structurally and biochemically expensive protection mechanisms. They seem to be able to divert the energy saved in this way into more rapid growth. This enables them, with the help of the ants as gardeners, to compete favourably in denser, wetter habitats than can other acacias which have not co-evolved with ants. In return for this protection, the acacia expends energy making the swollen stipules, nectar and Beltian bodies with which to house and feed the ants. The benefits to ants and acacias of this co-evolutionary relationship evidently far outweigh the costs.

There can be simpler associations between ants and *Acacia*. Wagner (1997) studied the ant *Formica perpilosa* which often nests underneath the shrub *Acacia constricta* in arid regions of the south-western United States. She found that plants with basal ant nests produced 1.9 times as many seeds on average than plants of similar size and location without ant nests. (The seeds from the two categories of *Acacia* were equally likely to germinate and didn't differ in size.) However, the ants do not appear to increase protection against herbivores. Plants with and without basal ant nests sustained similar levels of damage to leaves and seeds. On the other hand, soil from beneath plants with ant nests contained significantly

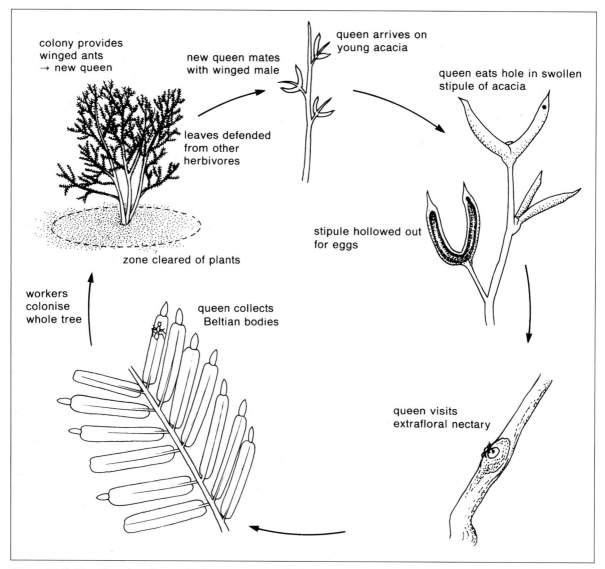

Figure 19.3 The phases in the establishment and later development of an ant colony on an acacia tree.

higher concentrations of nitrate, ammonia, phosphorus and water. The precise reasons for this are unclear but it may be that the processing, storage and discarding of food by ants in and around the nest contributes to the high concentrations of nitrogen and phosphorus.

19.3 Diffuse co-evolution

19.3.1 Co-evolution between groups of species

Where more than two species are involved then **diffuse co-evolution** is said to take place. This is a much more common situation than the close pair-wise evolutionary relationships described in Section 19.2. All species evolve within a community of some

sort and so are influenced by the many species around them. Most plant species, for example, are eaten by a variety of herbivores. Consequently thorns, tough leaves, toxins and distasteful chemicals may be responses to general herbivore activity, rather than to a single important species. Many plants use the same, or very similar, compounds in their defence against herbivores. So a single insect species may be selected to evolve tolerance to a number of different food sources. The same can be said for the evolution of basal meristems and silica in grass species, as a response to grazing by vertebrates, and the development of tall flat-topped teeth in the mammals which graze on such grasses. Another example of diffuse co-evolution is the evolution of predator avoidance mechanisms such as speed or camouflage in prey and corresponding speed and behavioural traits in their predators.

Like pairwise co-evolution, diffuse co-evolution can be either mutualistic or antagonistic. As we looked at an example of mutualism in Section 19.2.2, the example of diffuse co-evolution given in Section 19.3.2 will be an antagonistic one: the relationship between predators and prey.

19.3.2 The mammalian predator–prey example

The major predator–prey group of the Mesozoic (see Table 18.1) was a large and diverse group of reptiles which had a wide range of life styles, including flying and marine predators, large ponderous herbivores, small speedy carnivores and a few large ferocious carnivores. At the end of the Cretaceous the balance shifted, with the extinction of most of the main evolutionary lines of reptile, leaving behind a number of small, fairly unspecialised mammals. These mammals were often agile tree dwellers with short, mobile limbs and long muscular tails which were used for balance or for gripping branches. Their paws had long fingers, so they could grasp branches. In general body form these early mammals resembled small opossums found today in the forests of South America (Figure 19.4).

During the early Tertiary (Table 18.1), these mammals diversified to produce the wide range of terrestrial and marine forms we recognise today. They took over most of the reptile life styles, though, except for whales, they never reached the great sizes of some dinosaurs. Diffuse co-evolution of mammals within their communities frequently occurred. Here we will look in detail at just one relationship: that between the grassland herbivores and their predators.

During the Tertiary, the temperate regions, which had been warm and wet, became cooler and drier.

Figure 19.4 Murine opossum (*Marmosa pusilla*) from the Valdes Peninsula, Argentina.

The temperate to subtropical forests opened up and were replaced by grassland. Many mammalian herbivores evolved to live in the new habitats which were forming. Herbivores began to eat grasses instead of the leaves of forest trees as savannah spread.

With the shift in vegetation from forest to grassland the mammals had to move from thick forest, where they could escape predators by hiding in the dense cover, into the open plains. Agility to clamber around in the branches became less useful, and speed across open ground more important. The predators which had relied on ambush and stealth in the forests also changed their techniques of hunting to long-distance pursuit or high-speed chases. The herbivores of the plains increased in size, so that hunting in packs also became more profitable as a single kill could feed several animals. Herbivores too began living in groups, presumably because there was more safety in numbers with some animals to act as lookouts for predators (see Section 8.2.1).

So the small woodland mammals of the early Tertiary evolved into the large grazers and predators of the grasslands. Today's herbivores come from two main evolutionary lines: the even-toed grazers, or **Artiodactyla**, such as deer, antelopes, llamas and camels, and the odd-toed grazers, or **Perissodactyla**, which include rhinos, horses, zebra and tapirs. The carnivores which hunt them include pack hunters like wild dogs, wolves, hyenas and lions and the more solitary cheetahs and leopards.

During the Tertiary, diffuse co-evolution resulted in evolutionary changes in both the herbivores and carnivores which increased their running ability in open country. Many of these changes were identical in both groups. The limbs became longer and thinner, the bones of the toes shortened and there was a tendency for outside toes to be lost, while the bones behind the toes, the metatarsals, lengthened to add length to the legs. Movement at joints also became more restricted so that agility in paws and limbs decreased except for movements needed in running (Bakker, 1983). The loss of toes was most extreme in the herbivores, resulting in only one or two functional toes with the claws broadened into hooves. This trend was much less obvious in the carnivores, possibly because they required more flexibility in the paws to grip prey while catching, killing and eating it, although Bakker suggests it may be a response to having to dig dens.

The pattern of co-evolution can be discovered from the fossil record of the predators and prey through the Tertiary. A good measure of the specialisation of limbs for running and speed (rather than agility) is the length of the metatarsals relative to the femur (the long bone in the leg). Figure 19.5 is a plot of how this ratio has changed during the Tertiary for the fossil ungulates and an early group of carnivores,

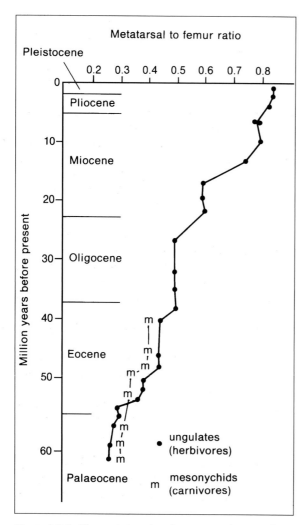

Figure 19.5 The metatarsal to femur ratio for ungulates and the carnivore mesonychids throughout the Tertiary. The ratio increases in both groups as they evolve to be faster runners indicating a co-evolutionary relationship between predator and prey. (Data from Bakker, 1983.)

the mesonychids. Most of the carnivore groups are more diverse than the grazers and have poorer fossil records making it harder to use them in comparisons. The retention in carnivores of a more basic paw structure also probably affects the metatarsal to femur ratio (M:F) so that the ever-increasing trend in the grazers is not so clearly seen. The M:F in recent carnivores is never as high as in the grazers, yet carnivores seem to be able to run, at least for short distances, at similar speeds to ungulates with much higher M:F. It has been suggested (Bakker, 1983) that this apparent increased efficiency in carnivores is a function of their diet. Ungulates have to have large, heavy stomachs and intestines to cope with their poor quality diet (Section 14.2.4). The higher quality diet of the carnivores allows them to be more streamlined, lighter in weight and to have a higher proportion of muscles in the legs. Nothing could be

more streamlined than the cheetah (Figure 6.8) which is the fastest mammal on land, capable of speeds up to 96 km/h. It is indeed a specialised short-distance running machine.

19.3.3 The Red Queen hypothesis

The example of the predator–prey changes in Section 19.3.2 shows an interesting phenomenon of co-evolution. In a herd of ungulates being chased by a predator, it is those ungulates with slightly longer limbs that are the most likely to survive, because they can run fastest. The predators, on the other hand, can only survive if they can catch enough to eat, so selection improves the speed with which predators can run. In other words the prey are constantly evolving better ways to escape, while the predators are evolving new ways to catch them. The situation is an arms race as mentioned earlier (Section 19.2.1). This phenomenon has been likened to a situation described in Lewis Carroll's Victorian children's book *Through the Looking Glass*:

> . . . [the Red Queen and Alice] were running hand in hand, and the Queen went so fast that it was all she could do to keep up with her . . . however fast they went, they never seemed to pass anything. . . . just as Alice was getting quite exhausted, they stopped . . .
>
> Alice looked round her in great surprise. 'Why, I do believe we've been under this tree the whole time! Everything's just as it was!'
>
> 'Of course it is,' said the Queen . . . 'here, you see, it takes all the running you can do, to keep in the same place.'

In other words species are constantly having to evolve just to keep up with other species in their community. A species which loses this evolutionary race will go extinct. These features of antagonistic diffuse co-evolution were encompassed in a hypothesis by Van Valen (1973) which he called, after the character in *Through the Looking Glass*, the Red Queen Hypothesis. This hypothesis was proposed as an explanation for the relatively constant extinction rates which are found throughout evolutionary time. Central to the hypothesis is the importance of antagonistic relationships between species.

19.4 Insect pollination

19.4.1 Angiosperm–pollinator relationships

There are several requirements a plant must fulfil to have an insect pollinator (to be **entomophilous**): the plant must attract the insect by producing a flower which advertises that the plant has provided a reason for the insect to visit, such as nectar. The insect benefits from visiting the flowers as it obtains nectar or pollen as food. The plant benefits from the visit by obtaining pollen from another flower (see the advantages of outbreeding in Section 6.3.2). To take advantage of an insect visitor a flower must have its reproductive organs in such a position that the insect collects pollen on to some part of its body and rubs this pollen off on to the stigma of another flower. This means the flower has to attract pollinators of the right size and shape to fit the flower. Ideally, the visiting insect should bring and deposit pollen only from another plant of the same species, and should carry the pollen it picks up to another compatible plant.

The insect likes ample food to make its visit profitable. It requires a flower where the food supply is not depleted. Thus, the insect needs a flower which can provide food to satisfy all visitors, or one which only a few other insects visit.

This complex interplay is the driving force for the diffuse co-evolution of insects and entomophilous angiosperms. The relationships are an intermix of mutualistic and antagonistic factors. The plant obviously wants to ensure pollination at minimum cost. It therefore produces as little pollen and nectar as possible. The pollinator wants to maximise the reward from its visits. If another species offers a better reward, then the pollinator may switch allegiance, so different flower species are competing among themselves in a community. Some ecologists have argued that flower species have co-evolved to avoid such competition by flowering at different times of the year, or by attracting different species of insect. Both these options are found in the tropical forests. Some species flower in a great burst of showy inflorescences. All the individuals of the species flower together and attract many different species of insect. When the plants have finished flowering, the insects move on to a different species which is coming into flower. Other tropical species produce flowers all year round. The flowers are often small and hidden among the leaves. They are visited by one faithful species of pollinator which hunts out the flowers and is kept alive throughout the year by their nectar production.

An entomophilous plant which has efficient pollinators does not need to produce much pollen as most of it will be carried to another flower. Thus entomophilous angiosperms produce much less pollen than anemophilous ones as the wind is much less accurate at hitting the stigma 'target' than a flying insect. This means that angiosperms relying on insects can be much more scattered throughout a community because the insects will hunt them out. Anemophilous plants tend to have to grow in uniform stands, with many individuals of the same species close together to ensure pollination. The diversity of the tropical forest is only possible because of the co-evolution of angiosperms with animal pollinators such as insects, birds, bats, other small mammals and even slugs.

Within the general background of diffuse co-evolution are examples of more specialised paired co-evolution between a plant and pollinator species. These presumably arose when one pollinator group began to predominate as the efficient pollinators for one species. More seed would have been set by flowers which attracted that group most strongly. This has led to an amazing range of flower shapes and colour. Non-specialised flowers which attract a range of insects tend to be open, flat cups, either single flowers, like those of the Magnoliaceae mentioned earlier, or as clusters of flowers as in the Asteraceae or Apiaceae (umbellifers). Bee-pollinated plants often have bilaterally symmetrical flowers, called **zygomorphic** flowers. These are frequently tubular, so the bee can orientate itself to the flower shape and then push its head or whole body into the flower. Bee flowers include many of the Orchidaceae. Moth-pollinated plants often have white flowers which are strongly scented and open at night (e.g. the night flowering catchfly, *Silene noctiflora*). Butterfly flowers usually have a long tubular corolla which matches the length of the pollinator's tongue: flowers visited by butterflies include thistles (*Cirsium* spp.) (Figure 19.6c) and knapweeds (*Centaurea* spp.).

Hummingbird-pollinated species often have bright red, tube-shaped flowers especially attractive to hummingbirds or sunbirds which have long beaks and can hover under the inflorescence while collecting nectar (see Figure 19.6b). Mammal-pollinated flowers are usually robust or small but in clusters held on a robust structure such as the stem of a tree, so that the weight of the animal can be supported while it collects pollen in its fur. Figure 19.6d shows a robust cactus flower being visited by a bat pollinator: the bat plunges its head into the flower and becomes covered with pollen. The pygmy-possum in Figure 19.6a is on a cluster of small flowers of an Australian banksia.

Despite all these specialisations by plants to improve the match between flower and pollinator, and therefore maximise the chances of successful pollination, there are some insects which break the rules and cheat. They are ahead in the evolutionary arms race, because they have learned to break into flowers by chewing a hole at the base of the tube of petals, so they can take nectar from a flower not designed for them. Other insects will then take advantage of the hole the robber has made and the flower may well not be pollinated as the anthers and stigma are bypassed. Short-tongued bees are often guilty of this sort of behaviour. Such insects gain the rewards of a pollinator without filling the desired role of pollen transfer. Perhaps the next step in this

Figure 19.6 Examples of animal pollinators: (a) Eastern pygmy-possum (*Cercartetus nanus*) on a banksia; (b) Costa's hummingbird (*Calyptae coata*) on a red tubular flower typical of hummingbird pollination: (c) Marbled white butterfly (*Melanargia galathea*) on a thistle; (d) Lesser long-nosed bat (*Leptonycteris curasoae*) on a saguaro cactus.

particular arms race is for the plants to evolve thicker bases to their corolla tubes, or to protect them with spines, enveloping sepals or bracts.

Although most flying insect groups, that is bees, butterflies, moths, hoverflies, beetles and so on, have co-evolved as pollinators of angiosperms, few species show tight, pairwise co-evolution with a single plant species. Fig trees (*Ficus* spp.) and their pollinating wasps are one group which do show tight, pairwise co-evolution. Generally, each of the 750 species of fig tree is pollinated by its own species of wasp which can only breed in the female flowers of its host (Anstett *et al.*, 1997). Usually plant species which do show

such tight one-on-one co-evolution are uncommon or rare in their particular community. If a species is widely scattered and has few flowers, it is unlikely to be pollinated very often if it is visited by generalists, as most of the time the generalists will be visiting other, more common species with the result that they will be loaded with other pollen types. It would be beneficial for such species to co-evolve with a single pollinator species prepared to search large distances for individuals (Feinsinger, 1983). However, if the species becomes too rare, it may be unable to maintain a population of the insects to pollinate it. Some rare species occur together in mixed populations and have very similar flowers so that they can share a pollinator. This way, two species can together maintain more individual pollinators, so that each flower may be visited more often by more insects, even though about half the pollen they deliver will be wasted as it comes from the other species. An example of shared pollinators is found in two rare species of ginger (*Costus* spp.) in Panama. These species have very different leaves, but extremely similar flowers; both are visited by the same euglossid bee which carries mixed pollen loads to each flower (Schemske, 1981).

Knowing which is the most effective pollinator for a plant is especially important for crops which require insect pollination to produce fruit. If a crop plant has been introduced into another country there may be no effective pollinator around. Such imports are usually done with little thought of what will happen without the co-evolved pollination partners. If yields are not satisfactory, the usual procedure is to pack the crop field with hives of honeybees as these are easy to move around, easy to keep and produce another cash crop, honey.

Sometimes the natural pollinator of a plant is not even known in its native country. This is the case for the Chinese gooseberry (*Actinidia chinensis*) which is grown as a fruit crop in several countries. The addition of honeybees to Chinese gooseberry crops may not result in an increase in fruit production. Chinese gooseberries, it appears, require pollination in a certain way, called **buzz pollination**. A bee visiting the flower buzzes loudly and the vibrations of the buzz shake dry pollen out of the male flowers and all over the bee. Unfortunately, honeybees are incapable of buzzing the flowers at the correct frequency to loosen the pollen in the anthers. The larger bumblebee species (*Bombus*) which do buzz-pollinate many species are probably a much better group to encourage in the *Actinidia* orchard (Corbet *et al.*, 1988). The tomato flower illustrated in Figure 19.1 is also a buzz-pollinated species which requires bumble bees to pollinate it or manual pollination achieved by tapping the flowers. Buzz pollination is so effective that Dutch tomato growers now set up bumblebee colonies in their greenhouses. In the 1980s, growers had to rely on humans equipped with electric vibrators, at an annual cost of some $20000 per hectare (Vines, 1997).

19.4.2 The early evolution of insect pollination

In almost every terrestrial community, plants can be found which are pollinated by insects. The ways in which angiosperms are pollinated by insects and other animals provide some of the most remarkable relationships within ecology. Such relationships are undoubtedly the result of co-evolution. The reason why angiosperms are so diverse today is because there are so many insects, while there are so many insects because the angiosperms are so diverse! The two groups have evolved and radiated together.

In the Jurassic and earlier, before angiosperms evolved (see Table 18.1), the diversity of insects was probably very much less, although, as insects are hardly ever preserved as fossils, we have a very meagre record of insect evolution. Carnivorous groups like dragonflies, and omnivores like cockroaches and other beetles, had been common since the Carboniferous. However, specialist pollinators like bees and hoverflies did not exist. Insects are thought to have been involved in spore transport ever since land plants became well established in the Devonian. Later, with the evolution of seed-bearing plants, they also carried pollen. Several fossil gymnosperm groups may have been insect pollinated by Jurassic times. Such relationships probably originated by accident, with foraging insects picking up and carrying pollen about in a fairly haphazard way. Pollen and spores are a valuable food source, if in rather small packages, and were probably eaten by insects from the Devonian onwards. As insects foraged more often for pollen, they became a more reliable transport system for plants and the co-evolution of insect pollinators and flowering plants had begun.

In the Jurassic, a group of gymnosperms called the cycadeoids had reproductive structures which could be called 'flowers'. The cycadeoids had thick stems with rosettes of leaves at the ends and strange reproductive structures (Figure 19.7). These were situated on the stem and were large and fleshy. Examination of the fossil 'flowers' reveals that some of them

Figure 19.7 A reconstruction of the cycadeoid *Williamsoniella* showing the 'fleshy flowers'. The model is in the Field Museum of Natural History, Chicago, USA.

Figure 19.8 *Rafflesia arnoldii* from West Sumatra, Indonesia, five days after the start of flowering. These are the largest flowers in the world, up to a metre across, yet are produced by a parasitic plant with no leaves.

appear to have been chewed. It is probable that these 'flowers' were beetle pollinated (Crepet, 1983). The fleshy structures may have attracted the beetles directly as a food source, or they may have been structures of deception which smelled like rotting dinosaur meat! Present-day angiosperms use flowers of both sorts, food providers and deceivers, to obtain insect pollinators. Most insect species visit flowers for nectar and pollen, but some visit flowers because they are tricked by the flower in some way. The huge fleshy flowers of *Rafflesia* look and smell like rotten meat and attract flies which feed or lay eggs on carcasses of animals (Figure 19.8). Even more surprising are the orchids which look and smell like a female insect to attract a male pollinator (see Section 19.4.3).

Although there is evidence of pollination by insects of gymnosperms in the fossil record, such **entomophily** in gymnosperms is extremely rare today. Nearly all gymnosperms are wind pollinated, that is **anemophilous**, and do not possess flower-like structures. It is the angiosperms, often called the 'flowering plants', that have flowers. Most angiosperm flowers are delicate and short-lived and so make poor fossils, but all the evidence suggests that both wind- and insect-pollinated angiosperm species existed from early Cretaceous times. The first entomophilous flowers in the Cretaceous were probably very simple and radially symmetrical. Flowers with a similar structure today, such as the cup-shaped flowers of the Magnoliaceae and Winteraceae, are pollinated by unspecialised small

insects like beetles (Gottsberger *et al.*, 1980) and this was probably true of the Cretaceous flowers too.

Because both flowers and insects fossilise only rarely we do not know a great deal about exactly how they have co-evolved. But it is clear that co-evolution has taken place from the huge number of species of angiosperms and the close relationships they have with their insect pollinators.

19.4.3 Orchids and Hymenoptera

As mentioned in Section 19.4.1 most angiosperms tempt their pollinators to visit them by offering nectar. Some, however, attract pollinators by deceit. Orchids of the genus *Ophrys* are a good example of flowers which cheat their insect pollinators. They attract members of the Hymenoptera, mainly bees, but also some wasps. Figure 19.9 shows a close-up of the flowering spike of *Ophrys vernixia*, a European species called a bee-orchid because of the similarity of the rounded velvety flower to a bee. This is the clue to the attractiveness of these flowers to their pollinators. *Ophrys* species attract insects from a small number of genera, mainly the bees *Eucera* and *Andrena*, and they always attract males. A male bee which lands on the flower appears to be trying to mate with it. In doing so, the bee comes into contact with two masses of pollen called pollinia and these detach from the flower and become stuck to the insect. When the bee visits another flower, the pollinia, which have a sticky outer surface, become attached to the downward curving stigma so that parts or all of the pollinia are rubbed off on to the stigma and the flower is pollinated.

An interesting part of this co-evolutionary relationship is that the flower does not only look like a female bee, it also smells like one! The flowers emit a large number of volatile chemicals, most notably terpenoids and fatty acid derivatives. These odours

Figure 19.9 A bee orchid (*Ophrys vernixia*) from Southern Spain being pollinated by two male wasps (*Campsoscolia ciliata*).

mimic those of female bees, or they may act as what is called a **super-stimulus**, that is, a signal which is even more obvious and encouraging than the natural one. (The huge gaping orange mouth of a cuckoo chick is another example of a super-stimulus, which is why such small birds as warblers and pipits chase themselves frantic in an effort to stuff it full of insects.) Different orchid species, and even different sub-species, emit a different mixture of scents and each attracts a different species or group of species of insect (Bergström, 1978).

Bergström suggests that, at an earlier evolutionary stage, a pre-*Ophrys* was visited by both males and females of a particular species in the Hymenoptera. This would have been followed by a transitional stage in the co-evolution of the present relationship. The change may well have been triggered by male behaviour: even a slight sexual stimulus presented by the flower would lead to males spending more time at the flowers than females. This would mean that the males had a greater chance of picking up the pollinia and of rubbing off pollen from pollinia already attached to them. Thus, flowers which stimulated the males the most would be more successful as they would be more likely to pass pollen on and be pollinated themselves. Evolution would then promote further improvements in the flower to attract males.

19.5 Introduced species

Diffuse co-evolution of all the species in a community will cause that community to fit together and function in ways that at times almost resemble a single organism. Populations may fluctuate with the changing abiotic environment or in response to fluctuations in other species, but one species rarely becomes so successful that it destroys the community or drives another species to extinction. Presumably this is because, in most cases, any advantage gained by one species in the evolutionary arms race will be only small and will rapidly (in evolutionary terms!) promote a response in the other species of the community. It is as if all species are boxed in on all sides by the constraints of the environment.

If a species is taken out of its box and introduced into a new situation, in a different community where it has not co-evolved, then various situations may occur. It may be that the species has no advantages in the new community, that the arms race there has reached a higher level, in which case the introduced species could go extinct. Alternatively, the community may be at a similar level of co-evolution, in which case the species may be able to survive with a similar level of success as in its own community. The third possibility is that the introduced species is well

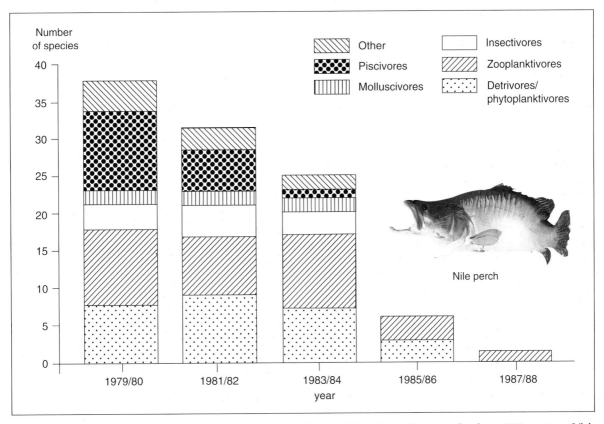

Figure 19.10 The introduction of the Nile perch to Lake Victoria led to the extinction of at least 200 species of fish that were endemic to the lake.

ahead of the new community in the arms race, in which case it will be able to increase with little or no check and will threaten individual species, or even the whole community, with extinction. A dramatic example of this has occurred since 1954 when a man emptied a bucket into a lake the size of Switzerland. The bucket contained Nile perch and the lake was Lake Victoria. Over the next 40 years at least 200 species of fish endemic to the lake went extinct, although the Nile perch does now support a huge fishing industry. Figure 19.10 shows that extinctions have affected all the various categories of fish in the lake.

It is often the case that the introduction of a mainland species into isolated communities, such as those on islands and detached continents, has the third devastating consequence (MacDonald & Cooper, 1995). Isolated oceanic islands and island chains are probably vulnerable because they have so few founding species (see Section 18.3.2) and because these founders subsequently undergo adaptive radiation to fill the niches in the available habitats with only very rare interference from other species (see Section 18.3.4). Many groups are absent on islands, and so co-evolution takes place without the influence of such species. In one sense the island arms races are taking place with a limited arsenal of weapons; introduce a new species with more efficient weapons and the result is a massacre. Large isolated continents, like Australia, and large islands, like Madagascar and New Zealand, suffer the same problem; they often lack major groups, and the species they have represent ancient evolutionary lines which have been surpassed in the intensive evolutionary rough and tumble of the larger continental landmasses.

There are many examples of how the introduction of species from otherwise unrepresented groups can disrupt or even destroy communities (e.g. Sutherland, 1995). Three examples are given here: the effect of ants in the Hawaiian Islands, and the introduction of rabbits and cacti into Australia.

Ants in the Hawaiian Islands

The Hawaiian islands have a unique native fauna (see Section 18.3) which does not include ants but which did, in about 1900, include an endemic flightless fly. At some time in the twentieth century a voracious predatory ant (*Pheidole megacephala*) which had been introduced into the islands wiped out the flightless fly. The ant now swarms in millions over the lower land of the islands where many other endemic Hawaiian insects once lived and can now no longer be found (Zimmerman, 1970). Many of Hawaii's endemic moths are being destroyed by introduced parasitic wasps, and several plant species are also at risk through predation of seeds and fruits by introduced rats. The Hawaiian Islands are an extreme example of a series of tightly co-evolved communities where the introduction of almost any mainland species seems to upset the balance and put the native biota at risk.

Rabbits in Australia

Rabbits have been introduced into several countries where their rapid rate of breeding and ability to graze the ground vegetation to almost bare earth has made them a considerable pest. In Britain, the rabbit (see Figure 16.7) was introduced by the Normans about 900 years ago but only became extremely numerous in the nineteenth century, possibly because of climatic changes, alteration of agricultural practices and the decline of predators due to sporting activities (Sumption & Flowerdew, 1985).

In Australia, the rabbit reached plague proportions after it had been introduced there in 1859. The rabbits destroyed the grazing land which was important for the Australian sheep and cattle ranchers. Huge rabbit drives were organised in an attempt to catch and kill rabbits, but these proved unsuccessful as a control method. Eventually, a virulent strain of the virus myxomatosis (see Section 16.4.3) was introduced into the countries, including Australia, where rabbits were out of control. At first the effect on rabbit populations was devastating as the virus caused the death of about 99% of the rabbits. Since then, the rabbit and myxomatosis have begun to co-evolve. Most of the rabbits which survived were those which were more resistant to the virus. The virus has also changed, so that it is now less virulent. This is presumably because the less virulent forms are less likely to completely wipe out a rabbit colony and therefore fail because of lack of hosts. The behaviour of the rabbits has changed too; they seem to spend more time on the surface of the ground instead of in burrows. This helps them avoid catching rabbit fleas which are the carrier of the virus.

In 1995, rabbits cost Australia around A$600 million a year in lost agricultural production. Rabbits also play havoc with the indigenous wildlife. They eat growing shoots and prevent many plants from regenerating. Native animals that depend on this vegetation for food and shelter may become locally extinct. For instance, the greater rabbit-eared bandicoot (*Macrotis lagotis*), classified by the World Conservation Union as a 'vulnerable' species, is now found only in rabbit-free areas (Anderson & Nowak, 1997).

In 1996, Australian scientists deliberately released rabbit calicivirus in an attempt to reduce rabbit numbers. The release has been controversial with some virologists saying that not enough is known about the biology of the virus to sanction such releases. Indeed, in July 1997 the New Zealand government rejected a request from farmers to import the virus from Australia to control New Zealand rabbits on the grounds that the epidemiology of the virus was still poorly understood (Anderson, 1997a). Despite this, farmers in New Zealand went ahead and introduced the calicivirus and by August 1997 dead rabbits were found in several locations (Anderson, 1997b).

In Australia, early indications are that the virus is killing around 80% of rabbits in infected areas. The virus causes blood to clot in the rabbits' lungs, heart and kidneys and the animals die from heart failure or asphyxiation, usually within a few days of infection. Although it is too early to be sure, there are some encouraging signs that the crash in rabbit numbers is leading to a rejuvenation of native species – both animal and plant (Drollette, 1997). Only time will tell if the attempt to control rabbits will be successful, like the following example, or another ecological disaster for Australia.

Figure 19.11 Before and after photographs of a site in Queensland, Australia. The first picture, taken in May 1928, shows thickets of *Opuntia*. In the second, taken in October 1929, the *Opuntia* has been attacked by *Cactoblastis cactorum* and is rotting.

Cacti in Australia

Introduced plant species can also become a threat to the native vegetation if they are able to grow and reproduce well in the climate. The Cactaceae are not native to Australia. Several species of cacti, notably *Opuntia inermis* and *O. stricta*, became severe economic pests in Australia. The *Opuntia*, or prickly pears, are all natives of America so they have not co-evolved with the Australian flora and fauna. *O. inermis* seems to have been introduced into Australia in 1839. By 1870 it was becoming a problem, forming dense impenetrable thickets on grazing land. By 1900 it had covered 10 million acres of land, and this rapidly increased to 60 million acres by 1920 (Dodd, 1929).

In the Americas, *Opuntia* species are attacked by a number of insects and diseases which were absent in Australia. By 1912, the *Opuntia* problem was so great that the possibility of introducing some cactus pest was considered in an attempt to limit cactus numbers. Such a method, using a natural predator, herbivore or disease to limit a species, is called **biological control**. The insect pests of *Opuntia* in America were studied in detail and breeding centres of various insects were set up to provide specimens in sufficient numbers to be worth introducing into Australia. Of 150 species investigated, about 50 were sent on to Australia for field trials to determine their usefulness as biological controls. Some insects which looked as if they might have been able to survive on other, Australian, species were rejected, as they might have become pests too. In the end about 12 species were successfully released as biological controls, but one species eclipsed all others in its devastating effect on the *Opuntia* species: *Cactoblastis*.

In 1925, 2750 eggs of the moth *Cactoblastis cactorum* were sent to Australia from Argentina. These were reared in breeding cages and an experimental introduction of 9 million eggs by 1927 showed that the insect was a promising contender for the control of the cacti. By 1930, 3000 million eggs had been distributed to the areas where prickly pears had become a pest (Dodd, 1940). The moth seems to have been aptly named, for blast the cactus it certainly did. By 1932 the moth had increased rapidly in number and the *Opuntia* populations crashed. The kind of effect the moth caused can be seen in the 'before' and 'after' photographs of a piece of land in Figure 19.11. The effect is dramatic and shows how good a biological control the moth was.

Cactoblastis lays its eggs in long stick-like clusters, each with about 80 eggs, on the surface of a cactus stem. When the larvae hatch they burrow down into the watery flesh of the cactus and eat the inside until they are ready to pupate. The control effects of the moth are thought to be helped by a bacterial disease which invades the plant tissue through scars left by grazing caterpillars. The bacteria causes the cacti to collapse and rot (Dodd, 1940). When the *Opuntia* populations crashed, there was also a crash in the number of *Cactoblastis*, so that in 1931–3 *Opuntia* regrew in various areas. By 1934 *Cactoblastis* had again increased and attacked the cacti to control the pest a second time. Since the 1930s, neither *Opuntia* nor *Cactoblastis* have become extinct in Australia, but the cactus now seems to be controlled and has not reached pest proportions since.

To be successful, a biological control must itself be controllable. It must be very host specific, so that it only attacks the pest it has been introduced to destroy. Considerable care has to be given to the investigation of the ecology of the control, to eliminate the possibility that it will move on to other, native species and itself become an ecological disaster. This host specificity is really another way of saying that the best biological controls are those which have arisen by one-on-one co-evolution in their native areas. A species which is only diffusely co-evolved with the pest may not be specific enough in its action. It will have other hosts in its natural ecosystem and will thus be much more likely to switch to some other species when its natural host declines in the area where it has been introduced.

In recent years biological control has become of

Figure 19.12 Starchy cassava tubers in Thailand. Biological control is helping to reduce pest damage of such valuable crops.

increasing importance. Many firms sell parasitic and other insects to horticulturists to enable them to control their pests. Kew Botanic Gardens, for example, now uses no pesticides, controlling any pests by the use of biological control agents. Of great importance to tropical agriculture is the attempt to use a predatory mite (*Typhlodromalus aripo*) to control the green spider mite (*Mononychellus tanajoa*). Steve Yaninek of the International Institute of Tropical Agriculture in Ibadan, Nigeria has spent more than a decade looking for a way to control the green spider mite which sometimes destroys a third of the crop of cassava – a staple food for some 500 million people (Figure 19.12). By 1997, the predatory mite had been successfully introduced in 12 countries of west and central Africa (Pearce, 1997d).

Summary

(1) When first introduced, the term co-evolution referred to the relationship between caterpillars and their food plants. Co-evolution was considered a subset of the general evolutionary relationships between species in communities which were called community evolution.

(2) Very close evolutionary relationships between individuals of two species are now known as pairwise or one-on-one co-evolution. Less specific evolutionary relationships are called diffuse co-evolution.

(3) Mutualism occurs where both species benefit from a symbiotic relationship. If the effects are deleterious then the relationship is antagonistic.

(4) Antagonistic relationships usually produce an arms race of chemical and mechanical attack and defence mechanisms.

(5) The Red Queen hypothesis suggests that in co-evolution, evolving species often simply maintain their position in the arms race in their community; species which cannot keep up in the arms race become extinct.

(6) Co-evolutionary relationships between flowering plants and pollinators are extremely varied. Most plants offer nectar or pollen in return for pollination, but some plants cheat by mimicking female insects or rotting meat. Some insects also cheat and steal nectar without collecting or depositing pollen.

(7) The introduction of species to islands or isolated continents where they are not in coevolutionary balance can have a devastating effect on the native flora and fauna.

(8) The best biological controls of introduced pests are generally ones which have undergone antagonistic pairwise co-evolution with the pest in its native land.

TWENTY

Conservation principles

20.1 Biology is not enough

This chapter is about the principles of conservation. It looks at the philosophical basis of conservation by addressing the question 'Why should we conserve?', and at the biology that underpins conservation. However, philosophy and biology are not enough. In Chapter 21 we will look at conservation in practice: at reasons why conservation sometimes works and at reasons why it often doesn't, even if the philosophy is sound and the biology correct.

20.2 The need for conservation

20.2.1 The pressure on wildlife

The number of people alive and the demands we make on the Earth's resources continue to increase. As the human population grows in size (see Box 4, p. 35), more and more land is brought under direct human control for agriculture and housing. In fact, around 40% of all the world's photosynthesis ends up being used by us. In consequence, the amount of natural vegetation diminishes and with it the space available for the species which live in such habitats. Although it is difficult to quantify with any precision the effects of human disturbance on the world's biomes (Hannah *et al.*, 1994), it is hard, if not impossible now, to find any habitat which is not affected by some change in species composition or balance due to introductions, disturbance or pollution. The vast expanses of tropical forest have become increasingly threatened in recent decades as large commercial companies back clearance schemes for cattle ranching and timber exploitation. Even in the oceans, fishing is so intensive that fish stocks are diminishing rapidly. In European waters most fish now caught for food, such as cod, are not fully grown. The large mature fish have already been caught and the smaller ones are captured before they can grow to replace them. We have become too efficient as predators.

Some animals are hunted not for food but for luxury items. African elephant numbers have been severely affected by poachers hunting for ivory (see Section 21.2.2). The various rhinoceros species are even closer to extinction as rhinoceroses are killed for their horns (Figure 20.1). Parrots and coral reef fish are captured in their millions for the pet trade. Most die before they ever reach the countries where they are to be sold. Some countries still use endangered species such as tigers and seahorses as part of their traditional medicine.

Figure 20.1 A 1996 study by TRAFFIC India, with the support of WWF India, reported that there were only about 1500 greater one-horned rhinoceros (*Rhinoceros unicornis*) left in the wild. Hunting has been banned since 1910 which allowed the populations to increase but poaching in the 1980s and 1990s has endangered the species. Demand for rhinoceros horn in Asian traditional medicines is a continued threat. These are horns taken from poachers in Kenya, Africa, where rhinoceros species are similarly endangered.

All this leads to a feeling that *Homo sapiens* is out of control. We have become a species that is no longer in co-evolved balance with its environment. **Conservation** is management of the Earth's resources in a way which aims to restore and maintain the balance between human requirements and the other species in the world. Those who support the active conservation of wildlife do so for a number of reasons which reflect the different ways people relate to the natural world (Section 20.3).

20.2.2 Maintaining biodiversity

The current pressure on ecosystems results from changes of land use by humans from either natural habitats or those exhibiting ancient farming practices. The cultivation of natural ecosystems, aggressive farming practices and increasing industrialisation now pose a considerable threat to local, and therefore also to global, **biodiversity**. The term 'biodiversity' was only coined in 1985 as a shortened form of 'biological diversity'. After the 1992 Rio Summit it became popular jargon: a buzzword for measuring the health of the planet. One measure of the biodiversity of an area is the number of species found there; however, the range of different life forms

(bacteria, fungi, ferns, flowering plants, nematodes, insects, birds and so on) is also important (Myers, 1995; Roy & Foote, 1997). Enormous effort is being made to provide better estimates of global biodiversity (Groombridge, 1992; Heywood, 1995).

One of the aims – perhaps the main aim – of conservation is to try to preserve the biodiversity we have inherited. Experiments and field studies suggest that a loss of diversity may lead to a loss of ecological stability (Section 14.4.3). Yet the biodiversity of some of our richest biomes (tropical rainforest and deep ocean) is still very incompletely known. Even in soils, the roles of the many species of bacteria, fungi and invertebrates are poorly understood (Groffman, 1997).

20.3 The philosophical basis for conservation

20.3.1 Ethical arguments

Some conservationists argue from ethical grounds that we have a duty to look carefully after the world. The world's various religions agree that we have, at the very least, a responsibility as to how we use the creation. It is true that a strong strand within some

---

Box 20

Ecofeminism

Ecofeminism represents a union of two concerns: **deep ecology** and **feminism** (Ruether, 1997). Deep ecologists insist that it is not enough to analyse the ecological damage to the Earth in terms of human social and technological use. We need to examine the ways in which humans have distanced themselves from nature and so claimed the ability to rule over it. Ecological healing requires that we recover the experience of being in communion with nature (Devall & Sessions, 1985). Feminism exists in many forms. Its principal aims are to transform human relationships and society through the exposure and analysis of inequalities, principally those resulting from the exploitation – whether or not conscious – of women by men.

Ecofeminism asserts that there is a fundamental connection in patriarchal societies between the domination of women and the domination of nature. The role of women as the caretakers of children, as gardeners, weavers, cleaners and cooks both exploits women and identifies them with an equally exploited natural world.

I remember standing in a market in Mexico in December looking hungrily at boxes of beautiful strawberries and wondering how I might sneak some back on the airplane through customs into the United States. A friend of mine, Gary McEoin, long-time Latin American liberation

journalist, standing next to me said softly, 'beautiful, aren't they ... and they are covered with blood'. To be an ecofeminist in my social context is to cultivate that kind of awareness about the goods and services readily available to me.

(Ruether, 1997, p.78)

It could be suggested that equating women with the exploited natural world is a way of denying any responsibility women may have for the present state of the Earth. Some ecofeminists seem almost fearful of suggestions that women might be 'closer to nature' than are men, others angry at the suggestion that this might not be the case. If women do identify strongly with nature, they become a powerful force with a pivotal role in conservation. For example, in affluent societies, such as western Europe and North America, women have purchasing power: the ability to choose which goods they buy. This empowers women to buy products that are eco-friendly and which support fellow women presently in exploited poverty through the promotion of fair trade and safe working conditions. Thus the strawberries in Mexico – a symbol of affluence but responsible for pesticide poisoning in those who grow them – become a representation of the choices available to us all.

Figure 20.2 The Garden of Eden is often portrayed as an idyllic paradise with friendly animals – an ideal no longer considered attainable in this world. Painted by Lucas Cranach the Elder (1472–1553) in 1526.

religions, particularly Judaism, Christianity and Islam, has viewed the created order as existing for humans to exploit. In *Genesis* Adam and Eve are told 'Be fruitful and increase, fill the earth and subdue it, rule over the fish in the sea, the birds of heaven, and every living thing that moves upon the earth' (*Genesis* 1.28). We cannot reconstruct with certainty how verses such as this one were first understood (Figure 20.2). There is little doubt, however, that the understanding that humans were created in 'the image of God' (*Genesis* 1.27) led many people to feel that they were set over nature and had authority to do with it pretty much as they liked. Christianity, in particular, has often been accused of a rapacious attitude towards nature:

> *Especially in its Western form, Christianity is the most anthropocentric religion the world has seen. In absolute contrast to ancient paganism and Asia's religions, it not only established a dualism of man and nature but also insisted that it is God's will that man exploit nature for his proper ends.*
>
> (White, 1967)

Certainly, the expectation, particularly within some Christian traditions, of life after death, an immanent Apocalypse and the advent of 'a new heaven and a new earth' (*Revelation* 21.1) made it easy for people to feel that they could ignore the needs of the rest of the created order and use this world for their own ends.

Despite this, the notion that humans can exploit the rest of the Earth for themselves is relatively uncommon in religions. More commonly, religions agree that humans have, at the very least, a responsibility as to how they use the creation (Tucker, 1997).

In Hinduism, the world's oldest major religion, all life is sacred. Visnu, as supreme being, endlessly creates the world and withdraws it into his existence time after time as the cycle of seasons endlessly repeats itself (Prime, 1992). In the Vedic literatures mother Earth is personified as the goddess Bhumi, or Prithvi, the abundant mother who showers her mercy on her children. It is not surprising that Hinduism views humanity as having a great responsibility towards the Earth.

In Judaism too there is a strong emphasis on the responsibilities that humans have towards nature. Agricultural land was supposed to lie fallow every seventh year as a 'Sabbath of sacred rest' (*Leviticus* 25.4). Further, every fifty years, on the Day of Atonement in the Jubilee Year, all land must return to its original owner. Because the Earth is the Lord's, no one has unconditional land rights (*Leviticus* 25.23). The Jewish scriptures also include a number of instructions relating to animal welfare, while some of the Wisdom writings argue that creation has a purpose beyond that of human benefit.

In Buddhism there is a very strong emphasis on how we should relate to the natural world; for example there is a prohibition on the taking of animal life. Although Buddhism exists in many different forms, human responsibility towards the creation is a common theme, though the word 'creation' is somewhat inappropriate as the Buddha taught that there is no creator God as the first cause, because there is no beginning. While Buddhism teaches that humans, unlike other creatures, have the opportunity to realise enlightenment, it does not teach that humanity is superior to the rest of the natural world. Indeed, the doctrine of 'emptiness' in Buddhism, as originally developed by the philosopher Nagarjuna, in asserting that all things are empty simply denies that anything can exist on its own (Batchelor & Brown, 1992).

The Chinese Buddhists, in particular, emphasised the intimate connections between all things. This philosophy found a more concrete expression in the Zen tradition. For instance, the Japanese Zen master Dogen says:

> *It is not only that there is water in the world, but there is a world in water. It is not just water. There is also a world of living things in clouds. There is a world of living things in the air. There is a world of living things in fire. There is a world of living things on earth. There is a world of living things in the phenomenal world. There is a world of living things in a blade of grass.*
>
> (Tanahashi, 1988, pp. 106–7)

The notion of **stewardship** is also a significant theme in recent writing on Christianity (World Council of Churches, 1990; DeWitt, 1994; Prance, 1996), Islam (Khalid & O'Brien, 1992) and other religions (Smart, 1989; Hinnells, 1991). Narrowly understood, stewardship still implies something of a 'them' (the rest of the natural world) versus 'us' (humans) situation. However, there is a more holistic view of our relationship to the rest of creation which can be described as one of mutuality. This reflects the ecological understanding that a relationship is mutual if it benefits both partners. The word 'mutuality' is therefore intended to avoid overtones of hierarchy and superiority but implies a beneficial influence of humans on other organisms which some people would consider hard to find. As some of the above quotations have indicated, humanity can be viewed as having a depth of inter-relatedness with the rest of the natural order that transcends an understanding which has humanity somehow set apart from, or above, the rest of creation.

Another interpretation of how humans relate to the rest of the natural world is that, in some sense, we are co-creators, co-workers or co-explorers with God (Peacocke, 1979; Barbour, 1990). Although this may at first sound blasphemous, or simply downright silly, the reasoning goes as follows. Our scientific understanding of the universe, in particular cosmology and biological evolution, indicates that creation has been an ongoing process for some fifteen thousand million years. Within the last few thousand years, humans have begun consciously to influence the course of that continued creation in a way never before attained by any species. Our influence changes the future of the world year by year, hour by hour, as we cause some species to become extinct and alter the genetic constitution of others by selective breeding as well as by the newer methods of genetic engineering. However, with the ability to change nature comes a responsibility not to misuse it.

Not all ethical arguments for conservation need have a religious basis (e.g. Elliot, 1995). Many who have no religious convictions still argue that we ought to try to save rare species and pristine habitats (Hågvar, 1994). For example, the American philosopher Paul Taylor, without reference to theological considerations, has argued that we should show respect for nature (Taylor, 1986). Taylor provides a theory of environmental ethics that rejects the notion that the natural environment exists only for human benefit. Instead, the creatures that make up the Earth's natural ecosystems are seen to possess inherent worth. The essential notion here is that living things have a value in themselves not simply because of what they can do to benefit humans.

20.3.2 Anthropocentric arguments

There are many **anthropocentric** (that is, human-oriented) arguments for conservation. One of the most common is that we should maintain global diversity as a resource. Many species have proved useful to us either in the past or today for food or as sources of medicinal compounds or building materials. The rate at which we are losing species or the wild races of our domesticated animals and plants is accelerating. It is extremely likely that there are wild species with genes for drought and disease resistance which could be bred into our crops. Such arguments – that species should be conserved because of their value to us – are called **instrumental**: they treat wildlife as an instrument for human purposes. One attempt to quantify the economic value of 'the world's ecosystem services and natural capital' produced a figure of US\$33 trillion per year – i.e. 3.3×10^{13} at 1997 prices (Costanza *et al.*, 1997). This is 1.8 times the 1997 global gross national product.

Anthropocentric arguments are not always to do with money, food or medicine. For example, it can be argued that tigers and giant redwood forests are beautiful and should be preserved for aesthetic reasons (Haldane, 1994). Certainly, increasing

Figure 20.3 Increasing numbers of people spend time in the countryside for recreation or personal fulfilment. Here walkers cross a marsh and river on duckboards in the Plitvice National Park, Croatia.

numbers of people enjoy visiting the countryside for recreation, for exercise or to give themselves the opportunity to grow spiritually (Figure 20.3). In addition, it can be argued that one reason why wild fish stocks should be preserved is that in their absence, the traditional way of life of fishing communities would end. Whatever our own attitudes to the environment, it is our descendants and future generations who will inherit the situation we create. Philosophers are increasingly trying to work out what precisely are our duties to those who come after us (Cooper & Palmer, 1995).

20.3.3 The role of ecology

The fact that we are destroying species, habitats and perhaps even the life-support systems of the planet by our irresponsible behaviour is a depressing thought. However, we can realise what we are doing and what we could do to try to put things right. This is where ecology fits into the picture. The autecology of rare or threatened species, the synecology of communities and the role of the abiotic environment in ecosystems are all relevant to conservation. The population dynamics of a threatened species and how it fits into the food web of an area indicate how likely it will be to recover once its habitat is secure. The setting up of protected areas or nature reserves is often a feature of conservation practice; theories of biogeography, an understanding of succession and the processes of nutrient cycling are all of importance in understanding how to establish and manage these areas effectively.

The principles of ecology can be applied to conservation at three levels: conservation of species, conservation of ecosystems and conservation of the biosphere. It is useful to make these distinctions as different aspects of ecology become more or less important at different levels of conservation.

20.4 Conservation of species

20.4.1 Why do species become extinct?

In geological time the fossil record shows that new species are continually evolving and others becoming extinct. Occasionally a major extinction event occurs when very large numbers of species disappear suddenly from the fossil record. These catastrophic events are thought to have been caused by such factors as changes in world climate, loss of habitat (for example when continental drift destroyed large areas of marine shelf), or the effects of giant meteors or volcanic eruptions. The best known of these extinctions happened some 65 million years ago when a giant meteor crashed into the present day north coast of the Yucatan Peninsula close to the Gulf of Mexico, causing the last of the dinosaurs to go extinct, ending the Cretaceous and heralding in the Tertiary period (Alvarez et al., 1992). However, the greatest known extinction event happened at the end of the Permian some 245 million years ago when around 95% of fossilisable marine species went extinct. The poor fossil record for terrestrial and soft-bodied marine species means that we will never know the precise extent of these extinction events.

Perhaps surprisingly, during many so-called mass extinctions probably only about 15–30 species went extinct each year. The reason for this is that most mass extinction 'events' take hundreds of thousands of years: it is just from the perspective of geological time that they appear instantaneous.

For the purposes of conservation we are obviously much more concerned with the species which were on the Earth when Homo sapiens evolved in Africa some 300 000 to 500 000 years ago (Clark et al., 1994; Bräuer et al., 1997). Since then, hundreds of mammal and flightless bird species have gone extinct and some have suggested that human hunting was responsible (Martin & Klein, 1984). The animals, of course, may have had their own unique parasites meaning that the total number of species that went extinct was even greater. Humans evolved as gatherers of plants and hunters of animals. Within the last 10 000 years a number of prey animals have gone extinct including the woolly mammoth (Figure

20.4a) and easy-to-catch flightless birds (Figures 20.4b & c). Predators too have gone extinct as a result of human persecution (Figure 20.4d).

However, it is probably only during the last few hundred years that extinction rates as a result of human activity have really rocketed. Every organism that goes extinct is the ending of a line that stretches back, unbroken, to the origins of life on Earth some 3500 million years ago. In the great majority of cases extinction really is forever. Genetic engineering, cloning and other Jurassic Park-inspired science fantasies are unlikely to bring more than a handful of species, if any, back from extinction. Over the last 400 years some 150 extinctions have been documented. Recorded extinctions are clearly only the tip of the iceberg. The dodo (Raphus cucullatus) (Figure 20.4c) of Mauritius is known to have been driven into extinction at the hands of Western sailors who found the docile birds an easy source of food. However, for every dodo-like documented species there must have been many others that went unnoticed. Estimates of total global species loss today range from 4000 to 300 000 per year, vastly higher than during even the great extinction events of the past (Pimm, 1995).

Extinctions by indirect human influence are almost certainly far more widespread than those which are a direct result of human activities such as poisoning and hunting (Figure 20.5). One common and indirect cause of extinctions results from the introduction of species, for example the predatory ant on the Hawaiian islands which has destroyed several endemic insect species as described in Section 19.5. Island biotas appear to be especially vulnerable. For one thing, islands are generally small so that the total number of individuals in a species is often smaller than it would be on a mainland. In addition, islands have distinctive patterns of colonisation and subsequent evolution (see Sections 18.3 and 19.5). Introductions of domestic animals, such as cats, dogs, goats and pigs, have caused much ecological damage to island ecosystems.

Undoubtedly the most important threat which humans pose to other species is by the destruction of suitable habitats. A few species, such as weeds (see Box 5, p. 49), certain birds (for example, the starling – Section 3.3) and other adaptable animals, have managed to thrive in the new environments produced by cultivation and urban life. Most species, however, have suffered a loss of available habitat. For many, this has reduced numbers to the point at which the species becomes extinct.

Actual extinction of a species obviously comes when the last individual dies. The process leading towards extinction may well, though, have started much earlier than this. The local extinctions of populations may finally lead to a situation where there is only one population left. Any isolated population is subject to accidents: freak weather, a virulent disease, the absence of a suitable host (if the species in question is a parasite), fire, flood and so on. If a species is widespread with many populations, the destruction of a single population produces a temporary decrease in numbers: emigrants from other populations will typically found a new population which, in effect, replaces the one that went extinct. If a species is rare with only a handful of populations, the loss of a single population may be a disaster from which the species never recovers.

If a species is rare and thought to be threatened with extinction, it is important for ecologists to ask 'Why is the species rare?' and 'Is it declining in numbers?'. Unfortunately, these questions require a

Figure 20.4 Extinct species. (a) The woolly mammoth (*Mammuthus primigenius*) was hunted by Palaeolithic humans. It became extinct about 8000 years ago. (b) The giant moa (*Diornis maximus*) stood over 3 m in height. It was one of about 22 moa species hunted to extinction in New Zealand between AD 900 and 1600. (c) The dodo (*Raphus cucullatus*) was first described in 1601 from the island of Mauritius. By 1681 it had become extinct as a result of hunting by Dutch colonists and egg predation by introduced pigs, dogs and monkeys. (d) The Tasmanian wolf (*Thylacinus cynocephalus*) was an agile marsupial carnivore. It became extinct on mainland Australia about 3000 years ago but survived in Tasmania. The last known specimen died in Hobart Zoo in 1936.

Figure 20.5 North American buffaloes (*Bison bison*) were hunted close to extinction by white colonists during the nineteenth century. In 1800 there were some 30 million of them; in 1870 some 20 million. By 1989 there were just one thousand. By 1994 the population had recovered to some 200 000.

good understanding of the autecology of the species. It is a sad fact that many of our rarest and most endangered species are ones we know least about. Many populations fluctuate considerably in size for all sorts of reasons. Untangling these reasons and determining whether or not the species really is undergoing a serious decline in abundance can take years – by which time it may be on the verge of extinction.

Species conservation should ideally begin when a species is known to be declining in numbers but is not yet threatened with extinction. The World Conservation Union – formerly known as the International Union for the Conservation of Nature and Natural Resources (**IUCN**) – keeps records and publishes detailed list of species at risk of becoming extinct. They recognise four categories of risk:

Rare species have small populations, usually within restricted geographical limits or localised habitats, or widely scattered individuals; they are at risk of becoming rarer but not of becoming extinct.

Vulnerable species are those which are rare or under threat or actually decreasing in number, or species which have been seriously depleted in the past and have not yet recovered.

Endangered species have very low population sizes and are in considerable danger of becoming extinct.

Extinct species are believed no longer to exist; they cannot be found in areas they once inhabited nor in other likely habitats.

The IUCN and, more recently, other conservation organisations produce what are called **Red Data Books** on species at risk in a number of groups including vascular plants, swallowtail butterflies, cetaceans, African primates and New World birds (Collar, 1996; Colston *et al.*, 1997). The Red Data Books for some of these groups list all known species in these four categories. The sheer scale of the task makes this impossible for most taxa, however. For example, it is estimated that around 25 000 plant species fall into one of the above four categories, that is, about 10% of all plant species. Most plants have probably been described, for other organisms our ignorance is enormous (Table 20.1). In particular, we have probably only named around 10% (one million) of all insect species and almost nothing is known of the population biology of the great majority of those we have named. Identifying the 'at risk' taxa under these circumstances is impossible.

20.4.2 Genetic diversity in rare species

In Chapter 6 we saw the importance of genetic variation in populations: how some species have very high degrees of heterozygosity while others are homozygous, often because of the effects of inbreeding. Empirical studies show that large population sizes are associated with greater genetic variation

Table 20.1 The approximate number of species known, and the total number thought to exist, for a number of different taxa.

Taxon	Number of species	
	Known	Estimated
Bacteria	4 000	1 000 000
Protists	42 000	200 000
Algae	40 000	300 000
Fungi	77 000	1 000 000
Viruses	5 000	400 000
Bryophytes	16 000	26 000
Pteridophytes	10 000	20 000
Seed plants	240 000	300 000
Nematodes	20 000	400 000
Molluscs	72 000	200 000
Crustaceans	43 000	150 000
Arachnids	80 000	750 000
Insects	980 000	8 000 000
Vertebrates	47 000	50 000

(Young *et al.*, 1996). However, even species with homozygous populations may preserve genetic variation if the various populations differ in their possession of alleles.

A species which has declined to such low numbers as to be considered endangered may encounter one of a number of genetic problems, including :

Complete loss of some alleles from the species resulting in a loss of genetic diversity with consequent inability to respond rapidly to selection.
Reduction of population breeding ability (resulting from increased relatedness between individuals) through the action of incompatibility mechanisms in plants or mate choice in animals.
Expression of deleterious alleles and increased homozygosity, increased mortality of young and inbreeding depression leading to reduced offspring fitness.

These problems may not be disastrous in the short term. As we saw (Section 6.4.2) the cheetah may have survived with a high degree of homozygosity for thousands of years. However, this illustrates the fact that once diversity is lost it takes a very long time – possibly thousands of generations – for evolution to build it up again. Thus it is very important during species conservation to preserve as much genetic diversity as possible, or the ability of the species to evolve in future environmental change will be reduced.

The loss of genetic diversity from a small population is affected both by the size of that population and by the number of generations that the population size remains at a low level. Consider a population of size N after t generations. The proportion of the genetic variation that remains among these N individuals after t generations is given by the formula:

$$\text{Variation} = \left(1 - \frac{1}{2N}\right)^t$$

The use of this equation can be seen in Table 20.2 which lists the variation remaining for different population sizes after breeding for up to 100 generations. From the table you can see that if the population is maintained at the extremely low level of just 10 individuals, then after only 10 generations only 60% of the genetic variation remains, i.e. 40% has been lost. Even with a larger population of 50 individuals, the passage of 100 generations means that only 36% of the genetic variation remains, i.e. almost two-thirds has been lost. As a rule of thumb, conservationists typically strive to maintain a population size of 500 as an absolute minimum (Soulé, 1980). However, many endangered species are far rarer than this. For example, in the vascular plant Red Data Book about 120 species are listed as endangered; 28 of these have population sizes of 20 or fewer individuals (IUCN, 1978). Some endangered vascular plants have only a single individual left. Some of these are self-compatible and so can set fertile seed, but others are doomed. One such doomed species is a member of the ebony family, a tree in the genus *Diospyros*. The single remaining known specimen lives on the island of Mauritius and is female so, unless a male plant can be found, the species' only chance of survival will be by **vegetative propagation** by botanists.

Table 20.2 The amount of genetic variation calculated to remain in a number of small populations of different sizes after various numbers of generations. Small populations which only interbreed for one or two generations lose much less variation than populations which remain small for many generations. (Data from Frankel & Soulé, 1981.)

Population size	Number of generations			
	1	5	10	100
2	75%	24%	6%	≪1%
10	95%	77%	60%	<1%
20	97.5%	88%	78%	8%
50	99%	95%	90%	36%
100	99.5%	97.5%	95%	60%

The number of individuals in a population is not the same as the number of individuals actually leaving offspring. Populations typically contain immature or senile individuals and sometimes the sex ratio is in imbalance. Further, social behaviour in animal populations such as dominance may prevent other individuals from breeding. In these cases the equation just given for genetic variation as a function of population size, N, and the number of generations, t, must be modified by the substitution of N_e, the **effective population size**, for N. The effective population size is always less than the actual population size that a survey would reveal. To maintain an effective population size of 500 typically requires around 2000 individuals (Soulé, 1980).

Although these guidelines of a minimum of 500 or 2000 individuals are useful, there are dangers in adopting them uncritically. For example, available data on extinct mammals on mountain-top islands show that smaller species need far larger population sizes if they are to avoid extinction (Remmert, 1994). This seems to be because, in general, the smaller a species is, the greater the fluctuations in its population size over time. This means that if both a small and a large species live in nature reserves that can only hold, say, 1000 of them, there is more chance of the smaller species suddenly dying out. It has been calculated that while a 100 kg mammal species requires a population size of around 320 individuals to maintain its genetic diversity over a long period of time, a 1 kg mammal species needs around 16 000 individuals (Remmert, 1994).

Figure 20.6 Speke's gazelle (*Gazella spekei*) in Saint Louis Zoo, Missouri, USA.

20.4.3 Captive breeding programmes

Captive breeding programmes are sometimes a last, but important, resort in the conservation of a species. It is rarely possible, however, for a single zoo or botanic garden to maintain populations which are large enough for the maintenance of genetic diversity, though Kew Gardens expects to collect and bank seed from 10% of the world's wild plants by the year 2010 (Pearce, 1996b). The problem is particularly acute for zoos because the public prefer to see large animals. No single zoo can keep 2000 individuals of a mammal or bird species – there simply isn't the space or money to feed them. It is feasible to keep 2000 individuals of an endangered insect or snail and some zoos are indeed beginning to do just this. However, even this costs money and too few people want to see any but the most exotic of invertebrates.

Captive breeding in zoos has improved significantly in recent years and zoos now collaborate with other zoos far more. Animals are shipped around the world to maximise outbreeding and so maintain genetic diversity in the species and the health of the individuals. Sometimes circumstances are such that inbreeding is impossible to avoid. Speke's gazelle (*Gazella spekei*) is found in the area of political unrest between Somalia and Ethiopia. In 1969 and 1972 animals were taken from the wild to start a captive herd in the USA (Figure 20.6). The herd was started with just one male and three females. Until 1976 the original male was bred with his daughters and granddaughters. Inbreeding depression rapidly resulted and infant mortality increased dramatically.

From 1976 a new breeding programme was implemented. The original male was only allowed to breed with the original females (to whom it was

presumed he was not closely related). Two sons born as a result of these crosses were used to breed with the other females which had been born to the group. These crosses of sons with half-sisters and cousins still constitute inbreeding – as this cannot be avoided given that only one original male was available – but inbreeding depression was minimised through the careful identification of the most suitable mates. By 1980 a total of 19 animals had been bred with two aims in view. One was to shift the balance of genes in the population so that equal proportions came from each of the four founding animals. This maximises the genetic variation retained in the population. The second aim was to try to maintain the health of the animals despite unavoidable inbreeding. This was done by, controversially, choosing animals which were healthy and fertile but inbred in preference to healthy, fertile animals which were not inbred. The logic behind this principle is that animals which have a high degree of homozygosity (i.e. are inbred) but are healthy and fertile (i.e. show no evidence of carrying deleterious alleles) may be better for breeding than heterozygous animals which may still carry deleterious, recessive alleles which weaken the population in later years (Templeton & Read, 1983).

20.4.4 Re-introductions

A captive breeding programme may be able to prevent a species from going extinct. Re-introducing that species to the wild is another matter (McKendrick, 1995; Sarrazin & Barbault, 1996). For a start, there may be no remaining habitats: they may have been destroyed by agriculture or building, for example. Even if the species' habitat does still exist, the composition of the other species in the community may have changed significantly. Flowering plants, for instance, may have lost their natural pollinators. Some animals face a different problem. They may have lost the ability to cope in the wild. Predators may lack the skills to catch prey; prey the ability to avoid predation. For all these reasons, re-introductions do not always go as smoothly as planned (e.g. Falk *et al.*, 1996).

Re-introductions are not only attempted when a species has gone extinct in the wild. They may also be done to boost numbers in an area or be made in an area where the species lived in the past but has gone locally extinct (e.g. Stewart, 1993). There can even be financial arguments in favour of re-introductions. It has been estimated that reintroducing wolves to Yellowstone National Park would cost $200 000 a year because of their predation on livestock and big game animals but would add $19 million a year to the local economy through increased tourism (Duffield, 1992).

Some of the most successful re-introductions to areas where species have become locally extinct have been for large birds. The sea eagle (*Haliaeetus albicilla*) has been re-introduced to the island of Rhum in Scotland from Norway; the bearded vulture (*Gypaetus barbatus*) has been successfully released into the European Alps; and the griffon vulture (*Gyps fulvus*) now breeds again in the Massif Central of France (Figure 20.7).

Figure 20.7 The griffon vulture (*Gyps fulvus*) has been successfully re-introduced into the Massif Central of France.

The success of these projects is probably due to several factors. All these birds are large, have few, if any, natural predators and nest in fairly inaccessible places on cliffs. Their original decline was directly to human interference rather than loss of habitat. They were shot by farmers and gamekeepers who believed the birds killed domestic animals. In fact, sea eagles take mostly fish, and vultures do not kill healthy animals but feed on the remains of animals which have already died. A third reason why these re-introductions have been successful is that in these species females often lay two eggs but only rear one chick to adulthood. Because of this, removal of an egg, or the smaller chick, from existing wild birds for hand rearing in captivity can be undertaken to increase population sizes.

Some species exist today only because they have been taken into captivity. The Hawaiian flora, like its fauna, is largely unique and very at risk. About 280 species and subspecies had already become extinct by the mid 1970s with some 800 being endangered (IUCN, 1978). In 1930 the last known example of a hibiscus tree (*Hibiscadelphus giffardianus*) died. The species' decline was probably due to a combination of factors including habitat damage by lava flows, extensive grazing, disturbance and the decline of pollinating honeycreepers (see Figure 18.13) – of the 39 species of honeycreepers on the Hawaiian islands, 30 are already extinct or listed in the Red Data Book. Fortunately, some *Hibiscadelphus giffardianus* individuals were being grown in cultivation and the species was re-introduced into its original locality in the Volcanos National Park. By 1968 the wild population had grown to 10 healthy mature trees with a number of seedlings.

Perhaps the species which has been preserved in undomesticated captivity longer than any other is Père David's deer (*Elaphurus davidiensis*). This deer seems to have become extinct in its native swamplands in China about 3000 years ago, but herds survived in the Emperor's Imperial Hunting Park near Beijing. In 1894 floods damaged the walls of the Imperial Park and the deer escaped only to be hunted for food by local peasants. Most of those that survived died in the 1900 rebellion and by 1921 the species was extinct in China. Fortunately a few animals had been taken to Europe in the nineteenth century and the Duke of Bedford built up a herd (Figure 20.8). By 1964 he had about 400 deer and was able to send four animals back to China (Usher, 1973). One day it is hoped that they may be re-introduced into a semi-wild situation back in their native lands.

Although there have been several successful re-introductions of captively bred species into the wild, there have been many failures. Often the ecology and behaviour of these species is poorly known (Curio, 1996). Without a knowledge of their true habitat

Figure 20.8 The Woburn Park herd of Père David's deer (*Elaphurus davidiensis*).

and the original reasons for their decline it is some-what optimistic to expect re-introductions to be successful (Cherfas, 1989). Captive breeding is a last resort, not the most satisfactory method of conservation. It can be expensive. For example, the successful project to breed the nearly extinct black-footed ferret, rediscovered in North America in 1981, cost $2 million in its first five years (May, 1989). Removing a significant number of individuals of a very endangered species from the wild into captivity is risky in itself. Take too few individuals and the captive breeding programme has little chance of success; take too many and you may effectively cause the species to go extinct in the wild. Even if captive breeding is successful there are problems of the loss of genetic diversity, unintended selection (e.g. for docile behaviour) and the difficulties of re-introductions. A far better solution is to preserve entire ecosystems in which rare species exist before they become endangered.

20.5 Conservation of ecosystems

20.5.1 *The importance of habitat conservation*

The destruction of existing natural ecosystems is the most important threat to species diversity. An idea of just how wide is the range of human activities which endangers species is illustrated in Table 20.3 which categorises the reasons why plants listed in the IUCN Red Data Book are endangered. At first sight, over-collecting appears most important, endangering 35

species. However, habitat destruction – i.e. land clearance, flooding, drainage and logging – is endangering 89 species, while habitat changes due to grazing, disturbance, introductions and changes in land use threaten another 62 species. Further, there is no doubt that Table 20.3 hugely underestimates the extent of the problem. If one thinks of a group like the insects, with their far greater diversity and abundance, habitat damage and destruction are vastly more threatening than any overcollecting.

Table 20.3 The major reasons why the plants recorded in the IUCN Red Data Book (1978) are endangered.

Threat	Number of species	Percentage
Collecting by gardeners/tourists/botanists	35	15
Browsing and overgrazing	33	14
Populations critically low for breeding	31	14
Clearance for agriculture	22	10
Industrial and urban growth	16	7
Logging in forests	12	5
Coastal development	11	5
Roads	9	4
Disturbance by cars or trampling	9	4
Fire	9	4
Effects of introduced plants	8	4
Dams and flooding	8	4
Changes in farming practice	7	3
Mining and quarrying	6	3
Drainage	5	2
Forestry	5	2
Pollution	2	1

In an attempt to protect at least some portions of natural ecosystems, many countries have set up national parks, nature reserves and other forms ofconservation area. How, then, should such areas be designed?

20.5.2 Design of nature reserves

Nature reserves are created as refuges for threatened species or as a means of preserving one or more ecosystems. Mostly, nature reserves are intended to preserve the biodiversity of an area, though they may also be used for recreation and tourism. It is virtually impossible to list all the species found in an area (Wilson, 1992). Fortunately, there is a surprisingly high correlation in many taxa between the number of families (or genera) in an area and the number of species. For instance, looking at 74 sites across the globe, the correlation between the number of genera of seed plants and the number of species is 0.96, so the more genera there are of seed plants, the more species there are. Good correlations also hold for such varied taxa as butterflies, reef fishes, frogs, birds and mammals (Lee, 1997). This means that it isn't necessary to know all the species in an area to obtain a good estimate of its biodiversity.

It is important to set up a reserve to minimise extinctions and maximise biodiversity. But how should this be done?

The first requirement is for the reserve to be large enough to hold viable populations of the species it is trying to conserve. The same principles apply as for the breeding of captive populations. Genetic variation within each species must be maintained as far as possible. This means an effective population size of at least 500, equal to an actual population size of around 2000 individuals (see Section 20.4.2).

If the reserve is surrounded by similar habitats then populations in the reserve and surrounding area may be large. However, it should be borne in mind that in future the surrounding area may change. If the natural surrounding ecosystem is destroyed, the reserve will be of limited value if it cannot maintain viable population sizes. If a species is migratory, a single reserve will not be able to protect it throughout the year, but reserves can be set up at feeding and breeding sites. For some species, especially large top carnivores such as hunting dogs and tigers, reserves will need to be very large. On the other hand, a rare orchid or localised insect may be afforded valuable protection by a reserve measured in square metres rather than thousands of square kilometres.

Further decisions which may need to be made, in addition to the size of a reserve, are its shape and whether a number of smaller reserves scattered through an area are better or worse than one large

one. Sometimes the shape of a reserve is predetermined if it is to include a natural feature such as a river bank, mountain top or lake. If the shape is not predetermined, then in most cases the best shape is circular. This is because a circular feature has the smallest amount of edge possible for a given area. It is generally supposed that edges are to be minimised because an edge zone is not representative of the ecosystem contained within the reserve: the edge of a wood, for example, may be lighter and windier at ground level than inside the wood.

Usually conservation areas are made from pre-existing sites, often the remaining fragments of natural ecosystems in amongst agricultural land. They will therefore already have well defined pre-existing edge zones. This gives an ecologist less opportunity to study the effects of edges on ecosystem conservation. One very interesting long-term study, the Minimum Critical Size of Ecosystems Project, is being carried out in the Amazon rainforest by the WWF and the National Institute for Amazon Research in Brazil (Lovejoy et al., 1986). Here the opportunity arose to create new reserves when large areas of the forest were cleared for grazing land between 1980 and 1984. Different-sized fragments of the original, undamaged forest were left standing as surrounding trees were cleared. Plots of 1 ha, 10 ha and 100 ha were left and their ecology compared with plots of the same sizes in uncleared forest.

For the 1 ha and 10 ha plots the edge effects reached right into their centres (Figure 20.9). These small plots were affected by increased light intensity and decreased humidity. This resulted in increased leaf fall, followed by increased tree mortality from 1.5% in the uncleared forest plots to 2.6%. When the experimental reserves were first set up, birds crowded into the remaining trees, but their numbers soon fell. Some birds took advantage of the denser edge growth, where insects were common, but this was only a temporary effect. Eventually only the centres of the 1 ha and 10 ha reserves maintained forest birds. In reserves of all three sizes there was a steady increase in the numbers of light-loving butterflies in a 200–300 m deep strip around the edges of the reserves but their activity disrupted the deep forest-dwelling species. Such studies indicate that, at least for tropical forest, a buffer zone several hundred metres deep of the same ecosystem needs to be set up around the reserve proper to take such edge effects into account.

The Minimum Critical Size of Ecosystems Project also studied the effects of reserve size. Even the 100 ha plots were too small to support certain animal species. Some small monkeys, including the golden-handed tamarin (*Sanguinus midas*) and the red howler monkey (*Alouatta seniculus*), appear to appreciate edge vegetation, and these species survived in 100 ha and

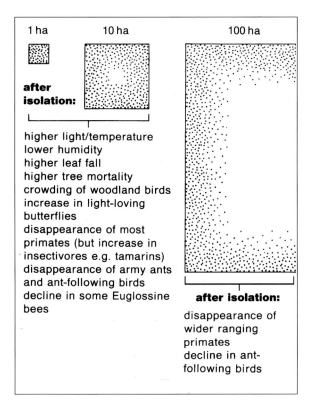

Figure 20.9 Summary of the experimental structure of the forest plots in the Minimum Critical Size of Ecosystems Project and some of the results which the experiments produced.

even some 10 ha reserves. Most primates, however, disappeared from the reserves within the first year of their isolation. Other species which persisted for only a limited period, even in the 100 ha reserves, were the predatory army ant *Eciton burchelli* and the various ant-following birds which forage around the ant columns of such species. There were also very significant reductions in the numbers of euglossine bees (Didham *et al.*, 1996). Male euglossines are important orchid pollinators; the females pollinate other species. A reduction in pollinators can have serious consequences not just for the various plant species but also for frugivores. Many tropical rainforest trees occur only at very low densities. Their populations can therefore extend over hundreds of square kilometres which makes them particularly susceptible to fragmentation (Chase *et al.*, 1996).

When the equilibrium theory of island biogeography (Section 18.3) was proposed, conservation biologists applied it with considerable enthusiasm to reserve design. Reserves, like islands, are isolated areas some distance apart and it seemed an obvious application of the theory. It was thought that small reserves (equivalent to small islands) would be less good than larger reserves (like large islands) because extinction rates would be higher. This would mean that the

equilibrium point for a small reserve would be for fewer species. In certain circumstances, though, it was thought that a series of reserves could be envisaged as optimal by analogy with a chain of islands on the grounds that species would hop from one to another (Wilson & Willis, 1975).

With time, however, it came to be accepted that the simple models derived from the theory of island biogeography may not necessarily suggest how best to distribute reserve land. There may, in fact, be no single 'best strategy' for apportioning land for conservation (Simberloff, 1986). One reason why large islands generally have more species than smaller ones may be that larger islands are more likely to have many available habitats than small islands. This would suggest that it is diversity of habitats, rather than overall size, which may be important in a reserve. In some cases several small reserves may be better than one large one because they cover a wider range of environmental conditions such as soil type and topography. A number of small reserves may also be beneficial in other ways: catastrophes such as fire, flood, diseases or hurricanes are unlikely to damage all the sites at the same time, so the risk of elimination of a rare species will be less. There are drawbacks though: edge effects will be more influential with a number of small reserves and each of them may be too small to maintain adequate populations of some species.

Rough guidelines for setting up reserves are summarised in Figure 20.10. These are only guidelines, not strict rules. Every case is different. The reserve may be set up to maintain one or two key species or an entire ecosystem; the species involved may have small area requirements or very large ones; they may need several habitats at different times of the year; and so on. The structure of the potential sites and the requirements of the communities involved must be considered individually for every potential reserve.

20.5.3 Maintenance of conservation areas

Once decisions about the size, shape and location of a reserve have been made, the design of the reserve may be over, but headaches about management will only just have begun!

The first important decision is whether to manage the reserve at all apart from marking boundaries and limiting human interference. If the site is likely to change without management so that its value as a reserve diminishes, then management will be necessary. That looks like a very simple statement but it is often difficult to decide what management is required to maintain the stability of a reserve. Management also needs to anticipate change; by the time the deterioration is noticed, it may be too late to correct it. An added complication occurs because community

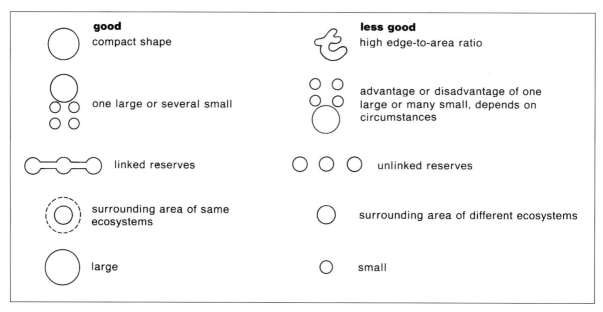

Figure 20.10 Guidelines for setting up a nature reserve.

structure changes naturally from year to year so that some observed changes will be either unimportant or important features maintaining the community.

A frequent problem with which conservation managers have to deal is that of succession. It is often necessary to prevent the invasion of another vegetation type into a reserve. This is especially important if the reserve is a diverted sere such as a grassland. We saw in earlier chapters how the invasion of scrub can irreversibly alter soil features such as pH and nutrient content. Under such circumstances, even if scrub that has invaded is cleared away, the desired vegetation may not return.

An example of this effect is provided by the story of the loss of chalk grassland species throughout much of southern England in the 1950s and 1960s. During this time a reduction in sheep grazing and the virtual disappearance of the rabbit as a result of myxomatosis meant that much chalk grassland became overgrown by bushes. Chalk grassland is the most species-rich form of vegetation in the United Kingdom, with a profusion of beautiful low plants and a number of rare orchids (Figure 20.11). Unfortunately, removal of the bushes, mostly hawthorn (*Crataegus monogyna*) and wild privet (*Ligustrum vulgare*), led not to the return of the much loved chalk grassland but to an unimpressive tangle of weeds including cleavers (*Galium aparine*) and even the common nettle (*Urtica dioica*). Almost 20 years of research by Peter Grubb, Barbara Kay and others finally sorted out what was going on and what to do about it (Stanier, 1993). The problem came down to the levels of phosphorus and nitrogen. These were much higher in the scrub soil than on original chalk grassland. Part of the problem is that the deeper roots

of the bushes absorb phosphorus from the subsoil. When the leaves, twigs and fruits of the bushes fall to the ground, this previously 'hidden' phosphorus is transferred to the topsoil. The solution is to sow upright brome (*Bromus erecta*) grass seed as soon as the shrubs have been removed from a patch of land. As the years go by the brome gradually reduces levels of nitrogen and phosphorus in the soil, allowing the desired chalk grassland plants to return.

Other studies have also shown the importance of low levels of phosphorus to the maintenance of species richness. Addition of phosphorus to old hay meadows dramatically reduces the abundance of the green-winged orchid (*Orchis morio*). In part this is because the addition of phosphorus encourages competitors. However, the effects of phosphorus on the green-winged orchid are so marked that it may be

Figure 20.11   The lizard orchid (*Himantoglossum hircinum*) flowering in its native chalk grassland.

that the element is toxic to it or to its mycorrhizal symbiont when added to excess (Silvertown *et al.*, 1994; see Figure 13.11).

A separate problem for managers of conservation areas is to do with natural regeneration if this is patchy and spasmodic. Suppose, for example, that the major elements of the community, such as forest trees, regenerate on a larger scale than the reserve size. If the normal pattern of events, for instance, is for one 10 000 ha area to blow down or burn down within a 100 000 ha forest every 20 years in an ever-changing mosaic, then maintaining a reserve of only 10 000 ha will require careful management. Such a reserve may be large enough for all the species to coexist in the short term but still not be a regenerating system unless the vegetation regenerates in smaller units than is naturally the case. It would obviously be disastrous for the whole reserve to burn down or blow down in one go.

The attitude of conservation managers to natural catastrophes in nature reserves has been an uneasy one. The aftermath of fire, for example, is unpleasant: much of the vegetation is destroyed and animals may be burnt alive or face starvation. Fires in the Yellowstone National Park, USA, were prevented or extinguished as soon as practicable for 100 years after its creation in 1872. This is now thought to have been a mistake for during that time the lack of fires had allowed a large amount of dry, combustible material to build up. In 1988 about half the park's 2 million acres burned. Fortunately, the sheer size of Yellowstone National Park means that this was not the disaster it might have been. The fire has given many plants the opportunity to germinate and grow,

Figure 20.12 Regeneration of *Pinus contorta* seedlings after fire in the Yellowstone National Park.

including the lodgepole pine (*Pinus contorta*) which, like several other pine species and Australian *Eucalyptus* gums, needs fire to trigger seed germination (Figure 20.12).

20.6 Conservation of the biosphere

We will leave our consideration of conservation of the biosphere to the next, the concluding chapter. Problems at this level of scale – such as those involved in tackling global warming, the destruction of the ozone layer and the damage from oil spills – require a particularly thorough integration of biology, politics and economics.

Summary

(1) The increasing demands of humans on the natural resources of the world threaten wildlife and, potentially, even the life support systems of the planet.

(2) Conservation is management of the Earth's resources in a way which aims to restore and maintain the balance between human requirements and the other species in the world.

(3) Ethical arguments for conservation may be based on religious faith but need not be.

(4) Wildlife may be conserved for human ends as well as for its own ends.

(5) Extinctions throughout geological time are due to natural disasters, habitat loss, climatic change, chance and evolutionary arms races.

(6) Within the last 10 000 years or so an increasing number of extinctions have resulted from human activity. Today, most extinctions are probably the result of human actions.

(7) Conservation can focus on individual species, individual ecosystems or the whole biosphere.

(8) If the number of individuals in a population drops to a small number (say, less than 500) for more than a few generations, a considerable amount of the genetic variation is lost.

(9) When taking an animal or plant species into captivity, it is important to have a breeding programme which maintains as much genetic variation as possible.

(10) Captive breeding can, occasionally, help conserve species.

(11) Re-introductions are sometimes successful but often not.

(12) Habitat destruction is probably the single most important threat to wildlife.

(13) Nature reserves should, if possible, be carefully designed. They are sometimes difficult to maintain due to succession or problems that arise from their not being large enough.

TWENTY ONE

Conservation in practice

21.1 The realities of attempting conservation

In Chapter 20 we looked at the principles of conservation. In this, our final, chapter, we look at conservation in practice. We shall examine why conservation sometimes succeeds and why it all too frequently fails, and ask what needs to be done in the future. This chapter is not, therefore, only about biology. It includes elements of economics, psychology and politics. The reason for this is that successful conservation demands much more than a good understanding of biology. It requires determination, political clout and an understanding of people and economic systems. Without such an understanding, conservationists risk being naive do-gooders unlikely ever to achieve much of lasting significance.

To help structure this chapter we will follow the same pattern as in Chapter 20, looking first at the conservation of species, then at ecosystems and lastly at the biosphere. It needs to be recognised, though, that the boundaries between these three levels of conservation are somewhat fluid. Effective species conservation nearly always involves habitat conservation and the distinction between habitat and ecosystem conservation is mainly one of scale.

21.2 Conservation of species

21.2.1 The golden lion tamarin – a successful re-introduction

The golden lion tamarin (*Leontopithecus rosalia*) provides an example of where captive breeding and re-introductions to the wild are playing a significant part in the conservation of a species. The golden lion tamarin is a small monkey (Figure 21.1) found in the Poco das Antas Reserve in Brazil and nearby. In the mid-1960s the reserve had a wild population of about 75–100 golden lion tamarins, which was about a quarter of the number of individuals existing in the wild. Fortunately this species breeds quite

rapidly in captivity, particularly if subordinates are separated from the rest of the group and bred as separate pairs. If kept in a group, only the dominant pair tends to breed: reproduction is suppressed in the subordinates (Curio, 1996). With the success of captive breeding, a plan to boost the wild population in the Poco das Antas Reserve by releasing captive bred animals has been undertaken by the WWF in association with the Brazilian Forestry Department and the Rio de Janeiro Primate Centre. An area of the reserve which did not have a resident family of monkeys was chosen as the site to release 15 captive-bred tamarins. Another six wild tamarins which had been

Figure 21.1 A golden lion tamarin (*Leontopithecus rosalia*).

rescued from an area of forest which was about to be destroyed were also released into the reserve. There was some mortality among the captive-bred individuals at first, but the majority of the introduced individuals survived and breeding has been successful. By 1997, the population had increased to around 800 (Mackinnon *et al.*, 1986; Bratley, 1997).

Captive breeding and re-introductions have been backed up by education programmes and public awareness schemes. Conservation organisations and local communities have worked together to conserve the fragmented ecosystem. However, only 2% of the tamarin's original habitat remains. Garo Batmanian, director of WWF–Brazil, wants to see that area doubled by the year 2025.

21.2.2 The African elephant – protective legislation

African elephants (Figure 21.2) provide an instructive case story of conservation in practice for several reasons. For a start, simply because elephants are so large and slow to reproduce, their numbers can be monitored with a reasonable degree of accuracy. We therefore know a fair amount about how their numbers have changed over the last hundred years or so. Secondly, the factors that have put increasing pressure on their numbers are several and include hunting for ivory and habitat destruction. This fact, together with the fact that they are found in several countries, cautions against easy solutions for their conservation. Thirdly, despite the fact that in some places elephant numbers are increasing and that

very considerable sums are being spent on protecting them, there is still some doubt as to whether the African elephant will be saved from extinction in the wild.

In the 1960s the African elephant (*Loxodonta africana*) was still thriving and there were around two million of them. During the 1970s, though, the price of ivory soared and more and more elephants began to be killed for their tusks. (The tusks are outgrowths of two of the teeth and are found in both males and females.) Between 1973 and 1987 between 85 and 89% of all the elephants in Kenya and Uganda were killed. Even the decline in numbers does not tell the whole tragedy. The animals with the largest tusks are the adults, so many mothers were killed leaving their youngsters behind. In some areas there are hardly any adults left and troops of young, delinquent elephants can be seen aimlessly roaming the savannah. They are not obtaining the normal teaching and guidance from older animals and it is unclear what they will be like as parents, if they survive to adulthood. An indication of what they may turn out to be like is afforded by the experience of Pilanesberg, a small wildlife reserve in South Africa. Here, young male elephants were imported into the park from the Kruger National Park in the early 1980s after the rest of their herd was culled as part of an attempt to control the Kruger's growing elephant population (Koch, 1996). Now these elephants have trampled to death tourists and professional hunters. They have even gored to death 19 white rhinoceroses and tried to mount rhinoceros cows (Figure 21.3).

Further problems resulting from a decline in

Figure 21.2 An old photograph of a herd of African elephants (*Loxodonta africana*). Such large groups with adults carrying well-grown tusks are now a rare sight.

Figure 21.3 African elephants brought up without the guidance of their elders may behave as delinquents. This orphaned bull elephant is attempting to mate with a white rhinoceros.

> ### Box 21.1
> ### *The Asian elephant*
>
> While public attention often focuses on the conservation status of the African elephant (*Loxodonta africana*), the Asian elephant (*Elephas maximus*) is actually far more endangered in the wild. Once Asian elephants roamed in their millions from Syria to China but by 1997 there were only some 35 000 untamed ones left, spread thinly over at least 13 countries in Asia (Fauna & Flora International, 1997). The main threat is forest clearance for agriculture and other human activities, though poaching for ivory is also a problem in some areas (Martin & Vigne, 1997).
>
> The Asian elephant has enjoyed an exceptionally close relationship with people for some 4000 years. Unlike its African counterpart, the Asian elephant can be tamed and trained. It has been used to carry great burdens, to provide pomp at festivals and to fight in battle. In most south and south-east cultures it is or has been worshipped as a god. Its extinction in the wild would be a terrible loss.

elephant numbers are due to the fact that African elephants are a **keystone species**: they play a pivotal role in the maintenance of their ecosystem and many other species depend on them. Elephants create and enlarge watering holes by trampling the mud and carrying it away after wallowing. In addition, their paths to the waterholes help funnel the rainwater to them. Also elephant dung – which is produced in large quantities – provides a habitat for dung beetles and other species, replenishes the fertility of the soil and distributes tree seeds. Elephants also open up scrubland, providing grassland for grazers such as gazelles and impala.

By the 1980s, most killing of African elephants was illegal poaching for ivory by poachers using guerrilla tactics and sophisticated automatic weaponry. With limited money to fund anti-poaching operations, exclusion of determined poachers was virtually impossible (Douglas-Hamilton, 1988). In an attempt to stem the demand for ivory, **CITES**, the Convention on International Trade in Endangered Species of Wild Fauna and Flora, agreed in 1989 to a total ban on ivory trade by placing the African elephant on its Appendix 1. Ivory prices fell sharply and poaching declined. Indeed, some ivory carvers started to use sub-fossil mammoth tusks instead (Swinbanks, 1989b). As a consequence of the drop in poaching, elephant numbers began to increase. By 1996, elephant cows in the South African Kruger National Park were even given long-term contraceptives in an attempt to stop them reproducing.

In 1995 there were probably around 580 000 African elephants – Tanzania, Congo (former Zaire),

Gabon, Zimbabwe and Botswana each having between 80 000 and 100 000 of them (Sharp, 1997). With the recovery in elephant numbers, Botswana, Namibia and Zimbabwe argued for a restoration of culling and/or trade in modern ivory. At their 1997 meeting, the 138 nations that belong to CITES agreed to remove elephant populations in these three countries from Appendix 1. However, before these countries are allowed to start a limited ivory trade again, they must convince a number of CITES-appointed monitoring committees that they have set up adequate controls to prevent the laundering (illegal sale) of ivory poached from other countries (Koch, 1997). Further, in an attempt to prevent a rise in poaching, trade will only be allowed from existing government stockpiles of ivory, tusks from which will be marked in some way for identification.

21.2.3 The tiger – teetering on the edge of extinction

At the beginning of the twentieth century, there were eight subspecies of tiger (*Panthera tigris*) in existence. By 1996, four of these were extinct leaving only the Indian tiger (*P. t. tigris*), the Indochinese tiger (*P. t. corbetti*), the Sumatran tiger (*P. t. sumatrae*) and the Siberian tiger (*P. t. altaica*) (Figure 21.4). Numbers are very difficult to determine. For instance, in 1996 the WWF estimated that there were between 415 and 475 Siberian tigers. This estimate relied on counting tiger tracks in the snow. However, a 1997 survey sponsored by the Tiger Protection Society in

Vladivostok put the number at only 250. This later survey used German shepherd dogs that had been trained to distinguish between the odours of different tigers (Jones, 1997). Whatever the precise numbers of the various subspecies, there is little doubt that all four subspecies are declining and that none has a population size of 1000.

Tiger populations have been reduced by habitat destruction, loss of wild prey, the authorised removal of 'problem animals', and hunting to supply the trade in skins, bones and other parts. The most important threats are probably habitat destruction and hunting. Estimating the extent of hunting for the medicine and souvenir trade is not easy. A 1995 study into the availability of tiger products in northern Sumatra, Indonesia revealed that of 88 medicine shops, souvenir stores and gold stores, 10 offered verified tiger products for sale (Plowden & Bowles, 1997). Prices (in equivalent US dollars) ranged from $7 to $125 per kilogram for bones, from $2 to $16 for a claw and from $34 to $68 for a tooth.

In India, the international, non-governmental organisation WWF has been prominent in the campaign to save the tiger. In the 1970s WWF persuaded the then prime minister, Indira Gandhi, to launch Project Tiger. Over the following 20 years, Project Tiger led to the establishment of 23 tiger reserves across India. WWF put in approximately £1 million

and the tiger population doubled (Pellew, 1997). WWF used a number of approaches: lobbying the government; law enforcement against traffickers and the illegal trade in animal parts; education to promote public awareness; and support for anti-poaching work.

In 1997, WWF initiated a new tiger conservation programme with a budget of £750 000 over three years. This programme focuses on three of India's premier tiger reserves – Corbett, Dudhwa and Manas. Radio communications are being improved, more anti-poaching patrols are being established and equipped and efforts are underway to improve the working conditions and morale of forest guards. The Indian parliament has voted more than adequate funds for the conservation of the tiger. However, there are difficulties in ensuring that central government releases these funds and in ensuring that state governments use the money for its intended purposes.

21.2.4 Northern spotted owl – habitat destruction

The northern spotted owl (*Strix occidentalis caurina*) (Figure 21.5) is a member of the late-successional forest community found in the Pacific Northwest of the USA (Franklin, 1995; Hunter, 1996). It preys on mammals such as northern flying squirrels

Figure 21.4 The Siberian tiger (*Panthera tigris altaica*) is very endangered. By 1996 there were probably only some 250 to 400 in the wild.

Figure 21.5 The northern spotted owl (*Strix occidentalis caurina*).

(*Glaucomys sabrinus*), redtree voles (*Arborimus longi-caudus*) and woodrats (*Neotoma* spp.). It competes with other owls, including the barred owl (*Strix varia*) and is even eaten by the great horned owl (*Bubo virginianus*). During the twentieth century, logging has reduced late-successional forests to about 15% of their original area and a number of species found only in them (including the northern spotted owl) are listed as threatened with extinction. USA legislation requires National Forests to maintain well-distributed populations of all vertebrate species and to maintain biodiversity in general. In addition, the northern spotted owl is protected under the US Endangered Species Act which means that its habitat must be preserved. However, logging is a billion dollar industry and supports the jobs of significant numbers of local people.

In the late 1980s the Interagency Scientific Committee to Address the Conservation of the Northern Spotted Owl was convened by the chief of the USDA Forest Service, under the leadership of wildlife biologist Dr Jack W. Thomas. Its conclusions were that large 'habitat conservation areas' should be created and that the exploited forest between these areas should be managed so as to improve the potential for successful owl dispersal. Each habitat conservation area should be large enough for at least 20 breeding pairs of owls. These areas should not be separated by more than 18 km and the intervening land should be subject to the '50–11–40' rule: 50% of this land should have forests with an average tree diameter of 11 inches at breast height and with a 40% canopy cover. It is the restrictions on the land between the habitat conservation areas that make this a revolutionary proposal. The 1992 northern spotted owl recovery plan by the United States Department of Interior accepted these recommendations.

Although this approach taken by the Interagency Scientific Committee has much to recommend it, it focuses exclusively on the northern spotted owl. From a publicity point of view there is something to be said for this – the owl is a 'flagship' species with which people can easily identify. To an ecologist, of course, and to many other people there is much more of value in this ecosystem than the owl. The most recent approach, devised by a Forest Ecosystem management Assessment Team, adopts a broader perspective. Its aim is to protect virtually all the species found in the virgin forest, including salmon and the other species associated with forest streams. However, even this plan, to the disappointment of many environmentalists, allows the thinning of stands in late-successional reserves. To the purist this may be regarded as a sell out. However, conservation is not about existence in an abstract realm of purity. It is about living in the real world where not everyone has the same value systems and where compromise

is needed for any movement forward to be made.

The above story applies only to the half of the remaining forest that is publicly owned. The other half is owned mainly by large timber corporations. Here virtually all of the forests have been cut and the emphasis is mainly on growing a single species, Douglas fir, on a 40–80 year cutting regime. Sites are clear cut and seedlings are planted and tended.

21.2.5 Spreading avens – habitat management

Spreading avens (*Geum radiatum*) is a flowering plant that is endemic to exposed rock outcrops on isolated mountain summits and high ridges in the Southern Appalachians, USA (Figure 21.6). Along with a number of other plants found in this habitat it has decreased in abundance over the last 100 years and is now classified as endangered on both state and federal criteria (Johnson, 1996). The main threats to spreading avens are visitor trampling, rock climbing, recreational and residential developments and changes in surrounding vegetation. At most of its present 11 sites, visitor trampling has caused severe decreases in abundance (Figure 21.7).

During the 1990s a conservation programme was mounted to benefit spreading avens and other plants found in this community at Craggy Gardens, Blue Ridge Parkway, North Carolina (Johnson, 1996). A combination of trail rerouting and the provision of seating and viewing areas for visitors reduced trampling by 97%. Restoration and reintroductions were carried out on 30 habitat patches with a total area of just 11 m². After considerable experimentation,

Figure 21.6 Spreading avens (*Geum radiatum*) – an endangered species with a chance of recovery thanks to active conservation measures.

Figure 21.7 Visitor trampling severely damages the habitat in which spreading avens (*Geum radiatum*) grows.

patches were constructed using woven coconut-hull fibre bags filled with native soil. The final plots were installed in 1993 and evaluation of first-year survival proved encouraging. Longer-term monitoring continues.

One encouraging feature of this project, as is often the case with plant conservation, is the relatively small cost, at least compared with the conservation of mammals and birds. The main sources of funding were $18 200 from the National Park Service for visitor management studies, restoration and trail design, $17 000 from the US Fish and Wildlife Service and the State of North Carolina for studies on *Geum radiatum*, and the provision of a graduate research assistantship from the University of Georgia. In addition, a number of organisations and individuals provided expertise free of charge.

21.2.6 Partula *snails – captive breeding*

On 1 January 1996 at 5.30 in the afternoon, *Partula turgida* went extinct. It was then that the last individual died in London Zoo (Morris, 1996a). Approximately 35 species of *Partula* are known, found in a number of Polynesian islands, some species being restricted in the wild to just a single valley. Although the snails were collected for their

---

Box 21.2
The good bulb guide

In May 1996 the Indigenous Propagation Project of the conservation organisation Fauna & Flora International celebrated its first bulb harvest (Figure 21.8). High in the Toros Mountains in Turkey, villagers harvested bulbs of snowdrops (*Galanthus elwesii*) and aconites (*Eranthis hyemalis*). The project had been launched five years earlier because of the massive conservation problems being caused by the collection of wild bulbs and corms in Turkey which has a wonderfully rich flora, including tulips, cyclamen and lilies. Entire hillsides were denuded and bulb populations were becoming greatly reduced to satisfy the horticultural markets of Europe and North America. The propagation project is providing local people with a greater, more reliable and sustainable income as well as taking the pressure off the wild populations and their habitats.

Fauna & Flora International has also been working in other ways to reduce the exports of wild-collected bulbs. For instance, retailers in Britain have been encouraged to pledge not to sell wild-collected bulbs and an education campaign in Turkey has warned of the dangers of over-collection (Fauna & Flora International, 1996). The success of the project is indicated by the fact that exports of wild-collected bulbs from Turkey fell by over 60% from 1989 to 1996 – a difference of more than 45 million bulbs (Morris, 1996b).

Figure 21.8 The first harvest of propagated bulbs in Turkey, May 1996

Figure 21.9 Dave Clarke, head keeper of the Invertebrate Department at London Zoo, with a Giant African land snail. It was the introduction of this snail to Polynesia that indirectly led to the extinction of a number of species of *Partula* snails.

shells which were used to make welcome necklaces, traditional snail collection had little impact. However, in the 1960s the Giant African land snail was introduced into Polynesia as a food (Figure 21.9). Unfortunately, as is so often the case with introduced species, the introduction went wrong. The snail escaped and started to ravage crops. In 1978 the carnivorous snail *Euglandina rosea* was imported as a biological control agent (London Zoo, 1994). However, *Euglandina* ignored the Giant African land snail, turning instead to the defenceless *Partula*. Within ten years, several species of *Partula* were wiped out.

London Zoo is one of 18 zoos and universities that have been trying to conserve *Partula* snails with financial backing from Fauna & Flora International. Fortunately, despite the loss of *P. turgida*, most of the species are easy to keep in captivity. Efforts are now being made on some Polynesian islands to provide enclosures from which *Euglandina* is excluded and into which various *Partula* species can be re-introduced. Early indications are that it may prove more difficult to exclude *Euglandina* than first thought. Three different *Partula* species were introduced in 1994 but when investigators returned in 1995, in every enclosure *Euglandina* had forced an entry and eaten almost all the *Partula*. The one encouraging finding was that the *Partula* had successfully bred before being eaten. This suggests that the re-introductions may prove successful provided *Euglandina* can be controlled.

21.3 Conservation of ecosystems

21.3.1 Different models of conservation

Different countries can have quite different aims for their nature reserves and other protected areas. For instance, in China, nature reserves are created simply for conservation; in Japan, they function for conservation and tourism (Ma & Numata, 1996). Countries also vary greatly in the protection that they give their wildlife habitats. The UK, for example, has practically no undisturbed areas of wildlife left, yet has a tremendous amount of legislation providing various degrees of protection to certain areas. However, as we shall see (Section 21.3.2), much of the UK legislation has exemptions and can legally be by-passed under certain circumstances. It can be argued that these features weaken the **statutory protection** (i.e. protection provided by the force of law) afforded to wildlife.

There can actually be conservation benefits, in some cases, in allowing what at first appear to be non-conservation interests to be taken into consideration. An **extractive reserve** is one in which local people continue to obtain limited resources such as nuts, fruits, rubber or fish from the reserve. One of the arguments in favour of such extractive reserves is that they help mobilise support amongst local peoples for the continuation of the reserve. It has been estimated that a 1 ha plot of Amazon forest in Peru can yield a profit of around $400 *each year* on fruit and rubber, while the same plot felled for timber yields a single crop of wood worth only around $1000. This compares with an income from cleared forest in Brazil used for cattle ranching of $150 per hectare per year (Peters *et al.*, 1989).

Extractive reserves fulfil the criterion of **sustainability** – one of the buzz words of conservation since the 1980s. A sustainable lifestyle or way of obtaining a resource is one which can, in principle, be continued indefinitely. Thus, obtaining energy from fossil fuels is not sustainable whereas energy from solar power, wind power and tidal power is. Cutting down trees in a forest can be sustainable providing the rate at which trees are cut down is less than or equals the rate at which they are replanted or naturally regenerate.

Deciding what precisely is sustainable can be complex. Conventional selective logging for mahogany – one of the most valuable tropical hardwoods – may not be sustainable from the point of view of the mahogany. However, mahogany is so rare – often there is only one tree every 5 ha – that, if the trees are extracted carefully, such logging can leave the ecosystem as a whole relatively undamaged. In the Chimanes forest of lowland Bolivia, for example, logging for mahogany destroys only 4–5% of forest cover. 'It looks almost like untouched forest' says Richard Rice of Conservation International (Holmes, 1997). However, when logging companies follow 'sustainable' management plans drawn up by environmentalists two problems arise. The first springs from the fact that these plans provide for regeneration of mahogany. Yet mahogany seedlings need large clearings if they are to survive. So to enable mahogany regeneration, logging companies have to clear more of the forest than the logging alone requires. The second problem is that these plans generally recommend that a number of tree species, rather than just mahogany, are felled. This damages more of the forest than when mahogany alone is taken. Over a 50-year period Rice has calculated that conventional selective logging for mahogany should lead to 2.5 times as much profit and 60% less damage than 'sustainable' logging.

Countries vary greatly in the extent to which they permit organisms to be collected from nature reserves. So-called 'butterfly ranching' is now an important aspect of habitat conservation in a number of tropical regions (New, 1994). Butterfly ranching occurs when unenclosed habitat patches are enriched with larval food plants and adult nectar sources. This makes the patches superattractive to certain butterflies. The butterflies are then caught and sold to collectors. Purists may object to such schemes but ranching is becoming increasingly popular because of its low capital outlay and because of the high prices some collectors are prepared to pay for perfect specimens. Incomes two or three times the minimum rural wage are common among butterfly farmers in the Papua New Guinea highlands.

21.3.2 UK legislation

In the UK, ecosystem protection is principally afforded by various pieces of legislation relating to National Parks, National Nature Reserves and Sites of Special Scientific Interest.

A **National Park** is an area in England or Wales of substantial size and outstandingly attractive scenery that is specially protected and reserved for public enjoyment. Some countries have had National Parks for over a hundred years but the first ones in the UK were established in 1949. Now there are ten in all as shown in Figure 21.10, covering 8% of the area of England and Wales. The Norfolk Broads and New Forest are also shown; although they are not technically National Parks, they have a very similar status. Scotland and Northern Ireland have a similar system to National Parks: in Scotland the protected areas are called **Regional Parks**, in Northern Ireland **Areas of Outstanding Natural Beauty**.

It needs to be realised that the main function of the National Parks of England and Wales is not real-ly the conservation of natural ecosystems. For one thing, the landscape of England and Wales has been so affected by human activity that almost no 'natural' ecosystem of any size remains. Then there is the fact that legislation permits farming, forestry and quarrying to take place in them.

National Nature Reserves (NNRs) are smaller and more numerous than National Parks. There are approximately 275 of them in England, Scotland and Wales. They are designated because of their importance in protecting biological diversity in Britain. However, they have functions in addition to conservation: research, demonstration and advice, education, and amenity and access.

Sites of Special Scientific Interest (SSSIs) in England, Scotland or Wales are areas of special interest by reason of their flora, fauna, geological or physiographic features. Northern Ireland has its own **Areas of Special Scientific Interest**. There are some 6000 SSSIs. Local planning authorities and landowners or occupiers must be informed of activities which might damage a SSSI. In theory this should protect them from damage. In practice, however, fully 5% of SSSIs are damaged each year by development for roads, housing or leisure amenities. Some government money is available to help protect SSSIs but almost no central funding exists to finance their management.

In addition to having National Parks, National

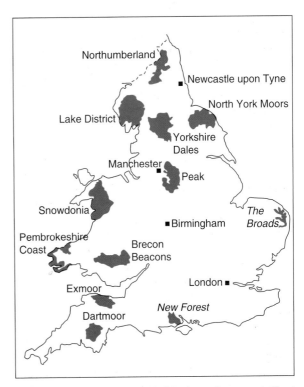

Figure 21.10 The ten National Parks and two equivalent areas (The Broads and New Forest) in England and Wales.

Nature Reserves and Sites of Special Scientific Interest, the UK has various other pieces of legislation that relate to ecosystem conservation. In 1987 the category of **Environmentally Sensitive Area (ESA)** was introduced. These areas are determined by the Minister of Agriculture according to the following criteria:

> Areas of national environmental significance whose conservation depends on the adoption, maintenance or extension of a particular farming practice; in which there have occurred, or there is a likelihood of, changes in farming practice which pose a major threat to the environment; which represent a discrete and coherent unit of environmental interest; and which would permit the economic administration of appropriate conservation aids.

There are 22 ESAs in England and Wales (Figure 21.11). Farmers in ESAs are paid to manage their land so as to conserve features created by traditional land-use management. For example, some farmers are paid to keep sheep, so maintaining grassland; others are paid not to drain their land, so that water meadows and other semi-aquatic habitats are preserved.

Other UK legislation concerned with the conservation of ecologically significant areas includes:

> **Nitrate Sensitive Areas** – areas where the application of fertilisers, slurry and manure is restricted and nitrate-absorbing crops are encouraged.
> **Heritage Coasts** – undeveloped coastline defined by the Countryside Commission and specified in local authority structure plans.
> **Farm Woodland Scheme** – a European Union (EU) scheme in which landowners are paid to plant trees.
> **Set-aside** – an EU scheme which compensates farmers who take at least 20% of their land out of arable production and leave the fields uncultivated.
> **Tree Preservation Orders** – which local planning authorities can put on important trees to protect them from being felled or damaged.

21.3.3 *Tropical rainforest*

There is still great uncertainty attached to estimates of deforestation rates for tropical rainforest (Grainger, 1993). Since the early 1970s, monitoring has been greatly helped by a growing number of satellites which provide continuous data on forest cover. The first of these remote-sensing satellites was Landsat 1, launched in 1972 by the US National Aeronautics and Space Administration (NASA). Despite the existence of remote sensing, however, considerable skill is needed in interpreting the findings. At present, the most accurate figure that can be given for global

Figure 21.11 Environmentally Sensitive Areas in England and Wales.

deforestation rates is of the order of 0.5% to 1% a year. That may not sound much – but is still the equivalent of an area the size of Belgium each year.

How concerned should we be at this loss of tropical rainforest? A typical Western response is to bewail the loss of biodiversity and worry about the loss of potentially life-saving drugs or threats to the global climate. However, remember that a few thousand years ago most of Western Europe was carpeted in thick temperate forest. The fact is, we have now lost over 90% of it. Similarly, we are also losing much of the forests that stretch from Canada and northern USA across to Siberia and China. Are the countries with tropical rainforest simply going through the same agricultural and industrial revolutions that much of the rest of the world has gone through? The simple answer is 'Yes, but do they have to?' Can some countries, such as Brazil, succeed where Western countries failed and combine increasing prosperity with significant conservation?

One positive sign is that rainforest destruction is not proceeding as fast as some environmentalists have claimed (de Selincourt, 1996; Fairhead & Leach, 1996). Alarmist predictions in the late 1980s and early 1990s that there would be no tropical rainforest within 20–30 years will hopefully be proved incorrect, although at times of drought, fires started for forest clearance can rage out of control as they did in Indonesia in 1997, when even nature reserves were damaged. Fairhead and Leach argue that rainforest loss is sometimes made worse as Western 'experts' with little knowledge of indigenous circumstances tell local people what to do. Sometimes this leads to the abandonment of traditional practices and to increased environmental damage. One encouraging feature is the increasing growth of indigenous conservation movements able to use Western expertise but not to rely on it exclusively (e.g. Gadgil & Guha, 1995).

Attempts to protect rainforests can be dangerous. Chico Mendes (Figure 21.12), the rubber tapper who was the founder and spokesman of the Union of Forest Peoples, was shot dead in December 1988. Mendes had spoken out for the rubber tappers and Amazon Indians against the large-scale destruction of forest for cattle and against the kind of funding from international banks lacking genuine conservation commitment to environmental protection.

It is too early to predict with any confidence the future of the tropical rainforests. There are different pressures in different countries. For example, in Brazil, deforestation is fuelled largely by inflation, foreign debt and land enclosures (Fearnside, 1993; Dore & Nogueira, 1994) whereas in Malaysia and Indonesia the international timber trade has been important (Vanclay, 1993). Writers differ in the extent to which they think the extraction of non-timber forest products will be significant (Grimes *et al.*, 1994; Adger *et al.*, 1995). During the late 1980s

Figure 21.13 Conservation is generally more effective if someone is prepared to pay for it.

some developing countries entered into 'debt-for-nature' swaps in which portions of their debts were cancelled in return for agreements to protect their forests. Certainly, while money cannot always solve everything, without it certain problems, including conservation, become far more severe (Figure 21.13).

Finally, it is perhaps surprising and encouraging just how many species can exist in even small remnants of tropical rainforest. Between 1845 and 1996 Singapore lost 99.8% of its primary rainforest. What survives is concentrated in the Bukit Timah Nature Reserve which has an area of just 164 ha. Primary rainforest accounts for some 50 ha of the reserve – the rest consisting of secondary forests of various types. Over the last 150 years Singapore has lost a substantial fraction of its native biota, including 26% of its vascular plants, 28% of its birds and 44% of its freshwater fish in addition to such large mammals as the tiger, leopard, banded langur, sambur and barking deer. Nevertheless, the small area of primary rainforest in the reserve contains nearly 900 plant species and has the appearance of natural tropical rainforest. Only one exotic plant has invaded – the shrub *Clidemia hirta* (Turner & Corlett, 1996).

21.3.4 Wetlands

Between 1900 and 1990, the world probably lost approximately half of its wetlands (Barbier, 1993). California has the honour of losing more of its wetlands – 91% – over the past 200 years than anywhere else (Anderson, 1996). Common instances of wetland destruction include the conversion of mangrove swamps to fishponds, diverting waters away

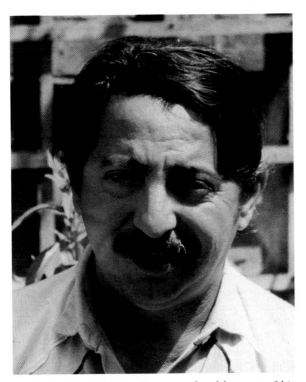

Figure 21.12 Chico Mendes – murdered because of his opposition to the destruction of the Amazonian rainforest.

Figure 21.14 Wicken Fen Nature Reserve, Cambridgeshire.

from river flood plains, draining wetlands for agriculture and other uses, and damage by pollution (Iannotta, 1996). In recent years, conservationists and economists have increasingly acknowledged the value of wetlands, not just in themselves but as sources of fish and other goods for human benefits.

In Section 20.5.2 we discussed the factors that should ideally govern the shape and size of conservation areas. In reality, such planning is rare, whether in wetland conservation or elsewhere. Reserves are established for a whole host of reasons; the scientific arguments generally play a minor role. Common sense rules of thumb are that 'The bigger the better and the less disturbed the better'. But you hardly need a Ph.D. in conservation biology to appreciate that.

A good example of the specifics of setting up and managing a small nature reserve are provided by the National Trust reserve at Wicken Fen in Cambridgeshire, England. Figure 21.14 shows what the reserve looks like today: a rather bleak flat landscape of grasses and bushes which is, frankly, uninspiring at first glance. The reserve is, however, of national importance because of its insect communities and management history. The site is also rich in bird life and has many rare and beautiful plants including several orchids, the fen violet (*Viola persicifolia*) and a relict population of 'stinging' nettles (*Urtica dioica*) which are stingless and have half the chromosome number of the common stinging nettle.

Wicken Fen is a tiny remnant (733 acres) of what was once a considerable area of about 1500 square miles of flat, water-logged, base-rich marsh and peat-land surrounding the Wash in eastern England. Parts of this fenland have been drained and cultivated for about 2600 years, but it was only in the seventeenth century that most of the habitat was altered by the systematic use of drainage ditches. As the rich peat was brought into cultivation it was dried out, burned off and ploughed up. These factors have caused the level of the soil to fall by about 4.5 m. As this happened, it became necessary to pump water out of the fens so as to lower the water table for agricultural purposes (Tansley, 1949). Only a few fragments of old fen now survive. The fen at Wicken was used as a source of sedge (*Cladium mariscus*) for thatching buildings from at least the 1650s. It became a nature reserve in fragments; the first part was bought in 1899 for entomologists to collect insects when there was a threat to drain the remaining sedge fen.

Several management problems arise in such a small area of fen as a whole range of autogenic and allogenic factors affect the site. As the ground around it has been drained and shrunk, Wicken now stands higher than the surrounding farmland. Consequently water drains from it readily and water has to be pumped into the fen, down the numerous channels (known locally as 'lodes') to maintain the best water balance for the various fen species. In addition, the entire reserve is now surrounded by a waterproof sheet to help retain the water. In the early part of the twentieth century, the drier parts of Wicken Fen were not cut for sedge and common reed (*Phragmites australis*) and this caused them to be invaded by scrub, especially alder buckthorn

Figure 21.15 In 1997 there were only 11 nesting pairs of bittern (*Botaurus stellaris*) in the United Kingdom. They breed in the reed beds and fens of East Anglia.

(*Frangula alnus*), common buckthorn (*Rhamnus catharticus*) and willows (*Salix* spp.). The scrub has increased the diversity of nesting sites for birds, but from a management point of view the scrub is less important than the sedge and reeds. In addition to their being cut, so obtaining a small amount of income for the running of the Fen, the sedge and reeds provide a habitat for many desirable species including bitterns – a nationally rare bird (Figure 21.15).

As Wicken was one of the earliest nature reserves in Britain, almost nothing was known about the principles of conservation and the fen suffered for many years from misguided management. The sedge was cut to continue the former human influence, which had maintained Wicken as a diverted sere, but unfortunately it was cut too often (every three years instead of every four) and often at the wrong time of year (autumn and winter instead of spring and summer). Both these practices led to a decline in the abundance of the sedge (Rowell, 1987). Even now the fen's problems are not over. In recent years, areas of the fen have become acidic due to leaching of the topsoil. This occurs because the fen soil surface is so much higher than the surrounding land. More acid-loving species such as bog myrtle (*Myrica gale*) and even *Sphagnum* mosses have begun to invade. Less well understood has been the appearance of a mysterious fungal disease which has wiped out much of the alder buckthorn, once again changing the appearance of the reserve.

On a more positive note, in 1994 the National Trust was able to purchase 128 acres of neighbouring farmland to extend the reserve by a quarter (Friday, 1997). This area has now been included within the water regime of the rest of the fen and its progress is being closely monitored. Hopefully it should soon provide a suitable habitat for at least some of the characteristic fen plants and animals. To accelerate the transition from commercial farmland to nature fen, the controversial decision was taken to sow the land with a mixture of native grasses and broad-leaved plants typical of fenland.

It is remarkable just how quickly wetlands can sometimes be restored to at least some of their previous glory (Edwards, 1997). In 1975 the former Communist leader of Romania, Nicolae Ceausescu, and his wife, Elena, dreamt up a plan to convert 180 000 hectares of Danube delta to agriculture, fisheries and forestry production. However, the plan was a disaster. The soil was often too sandy or too salty for crops and even the fish farms were a flop: the fish cost more to produce than their weight in caviar! With the collapse of the Ceausescu regime in 1989, plans began to restore the area. In 1994 a 2100 hectare section had its dikes breached to let the water in again. The recovery has been astonishing (Figure 21.16). Within three years, 18 species of wild fish were breeding in an area where there had been none and the number of bird species seen has increased from 34 to 72, of which 28 are breeding. In addition there are frogs, butterflies, dragonflies, racoon dogs, foxes, roe deer and wild boar.

Figure 21.16 Reflooded area of the Danube delta, showing thriving rooted aquatics.

21.4 Conservation of the biosphere

21.4.1 The greenhouse effect

In Section 13.3 we saw how rising atmospheric concentrations of carbon dioxide and certain other gases including CFCs, methane and nitrous oxide, have very probably led to an increase in average global temperatures of around 0.5 °C over the last 100 years. Although many climate modellers are working on the question, it is difficult to be certain what is going to happen as the levels of these greenhouse gases continue to rise (Mitchell *et al.*, 1989; Trenberth, 1992; Tans & Bakwin, 1995). There is general agreement that the Earth will continue to get warmer (Jacoby *et al.*, 1996). However, the extent of this warming is still highly contentious. When we wrote the first edition of this book in 1990, the consensus was that the twenty-first century would see an increase in world temperature of between 2 and 8 °C. By now (1997), the predictions are less extreme. In 1996, the Intergovernmental Panel on Climate Change (**IPCC**) predicted an increase of between 1 and 3.5 °C with a best estimate of 2 °C.

It is important to stress the genuine scientific uncertainty that surrounds such predictions. One problem is that we cannot produce a realistic physical model of the Earth with its oceans and atmosphere. In the absence of such a model – which, if it existed, might allow us experimentally to alter the atmospheric concentrations of carbon dioxide and other gases and so see their effects on climate – we have to rely on computer models and extrapolations from present-day changes. This approach becomes more reliable with the gathering of more data and the building of more powerful computers, but ultimately relies on scientists understanding what determines climate. Even slight changes in the height and extent of cloud in these models can produce quite different results. Most climatologists would admit that there is still much we do not fully understand about climate nor about how the biota and climate would interact in the event of climate change.

For example, it is known that increasing levels of atmospheric carbon dioxide have led to the faster growth of trees, especially in boreal forests (MacKenzie, 1994). It has been assumed that this would act to reduce the greenhouse effect as carbon dioxide is absorbed from the atmosphere. Unfortunately, there is now evidence that increased levels of carbon dioxide cause wetlands to produce significantly more methane (Dacey *et al.*, 1994), itself a greenhouse gas (Table 13.1). Then, there is the fact that it has recently been realised that cement production already produces 7% of global carbon dioxide emissions (Pearce, 1997b). Globally, cement production is rising by 5% a year. A different problem

– the significance of which is still unclear – is the release by aircraft of an estimated 230 million tonnes of water vapour into the atmosphere each year (Pearce, 1997c). Aircraft vapour trails seems to create cirrus clouds. These can have a strong warming effect on the atmosphere. It has been estimated that as many as 10% of the cirrus clouds over central Europe may have been produced by aircraft. Global air traffic is forecast to double within 15 years.

What are the likely consequences of global warming? Sea levels have already risen by about 15 cm in the last 100 years, most of this being the result not of the melting of ice but of the thermal expansion of surface water as the oceans have warmed slightly. This 15 cm may not sound very much until one realises that many of the world's cities are within a few metres of sea level. The 1996 IPCC report's best estimate of the rise by 2100 is 48 cm. That would place an extra 9 million people at serious risk of flooding. Should global warming ever lead to the West Antarctic ice sliding into the sea, sea levels would rise by around 5 m. (The melting of *floating* ice has no effect on sea levels – a consequence of Archimedes' principle.)

Global warming is also likely to alter the climate in other ways. It is possible that extremes of weather will become more frequent – as may indeed already be happening. That means that we will see more droughts, more floods and more hurricanes. It is also possible that there will be some significant global shifts in rainfall distributions: it may be that subtropical latitudes will receive less rain than previously. The effects of global warming on shifts in oceanic currents are particularly difficult to predict. Evidence from the Greenland ice sheet and ocean deposits shows there were sudden and extreme shifts in temperature in the northern hemisphere at the end of the last ice age. This suggests that there can be violent swings in climate as meltwater disrupts oceanic currents (Rahmstorf, 1997). The Gulf Stream keeps European winters warmer than their counterparts in America. If melting ice were to prevent the Gulf Stream flowing so far north, winter temperatures in north-western Europe might be 5–10 °C *colder* than at present while the ice melts, a process which could take hundreds of years!

So far we have only considered the effect of global warming on humans, yet global warming may have a profound effect on many other species (MacGillivray & Grime, 1995; Whittaker & Tribe, 1996; Condit, 1997). The earlier onset of spring in temperate regions may already be altering spring flowering times and nesting times in birds. Global warming may already have been responsible for the loss of the burbot (*Lota lota*) from Britain. This fish spawns in mid-winter and requires temperatures only slightly below freezing (Anon, 1997). Warmer temperatures are also thought to have been responsible for the decrease in the Adelie penguin: the population fell from 15 200 breeding pairs in 1975 to 9200 in 1997 (Greenpeace, 1997b). Part of the reason is a decrease in winter sea ice. The penguins feed on krill which in turn feed on algae which grow on the underside of the ice. Warmer temperatures mean less ice, fewer algae, fewer krill and so fewer penguins.

Looking to the future, a 1 °C rise in mean temperatures is likely to mean that vegetation zones will move 100–160 km polewards (Dobson *et al.*, 1989). If average temperatures increase by 2 °C during the twenty-first century, vegetation zones are predicted to move polewards at about 2–3 km a year. At the end of the last ice age, many species migrated polewards, often at speeds of half a kilometre a year. This was across continuous expanses of natural vegetation with the first tree species invading open grasslands. Our present flora and fauna would have to move over a dissected landscape with great areas of cultivated land separating small areas of 'natural' vegetation. Conservation might end up as a bizarre shuttle service, moving species from one reserve to the next in an attempt to keep up with the changing climate!

Recently, ecologists have begun experimentally to investigate the effect of raised carbon dioxide concentrations on organisms (e.g. Spring *et al.*, 1996). As yet it is difficult to generalise. For example, Salt *et al.* (1995) found that two common species of docks (*Rumex crispus* and *R. obtusifolius*) grew faster, as one might expect, when given carbon dioxide levels of 600 p.p.m. However, the leaf-mining insect *Pegomya nigritarsis* that feeds on them produced significantly bigger mines. This means that the overall effect of raised carbon dioxide concentrations on dock growth is likely to depend on the level of infestation of the leaf-mining insect: at low infestation rates the docks are predicted to grow better; at high infestation rates the opposite is predicted to be the case.

So what should we do about the threat of global warming? The uncertainty about its likely consequences has, not surprisingly, lead some politicians to suggest we do nothing. To the fury of many environmentalists such 'do nothing' politicians have received qualified support from some highly respected scientists. In particular, Tom Wigley, of the University Corporation for Atmospheric Research in Boulder, Colorado, has argued that it makes sense to delay attempts to reduce global warming (Wigley *et al.*, 1996). Such a delay is likely to save a great deal of money, giving time for cleaner technologies to be developed as we move away from our present dependence on fossil fuels. Wigley accepts the IPCC recommendation that atmospheric carbon dioxide levels should not be allowed to rise above 550 p.p.m. However, he calculates that a 30-year delay in doing anything about global warming will only lead to an

increase in global temperatures of about 0.2 °C.

The 'wait and see' approach may be dangerous if there is a threshold effect in global warming: a point where a sudden and rapid change in climate occurs which is then very hard to reverse. There certainly seems to have been such a threshold effect in operation in the northern hemisphere at the end of the last ice age. Sedimentary and ice core evidence suggests that temperature fluctuations, including a 1000-year cold 'snap', occurred. The transition involved changes of around 5 °C and yet took only about five years (Taylor *et al.*, 1993; Landmann *et al.*, 1996). These abrupt transitions seem to indicate a sudden switching of atmospheric and/or ocean circulation patterns. The most likely scenario is that rapid changes in North Atlantic ocean circulation, influenced by meltwater from ice sheets, affected atmospheric temperatures and further ice-melting (Lehman & Keigwin, 1992).

The 1992 Earth Summit in Rio de Janeiro recommended, but did not legally require, that developed countries have greenhouse gas emissions in the year 2000 equal to those in the year 1990. In 1996, most of the world's industrial countries (except for Russia, Australia and Saudi Arabia) agreed to set legally binding targets to cut greenhouse gas emissions from the year 2000. Countries vary in the extent to which they are likely to meet these targets. Emissions can be cut. Britain's peaked back in 1973, just before the 1974 oil crisis, when they were equivalent to 178 million tonnes of carbon a year. In the 1970s, global atmospheric methane levels were increasing at 1.1% a year; in the late 1980s, at 0.6% a year and in the early-1990s they almost stabilised (Anon, 1993).

If global levels of greenhouse gases are to be prevented from rising as quickly as would otherwise be the case, a whole range of approaches will be needed, particularly as more and more countries increase their standard of living. China, for example, has 1.2 billion people and experienced annual economic growth rates of around 10% throughout the 1980s and 1990s. In the West, economic growth has always been accompanied by carbon dioxide pollution. A number of proposals have been made as to how the global growth in greenhouse emissions might be curbed:

Move from fossil fuels to solar power, wind power and nuclear power. France reduced its carbon dioxide emissions during the 1980s by making a major shift in electricity generation towards nuclear fuels which now supply 70% of the country's electricity.
Continue to plant more trees (Moffat, 1997).
Introduce more energy-saving measures. Though hardly headline news, this approach is extremely effective. If Britain's houses had the insulation of

those in Scandinavia, heating bills would fall by a third.

Tax individuals according to how many greenhouse gases they produce (Hanna, 1995). On a global scale there is massive inequality with respect to energy consumption in different countries (Figure 21.17).

Educate people so that they have reduced expectations about their standards of living.

More speculatively, *seed the oceans with iron*, so causing more photosynthesis by phytoplankton (Coale *et al.*, 1996; Behrenfeld *et al.*, 1996).

Even more speculatively, *store carbon dioxide* on the Earth's surface as dry ice (the solid form of carbon dioxide that forms at −78 °C) or pump it to the ocean depths where it should remain for hundreds of years (Pearce, 1993).

21.4.2 Conserving the seas

It is easy to forget, when thinking about conservation, that two-thirds of the Earth's surface is water and that the oceans are just beginning to reveal their fascinating biodiversity. Here we will briefly look at two aspects of marine conservation – oil pollution and overfishing. Both illustrate the need for international co-operation. Table 21.1 shows selected major environmental treaties, a number of which relate to marine conservation.

Marine oil pollution is of two main types. The better known is when an oil tanker runs aground or breaks up at sea. Tens, sometimes hundreds, of thousands of tonnes of oil are released into the sea; tens, sometimes hundreds of journalists descend and the event is widely reported as being an ecological disaster. Less frequently noted, but possibly as serious for marine wildlife, is the continual release of small amounts of oil into the seas by very large numbers of vessels. Some of this release is almost unavoidable, but much is still due to the illegal practice in which certain ships flush out their oil storage tanks while still at sea.

Major oil spills are surprisingly frequent with an average of around one every two years (Figure 21.18). However, their long-term ecological consequences remain unclear. One reason for this is that, perhaps unsurprisingly, though somewhat disappointingly, scientists' conclusions tend to depend on who is paying them. The most thoroughly researched oil spill has been that of the *Exxon Valdez* which spilled some 38 000 tonnes of oil in Prince William Sound, Alaska in March 1989. Scientists hired by Exxon produced study after study which contradicted the findings of government researchers (Pain, 1993). Exxon scientists concluded that within four years of the spill, recovery was almost complete. The National Oceanic and Atmospheric Administration, on the

Table 21.1 Selected major environmental treaties.

Year of treaty	Name and nature of treaty
1946	*The International Convention for the Regulation of Whaling.* Establishes the International Whaling Commission to regulate whaling. (This has led to significantly less whaling.)
1963	*The Treaty Banning Nuclear Weapon Tests in the Atmosphere, in Outer Space, and Under Water.* Prohibits tests that could distribute radioactivity across national boundaries.
1971	*The Convention on Wetlands of International Importance Especially as Waterfowl Habitat* (often known as *The Ramsar Convention*). Promotes protection of important wetlands.
1972	*The Convention Concerning the Protection of World Cultural and Natural Heritage.* Establishes a system of World Heritage Sites.
1973	*The Convention on International Trade in Endangered Species of Wild Fauna and Flora (CITES).* Controls the trade in endangered species. Species on Appendix I cannot be traded internationally for commercial purposes.
1987	*The Protocol on Substances that Deplete the Ozone Layer.* Requires reductions in emissions of chlorofluorocarbons and halons that deplete the ozone layer.
1992	*The Convention of Biodiversity.* Promotes protection of biodiversity and its sustainable use and allows nations to benefit financially from the commercial exploitation of genes or species obtained from them.
1992	*The Convention on Climate Change.* Requires stabilisation of the concentrations of carbon dioxide, methane and other greenhouse gases.

other hand, described Exxon's findings as being based on 'misinterpretations' and 'erroneous interpretation' (Pain, 1993, p. 4). The most obvious ecological effects of oil spills are oiled seas birds and mammals. More difficult to monitor are the effects on phyto- and zooplankton and the long-term consequences of oil in the sediment. Even the various 'clean-up' attempts have been accused of doing more harm than good in some circumstances.

While the precise consequences of oil pollution are imperfectly understood, more is known about the consequences of fishing. The take-home message is depressingly straightforward: we are seriously overfishing the world's seas. Species after species of fish has been overfished. Future projections predict a steadily widening gap between the world's demand for fish and the ability of the oceans to meet it (Masood, 1997). Take cod, for example. In the early 1990s the cod fishing industry in Canada collapsed

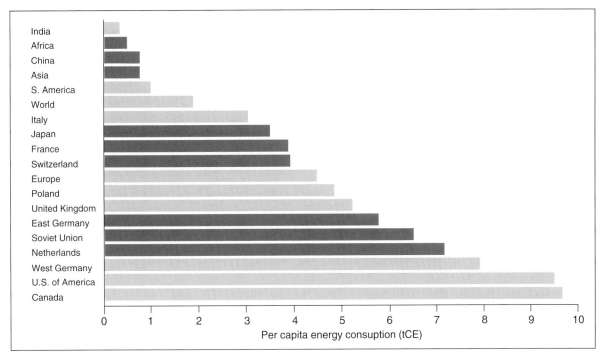

Figure 21.17 Per capita energy consumption in 1986 in various places. Figures are given in tonnes of coal equivalents (tCE).

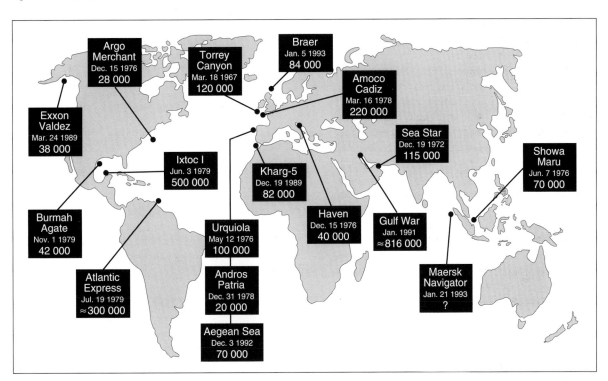

Figure 21.18 Major oil spills from 1967 to 1996. Figures show the loss of oil in tonnes.

as it already had in the north-east Arctic and Iceland. The Canadian fishing area has remained closed since then with the loss of 40 000 jobs (Spurgeon, 1997). The evidence suggests that exactly the same may happen soon in the North Sea (Cook *et*

al., 1997). Cod can live for many years and only reach reproductive maturity in significant numbers by the age of four. However, mesh sizes are such that fish are currently caught when as young as one year old. Cook, Sinclair and Stefánsson (fishery scientists from Scotland, Canada and Iceland, respectively) conclude that 'Without a substantial reduction in the rate of fishing, the North Sea cod stock may well collapse' (Cook *et al.*, 1997). Sadly, while the scientists recommend fishing quotas, it is politicians who set them. In too many countries history shows that, year after year, politicians set higher quotas than those recommended by scientists.

It is easy to blame politicians, but the reality is that scientists can rarely be certain that their recommendations are the best options. No doubt there have been occasions when scientists recommended quotas that were lower than necessary. One useful guiding principle when such uncertainty exists is the **precautionary principle**. Essentially, the precautionary principle states that when there aren't enough scientific data to be certain, it is better to play safe. It is enshrined in the 1992 Rio declaration as principle 15:

> In order to protect the environment, the precautionary approach shall be widely applied by States according to their capabilities. Where there are threats of serious or irreversible damage, lack of full scientific certainty shall not be used as a reason for postponing cost-effective measures to prevent environmental degradation.

The precautionary principle is beginning to be used increasingly in areas when there are biological uncertainties about the environmental consequences of different courses of action (Kaiser, in press).

21.5 What can individuals do?

These discussions of nature reserves, endangered species, tropical rainforests and overfishing may seem rather distant from you as an individual. You may feel like asking: 'Well, what can *I* do to conserve wildlife?'. The answer is that you can do quite a lot.

First, you can try to understand the principles of ecology, and see how they can be applied to the conservation of species, ecosystems and the biosphere. If you have read much of this book, then you will probably already have achieved this. Once you can look at the world through the eyes of an ecologist, you are in an ideal position to explain the problems and ideals of conservation to others.

More practically, you can make sure your local environment is as favourable for wildlife as possible. Precisely what you can do will depend on your particular circumstances. If you or your family have a garden or access to land, encourage plants which are valuable for animals such as bumblebees and butterflies. Table 21.2 gives a list of useful plants for British species. Leave undisturbed areas for insects (such as ants and those which help break down wood) and small mammals. Grow plants with berries or seeds which birds can eat during the winter

Table 21.2 Plants which can be grown in a garden to increase its ecological value by providing food for insects and birds. (Caterpillar data from Brooks & Knight, 1982.)

Food plants for butterfly caterpillars

Plant	Butterfly species
Dog violet (*Viola canina*)	Fritillaries (*Boloria* spp. and *Argynnis* spp.)
Nettle (*Urtica dioica*)	Red admiral (*Vanessa atalanta*) Small tortoiseshell (*Aglais urticae*) Peacock (*Inachis io*) Comma (*Plygonia c-album*)
Currants (*Ribes*), Hops (*Humulus lupulus*)	Comma (*Plygonia c-album*)
Honeysuckle (*Lonicera* spp.)	White admiral (*Ladoga camilla*)
Horseshoe vetch (*Hippocrepis comosa*), kidney vetch (*Anthyllis vulneraria*), trefoil (*Lotus corniculatus*)	Small blues (*Lycaenidae*)
Docks (*Rumex* spp.)	Coppers (*Lycaena* spp.)
Crucifers	Orange tip (*Anthocharis cardamines*)
Long meadow grasses (e.g. *Dactylis glomerata*, *Poa annua*)	Gatekeeper (*Pyronia tithonus*) Ringlet (*Aphantopus hyperantus*) Meadow brown (*Maniola jurtina*)
Buckthorns (*Rhamnus catharticus*, *Frangula alnus*)	Brimstone (*Gonepteryx rhamni*)

Plants for bees and butterflies

Labiates (such as *Lamium*, *Salvia* and *Ballota nigra*), thistles (such as *Cirsium* spp.), dog rose (*Rosa canina*), honeysuckles (*Lonicera* spp.), *Buddleia*, goldenrod (*Solidago*), centauries (*Centaurea* spp.), dandelion (*Taraxacum* agg.).

Plants with seeds for birds

Teasles (*Dipsacus fullonum*), thistles, rowan (*Sorbus aucuparia*), blackberry and raspberry (*Rubus* spp.), elder (*Sambucus nigra*).

months (see Table 21.2) and leave the snails for birds such as thrushes. Dig a pond, if you have space (see Box 21.3 for how to construct a simple pond). It is amazing how many species are attracted to a pond – you will see many interesting animals. Most important of all, avoid using pesticides if you can as they enter the food chains and may harm many of the species you are trying to encourage.

You may then be able to widen your ecological involvement by joining any local ecological societies or conservation organisations. Visit reserves to see how they are managed. You may even be able to persuade local councils or equivalent organisations to take a conservationist approach to the management of roadside verges, churchyards and other local areas (e.g. Cooper, 1995). Consider protesting at changes to the environment with which you disagree. In many countries, protest groups are wielding increasing political power. A surprising amalgam of women's

Box 21.3
Garden ponds

However conscious one is about the need for conservation, it is easy to feel that there is little one can do to create new habitats of ecological value to organisms. We can't go planting tropical rainforests, for example. However, one habitat that is relatively easy to create, provided you have a garden or other area of flat land, is a pond.

In many countries, ponds are becoming scarcer. In England and Wales, the $2\frac{1}{2}''$ Ordnance Survey maps of the 1920s indicate a total of 340 000 ponds, lakes and wet pits over 20 feet in diameter (Rackham, 1986). This works out at an average of 5.8 per square mile. It is difficult to be sure, but Rackham (1986) suggests that the average pond lasts about 100 years. Ponds disappear for a variety of reasons. They get filled in with silt and dead leaves (see Section 16.3.2). Once the water is shallower than about two feet, many plants start to root in the bottom and the pond soon becomes filled in. As well as disappearing as a result of this succession, many ponds are drained. Sometimes this is done for health reasons – to prevent the breeding of mosquitoes, for instance. Often ponds are filled in so that the land can be used for building or agriculture (Baird, 1996).

Artificial ponds are relatively easy to construct and are soon colonised by a variety of animals. In most countries, only a very small percentage of the available land is covered by fresh water. By constructing a pond, therefore, you will be helping to reverse the loss of this important and rare habitat.

If you have access to a soil with a very high clay content, it may be possible simply to dig out a pond and fill it up with water. Usually, however, you will need to provide a lining for a pond. By far the best substance for this is **butyl**. This is a plastic which is expensive, but is more or less totally resistant to sunlight and frost. The bigger the pond the better but 4 m by 2 m is a good size and will allow many amphibians to breed. Even a pond as small as 1.5 m by 1 m will allow some animals to breed and provide a great deal of enjoyment. The pond should be in a position where it will get lots of sun.

There is no need for a pond to be deeper than 60–80 cm at its deepest point. (This is deep enough to ensure that the pond will not freeze solid in winter.) It is best to make the pond irregular in outline. This makes it look more natural. It is a good idea to have a variety of depths in the pond (Figure 21.19). Different animals and plants prefer to live in water of different depths. Digging the pond may be hard work, but requires no particular skill. Before the lining (whether plastic, fibreglass, concrete or clay) is put in, it is a good idea to remove any stones which would otherwise be in contact with the lining. If you can, put 1–3 cm of sand between the soil and the lining.

It does not really matter at what time of year ponds are dug, though concrete ones should not be made while there is a danger of frost. On all but the shallowest shelves in the pond, it is a good idea to put in about 2–3 cm of sand and garden soil. Tap water should then be added until the pond is full. At various times of year it may be necessary to top the pond up with water. This is not essential, though. Indeed, the ecology of temporary waters is a subject in its own right (Dudley Williams, 1987).

In many countries suitable plants may be purchased from garden centres. Try to buy ones that are native to your country. It may be illegal to obtain plants, and animals, from the wild. However, friends may let you have some if they have ponds. It is surprising, though, how quickly organisms will colonise a pond. Some people prefer simply to start with a bare pond, and let succession take its natural course. This does mean, though, that it may take quite a few years for the pond to look interesting. It is worth ensuring that a variety of habitats are available immediately adjacent to the pond. Some tall grass, for instance, will provide valuable cover for amphibians.

If you do make a pond, keep a record of what organisms you introduce and which colonise of their own accord. If you want your pond to have good variety of invertebrates, do not put in any fish, especially ornamental ones like goldfish, as these will eat many of the native invertebrates.

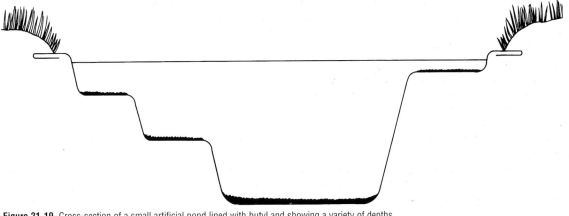

Figure 21.19 Cross-section of a small artificial pond lined with butyl and showing a variety of depths.

groups, local landowners and young people are protesting against such things as the building of new motorways and the destruction of local woods (Staheli, 1995). Protests succeeded in persuading Shell in June 1995 not to dump the 14 500 tonne Brent Spar North Sea oil platform at sea (Melchett, 1995). Whether that was the best environmental decision is uncertain. What the episode did demonstrate though was that, even when a course of action is perfectly legal and widely supported by industry, pressure from environmental organisations and consumers can effect change. Greenpeace spent $2.2 million on their Brent Spar campaign and across Europe sales from Shell petrol stations fell by around 10%.

Making an impact on a global scale is much harder for an individual. However, there are many choices which, if taken by many individuals, will benefit our planet (Myers, 1994). When shopping, choose products which minimise deleterious effects on the environment. These include organically grown food, CFC-free products (products that release CFCs are already virtually impossible to buy in many countries) and recycled paper, glass and plastic. Buying unbleached paper and cotton reduces river pollution. There are many books about **environmentally friendly products** (e.g. Elkington & Hailes, 1988).

More and more of us live in a 'throw away' consumer culture where we waste an enormous amount. From an environmental perspective, reusing products (e.g. plastic bags) is even better than recycling them. Countries vary greatly in how much they recycle or reuse. In 1995, less than 5% of household waste in Britain was recycled; in Germany the figure was 30% (Edwards, 1995). Another way we can help is by conserving energy. The less oil, gas and petrol and electricity used, the slower the rate of release of carbon dioxide into the atmosphere. Walking, cycling and using public transport rather than cars makes a difference.

Finally, by way of encouragement, the possibility of undoing some of the world's environmental damage as a result of campaigning by individuals should not be underestimated. In 1996, and after two years of intensive lobbying by people living near the dam on the Clyde River in Newport, Vermont, the dam known as Newport number 11 was demolished at a cost of $1 million. The local people argued that they would rather have fish in their river than the 2 megawatts of power that the dam generated. 'If I could catch a salmon', 10-year-old Kate Grim told the assembled adults, 'I'd turn my television off, my electric blanket and my stereo, so that we could save electricity' (Chatterjee, 1997). Within a few weeks of the dam's demolition, the salmon were back. This story illustrates some of the problems of conservation. Hydroelectric power is clean as it does not release additional carbon dioxide (apart from the concrete used in a dam's construction). Yet damming all our rivers would have devastating costs for many species and be aesthetically most undesirable. Balancing costs and benefits is rarely straightforward.

World-wide there are approximately one million dams. Many are now approaching the end of their useful lives. While demolition is expensive, so is repair. In a significant number of cases, removing the dams can improve the local economy, especially when income from fishing and tourism is taken into account. In France, the environmental group, SOS Loire, has persuaded the government to remove the Saint Etienne du Vigan Dam on the upper Allier, and the Maisons-Rouges on the Vienne River, to restore passage for fish through the upper Loire Valley. In some cases, though, restoration of river ecosystems can be very expensive. In 1996 a feasibility study reported that rejuvenating the Elwha River on the isolated Olympic peninsula in Washington State will cost $113 million and take up to 20 years. The Elwha S'Klallam, a tribe of Native Americans, have campaigned for fishing to be restored since the building of two huge dams, one 100-foot high, the other twice that height, early this century led to the disappearance of Chinook salmon, some of which weighed in at 45 kg.

Summary

(1) Successful conservation demands much more than a good understanding of biology. It requires determination, political clout and an understanding of people and economic systems.
(2) Conservation has helped some species to survive in the wild (e.g. the golden lion tamarin) but many other species remain endangered (e.g. the tiger) and many have gone extinct (e.g. some *Partula* snails).
(3) There are arguments for and against the use of extractive reserves.
(4) Deciding what precisely is sustainable can be complex.
(5) Tropical rainforest destruction continues but the extent of the damage has probably been overestimated.
(6) Wetland destruction has been extensive but can be reversed in some cases.
(7) The consequences of the greenhouse effect for humans and other species remains unclear though global warming will become increasingly significant during the twenty-first century.
(8) Oil pollution and overfishing continue to have a significant impact on marine ecosystems.
(9) Individuals can do a great deal to promote conservation.

Glossary

Abiotic environment Non-biological surroundings of an organism, such as temperature, light intensity and rainfall.

Absorption Uptake from the gut of salts, vitamins and digested food.

Acid rain Rainwater with a pH of less than about 4.5.

Adaptive radiation Evolution of a number of species from a single ancestral species to fill a variety of niches.

Aflatoxin Toxin made by grain-inhabiting fungi.

Alleles Alternative forms of a gene found at the same locus.

Allogenic succession Changes in the species that make up a community as a result of changes in the external environment.

Allometric equation Equation of the form $y = ax^b$.

Alpine Mountainous.

Alternative strategy Two or more responses to the same problem.

Altruism Helping behaviour.

Ammonification Conversion of nitrogenous material to ammonium ions by decomposers.

Anabolism Metabolic reactions involving the synthesis of macromolecules.

Anemophilous Wind pollinated.

Annual net assimilation see **Annual net productivity**.

Annual net productivity Yearly difference for a plant between its photosynthesis and its respiration.

Annual primary productivity Amount of carbon fixed by plants in a year.

Antagonistic co-evolution Co-evolution in which the relationship harms one of the species.

Anthropocentric arguments for conservation Reasons that focus on the advantages to humans of conservation.

Apomixis Asexual reproduction in plants in which cells develop into embryos without undergoing meiosis and fertilisation.

Area of Outstanding Natural Beauty The Northern Irish equivalent of a National Park.

Area of Special Scientific Interest The Northern Irish equivalent of a Site of Special Scientific Interest.

Arms race Retaliatory relationship between species over evolutionary time.

Artiodactyla Mammalian herbivores with hooves and an even number of toes, such as antelopes and deer.

Asexual reproduction Production of new individuals without the formation of haploid cells by meiosis.

Assimilation Uptake by cells of the products of absorption which are then synthesised into macromolecules.

Assimilation efficiency Percentage of the energy an organism ingests that is assimilated rather than egested.

Autecology Study of the ecological relationships of a single species.

Autogenic succession Change in the species that make up a community as a result of effects of the organisms themselves.

Autotroph Organism, such as a green plant, that can synthesise its organic compounds from inorganic ones.

Basal metabolic rate Minimum amount of energy an organism needs per unit time just to remain alive.

Batesian mimicry Mimicry in which a palatable mimic imitates a harmful or distasteful species to avoid predation.

Behaviour Activities of an organism, such as mate location, parental care and predator avoidance.

Behavioural ecology Science which studies how the behaviour of organisms is related to their ecology.

Benthic Living on the bottom of the ocean.

Bet-hedging Evolutionary strategy in which an individual adopts two alternative tactics so as to maximise the probability of survival or reproduction.

Binary fission Reproduction in which individuals divide in two asexually.

Biocoenose Ecosystem.

Biodegradable Quickly decomposed.

Biodiversity A measure of the number of species and range of life forms found in an area.

Biogeographic regions Major areas of the world with biomes determined by the past and present positions of the Earth's major land masses.

Biogeography Study of the geographical distribution of species.

Biological control Reduction in population size of an unwanted organism by means of another organism.

Biological index Use of the presence or absence of certain organisms to indicate something about the environment, such as the level of pollution.

Biological oxygen demand Uptake of oxygen in milligrams by a sealed dm^3 of water kept in the dark at $20\ °C$ for five days.

Biome Major regional community extending over a large area.

Biotic environment Influence of other organisms on an organism.

Birth Production of new individuals or release of fertilised eggs by a female.

Blanket bog Wetland that receives more than about 1000 mm of rainfall a year, resulting in the build-up of deep peat.

BOD see **Biological oxygen demand**.

Bog Wetland that only receives water from rainfall.

Boreal forest Biome, also known as taiga, found at high latitudes and dominated by evergreen conifers.

Bottleneck Very small population size from which the

population subsequently recovers, but with some loss of genetic variability.

Bourgeois Strategy for an animal of fighting in a conflict only when it is in its own territory.

Brown earth Fertile brown soil type with a characteristic soil profile.

Browser Herbivore that feeds on the leaves, young shoots and fruit of shrubs and trees.

Butyl Plastic often used as a pond liner as it is very resistant to sunlight and frost.

Buzz pollination Pollination technique in which the vibrations of a buzzing bee shake dry pollen from anthers on to the bee.

Calcicole Plant associated with alkaline soils rich in calcium.

Calcifuge Plant associated with acidic soil.

Capillary action Ascent of water a short distance up very narrow tubes or spaces.

Carcinogen Substance which can cause cancer.

Carnivore Organism, such as a polar bear, that usually feeds on prey smaller than itself.

Carrying capacity Number of individuals in a population which the habitat can support.

Caste Group of individuals within a colony of a social animal that are recognisably different from other individuals in the colony and perform specific tasks.

CFCs see **Chlorofluorocarbons**.

Chaos Unpredictability.

Chaparral Temperate shrubland found in California.

Character displacement Divergence over evolutionary time between competing species in some character such as beak length.

Cheat Term used in games theory for an individual that breaks one of the rules, e.g. an individual that fails to reciprocate help in a situation where reciprocation is the norm.

Chemical weathering Breakdown of rocks caused by the influence of carbonic acid and other substances.

Chemoautotroph Organism that gets its energy from chemical reactions involving only inorganic substrates: for example, some bacteria obtain their energy from the oxidation of Fe^{2+} to Fe^{3+}.

Chemosynthetic organism see **Chemoautotroph**.

Chernozem Fertile, deep, black soil with a characteristic soil profile.

Chlorofluorocarbons Chemicals which are responsible for the hole in the ozone layer and contribute to the greenhouse effect.

Chromosome Eukaryotic assemblages of DNA and protein that carry genes in a fixed order.

CITES Convention on International Trade in Endangered Species of Wild Fauna and Flora.

Cleaning symbiosis Relationship in which one species removes ectoparasites from another species.

Cleistogamy Fertilisation within unopened flowers.

Climate Seasonal variations in the temperature, rainfall, etc. of an area.

Climatic climax Community at the end of a succession, whose composition is determined by the climate.

Climax vegetation Final plant community at the end of a succession.

Climber Plant that grows up another plant.

Clone Group of organisms formed from a single individual and thus containing the same genetic material, for example, the many individuals in a large fragmented

clump of bracken.

Co-evolution Genetic change of two or more species in response to the evolutionary pressures they exert on each other.

Cohort Individuals in a population all of approximately the same age.

Cohort life table Life table produced using data from a single cohort.

Colonial growth Growth shown in organisms such as corals where many genetically identical multicellular subunits remain attached to one another.

Commensalism Association between two organisms in which one benefits and the other neither gains nor loses.

Communal suckling Behaviour in which offspring obtain milk from one of several females in a colony, not just from their own mothers.

Community Interacting collection of species found in a common environment or habitat.

Community evolution Evolutionary interactions found among different species of organisms where exchange of genetic information among the species is assumed to be minimal or absent.

Compensation point Light intensity at which the rate of carbohydrate manufacture in photosynthesis equals the rate at which carbohydrates are lost in respiration.

Competition Utilisation of a resource in short supply by two or more organisms with the result that at least one of the organisms grows or reproduces less.

Competitive exclusion Absence of one species due to presence of another species utilising the same resources.

Competitor Individual whose growth or survival is affected by its neighbours.

Conservation Management of the Earth's resources in a way which restores and maintains the balance between human requirements and the other species in the world.

Constantly limiting factor Environmental factor that is always in short supply but varies little over time so that a population is limited to a more or less fixed size by the factor.

Continental drift Slow movement of large plates which make up the surface of the globe.

Control Experiment designed to show what happens in the absence of the factor under investigation.

Coppice Trees cut periodically so they produce straight poles from their stumps.

Cross-fertilisation Sexual reproduction in which the two gametes come from different individuals.

Daily energy expenditure Amount of energy a free-living organism needs over 24 hours.

Death End of life.

Decomposer Organism that lives off dead organic matter which it breaks down to simpler compounds.

Deep ecology The belief that we need to examine the ways in which humans have distanced themselves from nature and recover the experience of being in communion with nature.

Deleterious allele Allele the possession of which (usually in the homozygous state) harms an individual.

Demography Study of changes in population sizes.

Denitrification Reduction of the nitrate ion by certain bacteria.

Density dependent Describes an environmental factor whose influence on population size increases as the population grows in size.

Density independent Describes an environmental factor

whose influence on population size is unrelated to the number of individuals in the population.

Desert Biome characterised by unpredictable rainfall usually averaging less than 50 mm a year.

Desertification The spread of deserts into semi-arid lands.

Detritivore Organism that ingests small particles of organic matter derived from the dead remains of animals or plants.

Detritus Small particles of organic matter derived from the remains of dead organisms.

Diffuse co-evolution Co-evolution among more than two species, for instance between predators and their prey.

Digestion Process by which animals and other heterotrophs break down their food into particles small enough to be absorbed.

Dioecious Plant species in which individuals are either male or female.

Diploid Cell or organism with chromosomes in pairs.

Dispersal Movement of an individual from one population to another.

Diversity Index of the number of species in a community.

DNA Deoxyribonucleic acid – the hereditary material in organisms.

Dominance hierarchy Linear order of individuals in a group such that individuals higher in the order have preferential access to a specified resource.

Dormancy State of almost complete absence of metabolism despite the organism or organ being alive.

Dove Strategy of never fighting in a conflict.

Ecocline Continuous variation in plant phenotype, and associated genotype, along an environmental gradient.

Ecofeminism The union of ecology and feminism.

Ecological compression Concentration by a species of its feeding efforts on only some of the available food due to the presence of other species.

Ecological isolation Coexistence of two or more species by virtue of differences in their ecology.

Ecologist Person who studies ecology.

Ecology Scientific study of the interactions that determine the distribution, abundance and characteristics of organisms.

Ecosystem A community of organisms and its physical environment.

Ecotype Morphologically distinct form within a species produced by natural selection.

Ectoparasite Parasite that lives outside its host.

Effective population size Formula for the breeding population size of a population which takes account of the fact that not all adults breed.

Efficiency of photosynthesis Percentage of energy in sunlight converted by a plant into carbohydrate.

Egestion Removal from an animal of ingested food that is never digested. Humans, for instance, egest cellulose in faeces.

Elfinwood Low-growing wood found high on certain tropical mountains.

Eluvial Leached.

Emigration Movement of an individual from a population.

Endoparasite Parasite that lives inside its host.

Endosymbiotic theory View that several of the important components of eukaryotic cells, such as mitochondria, were once free-living organisms.

Energy budget Table which shows the percentage of the energy gathered by an organism that is diverted to various metabolic activities such as maintenance, growth and reproduction.

Entomophilous Insect pollinated.

Entomophily Insect pollination.

Environment Surroundings of an organism, including other organisms and physical features.

Environmentally friendly product Product whose manufacture and use causes the minimum damage to the environment.

Environmentally Sensitive Area (ESA) In England and Wales, an area of national environmental significance whose conservation depends on the adoption, maintenance or extension of a particular farming practice.

Ephemeral Short-lived.

Epilimnion Layer of warm water in a lake.

Epiphyte Plant that grows on another plant.

Equilibrium Balance between two counteracting features which produces a lack of overall change.

Equilibrium theory of island biogeography Notion that the number of species on an island is determined by a balance between the rate at which species arrive at the island and the rate at which they go extinct.

ESS see **Evolutionarily stable strategy**.

Eusocial Species in which there is co-operation in looking after the young, some individuals are permanently sterile and there is an overlap of at least two generations contributing to colony labour.

Eutrophic Nutrient rich.

Eutrophication Release of large amounts of organic matter or phosphate and nitrate into water resulting in a lowering of dissolved oxygen concentration.

Evapotranspiration Sum of evaporation and transpiration from an area of land and its vegetation.

Evolutionarily stable strategy Set of tactics which, when adopted by most members of a population, cannot be bettered.

Evolutionary arms race Contest over evolutionary time between two species which causes them to evolve in response to each other.

Excretion Process in which the waste products of metabolism are expelled from an organism.

Exploitation efficiency Percentage of the production of one trophic level that is ingested by the trophic level above it.

Exponent b in the allometric equation $y = ax^b$.

Exponential growth Increase in population size by a constant factor each unit of time.

Extractive reserve One in which local people continue to obtain limited resources such as nuts, fruits and rubber.

Farm Woodland Scheme A European Union scheme in which financial incentives are provided to encourage landowners to plant trees.

Feminism A movement whose principal aims are to transform human relationships and society through the exposure and analysis of inequalities, principally those resulting from the exploitation of women by men.

FFPS Fauna and Flora Preservation Society.

Field work Observation and recording of organisms in their natural environment.

Folivore Herbivore that feeds on the leaves of shrubs or trees.

Food chain Diagrammatic representation of the fate of a single carbon atom in a community.

Food cycle Passage of nutrients round an ecosystem.

Food web Diagrammatic representation of the energy and nutrient flow through a community.

Founder effect Phenomenon in which only a few

individuals give rise to a population.

Frass Faecal pellets produced by small soil invertebrates.

Frugivore Herbivore that feeds on fruit.

Fundamental niche Total niche an organism could occupy in the absence of competition and predation by other species.

Fungicide Chemical designed to kill fungi.

Games theory Theoretical approach which investigates the strategies that individuals should adopt in situations when the consequence of an individual's behaviour depends on what other individuals are doing.

Gamete Haploid reproductive cell.

Gashed pyramid Population pyramid in which certain cohorts are missing or rare.

Gause's competitive exclusion principle Belief that no two species can exist together if they occupy the same niche.

Gene Length of DNA that codes for a polypeptide.

Generalist feeder Organism that can eat a variety of foods.

Genet Plant that originates from a single seed.

Genetic drift Random change in allele frequencies at polymorphic genes.

Genotype Genetic make-up of an individual.

Geophyte Perennial plant that survives drought by living underground as a bulb or corm.

Glacial till Debris left by retreating glacier.

Gradient analysis Technique that allows communities to be studied across a range of environments.

Granivore Herbivore that feeds on grain or seeds.

Grazer Herbivore that feeds on herbs and grasses.

Greenhouse effect Trapping of heat within the Earth's atmosphere.

Gross primary productivity Dry weight of carbohydrate made by a plant in photosynthesis.

Group selection Form of natural selection in which the differential survival and reproduction of groups leads to the evolution of characteristics expressed by individuals.

Growth efficiency Percentage of the energy assimilated that an organism invests in growth.

Haber-Bosch process Industrial process in which atmospheric nitrogen reacts with hydrogen at high temperatures and pressures in the presence of inorganic catalysts for the production of nitrogen-rich fertilisers.

Habitat Place where an organism lives.

Haplodiploid Species in which males have one set of chromosomes and females two.

Haploid Cell or organism with only one set of chromosomes.

Hawk Strategy of always fighting in a conflict.

Heathland Biome characterised by the presence of evergreen dwarf shrubs, especially those in the Ericaceae.

Hemiparasite Organism that obtains some but not all of its nutrients parasitically.

Herbicide Chemical designed to kill plants.

Herbivore Organism, such as a camel, that feeds on vegetation.

Heritage Coast In the UK, undeveloped coastline defined by the Countryside Commission and specified in local authority structure plans.

Hermaphrodite Individual that can produce both male and female gametes.

Heterosis Hybrid vigour shown when an organism has a high degree of heterozygosity.

Heterotroph Organism, such as an animal, that needs to take in organic compounds from its environment.

Heterozygous Gene which carries two different alleles in a diploid individual.

Holozoon Organism, such as a wolf or rat, which feeds on relatively large pieces of dead organic matter.

Home range area Ground over which an organism travels to obtain its food.

Homoiothermic animals Animals which can regulate and maintain their body temperature at a higher level than the environment by altering their metabolic rate. Also called endotherms.

Homozygous Gene which carries two identical alleles in a diploid individual.

Horizon Soil layer.

Host Organism suffering parasitism.

Humification Production of humus during the decay of biological materials.

Humus Organic matter resulting from the partial breakdown of organic material in the soil.

Hydrosere Series of communities found in an aquatic habitat as a result of primary succession.

Hypocotyl Part of seedling stem below cotyledons.

Hypolimnion Cool layer of water at the bottom of a lake.

Ice age Period during the last two million years characterised by cold climate with glaciers and little vegetation.

Immigration Movement of an individual into a population.

Inbreeding Production of offspring by two closely related individuals.

Inbreeding depression Decrease in viability or fertility of an individual as a result of a high degree of homozygosity of its genes due to its parents being closely related.

Indeterminate growth Growth, as occurs in most vascular plants, where the organism continues to increase in size until resources run out.

Infanticide Killing of offspring by other individuals in the same species.

Ingestion Process by which heterotrophs such as animals take in their food.

Insecticide Chemical designed to kill insects.

Instrumental arguments for conservation Reasons that focus on the benefits that humans may obtain from the use of conserved species or habitats.

Interglacial Warmer interval between ice ages.

Interspecific Between species.

Intertidal zone Land exposed by tides.

Intraspecific Within a species.

Intrinsic rate of increase Maximum number of offspring born per individual in a population.

Inverted pyramid of numbers Situation in which more individuals at one trophic level exist than at the trophic level beneath.

IPCC The Intergovernmental Panel on Climate Change.

Iteroparous Species in which individuals reproduce several times during their lives.

IUCN The World Conservation Union – formerly known as the International Union for the Conservation of Nature and Natural Resources.

K-selected Describes a species adapted by natural selection to living at relatively constant population levels.

Key group Critical taxon, the presence or absence of which indicates something about the environment, such as the presence of pollution.

Keystone species A species that plays a pivotal role in the maintenance of an ecosystem and on which many other species depend.

Kin selection Form of helping behaviour favoured by natural selection through the benefit obtained by relatives.

Late successional vegetation Mature communities that develop at the end of a succession.
Laterite Reddish deeply weathered infertile soil.
Law of conservation of energy Statement that energy can neither be created nor destroyed.
Leaching Loss of minerals as water drains through a soil.
Leaf area index The area of leaves found over a unit area of ground in the vegetation.
Leghaemoglobin Haemoglobin-like protein in leguminous root nodules which functions in nitrogen fixation.
Lek Grouping of adult males allowing females to choose with which males to breed.
Lethal allele Allele the possession of which (usually when in the homozygous state) is fatal to an individual.
Levée Sandy bank built up at the side of a river.
Life table Presentation of numerical data collected during a population study to show age-specific survival and reproduction.
Lifetime individual reproductive success see **Lifetime reproductive success**.
Lifetime reproductive success Total number of offspring an individual has over its life.
Limiting nutrient Nutrient which if present in greater amounts would lead to an increase in the growth or reproduction of an organism.
Lincoln index Estimate of population size calculated by the mark, release and recapture of organisms.
Lindeman's law of trophic efficiency Belief that the efficiency of energy transfer from one trophic level to the next is about 10%.
Lithosere Series of communities that develops on rock as a result of primary succession.
Litter Dead organic matter added to the soil in the form of leaves, etc.
Little ice age Period from AD 1540 to 1700 notable for many long cold winters and cool summers.
Littoral Occurring at the border of the land and sea.
Loess Wind-blown soil deposits originating from debris left by retreating glaciers.
Logistic curve Shape of the graph of the number of individuals in a population as a function of time when the growth in the size of the population follows the logistic equation and the initial number of individuals in the population is very low.
Logistic equation Idealised growth equation:
$dN/dt = rN(1 - N/K)$, where dN/dt = rate of growth of the population, r = intrinsic rate of increase, K = carrying capacity, and N = number of individuals in the population at a given time.

Macrofauna Large soil invertebrates, such as earthworms.
Maquis Temperate shrubland found in the Mediterranean region.
Mark, release, recapture Technique used for determining population sizes of mobile species.
Marsh Permanently flooded ecosystem dominated by grasses and reeds.
Marsupial Mammal that gives birth to young in a very immature state and usually possesses a pouch over her teats.
Mechanical weathering Breakdown of rocks due to the action of water, heat and wind.
Metabolic rate Amount of energy an organism expends per unit time.
Metalimnion Region in a lake where sudden changeover in temperature occurs.
Metapopulation A network of populations with occasional movements of individuals between them.
Microhabitat Very small habitat, such as the petiole of an oak leaf or gap under a stone.
Microparasite Small disease-causing organism.
Migration Regular movements by animals over quite long distances.
Mixed strategy Situation when the optimal solution to a problem involves more than one response.
Modular organism Organism, such as a tree or coral, which is made up of an indeterminate number of repeatable units.
Monotreme Mammal that lays eggs and lacks mammary glands.
Montreal Protocol 1987 agreement signed by over 30 countries that led to significant reductions in CFC emissions.
Mor Thin, sticky, structureless layer of organic soil material.
Mull Crumbly, dark, fertile soil layer.
Müllerian mimicry Mimicry in which both the model and the mimic are unpalatable or potentially harmful to a prospective predator.
Mutualism Relationship between two species from which each species benefits.
Mycorrhiza Association between a non-pathogenic or weakly pathogenic fungus and living plant root cells.

n-dimensional hypervolume Complete description of the niche of an organism.
Natal group Collection of individuals of the same species into which an individual is born.
National Nature Reserve (NNR) In the UK, an area protected because of its importance for biological diversity.
National Park An area of substantial size and outstandingly attractive scenery or wildlife significance that is specially protected.
Natural selection A major mechanism of evolution where the individuals which survive pass on their advantageous traits to their offspring.
Nature reserve Area created as a refuge for threatened species or ecosystems.
Nekton Aquatic animals that can swim.
Net primary productivity Overall gain in dry weight of a plant or community, equal to gross primary productivity minus respiration.
Niche Complete account of how an organism uses its environment.
Nitrate Sensitive Area In the UK, an area where the use of fertilisers, slurry and manure is restricted and nitrate-absorbing crops are encouraged.
Nitrification Oxidation of ammonium ions to nitrite and nitrate ions.
Nitrifying bacteria Bacteria that oxidise ammonium ions to nitrite and/or nitrate ions.
Nitrogen fixation Reduction of atmospheric nitrogen to ammonium ions by certain prokaryotes.
Nitrogenase Enzyme responsible for nitrogen fixation.
Non-equilibrium theory of species coexistence Idea that competition between species is relatively unimportant in shaping communities.
Nucleation Birth of an ice crystal around a small impurity.

Nutrient cycling Movement of substances from organism to organism and between organisms and the physical environment, allowing the same substances to be available to organisms time and time again.

Obligate anaerobe Organism that cannot utilise oxygen, indeed which may be poisoned by it.

Oestrus Condition of a female mammal when she is prepared to mate. Humans are not said to show oestrus because women are sexually receptive throughout their reproductive cycle.

Oligotrophic Nutrient poor.

Omnivore Animal which feeds on a mixture of plant and animal foods.

One-on-one co-evolution Tight co-evolution between two species.

Optimal foraging Way in which an organism searches for food so as to maximise energy and nutrient intake.

Optimal group size Preferred number of individuals in a group; value may not be the same for all individuals in the group.

Optimisation theory Notion that for any situation there is an optimum or best solution.

Organism Individual from any of the five kingdoms: Prokaryotae, Plantae, Protoctista. Fungi, Animalia.

Outbreeding Sexual reproduction of offspring by two unrelated individuals.

Outcrossing Sexual reproduction in which the two gametes come from different individuals.

Overgrazing The removal of excessive amounts of vegetation as a result of too high a density of herbivores. Overgrazing normally results from the introduction by humans of domestic animals.

Ovoviviparous Producing eggs which hatch within the mother so that the young are born having already left their eggs.

Oxbow lake Small crescent-shaped lake formed when part of a river is cut off.

Pair bond Prolonged association between a male and a female of the same species for the purposes of reproduction.

Pairwise co-evolution Tight co-evolution between two species.

Palaeobiogeography Biogeography of past continents.

Pampas Temperate grassland found in Argentina.

Parasite Individual that feeds off an organism which is still alive.

Parasite load Heavy infestation of parasites.

Parent material Underlying substrate that gives rise to a soil.

Parthenogenesis Asexual reproduction in animals when sex cells start to develop into an embryo without undergoing meiosis and fertilisation.

Pathogen Organism that causes disease.

Payoff Evolutionary consequence of a behaviour to the individual that performs it.

Payoff matrix Table which shows the evolutionary consequences of all possible behaviours in a situation to the various individuals involved.

Ped Clump of soil.

Pelagic Inhabiting the open ocean.

Perissodactyla Mammalian herbivores with hooves and an odd number of toes, such as horses and tapirs.

Permafrost Where the lower parts of the soil profile are always frozen.

Pesticide Chemical designed to kill unwanted organisms.

Phenotype Outward expression of a genotype.

Pheromone Airborne chemical produced by organisms and which alters the behaviour of other individuals in the same species.

Photic Reached by light, e.g. the upper levels of a lake.

Photoautotroph Organism, such as a green plant, that gets its energy from sunlight.

Photosynthesis Manufacture of organic compounds by organisms using the energy from sunlight.

Photosynthetic efficiency Percentage of available sunlight that is used in photosynthesis.

Physical environment Inorganic surroundings of an organism, such as temperature, salinity and pH.

Phytoplankton Small photosynthetic organisms in lakes, oceans or seas which drift almost passively.

Phytosociology Classification of plant communities.

Plankton Small organisms in lakes, oceans or seas which drift almost passively.

Plate tectonics Mechanism to explain continental drift.

Pneumatophore Stick-like structure that projects from the mud in mangrove forests and obtains oxygen for the mangrove trees.

Poikilothermic animals Animals whose body temperature changes depending on the environmental temperature. Also called ectotherms.

Pollutant Substance that causes pollution.

Pollution Consequence of the release into the environment of harmful amounts of a substance as a result of human activity.

Polygyny Mating system in which one male may mate with several females.

Polymorphic Existing in several forms; e.g. a polymorphic gene has several alleles.

Population Group of organisms of the same species living together within a common area at the same time.

Population cycle Regular pattern of change in population size over several years found in some organisms such as lemmings.

Population pyramid Representation of demographic data in which the ages of the individuals in a population are plotted against their abundance.

Population regulation Control of population size.

Pore space Volume between soil particles.

Prairie Temperate grassland found in central North America.

Precautionary principle The notion that when uncertainty exists about the likely consequences of a course of action, it is better to play safe.

Pride Group of lions containing adult females.

Primary consumer Organism that eats only plants.

Primary production Synthesis of organic compounds by photoautotrophs and chemoautotrophs.

Primary succession Succession that begins on bare uncolonised ground or in newly formed lakes.

Producer Organism (photoautotroph or chemoautotroph) which exists without needing to take in organic compounds.

Production Sum of the resources an organism devotes to growth and reproduction.

Production efficiency Percentage of energy assimilated that is devoted to growth or reproduction.

Prothallus Small flat sheet of haploid cells produced by club mosses, horsetails and ferns.

Putrefaction Rotting of food by microorganisms.

Pyramid of biomass Diagrammatic representation of the

weight or mass of organisms at the different trophic levels in a community.

Pyramid of energy Diagrammatic representation of the flow of energy from a trophic level of a community to the ones above.

Pyramid of numbers Diagrammatic representation of the numbers of organisms at the different trophic levels in a community.

Quaternary consumer Organism that eats tertiary consumers.

r-selected Describes a species which natural selection has caused to maximise its rate of increase.

Raised bog Wetland ecosystem that receives its water only in rainfall and is characterised by the formation of huge domes of *Sphagnum* peat.

Ramet Morphological unit of a plant with the capability of independent existence.

Realised niche Actual niche of an organism, as opposed to its fundamental niche.

Reciprocal altruism Helping behaviour in which individuals subsequently help each other back.

Red Data Books Books produced by the IUCN that list species at risk of becoming extinct.

Red Queen hypothesis Notion that organisms are constantly having to evolve to keep up with other organisms.

Regeneration niche Set of germination requirements for a seed.

Regional Park The Scottish equivalent of a National Park.

Regulation Control.

Rendzina Shallow, stony, fertile soils formed over limestone or chalk.

Replicate Repeat investigation to improve the confidence with which a conclusion can be drawn from a study.

Reproductive efficiency Percentage of the energy assimilated that an organism invests in reproduction.

Reproductive effort Measure of how much of its resources an organism allocates to reproduction.

Reproductive potential Maximum number of offspring that can be born to an individual during its lifetime.

Reproductive success Contribution of an individual to the next generation.

Resource axis Axis showing variation in some environmental variable, such as food size or ambient temperature.

Resource partitioning Division of a resource such as food among two or more species so that each species has access to a different part of the resource.

Rhizome Underground root-like stem of a vascular plant.

Rhizosphere Transition zone between the root and the soil.

Roost Place where birds settle at night.

Root nodule Swelling on the roots of plants containing symbiotic nitrogen fixing bacteria.

RSNC Royal Society for Nature Conservation.

RSPB Royal Society for the Protection of Birds.

Rumen Enlarged pouch in the foregut of ruminants that holds huge numbers of cellulose-digesting micro-organisms.

Ruminant Mammal, such as a sheep or deer, that possesses a rumen.

Rut Short breeding season of deer.

Salt marsh Tidal ecosystem at the edge of land and sea consisting of low vegetation cut by creeks.

Saprotroph Organism, such as an earthworm, that feeds off dead organic matter which is either absorbed in solution or ingested as very small pieces.

Satellite Male that remains inconspicuously in the territory of another male, hoping to be able occasionally to mate.

Savannah Tropical grassland.

Scavenger Organism which feeds off quite large pieces of dead organic matter which it has not killed itself.

Secondary consumer Organism that eats only herbivores.

Secondary succession Succession that begins in areas that have previously supported vegetation and retain at least some of the effects of this previous colonisation.

Seed apomixis Asexual reproduction in plants when a sex cell or nearby cell develops into an embryo without undergoing meiosis and fertilisation.

Seed bank Store of seeds in the soil.

Self-fertilisation Sexual reproduction in which both gametes come from the same individual.

Self-thinning Regulation of plant numbers in a population as the individuals grow in size and compete for resources.

Semelparous Species in which individuals reproduce on only one occasion during their lives.

Semi-desert Biome intermediate between desert and grassland or scrub in the amount of rainfall received.

Sere The sequence of vegetation types that occurs during succession.

Set-aside A European Union scheme in which grants are made to farmers to take at least 20% of their land out of arable production and leave the fields uncultivated.

Sexual reproduction Production of offspring by the fusion of two haploid cells, typically from two different individuals.

Shade intolerance Inability of a plant to grow well in the shade.

Sigmoid curve Shape of the graph of the number of individuals in a population as a function of time when population growth follows the logistic equation and the initial number of individuals is very low.

Sink Compartment in a nutrient cycle where the nutrient accumulates.

Site of Special Scientific Interest (SSSIs) In England, Scotland and Wales, an area of land of special interest by reason of its flora, fauna, geological or physiographic features.

Sociobiology Study of the reasons for social behaviour.

Soil erosion The loss of soil from an area, often as an indirect result of a human activity such as the removal of trees or the introduction of domestic herbivores.

Soil profile Series of soil horizons.

Solar radiation Sunlight, including the non-visible part of the spectrum.

Specialist feeder Organism that lives on a narrow range of foods.

Standing crop Mass of living organisms in an area.

Statutory protection Protection afforded, for example to a species or area, by law.

Steppe Temperate grassland found in eastern Europe and Asia.

Stewardship The notion that humans have a responsibility, as stewards, to look after the Earth.

Stochastic Discontinuous.

Stolon Horizontal stem that produces roots at its nodes.

Strategy Set of tactics favoured by natural selection for an organism's behaviour or ecology.

Suboptimal Structure or behaviour that could be improved on.

Succession Natural change in the structure and species composition of a community.

Succulent Plant which minimises the loss of water through the possession of a thick cuticle, a very low surface area to volume ratio and sunken stomata.

Super-stimulus Signal that is larger than the one an animal is used to and gives rise to an excessive behavioural response.

Survivorship curve Graphical representation of the probability of individuals living to a given age.

Sustainability The principle that a way of life or method of obtaining a resource should be capable of being continued indefinitely.

Swamp Wetland ecosystem dominated by trees.

Symbiosis Any close relationship between two species.

Synecology Study of the ecological relationships of a community.

Taiga Boreal forest dominated by evergreen coniferous trees.

Taxonomy Science of the classification of organisms.

Territory Relatively fixed area defended by an animal because it contains a valuable resource.

Tertiary consumer Organism that eats only secondary consumers.

Thermal pollution Release of excessive amounts of heated water.

Thermocline Rapid changeover in the temperature of a lake between warmer and cooler layers.

Tolerant ecotype Form of a species that has evolved resistance to a particular chemical, such as a pollutant.

Topography Physical features of the environment such as altitude, slope and aspect.

Trace element Element required by an organism only in very small amounts.

Tree line Altitude or latitude beyond which no trees grow.

Tree Preservation Order In the UK, a protection which can be given to an important tree by a local planning authority to prevent it from being felled or damaged.

Trophic efficiency Efficiency of energy transfer from a trophic level to the one above.

Trophic level Position at which an organism feeds in a community; for instance, carnivores feed off herbivores.

Tundra Treeless vegetation found at higher latitudes than the tree line.

Turnover time Measure of how long a typical atom spends in a particular compartment in a nutrient cycle.

Unpredictable factor Environmental factor that changes in an irregular manner.

Variably limiting factor Environmental factor that changes with time but in a predictable fashion.

Vector Species that carries disease from one species to another.

Vegetative propagation Asexual production of new plants by gardeners or botanists.

Vegetative reproduction Asexual production of new plants without the involvement of seeds.

Weather Unpredictable hour-to-hour or day-to-day variations in the temperature, rainfall, etc. of an area.

Weathering Breakdown of rocks into smaller particles.

Worker sterility Existence, in a colonial species, of healthy adult individuals that never mate.

WWF World Wide Fund for Nature, formerly the World Wildlife Fund.

Xerosere Series of communities which develop on dry land as a result of primary succession.

Zonation Clear orientation of communities along a steep environmental gradient, such as occurs at the sea shore.

Zooplankton Small animals in lakes, oceans or seas which drift almost passively.

Zygomorphic Bilaterally symmetrical.

– 3/2 power law Ecological law stating the slope of the graph of log (individual plant mass) as a function of log (density).

Bibliography

Abrams, P. A. (1986). [Reply] *Trends in Ecology and Evolution* **1**. 131–2.

Adams, J. M. (1989). 'Species diversity and productivity of trees.' *Plants Today* Nov–Dec. 183–7.

Adger, W. N. & Brown, K. (1993). 'A UK greenhouse gas inventory: on estimating anthropogenic and natural sources and sinks.' *Ambio* **22**. 509–17.

Adger, W. N., Brown, K., Cervigni, R. & Moran, D. (1995). 'Total economic value of forests in Mexico.' *Ambio* **24**. 286–96.

Alcock, J. (1979). 'The evolution of intraspecific diversity in male reproductive strategies in some bees and wasps.' In Blum, M. S. & Blum, N. A. (Eds), *Sexual Selection and Reproductive Competition in Insects*, Academic Press, London. pp. 381–402.

Alexander, R. D. (1974). 'The evolution of social behaviour.' *Annual Review of Ecology and Systematics* **5**. 325–83.

Alexander, R. McN. (1982). *Optima for Animals*, Edward Arnold, London.

Al-Joborae, F. F. (1979). *The Influence of Diet on the Gut Morphology of the Starling* (Sturnus vulgaris) *L1758*. D.Phil. thesis, University of Oxford, Oxford.

Alvarez, W., Smit, J., Lowrie, W., Asaro, F., Margolis, S. V., Claeys, P., Kastner, M. & Hildebrand, A. R. (1992). 'Proximal impact deposits at the Cretaceous–Tertiary boundary in the Gulf of Mexico: a restudy of DSDP Leg 77 Sites 536 and 540.' *Geology* **20**. 697–700.

Anderson, I. (1987). 'Epidemic of bird deformities sweeps US.' *New Scientist* 3 September. 21.

Anderson, I. (1989). 'Biosphere II: a world apart.' *New Scientist* 18 March. 34–5.

Anderson, I. (1996). 'World's wetlands sucked dry.' *New Scientist* 30 March. 9.

Anderson, I. (1997a). 'Border alert against virus smugglers.' *New Scientist* 12 July. 12.

Anderson, I. (1997b). 'Viral renegades.' *New Scientist* 6 September. 10.

Anderson, I. & Nowak, R. (1997). 'Australia's giant lab.' *New Scientist* 22 February. 34–7.

Andrewartha, H. G. & Birch, L. C. (1954). *The Distribution and Abundance of Animals*, University of Chicago Press, Chicago.

Anon (1982). 'Pandas eat a lot, but digest little.' *New Scientist* 2 September. 620.

Anon (1989). 'Japan Targets the Amazon.' *Oryx* **23** (3). 169–70.

Anon (1993). 'Build-up of greenhouse gas coming to a halt.' *New Scientist* 9 October. 10.

Anon (1997). 'Too hot for a homecoming.' *Nature* **386**. 108.

Anstett, M. C., Hossaert-McKey, M. & Kjellberg, F. (1997). 'Figs and fig pollinators: evolutionary conflicts in a coevolved mutualism.' *Trends in Ecology and Evolution* **12**. 94–9.

Archibold, O. W. (1995). *Ecology of World Vegetation*, Chapman & Hall, London.

Arthur, W. (1982). 'The evolutionary consequences of interspecific competition.' *Advances in Ecological Research* **12**. 127–87.

Attenborough, D. (1979). *Life on Earth*, Collins, London.

Attenborough, D. (1984). *The Living Planet: A Portrait of the Earth*, Collins, London.

Auclair, A. N. & Cottam, G. (1971). 'Dynamics of the black cherry (*Prunus serotina* Erhr.) in Southern Wisconsin oak forests.' *Ecological Monographs* **41** (2). 153–77.

Baird, N. (1996). 'Undercurrents.' *The Conserver* **25**. 15–7.

Baker, R. J., Van Den Bussche, R. A., Wright, A. J., Wiggins, L. E., Hamilton, M. J., Reat, E. P., Smith, M. H., Lomakin, M. D. & Chesser, R. K. (1996). 'High levels of genetic change in rodents of Chernobyl.' *Nature* **380**. 707–8.

Bakker, R. T. (1983). 'The deer flees, the wolf pursues: incongruencies in predator–prey coevolution.' In Futuyma, D. J. & Slatkin. M. (Eds) *Coevolution*, Sinauer Associates Inc., Massachusetts. pp. 350–82.

Barbier, E. B. (1993). 'Sustainable use of wetlands. Valuing tropical wetland benefits: economic methodologies and applications.' *The Geographical Journal* **159**. 22–32.

Barbour, I. G. (1990). *Religion in an Age of Science: The Gifford Lectures 1989–1991 Volume 1*, SCM Press, London.

Barnes, R. S. K. & Hughes, R. N. (1988). *An Introduction to Marine Ecology (2nd Edn)*, Blackwell Scientific Publications, Oxford.

Barnola, J. M., Raynaud, D., Neftel, A. & Oeschger, H. (1983). 'Comparison of CO_2 measurements by two laboratories on air from bubbles in polar ice.' *Nature* **302**. 410–13.

Barrett, J. & Yonge. C. M. (1958). *Collins Pocket Guide to the Sea Shore*, Collins, London.

Barrow, C. J. (1991). *Land Degradation: Development and Breakdown of Terrrestrial Environments*, Cambridge University Press, Cambridge.

Batchelor, M. & Brown, K. (Eds) (1992). *Buddhism and Ecology*, Cassell, London.

Bates, J. W. (1993). 'Regional calcicoly in the moss *Rhytidiadelphus triquetrus*: survival and chemistry of transplants at a formerly SO_2-polluted site with acid soil.' *Annals of Botany* **72**. 449–55.

Bates, J. W., McNee, P. J. & McLeod, A. R. (1996). 'Effects of sulphur dioxide and ozone on lichen colonization of conifers in the Liphook Forest Fumigation Project.' *New Phytologist* **132**. 653–60.

Bauchop, T. (1978). 'Digestion of leaves in vertebrate arboreal folivores.' In Montgomery, G. G. (Ed) *The Ecology of Arboreal Folivores*, Smithsonian Institute, Washington, USA. pp. 193–204.

Beattie, A. J. (1985). *The Evolutionary Ecology of Ant–Plant Mutualisms*, Cambridge University Press, Cambridge.

Begon, M., Harper, J. L. & Townsend, C. R. (1996). *Ecology: Individuals, Populations and Communities (3rd Edn)*, Blackwell Scientific Publications, Oxford.

Begon, M. & Mortimer, M. (1996). *Population Ecology: A Unified Study of Animals and Plants (3rd Edn)*, Blackwell Scientific Publications, Oxford.

Behrenfeld, M. J., Bale, A. J., Kolber, Z. S., Aiken, J. & Falkowski, P. G. (1996). 'Confirmation of iron limitation of phytoplankton photosynthesis in the equatorial Pacific Ocean.' *Nature* **383**. 508–11.

Bell, P. R. F. & Elmetri, I. (1995). 'Ecological indicators of large-scale eutrophication in the Great Barrier Reef Lagoon.' *Ambio* **24**. 208–15.

Bell, R. H. V. (1971). 'A grazing system in the Serengeti.' *Scientific American* **225** (1). 86–93.

Bennett, P. M. & Harvey, P. H. (1987). 'Active and resting metabolism in birds: allometry, phylogeny and ecology.' *Journal of Zoology, London* **213**. 327–63.

Bergamini, D. (1965). *The Land and Wild-life of Australasia*, Time-Life International.

Bergström, G. (1978). 'Role of volatile chemicals in *Ophrys*–pollinator interactions.' In Harborne, J. B. (Ed) *Biochemical Aspects of Plant and Animal Evolution*, Academic Press, London. pp. 207–32.

Bernays, E. & Graham, M. (1988). 'On the evolution of host specificity in phytophagous arthropods.' *Evolution* **69** (4). 886–92.

Berry, R. J. (1964). 'The evolution of an island population of the house mouse.' *Evolution* **18**. 468–83.

Berry, R. J. (1977). *Inheritance and Natural History*, Collins, London.

Berryman, A. A. & Millstein, J. A. (1989). 'Are ecological systems chaotic – and if not, why not?' *Trends in Ecology and Evolution* **4** (1). 26–8.

Bertram, B. C. R. (1978a). 'Living in groups: predators and prey.' In Krebs, J. R. & Davies, N. B. (Eds) *Behavioural Ecology: An Evolutionary Approach*, Blackwell Scientific Publications, Oxford. pp. 64–96.

Bertram, B. (1978b). *Pride of Lions*, J. M. Dent, London.

Bertram, B. C. R. & Moltu, D. (n.d.). *The Reintroduction of Red Squirrels into Regent's Park*, Zoological Society of London, London.

Betzig, L., Borgerhoff Mulder, M. & Turke. P. (Eds) (1988). *Human Reproductive Behaviour: A Darwinian Perspective*, Cambridge University Press, Cambridge.

Birks, H. J. B. (1980). 'British trees and insects: a test of the time hypothesis over the last 13 000 years.' *American Naturalist* **115**. 600–605.

Bishop, K. H. & Hultberg, H. (1995). 'Reversing acidification in a forest ecosystem: the Gårdsjön covered catchment.' *Ambio* **24**. 85–91.

Biswas, M. R. (1994). 'Agriculture and environment: a review, 1972–1992.' *Ambio* **23**. 192–7.

Black, J. N. (1958). 'Competition between plants of different initial seed sizes in swards of subterranean clover (*Trifolium subterraneum* L.) with particular reference to leaf area and the light microclimate.' *Australian Journal of Agricultural Research* **9**. 299–318.

Black, J. N. (1960). 'An assessment of the role of planting density in competition between red clover (*Trifolium pratense* L.) and lucerne (*Medicago sativa* L.) in the early vegetative stage.' *Oikos* **11**. 26–42.

Blest, A. D. (1963). 'Longevity, palatability and natural selection in five species of New World saturniid moths.' *Nature* **197**. 1183–6.

Bolin, B. (1983). 'The carbon cycle.' In Bolin, B. & Cook, R. B. (Eds) *The Major Biogeochemical Cycles and Their Interactions. Scope 21*, John Wiley. Chichester. pp.41–5.

Boone, J. L. III (1988). 'Parental investment, social subordination, and population processes among the 15th and 16th century Portuguese nobility.' In Betzig, L., Borgerhoff Mulder, M. & Turke, P. (Eds), *Human Reproductive Behaviour: A Darwinian Perspective*, Cambridge University Press. Cambridge. pp. 201–19.

Bothwell, M. L., Sherbot, D. M. J. & Pollock, C. M. (1994). 'Ecosystem response to solar ultraviolet-B radiation: influence of trophic-level interactions.' *Science* **265**. 97–100.

Bradshaw, A. D. (1952). 'Populations of *Agrostis tenuis* resistant to lead and zinc poisoning.' *Nature* **169**. 1098.

Bradshaw, A. D. & Chadwick, M. J. (1980). *The Restoration of Land: The ecology and reclamation of derelict and degraded land*, Blackwell Scientific Publications, Oxford.

Bradshaw, A. D. & McNeilly, T. (1981). *Evolution and Pollution*, Edward Arnold, London.

Bratley, A. (1997). 'Tamarin goes from strength to strength.' *WWF News* Summer. 16.

Bräuer, G., Yokoyama, Y., Falguères, C & Mbua, E. (1997). 'Modern human origins backdated.' *Nature* **386**. 337–8.

Braun-Blanquet, J. (1932). *Plant Sociology: The study of plant communities*. Translated, revised & edited by Fuller, G. D. & Conard. H. S., McGraw-Hill Inc., New York.

Brian, M. V. (1983). *Social Insects: Ecology and Behavioural Biology*, Chapman & Hall. London.

Briggs, D. & Walters, S. M. (1984). *Plant Variation and Evolution (2nd Edn)*, Cambridge University Press, Cambridge.

Briggs, D. & Walters, S. M. (1997). *Plant Variation and Evolution (3rd Edn)*, Cambridge University Press, Cambridge.

Bright, P. (1987). 'To let: small des. res. for dormouse.' *Natural World*, Autumn Issue, Royal Society for Nature Conservation. 12–15.

Brookes, M. & Knight, C. (1982). *A Complete Guide to British Butterflies*, Jonathan Cape. London.

Brown, C. R. (1986). 'Cliff swallow colonies as information centers.' *Science* **234**. 83–5.

Brown, J. H. (1975). 'Geographical ecology of desert rodents.' In Cody. M. L. & Diamond. J. M. (Eds), *Ecology and Evolution of Communities*, Belknap Press of Harvard University Press, Cambridge, Massachusetts. pp. 315–41.

Brown, J. H. (1984). 'On the relationship between abundance and distribution of species.' *American Naturalist* **124**. 255–79.

Brown, P. (1996). 'How the parasite learnt to kill.' *New Scientist* 16 November. 32–6.

Bruun, M., Sandell, M. I. & Smith, H. G. (1997). 'Polygynous male starlings allocate parental effort according to relative hatching date.' *Animal Behaviour* **54**. 73–9.

Burdon, J. J. (1987). *Diseases and Plant Populations in Biology*, Cambridge University Press, Cambridge.

Calhoun, J. B. (1962). 'Population density and social pathology.' *Scientific American* **206** (2). 139–48.

Calow, P. (1977). 'Conversion efficiencies in heterotrophic organisms.' *Biological Reviews of the Cambridge Philosophical Society* **52**. 385–409.

Cammen. L. M. (1980). 'Ingestion rate: an empirical model for aquatic deposit feeders and detritivores.' *Oecologia (Berl.)* **44**. 303–10.

Carlquist. S. (1974). *Island Biogeography*, Columbia University Press, New York.

Chagnon. N. A. & Irons. W. (Eds) (1979). *Evolutionary Biology and Human Social Behavior: An Anthropological Perspective*, Duxbury Press, North Scituate, Massachusetts.

Chaloner, W. G. & Lacey, W. S. (1973). 'The distribution of late Palaeozoic floras.' *Special Papers in Palaeontology* **12**. 271–89.

Chapman, F. M. (1895). *Handbook of Birds of Eastern North America*, Appleton, New York.

Charnov, E. L. (1984). 'Behavioural ecology of plants.' In Krebs, J. R. & Davies, N. B. (Eds), *Behavioural Ecology: An Evolutionary Approach (2nd Edn)*, Blackwell Scientific Publications, Oxford. pp. 362–79.

Chase, M. R., Moller, C., Kesseli, R. & Bawa, K. S. (1996). 'Distant gene flow in tropical trees.' *Nature* **383**. 398–9.

Chatterjee, P. (1997). 'Dam busting.' *New Scientist* 17 May. 34–7.

Cherfas, J. (1989). 'Return of the native.' *New Scientist* 11 March. 50–53.

Chitty, D. (1977). 'Natural selection and the regulation of density in cyclic and non-cyclic populations.' In Stonehouse, B. & Perrins. C. (Eds), *Evolutionary Ecology*, Macmillan, London. pp. 27–32.

Christiansen, F. B. & Fenchel, T. M. (1979). 'Evolution of marine invertebrate reproductive patterns.' *Theoretical Population Biology* **16**. 267–82.

Clark, J. D. *et al.* (1994). 'African *Homo erectus*: old radiometric ages and Young Oldowan assemblages in the Middle Awash Valley, Ethiopia.' *Science* **264**. 1907–10.

Clausen, J., Keck, D. D. & Hiesey, W. M. (1940). *Experimental Studies on the Nature of Species I. Effect of Varied Environments on Western North American Plants*, Carnegie Institution, Washington.

Clements, F. E. (1916). *Plant Succession*, Carnegie Institute, Washington.

Clutton-Brock, T. H. (Ed) (1988). *Reproductive Success: Studies of Individual Variation in Contrasting Breeding Systems*, University of Chicago Press, Chicago.

Clutton-Brock, T. H., Guinness. F. E. & Albon, S. D. (1982). *Red Deer: Behaviour and Ecology of Two Sexes*, University of Chicago Press, Chicago.

Coale, K. H. *et al.* (1996). 'A massive phytoplankton bloom induced by an ecosystem-scale iron fertilization experiment in the equatorial Pacific Ocean.' *Nature* **383**. 495–501.

Cody, M. L. (1986). 'Roots in plant ecology.' *Trends in Ecology and Evolution* **1**. 76–8.

Cody, M. L. & Overton, J. McC. (1996). 'Short-term evolution of reduced dispersal in island plant populations.' *Journal of Ecology* **84**. 53–61.

Coghlan, A. (1996). 'Doomsday has been postponed.' *New Scientist* 5 October. 8.

Cohen, J. E. (1995). *How Many People Can the Earth Support?*, W. W. Norton, New York.

Cohen, M. N. (1980). 'Speculations on the evolution of density measurement and population regulation in *Homo sapiens*.' In Cohen, M. N., Malpass. R. S. & Klein, H. G. (Eds), *Biosocial Mechanisms of Population Regulation*, Yale University Press. pp. 275–303.

Colbert, E. H. (1973). 'Continental drift and the distributions of fossil reptiles.' In Tarling, D. H. & Runcorn, S. K. (Eds), *Implications of Continental Drift to the Earth Sciences Vol. 1*, Academic Press, London. pp. 395–412.

Coles, B. & Coles, J. (1986). *Sweet Track to Glastonbury*, Thames & Hudson, London.

Colinvaux, P. (1980). *Why Big Fierce Animals are Rare*, Penguin Books, Harmondsworth. Middlesex.

Colinvaux, P. (1993). *Ecology (2nd Edn)*. John Wiley, New York.

Colinvaux, P. A., De Oliveira, P. E., Moreno, J. E., Miller, M. C. & Bush, M. B. (1996). 'A long pollen record from lowland Amazonia: forest and cooling in glacial times.' *Science* **274**. 85–8.

Collar, N. J. (1996). 'The reasons for Red Data Books.' *Oryx* **30**. 121–30.

Colston, A., Gerrard, C. & Parslow, R. (1997). *Cambridgeshire's Red Data Book including Huntingdonshire, Old Cambridgeshire & The Soke of Peterborough*, The Wildlife Trust for Cambridgeshire, Cambridge.

Condit, R. (1997). 'Forest turnover, diversity, and CO_2.' *Trends in Ecology and Evolution* **12**. 249–50.

Connell, J. H. (1975). 'Some mechanisms producing structure in natural communities.' In Cody, M. L. & Diamond, J. M. (Eds), *Ecology and Evolution of Communities*, Belknap Press of Harvard University Press, Cambridge, Massachusetts. pp. 460–90.

Connor, E. F. (1986). 'The role of Pleistocene forest refugia in the evolution and biogeography of tropical biotas.' *Trends in Ecology and Evolution* **1**. 165–8.

Cook, R. M., Sinclair, A. & Stefánsson, G. (1997). 'Potential collapse of North Sea cod stocks.' *Nature* **385**. 521–2.

Cooper, D. E. & Palmer, J. A. (Eds) (1995). *Just Environments: Intergenerational, International and Interspecific Issues*, Routledge, London.

Cooper, N. S. (1995). 'Wildlife conservation in churchyards: a case-study in ethical judgements.' *Biodiversity and Conservation* **4**. 916–28. Reprinted in Cooper, N. S. & Carling, R. C. J. (Eds) *Ecologists and Ethical Judgments*, Chapman & Hall, London. pp. 137–49.

Cooper-Driver, G. (1976). 'Chemotaxonomy and phytochemical ecology of bracken.' *Botanical Journal of the Linnean Society* **73**. 35–46.

Corbet, G. B. (1977). 'Common dormouse *Muscardinus avellanarius*.' In Corbet, G. B. & Southern, H. N. (Eds), *The Handbook of British Mammals (2nd Edn)*, The Mammal Society, Blackwell Scientific Publications, Oxford. 250–55.

Corbet, S. A., Chapman, H. & Saville, N. (1988). 'Vibratory pollen collection and flower form: bumble-bees on *Actinidia, Symphytum, Borago* and *Polygonatum*.' *Functional Ecology* **2**. 147–56.

Cossins, A. R. & Roberts, N. (1996). 'The gut in feast and

famine.' *Nature* **379**. 23.

Costantino, R. F., Desharnais, R. A., Cushing, J. M. & Dennis, B. (1997). 'Chaotic dynamics in an insect population.' *Science* **275**. 389–91.

Costanza, R. *et al.* (1997). 'The value of the world's ecosystem services and natural capital.' *Nature* **387**. 253–60.

Cousens, R. & Mortimer, M. (1995). *Dynamics of Weed Populations*, Cambridge University Press, Cambridge.

Cousins, S. (1987). 'The decline of the trophic level concept.' *Trends in Ecology and Evolution* **2**. 312–16.

Cox, C. B. & Moore, P. D. (1985). *Biogeography: An Ecological and Evolutionary Approach (4th Edn)*, Blackwell Scientific Publications, Oxford.

Crawley, M. J (1986). 'Life history and environment.' In Crawley, M. J. (Ed), *Plant Ecology*, Blackwell Scientific Publications. Oxford. pp. 253–90.

Crawley, M. J. (1990). 'The population dynamics of plants.' *Philosophical Transactions of the Royal Society of London B* **330**. 125–40.

Creber, G. T. (1977). 'Tree rings: a natural data-storage system.' *Biological Reviews of the Cambridge Philosophical Society* **52**. 349–83.

Crepet, W. L. (1983). 'The role of insect pollination in the evolution of the angiosperms.' In Real. L. (Ed). *Pollination Biology*, Academic Press, London. pp. 29–50.

Cronin, T. M. & Raymo, M. E. (1997). 'Orbital forcing of deep-sea benthic species diversity.' *Nature* **385**. 624–7.

Cronk, Q. C. B. (1987). 'The history of the endemic flora of St Helena: a relictual series.' *New Phytologist* **105**. 509–20.

Curio, E. (1996). 'Conservation needs ethology.' *Trends in Ecology and Evolution* **11**. 260–3.

Dacey, J. W. H., Drake, B. G. & Klug, M. J. (1994). 'Simulation of methane emission by carbon dioxide enrichment of marsh vegetation.' *Nature* **370**. 47–9.

Daday, H. (1954). 'Gene frequencies in wild populations of *Trifolium repens*. 1. Distribution by latitude.' *Heredity, London* **8**. 61–78.

Darwin, C. (1859). *The Origin of Species by Means of Natural Selection or the Preservation of Favoured Races in the Struggle for Life*, John Murray, London.

Davies, N. B. (1978). 'Territorial defence in the speckled wood butterfly (*Parage aegeria*): the resident always wins.' *Animal Behaviour* **26**. 138–47.

Dawkins, R. (1976). *The Selfish Gene*, Oxford University Press, Oxford.

Dawkins, R. (1982). *The Extended Phenotype: The Gene as the Unit of Selection*, Oxford University Press, Oxford.

Dawkins, R. & Krebs, J. R. (1979). 'Arms races between and within species.' In Maynard Smith, J. & Holliday, R. (Eds), *Evolution and Adaptation by Natural Selection*, The Royal Society of London, London. pp. 55–78.

Deevey, E. S. (1947). 'Life tables for natural populations of animals.' *Quarterly Review of Biology* **22**. 283–314.

Deevey, E. S. (1960). 'The human population.' *Scientific American* **203**, September. 194–204.

Delvingt, W. (1961). 'Les dortoirs d'Etourneaux *Sturnus vulgaris* L. de Belgique en 1959–1960.' *Le Gerfaut* **51**. 1–27.

den Boer, P. J. (1986). 'The present status of the competitive exclusion principle.' *Trends in Ecology and Evolution* **1**. 25–8.

de Selincourt, K. (1996). 'Demon farmers and other myths.' *New Scientist* 27 April. 36–9.

De Smet, K. (1993). 'Cheetahs teetered on brink in ice-age.' *New Scientist* 29 May. 16.

Devall, B. & Sessions, G. (1985). *Deep Ecology: Living as if Nature Mattered*, Peregrine Smith, Salt Lake City.

DeWitt, C. B. (1994). *Earth-Wise: A Biblical Response to Environmental Issues*, CRC Publications, Grand Rapids, Michigan.

Dickman, C. R. (1985). 'Sloths.' In Macdonald. D. (Ed). *The Encyclopaedia of Mammals: 2*, Guild Publishing, London. pp. 776–9.

Didham, R. K., Ghazoul, J., Stork, N. E. & Davis, A. J. (1996). 'Insects in fragmented forests: a functional approach.' *Trends in Ecology and Evolution* **11**. 255–60.

Dobson, A., Jolly, A. & Rubenstein, D. (1989). 'The green house effect and biological diversity.' *Trends in Ecology and Evolution* **4**. 64–8.

Dodd, A. P. (1929). *The Progress of Biological Control of Prickly-Pear in Australia*, The Commonwealth Prickly-Pear Board, A. J. Cumming, Brisbane.

Dodd, A. P. (1940). *The Biological Campaign against Prickly-Pear*, The Commonwealth Prickly-Pear Board, A. H. Tucker, Brisbane.

Dore, M. H. I. & Nogueira, J. M. (1994). 'The Amazon rain forest, sustainable development and the biodiversity convention: a political economy perspective.' *Ambio* **23**. 491–6.

Douglas-Hamilton. I. (1988). 'The great East African elephant disaster.' *Swara* **11** (2). 8–11.

Downhower, J. F. & Armitage, K. B. (1971). 'The yellow-bellied marmot and the evolution of polygamy.' *American Naturalist* **105**. 355–70.

Drent, P. J. & Woldendorp. J. W. (1989). 'Acid rain and eggshells.' *Nature* **339**. 431.

Drollette, D. (1997). 'Wide use of rabbit virus is good news for native species.' *Science* **275**. 154.

Dudley Williams, D. (1987). *The Ecology of Temporary Waters*, Croom Helm. London.

Duffield, J. W. (1992). 'An economic analysis of wolf recovery in Yellowstone: park visitor attitudes and values.' In Varley, J. D., Brewster, W. G., Broadbent, S. E. & Evanoff, R. (Eds), *Wolves for Yellowstone?*, National Park Services, Yellowstone National Park. pp. 25–28.

Dunbar, R. I. M. (1984). *Reproductive Decisions: An Economic Analysis of Gelada Baboon Social Strategies*, Princeton University Press, Princeton.

Eady, R., Robson, R. & Postgate, J. (1987). 'Vanadium puts nitrogen in a fix.' *New Scientist* 18 June. 59–62.

Edwards, E. (1995). 'Waste not, want not.' *New Scientist* 23/30 December. 64–5.

Edwards, J. S. (1988). 'Life in the allobiosphere.' *Trends in Ecology and Evolution* **3**. 111–14.

Edwards, R. (1996). 'Mutation rate doubled in Chernobyl's children.' *New Scientist* 27 April. 6.

Edwards, R. (1997). 'Return of the pelican.' *New Scientist* 29 March. 32–5.

Ehrlich, P. R. & Ehrlich, A. H. (1981). *Extinction: The Causes and Consequences of the Disappearance of Species*, Random House, New York.

Ehrlich, P. R. & Raven. P. H. (1964). 'Butterflies and plants: a study in coevolution.' *Evolution* **18**. 586–608.

Elgar, M. A. & Harvey, P. H. (1987a). 'Basal metabolic rate in mammals: allometry, phylogeny and ecology.' *Functional Ecology* **1**. 25–36.

Elgar, M. & Harvey, P. (1987b). 'Colonial information

centres.' *Trends in Ecology and Evolution* **2**. 34.

Elkinton, J. & Hailes, J. (1988). *The Green Consumer Guide*, Victor Gollancz, London.

Elliot, R. (Ed) (1995). *Environmental Ethics*, Oxford University Press, Oxford.

Elliot. S. (1989). 'Logging banned in Thailand.' *Oryx* **23**. 122–3.

Elton, C. (1927). *Animal Ecology*, Sidgwick & Jackson. Quotes taken from the 1971 reprint by Methuen.

Emson, R. H. (1985). 'Life history patterns in rock pool animals.' In Moore, P. G. & Seed, R. (Eds), *The Ecology of Rocky Coasts*, Hodder & Stoughton, London. pp. 220–22.

Engel, L. (1963). *The Sea*, Time-Life International.

Estes, J. A. (1995). 'Top-level carnivores and ecosystem effects: questions and approaches.' In Jones, C. G. & Lawton, J. H. (Eds), *Linking Species & Ecosystems*, Chapman & Hall, New York. pp. 151–58.

Evans, I. A. (1986). 'The carcinogenic, mutagenic and teratogenic toxicity of bracken.' In Smith, R. T. & Taylor, J. A. (Eds), *Bracken: Ecology, Land Use and Control Technology*, Parthenon Publishing, Carnforth, England. pp. 139–46.

Evans, P. G. H. (1980). 'Population genetics of the European starling *Sturnus vulgaris*.' D.Phil. thesis, University of Oxford, Oxford.

Evans, W. C. (1986). 'The acute diseases caused by bracken in animals.' In Smith, R. T. & Taylor, J. A. (Eds), *Bracken: Ecology, Land Use and Control Technology*, Parthenon Publishing, Carnforth, England. pp. 121–32.

Everett, K. R., Vassilijevskaya, V. D., Brown, J. & Walker. B. D. (1981). 'Tundra and analagous soils.' In Bliss, L. C., Heal, O. W. & Moore, J. J. (Eds), *Tundra Ecosystems: A Comparative Analysis*, Cambridge University Press, Cambridge. pp. 139–79.

Fairhead, J. & Leach, M. (1996). *Misreading the African Landscape: Society and Ecology in Forest–Savanna Mosaic*, Cambridge University Press, Cambridge.

Falk, D. A., Millar, C. I. & Olwell, M. (Eds) (1996). *Restoring Diversity: Strategies for Reintroduction of Endangered Plants*, Island Press, Washington DC.

Falla, R. A., Sibson, R. B. & Turbott. E. G. (1979). *Birds of New Zealand*, Collins, London.

Farlow, J. O. (1976). 'A consideration of the trophic dynamics of a Late Cretaceous large-dinosaur community (Oldham Formation).' *Ecology* **57**. 841–57.

Farman, J. (1987). 'What hope for the ozone layer now?' *New Scientist* 12 November. 50–54.

Faulkes, C. G. & Abbott, D. H. (1993). 'Evidence that primer pheromones do not cause social suppression of reproduction in male and female naked mole-rats (*Heterocephalus glaber*).' *Journal of Reproduction and Fertility* **99**. 225–30.

Fauna & Flora International (1996). *Good Bulb Guide*, Fauna & Flora International, Cambridge.

Fauna & Flora International (1997). *Asian Elephant Conservation Programme*, Fauna & Flora International, Cambridge.

Feare, C. J. (1984). *The Starling*, Oxford University Press, Oxford.

Feare, C. J. & Inglis. I. R. (1979). 'The effects of reduction of feeding space on the behaviour of captive starlings *Sturnus vulgaris*.' *Ornis Scandinavica* **10**. 42–7.

Fearnside, P. M. (1993). 'Deforestation in Brazilian Amazonia: the effect of population and land tenure.' *Ambio* **22**. 537–40.

Feder, H. M. (1966). 'Cleaning symbioses in the marine environment.' In Henry, S. M. (Ed), *Symbioses, Volume 1*, Academic Press, New York. pp. 327–80.

Feinsinger, P. (1983). 'Coevolution and pollination.' In Futuyma, D. J. & Slatkin, M. (Eds), *Coevolution*, Sinauer Associates Inc., Massachusetts. pp. 282–310.

Fenner, M. (1987). 'Seed characteristics in relation to succession.' In Gray, A. J., Crawley, M. J. & Edwards, P. J. (Eds), *Colonisation, Succession & Stability*, Symposium of the British Ecological Society **26**. 103–114.

Finerty, J. P. (1980). *The Population Ecology of Cycles of Small Mammals*, New Haven, Yale University Press.

Fitzpatrick. E. A. (1986). *An Introduction to Soil Science (2nd Edn)*, Longman Scientific and Technical, Harlow, Essex.

Flowerdew, J. R. (1977). 'Bank vole *Clethrionomys glareolus*.' In Corbet, G. B. & Southern, H. N. (Eds), *The Handbook of British Mammals (2nd Edn)*, Blackwell Scientific Publications, Oxford. pp. 173–84.

Ford, M. J. (1982). *The Changing Climate: Responses of the Natural Fauna and Flora*, George & Unwin Ltd., London.

Fosberg. F. R. (1948). 'Derivation of the flora of the Hawaiian Islands.' In Zimmerman. E. C. (Ed), *Insects of Hawaii Volume 1: Introduction*, Honolulu, University of Hawaii Press. pp. 107–19.

Fox, L. R. (1988). 'Diffuse coevolution within complex communities.' *Ecology* **69**. 906–7.

Frankel, O. A. & Soulé, M. E. (1981). *Conservation and Evolution*, Cambridge University Press, Cambridge.

Franklin, J. F. (1995). 'Why link species conservation, environmental protection, and resource management?' In Jones, C. G. & Lawton, J. H. (Eds), *Linking Species & Ecosystems*, Chapman & Hall, New York. pp. 326–35.

Franks, F. (1983). 'Cold stress and resistance in plants.' *What's New in Physiology?* **14**. 37–40.

Freed, L. A., Conant, S. & Fleischer, R. C. (1987). 'Evolutionary ecology and radiation of Hawaiian passerine birds.' *Trends in Ecology and Evolution* **2**. 196–203.

Freeman, D. C., Harper, K. T. & Charnov, E. L. (1980). 'Sex change in plants: old and new observations and new hypotheses.' *Oecologia* **47**. 222–32.

Friday, A. & Ingram, D. S. (Eds) (1985). *The Cambridge Encyclopedia of Life Sciences*, Cambridge University Press. Cambridge.

Friday, L. E. (Ed) (1997). *Wicken Fen: The Making of a Wetland Nature Reserve*, Harley Books, Colchester.

Gadgil, M. & Guha, R. (1995). *Ecology and Equity: The Use and Abuse of Nature in Contemporary India*, Routledge, London.

Gamlin, L. (1987). 'Rodents join the commune.' *New Scientist* 30 July. 40–47.

Garrells, R. M. & MacKenzie F. T. (1971). *Evolution of Sedimentary Rocks*, Norton, New York.

Garrells, R. M., MacKenzie, F. T. & Hunt, C. (1975). *Chemical Cycles and the Global Environment*, Kaufmann, Los Altos, California.

Gass, C. L., Angehr, G. & Centa, J. (1976). 'Regulation of food supply by feeding territoriality in the rufus hummingbird.' *Canadian Journal of Zoology* **54**. 2046–54.

Gatsuk, E., Smirnova, O. V., Vorontzova, L. I., Zavgolnova, L. B. & Zhukova. L. A. (1980). 'Age states of plants of various growth forms: a review.' *Journal of Ecology* **68**. 675–96.

Gause. G. F. (1934). *The Struggle for Existence*, Williams & Wilkins, Baltimore.

Gemmell, R. P. (1977). *Colonization of Industrial Wasteland*, Edward Arnold, London.

Gessaman, J. A. (1973). 'Methods of estimating the energy cost of free existence.' In Gessaman, J. A. (Ed), *Ecological Energetics of Homeotherms: A View Compatible with Ecological Modeling*, Utah State University Press, Logan, Utah. pp. 3–31.

Ghiorse, W. C. (1997). 'Subterranean life.' *Science* **275**. 789.

Gilbert, F. S. (1980). 'The equilibrium theory of island biogeography: fact or fiction?' *Journal of Biogeography* **7**. 209–35.

Gilbert. F. S. (1985). 'Ecomorphological relationships in hoverflies (Diptera, Syrphidae).' *Proceedings of the Royal Society of London B* **224**. 91–105.

Gilbert. F. S., Harding, E. F., Line J. M. & Perry, I. (1985). 'Morphological approaches to a community structure in hoverflies (Diptera. Syrphidae).' *Proceedings of the Royal Society of London B* **224**. 115–30.

Giller, P. (1986). [Reply] *Trends in Ecology and Evolution* **1**. 132.

Gimingham, C. H. (1972). *Ecology of Heathlands*, Chapman and Hall, London.

del Giorgio, P. A., Cole, J. J. & Cimbleris, A. (1997). 'Respiration rates in bacteria exceed phytoplankton production in unproductive aquatic systems.' *Nature* **385**. 148–51.

Glantz, M. H. & Krenz, J. H. (1992). 'Human components of the climate system' In Trenberth, K. E. (Ed.), *Climate System Modellng*, Cambridge University Press, Cambridge, pp. 27–49.

Gleason, H. A. (1927). 'Further views on the succession concept.' *Ecology* **8**. 299–326.

Godwin, H. (1960). 'The history of weeds in Britain.' In Harper, J. L. (Ed), *The Biology of Weeds*, Blackwell, Oxford.

Gooday, G. W. (1988). 'The potential of the microbial cell and its interaction with other cells.' In Lynch, J. M. & Hobble, J. E. (Eds), *Micro-organisms in Action: Concepts and Applications in Microbial Ecology*, Blackwell Scientific Publications, Oxford. pp. 7–32.

Gorham, E. (1996). 'Lakes under a three-pronged attack.' *Nature* **381**. 109–10.

Gottsberger, G., Silberbauer-Gottsberger, I. & Ehrendorfer, F. (1980). 'Reproductive biology in the primitive relic angiosperm *Drimys brasiliensis* (Winteraceae).' *Plant Systematics and Evolution* **135**. 11–39.

Gould. L. L. & Heppner, F. (1974). 'The vee formation of Canada geese.' *Auk* **91**. 494–506.

Gould, S. J. & Lewontin, R. (1979). 'The spandrels of San Marco and the Panglossian paradigm: a critique of the adaptationist programme.' *Proceedings of the Royal Society of London B* **205**. 581–98.

Grady, M. M., Wright, I. P. & Pillinger, C. T. (1997). 'Microfossils from Mars: a question of faith?' *Astronomy & Geophysics* **38**. 26–9.

Grainger, A. (1993). 'Rates of deforestation in the humid tropics: estimates and measurements.' *The Geographical Journal* **159**. 33–44.

Grant, P. R. (1986). *Ecology and Evolution of Darwin's Finches*, Princeton University Press. Princeton.

Greenpeace (1997a). 'Bears under threat as Canada's rainforests move closer to the brink.' *Campaign Report* **27**.

Greenpeace (1997b). 'Decline in Adelie penguins.' *Campaign Report* **26**.

Gribbin, J. & Gribbin, M. (1996). 'The greenhouse effect.' *New Scientist Inside Science* 6 July. 1–4.

Grime, J. P. (1979). *Plant Strategies and Vegetation Processes*, John Wiley, Chichester.

Grimes, A. et al. (1994). 'Valuing the rain forest: the economic value of nontimber forest products in Ecuador.' *Ambio* **23**. 405–10.

Grinnell, J. (1924). 'Geography and evolution.' *Ecology* **5**. 225–9.

Groffman, P. M. (1997). 'Global biodiversity: is it in the mud and the dirt?' *Trends in Ecology and Evolution* **12**. 301–2.

Groombridge, B. (Ed) (1992) *Global Biodiversity: Status of the Earth's Living Resources*, Chapman & Hall, London.

Groot, P. de (1980). 'Information transfer in a socially roosting weaver bird (*Quelea quelea*; Ploceniae): an experimental study.' *Animal Behaviour* **28**. 1249–54.

Gross, M. R. & Shine, R. (1981). 'Parental care and mode of fertilisation in ectothermic vertebrates.' *Evolution* **35**. 775–93.

Grubb, P. J. (1977). 'The maintenance of species-richness in plant communities: the importance of the regeneration niche.' *Biological Reviews of the Cambridge Philosophical Society* **52**. 107–45.

Grubb, P. J. (1986). 'The ecology of establishment.' In Bradshaw, A. D., Goode. D. A. & Thorpe, E. (Eds), *Ecology and Design in Landscape*, Symposium of the British Ecological Society **24**. 83–97.

Grubb, P. J. (1987). 'Global trends in species-richness in terrestrial vegetation: A view from the northern hemisphere.' In Gee, J. H. R. & Giller, P. S. (Eds), *Organisation of Communities Past & Present*, Blackwell Scientific Publications, Oxford. pp. 99–118.

Grubb, P. J. (1996). 'Rainforest dynamics: the need for new paradigms.' In Edwards, D. S., Booth, W. E. & Choy, S. C. (Eds), *Tropical Rainforest Research – Current Issues*, Kluwer, Dordrecht. pp. 215–33.

Grubb, P. J., Green, H. E. & Merrifield. R. C. J. (1969). 'The ecology of chalk heath: its relevance to the calcicole–calcifuge and soil acidification problems.' *Journal of Ecology* **57**. 175–212.

Haeckel, E. (1869, pub. 1870). 'Ueber Entwickelungsgang und Aufgabe der Zoologie.' *Senaische Z.***5**. 353–70.

Hågvar, S. (1994). 'Preserving the natural heritage: the process of developing attitudes.' *Ambio* **23**. 515–8.

Haldane, J. (1994). 'Admiring the high mountains: the aesthetics of environment.' *Environmental Values* **3**. 97–106.

Hamilton. W. A. (1988). 'Microbial energetics and metabolism.' In Lynch, J. M. & Hobbie, J. E. (Eds), *Micro-organisms in Action: Concepts and Applications in Microbial Ecology*, Blackwell Scientific Publications, Oxford. pp. 75–100.

Hamilton, W. D. (1964). 'The genetical evolution of social behaviour. I & II.' *Journal of theoretical Biology* **7**. 1–16; 17–52.

Hamilton, W. D. (1971). 'Geometry for the selfish herd.' *Journal of theoretical Biology* **31**. 295–311.

Hamilton, W. J. & Gilbert, W. M. (1969). 'Starling dispersal from a winter roost.' *Ecology* **50**. 886–98.

Hammond, K. A. & Diamond, J. (1997). 'Maximal sustained energy budgets in humans and animals.' *Nature* **386**. 457–62.

Hanna, J. (1995). 'Towards a single carbon currency.' *New Scientist* 29 April. 50–1.

Hannah, L., Lohse, D., Hutchinson, C., Carr, J. L. & Lankerani, A. (1994). 'A preliminary inventory of human disturbance of world ecosystems.' *Ambio* **23**. 246–50.

Hanski, I. & Gyllenberg, M. (1997). 'Uniting two general patterns in the distribution of species.' *Science* **275**. 397–400.

Harland, W. B. *et al.* (1990). *A Geological Time Scale 1989*, Cambridge University Press, Cambridge.

Harper, J. L. (1977). *Population Biology of Plants*, Academic Press, London.

Harper, J. L. & McNaughton, I. H. (1962). 'The comparative biology of closely related species living in the same area. VII. Interference between individuals in pure and mixed populations of *Papaver* species.' *New Phytologist* **61**. 175–88.

Harper, J. L. & Ogden, J. (1970). 'The reproductive strategy of higher plants. I. The concept of strategy with special reference to *Senecio vulgaris* L.' *Journal of Ecology* **58**. 681–98.

Harper, J. L., Lovell, P. H. & Moore, K. G. (1970). 'The shapes and sizes of seeds.' *Annual Review of Ecology and Systematics* **1**. 327–56.

Hassan, F. A. (1980). 'The growth and regulation of human population in prehistoric times.' In Cohen, M. N., Malpass, R. S. & Klein, H. G. (Eds), *Biosocial Mechanisms of Population Regulation*, Yale University Press. pp. 305–19.

Hatcher, B. G. (1988). 'Coral reef primary productivity: a beggar's banquet.' *Trends in Ecology and Evolution* **3**. 106–11.

Hawksworth, D. L. & Rose, F. (1970). 'Qualitative scale for estimating sulphur dioxide air pollution in England and Wales using epiphytic lichens.' *Nature* **227**. 145–8.

Hawksworth, D. L. & Rose, F. (1976). *Lichens as Pollution Monitors*, Edward Arnold, London.

Hays. J. D., Imbrie, J. & Shackleton, N. J. (1976). 'Variations in the Earth's orbit: Pacemaker of the Ice Ages.' *Science* **194**. 1121–32.

Heath, I. B. (1988). 'Gut Fungi.' *Trends in Ecology and Evolution* **3**. 167–71.

Hedberg, O. (1986). 'Origins of the Afroalpine flora.' In Vuilleumier, F. & Monasterio, M. (Eds), *High Altitude Tropical Biogeography*, Oxford University Press, Oxford. pp. 443–68.

Heimsath, A. M., Dietrich, W. E., Nishiizumi, K. & Finkel, R. C. (1997). 'The soil production and landscape equilibrium.' *Nature* **388**. 358–61.

Hemmingsen, A. M. (1960). 'Energy metabolism as related to body size and respiratory surfaces, and its evolution.' *Reports of the Steno Memorial Hospital and Nordinsk Insulin Laboratorium* **9** (2). 7–110.

Hessler. R., Lonsdale, P. & Hawkins, J. (1988). 'Patterns on the ocean floor.' *New Scientist* 24 March. 47–51.

Heywood, V. H. (Ed) (1995). *Global Biodiversity Assessment*, Cambridge University Press, Cambridge.

Hinnells, J. R. (1991). *A Handbook of Living Religions*, Penguin Books, London.

Hirayama. T. (1979). 'Diet and Cancer.' *Nutrition and Cancer* **1**. 67–81.

Hofmann, D. J. (1996). 'Recovery of Antarctic ozone hole.' *Nature* **384**. 222–3.

Holland, H. D. (1995). 'Atmospheric oxygen and the biosphere.' In Jones, C. G. & Lawton, J. H. (Eds), *Linking Species & Ecosystems*, Chapman & Hall, New York. pp. 127–136.

Hollingsworth, T. H. (1969). *Historical Demography*, The Sources of History Limited, Cambridge University Press, Cambridge.

Holmes, A. (1978). *Principles of Physical Geology (3rd Edn)*, Nelson, London.

Holmes, B. (1997). 'Chop down a tree to save the forest.' *New Scientist* 22 February. 10.

Hong, S., Candelone, J.-P., Patterson, C. C. & Boutron, C. F. (1996). 'History of ancient copper smelting pollution during Roman and medieval times recorded in Greenland ice.' *Science* **272**. 246–8.

Howard, R. D. (1978). 'The evolution of mating strategies in bullfrogs, *Rana catesbeiana*.' *Evolution* **32**. 850–71.

Howarth, F. G. (1987). 'Evolutionary ecology of aeolian and subterranean habitats in Hawaii.' *Trends in Ecology and Evolution* **2**. 220–23.

Howarth, R. W., Giblin, A., Gale, J., Peterson, B. J. & Luther, G. W. III (1983). 'Reduced sulfur compounds in the pore waters of a New England salt marsh.' *Environment, Biogeochemistry and Geology Bulletin* (Stockholm) **35**. 135–52.

Howarth, S. E. & Williams, J. T. (1972). 'Biological Flora of the British Isles: *Chrysanthemum segetum*'. *Journal of Ecology* **60**. 573–84.

Hughes. R. D. & Walker, J. (1970). 'The role of food in the population dynamics of the Australian bushfly.' In Watson, A. (Ed), *Animal Populations in Relation to their Food Sources*, British Ecological Symposium 10, Blackwell Scientific Publications, Oxford. pp. 255–69.

Humphreys, W. F. (1979). 'Production and respiration in animal communities.' *Journal of Animal Ecology* **48**. 427–53.

Hunter, M. L. (1996). *Fundamentals of Conservation Biology*, Blackwell Science, Cambridge, Massachusetts.

Husband, B. C. & Barrett, S. C. H. (1996). 'A metapopulation perspective in plant population biology.' *Journal of Ecology* **84**, 461–9.

Hutchinson, G. E. (1958). 'Concluding remarks.' *Cold Spring Harbour Symposium on Quantitative Biology* **22**. 415–27.

Hutchinson. G. E. (1959). 'Homage to Santa Rosalia or Why are there so many kinds of animals?' *American Naturalist* **93**. 145–59.

Iannotta, B. (1996). 'Mystery of the Everglades.' *New Scientist* 9 November. 34–7.

IUCN (1978). *The IUCN Plant Red Data Book*. Compiled by Lucas, G. & Synge, H. for the Threatened Plants Committee of the Survival Service Commission of the IUCN.

Jacoby, G. C., D'Arrigo, R. D. & Davaajamts, T. (1996). 'Mongolian tree rings and 20th-century warming.' *Science* **273**. 771–3.

Janzen. D. H. (1966). 'Coevolution of mutalism between ants and acacias in Central America.' *Evolution* **20**. 249–75.

Janzen, D. H. (1975). *Ecology of Plants in the Tropics*, Edward Arnold, London.

Janzen, D. H. (1979). 'Why food rots.' *Natural History* **88**

(6). 60–65.

Janzen, D. H. (1980). 'When is it coevolution?' *Evolution* **34**. 611–12.

Jarvis, J. U. M. (1981). 'Eusociality in a mammal: cooperative breeding in naked mole-rat colonies.' *Science* **212**. 571–3.

Jarvis, J. U. M. (1985). 'African mole-rats.' In Macdonald. D. (Ed), *The Encyclopaedia of Mammals: 2*, Guild Publishing, London. pp. 708–11.

Jefferies, R. L. & Maron, J. L. (1997). 'The embarrassment of riches: atmospheric deposition of nitrogen and community and ecosystem processes.' *Trends in Ecology and Evolution* **12**. 74–8.

Johanson, U., Gehrke, C., Björn, L. O., Callaghan, T. V. & Sonesson, M. (1995). 'The effects of enhanced UV-B radiation on a subarctic heath ecosystem.' *Ambio* **24**. 106–11.

Johnson, B. R. (1996). 'Southern Appalachian rare plant reintroductions on granite outcrops.' In Falk, D. A., Millar, C. I. & Olwell, M. (Eds), *Restoring Diversity: Strategies for Reintroduction of Endangered Plants*, Island Press, Washington DC. pp. 433–43.

Johnson, K. H., Vogt, K. A., Clark, H. J., Schmitz, O. J. & Vogt, D. J. (1996) 'Biodiversity and the productivity and stability of ecosystems.' *Trends in Ecology and Evolution* **11**. 372–77.

Jones, A. E. & Shanklin, J. D. (1995). 'Continued decline of total ozone over Halley, Antarctica, since 1985.' *Nature* **376**. 409–11.

Jones, A. K. G. (1986). 'Parasitological investigation on Lindow Man.' In Stead, I. M. *et al.* (Eds), *Lindow Man: The Body in the Bog*, Guild Publishing, London. pp. 136–9.

Jones, L. (1997). 'The scent of a tiger.' *New Scientist* 12 July. 18.

Jordan, C. F. (1985). *Nutrient Cycling in Tropical Forest Ecosystems*, John Wiley, Chichester.

Juday. C. (1940). 'The annual energy budget of an inland lake.' *Ecology* **21**. 438–50.

Kaiser, M. (in press). 'Fish-farming and the precautionary principle: context and values in environmental science for policy'. *Foundations of Science*.

Kaneshiro, K. Y. (1995). 'Evolution, speciation and the genetic structure of island populations.' In Vitousek, P. M., Loope, L. L. & Adsersen, H. (Eds) (1995) *Islands: Biological Diversity and Ecosystem Function*, Springer, Berlin. pp. 23–33.

Keeling, C. D., Guenther, P. R. & Whorf, T. P. (1986). *An Analysis of the Concentrations of Atmospheric Carbon Dioxide at Fixed Land Stations and over the Oceans and Daily Averaged Continuous Measurements*, Scripps Institute of Oceanography, La Jolla.

Keeling, C. D., Moss, D. J., & Whorf, T. P. (1987). *Measurements of the Concentration of Atmospheric Carbon Dioxide at Mauna Loa Observatory, Hawaii, 1958–1986*, Final Report for the Carbon Dioxide Information and Analysis Centre, Martin-Marietta Energy Systems Inc., Oak Ridge, Tennessee.

Kelly, F. J. (1996). 'Air pollution: an old problem in a new guise.' *Biologist* **43**. 102–5.

Kempe, S. (1979). 'Carbon in the rock cycle.' In Bolin, B., Degens, E. T., Duvigneaud, P. & Kempe, S. (Eds), *The Global Carbon Cycle, Scope 13*, John Wiley, Chichester. pp. 343–78.

Kenward, R. E. (1978). 'Hawks and doves: factors affecting success and selection in goshawk attacks on woodpigeons.' *Journal of Animal Ecology* **47**. 449–60.

Kerr, J. T. & Packer, L. (1997). 'Habitat heterogeneity as a determinant of mammal species richness in high-energy regions.' *Nature* **385**. 252–4.

Khalid, F. & O'Brien, J. (Eds) (1992). *Islam and Ecology*, Cassell, London.

King, J. R. (1974). 'Seasonal allocation of time and energy resources in birds.' In Paynter, R. A. Jr. (Ed). *Avian Energetics*, Publications of the Nuttall Ornithological Club, Cambridge, Massachusetts. pp. 4–70.

Klieber, M. (1947). 'Body size and metabolic rate.' *Physiological Review* **27**. 511–41.

Kleiner, K. (1994). 'Ozone hole could be killing amphibians.' *New Scientist* 5 March. 7.

Koch, E. (1996). 'Orphan elephants go on the rampage.' *New Scientist* 20 July. 5.

Koch, E. (1997). 'Ivory plan splits elephant experts.' *New Scientist* 28 June. 4.

Kodric-Brown, A. & Brown, J. H. (1978). 'Influence of economics, interspecific competition and sexual dimorphism on territoriality of migrant rufous hummingbirds.' *Ecology* **59**. 285–96.

Kokwaro, J. O. & Beck, E. (1987). 'The animal threat to Mount Kenya's Afro-alpine plants.' *Swara* **10** (1). 30–31.

Krebs, C. J. (1964). 'The lemming cycle at Baker Lake, Northwest Territories, during 1959–1962.' *Arctic Institute of North America, Technical Paper* **15**.

Krebs, C. J. (1978). *Ecology: The Experimental Analysis of Distribution and Abundance (2nd Edn)*, Harper & Row, New York.

Krebs, C. J. (1985). *Ecology: The Experimental Analysis of Distribution and Abundance (3rd Edn)*, Harper & Row, New York.

Krebs, C. J. (1988). *The Message of Ecology*, Harper & Row, New York.

Krebs, J. R. (1971). 'Territory and breeding density in the great tit *Parus major* L.' *Ecology* **52**. 2–22.

Krebs, J. R. & Davies, N. B. (1993). *An Introduction to Behavioural Ecology (3rd Edn)*, Blackwell Scientific Publications, Oxford.

Kronzucker, H. J., Siddiqi, M. Y. & Glass, A. D. M. (1997). 'Conifer root discrimination against soil nitrates and the ecology of forest succession.'*Nature* **385**. 59–61.

Kruuk, H. (1972). *The Spotted Hyena*, University of Chicago Press, Chicago.

Kurihara, Y. & Kikkawa. J. (1986). 'Trophic relations of decomposers.' In Kikkawa, J. & Anderson, D. J. (Eds), *Community Ecology: Pattern and Process*, Blackwell Scientific Publications. Melbourne. pp. 127–60.

Kurland, J. A. (1979). 'Paternity, mother's brother, and human sociality.' In Chagnon. N. A. & Irons. W. (Eds), *Evolutionary Biology and Human Social Behavior: An Anthropological Perspective*, Duxbury Press, North Scituate, Massachusetts. pp. 145–80.

Lacey. J. (1988). 'Aerial dispersal and the development of microbial communities.' In Lynch, J. M. & Hobbie, J. E. (Eds), *Micro-organisms in Action: Concepts and Applications in Microbial Ecology*, Blackwell Scientific Publications. Oxford. pp. 207–37.

Lack, D. (1947). *Darwin's Finches*, Cambridge University Press, Cambridge.

Lamb. H. H. (1977). *Climate Present, Past and Future*.

Volume 2: Climatic History and the Future, Methuen, London.

Lamb, H. H. (1982). *Climatic History and the Modern World*, Methuen. London.

Landmann, G., Reimer, A., Lemcke, G. & Kempe, S. (1996). 'Dating Late Glacial abrupt climate changes in the 14,570 yr long continuous varve record of Lake Van, Turkey.' *Palaeogeography, Palaeoclimatology, Palaeoecology* **122**. 107–18.

Larson, J. S., Bedinger, M. S., Bryan. C. F., Brown, S., Huffman, R. T., Miller, E. L., Rhodes, D. G. & Touchet, B. A. (1981). 'Transition from wetlands to uplands in southeastern bottomland hardwood forests.' In Clark, J. R. & Benforado, J. (Eds). *Wetlands of Bottomland Hardwood Forests*, Elsevier, Amsterdam. pp. 225–73.

Laurenson, M. K., Caro, T. M., Gros, P. & Wielebnowski, N. (1995). 'Controversial cheetahs?' *Nature* **377**. 392.

Lawton, J. H. (1976). 'The structure of the arthropod community on bracken.' *Botanical Journal of the Linnean Society* **73**. 187–216.

Lawton, J. H. (1984). 'Non-competitive populations, non-convergent communities, and vacant niches: the herbivores of bracken.' In Strong, D. R. *et al.* (Eds), *Ecological Communities: Conceptual Issues and the Evidence*, Princeton University Press, Princeton. pp. 67–101.

Lawton, J. H. (1986). 'Surface availability and insect community structure: the effects of architecture and fractal dimension of plants.' In Juniper, B. & Southwood, R. (Eds), *Insects and the Plant Surface*, Edward Arnold, London. pp. 317–31.

Lawton, J. H., Heads, P. A., Hefin Jones, T. & Thompson, L. J. (1996). 'The Ecotron: building model ecosystems.' *Biologist* **43**. 19–21.

Lee, J. (1988). 'Acid rain.' *Biological Sciences Review* **1**. 15–18.

Lee, M. S. Y. (1997). 'Documenting present and past biodiversity: conservation biology meets palaeontology.' *Trends in Ecology and Evolution* **12**. 132–3.

Lehman, S. J. & Keigwin, L. D. (1992). 'Sudden changes in North Atlantic circulation during the last deglaciation.' *Nature* **356**. 757–62.

Lever, C. (1977). *The Naturalized Animals of the British Isles*, Hutchinson, London.

Lewin, R. (1996). 'A strategy for survival?' *New Scientist* 17 February. 14–5.

Likens, G. E., Driscoll, C. T. & Buso, D. C. (1996). 'Long-term effects of acid rain: response and recovery of a forest ecosystem.' *Science* **272**. 244–6.

Limbaugh. C. (1961). 'Cleaning symbioses.' *Scientific American* **205** (2). 42–9.

Lind, E. M. & Morrison, M. E. S. (1974). *East African Vegetation*, Longman Group Ltd., London.

Lindeman, R. L. (1942). 'The trophic-dynamic aspect of ecology.' *Ecology* **23**. 399–417.

Little, C. T. S., Herrington, R. J., Maslennikov, V. V., Morris, N. J. & Zaykov, V. V. (1997). 'Silurian hydrothermal-vent community from the southern Urals, Russia.' *Nature* **385**. 146–8.

London Zoo (1994). *Your Guide to London Zoo: Conservation in Action (2nd Edn)*, Zoological Society of London, London.

Longhurst, A. R. & Pauly, D. (1987). *Ecology of Tropical Oceans*, Academic Press, San Diego.

Louw, G. N. & Seely, M. K. (1982). *Ecology of Desert Organisms*, Longman, London.

Lovejoy, T. E. *et al.*, (1986). 'Edge and other effects of isolation on Amazon Forest fragments.' In Soulé, M. E. (Ed), *Conservation Biology: The Science of Scarcity and Diversity*, Sinauer Associates Inc., Massachusetts. pp. 257–85.

Lovelock, J. (1988). *The Ages of Gaia*, Oxford University Press, Oxford.

Lucas, M. C., Johnstone, A. D. F. & Priede, I. G. (1993). 'Use of physiological telemetry as a method of estimating metabolism of fish in the natural environment.' *Transactions of the American Fisheries Society* **122**. 822–33.

Luke, A. (1995). 'Lead kills Spanish birds as hunters shoot wild.' *New Scientist* 3 June. 7.

Lüscher, M. (1961). 'Air-conditioned termite nests.' *Scientific American* **205** (1). 138–45.

Lynch, J. M. (1988). 'The terrestrial environment.' In Lynch, J. M. & Hobbie, J. E. (Eds), *Micro-organisms in Action: Concepts and Applications in Microbial Ecology*, Blackwell Scientific Publications, Oxford. pp. 103–31.

Ma, N. & Numata, M. (1996). 'Comparative studies of nature conservation areas in China and Japan.' *Natural History Research* **4**. 1–10.

Mabberley, D. J. (1986). 'Adaptive syndromes of the Afroalpine species of *Dendrosenecio*.' In Vuilleumier, F. & Monasterio. M. (Eds), *High Altitude Tropical Biogeography*, Oxford University Press, Oxford. pp. 81–102.

MacArthur, J. (1975). 'Environmental fluctuations and species diversity.' In Cody, M. L. & Diamond, J. M. (Eds). *Ecology and Evolution of Communities*, Belknap Press of Harvard University Press, Cambridge, Massachusetts. pp. 74–80.

MacArthur, R. H. (1958). 'Population ecology of some warblers of northeastern coniferous forests.' *Ecology* **39**. 599–619.

MacArthur, R. H. & Pianka, E. R. (1966). 'On optimal use of a patchy environment.' *American Naturalist* **100**. 603–9.

MacArthur, R. H. & Wilson, E. O. (1967). *The Theory of Island Biogeography*, Princeton University Press, Princeton.

Macdonald, D. (Ed) (1985). *The Encyclopaedia of Mammals I & II*, Guild Publishing, London.

MacDonald, I. A. W. & Cooper, J. (1995). 'Insular lessons for global biodiversity conservation with particular reference to alien invasions.' In Vitousek, P. M., Loope, L. L. & Adsersen, H. (Eds), *Islands: Biological Diversity and Ecosystem Function*, Springer, Berlin. pp. 189–203.

MacGillivray, C. W. & Grime, J. P. (1995). 'Genome size predicts frost resistance in British herbaceous plants: implications for rates of vegetation response to global warming.' *Functional Ecology* **9**. 320–5.

MacKenzie, D. (1994). 'Where has all the carbon gone?' *New Scientist* 8 January, 30–3.

MacKenzie, D. (1995). 'Killing crops with cleanliness.' *New Scientist* 23 September. 4.

Mackinnon. J., Mackinnon, K., Child, G. & Thorsell, J. (1986). *Managing Protected Areas in the Tropics*, IUCN, Switzerland.

Macilwain, C. (1996). 'Biosphere 2 begins fight for credibility.' *Nature* **380**. 275.

Maiorana, V. C. (1978). 'An explanation of ecological and developmental constants.' *Nature* **273**. 375–7.

Makhijani, A. & Gurney, K. R. (1995). *Mending the Ozone*

Hole: Science, Technology and Policy, MIT Press, Cambridge, Massachusetts.

Major. J. (1963). 'A climatic index to vascular plant activity.' *Ecology* **44**. 485–98.

Malcolm, J. R. & van Lawick, H. (1975). 'Notes on wild dogs (*Lycaon pictus*) hunting zebras.' *Mammalia* **39**. 231–40.

Mandelbrot, B. (1967). 'How long is the coast of Britain? Statistical self similarity and fractal dimensions.' *Science* **156**. 636–8.

Manley, G. (1974). 'Central England temperature: monthly means 1659 to 1973.' *Quarterly Journal of the Royal Meteorological Society* **100**. 389–405.

Mapstone, B. D. & Fowler, A. J. (1988). 'Recruitment and the structure of assemblages of fish on coral reefs.' *Trends in Ecology and Evolution* **3**. 72–7.

Margulis, L. (1981). *Symbiosis in Cell Evolution*, W. H. Freeman & Co., San Francisco.

Margulis, L. & Schwartz, K. V. (1982). *Five Kingdoms*, W. H. Freeman & Co., San Francisco.

Marples, B. J. (1934). 'The winter starling roosts of Great Britain. 1932–1933.' *Journal of Animal Ecology* **3**. 187–203.

Martin, E. & Vigne, L. (1997). 'Expanding ivory production in Myanmar threatens wild elephants.' *Oryx* **31**. 158–60.

Martin, M. H. (1968). 'Conditions affecting the distribution of *Mercurialis perennis* L. in certain Cambridgeshire woodlands.' *Journal of Ecology* **56**. 777–93.

Martin, P. S. & Klein, R. G. (Eds) (1984) *Quaternary Extinctions*, University of Arizona Press.

Mason, C. F. (1977). *Decomposition*, Edward Arnold, London.

Masood, E. (1997). 'Fisheries science: all at sea when it comes to politics?' *Nature* **386**. 105–6.

May, R. M. (1979). 'Production and respiration in animal communities.' *Nature* **282**. 443–4.

May, R. M. (1989). 'Black-footed ferret update.' *Nature* **339**. 104.

May, R. M. & MacArthur, R. H. (1972). 'Niche overlap as a function of environmental variability.' *Proceedings of the National Academy of Sciences*, USA **69**. 1109–13.

Maynard Smith, J. (1964). 'Group selection and kin selection.' *Nature* **201**. 1145–7.

Maynard Smith, J. (1976). 'Group selection.' *Quarterly Review of Biology* **51**. 277–83.

Maynard Smith, J. & Price, G. R. (1973). 'The logic of animal conflict.' *Nature* **246**. 15–18.

McKendrick. S. L. (1995). 'The effects of herbivory and vegetation on laboratory-raised *Dactylorhiza praetermissa* (Orchidaceae) planted into grasslands in Southern England.' *Biological Conservation* **73**. 215–20.

McKenzie, D. P. & Parker, R. L. (1967). 'The North Pacific: an example of tectonics on a sphere.' *Nature* **216**. 1276–80.

McNaughton, S. J. (1979). 'Grassland–herbivore dynamics.' In Sinclair. A. R. E. & Norton-Griffiths, M. (Eds), *Serengeti: Dynamics of an Ecosystem*, University of Chicago Press, Chicago. pp. 46–81.

Medina, E. & Klinge, H. (1983). 'Productivity of tropical forests and tropical woodlands.' In Lange, O. L., Nobel, P. S., Osmond, C. B. & Zeigler, M. H. (Eds), *Physiological Plant Ecology, vol. IV, Ecosystem Processes: Mineral Cycling, Productivity and Man's Influence,*

Springer-Verlag, Berlin. pp. 281–303.

Melchett, P. (1995). 'Green for danger.' *New Scientist* 23/30 December. 50–1.

Melillo, J. M. & Gosz, J. R. (1983). 'Interactions of biogeochemical cycles in forest ecosystems.' In Bolin, B. & Cook, R. B. (Eds), *The Major Biogeochemical Cycles and Their Interactions, Scope 21*, John Wiley, Chichester. pp. 177–222.

Mellanby, K. (1972). *The Biology of Pollution*, Edward Arnold, London.

Mertens, J. A. L. (1969). 'The influence of brood size on the energy metabolism and water loss of nestling great tits *Parus major major*.' *Ibis* **111**. 11–16.

Mestel, R. (1997). 'Let's make nodules.' *New Scientist* 11 January. 22–5.

Meyer, W. H. (1938). 'Yield of even aged stands of Ponderosa pine.' *USDA Technical Bulletin* **630**.

Mitchell, J. F. B., Senior, C. A. & Ingram, W. J. (1989). 'CO_2 and climate: a missing feedback?' *Nature* **341**. 132–4.

Mitsch, W. J. & Grosselink, J. G. (1986). *Wetlands*, Van Nostrand Reinhold Co., New York.

Möbius. K. (1877). *Die Auster und die Austernwirtschaft*, Wiegundt, Hampel & Parey, Berlin.

Moen, A. N. (1973). *Wildlife Ecology*, W. H. Freeman & Co., San Francisco.

Moffat, A. S. (1992). 'An intimate look at nitrogen's biopartner.' *Science* **257**. 1624–5.

Moffat, A. S. (1997). 'Resurgent forests can be green house gas sponges.' *Science* **277**. 315–6.

Montgomery, G. G. & Sunquist, M. E. (1978). 'Habitat selection and use by two-toed and three-toed sloths.' In Montgomery, G. G. (Ed), *The Ecology of Arboreal Folivores*, Smithsonian Institution Press, Washington, D.C. pp. 329–59.

Montzka, S. A., Butler, J. H., Myers, R. C., Thompson, T. M., Swanson, T. H., Clarke, A. D., Lock, L. T. & Elkins, J. W. (1996). 'Decline in the tropospheric abundance of halogen from halocarbons: implications for stratospheric ozone depletion.' *Science* **272**. 1318–22.

Morris, J. (1996a). 'Extinction of one snail but hope for others.' *Oryx* **30**. 81.

Morris, J. (1996b). 'A conservation success.' *Oryx* **30**. 225.

Moss, B. (1988). *Ecology of Fresh Waters: Man and Medium (2nd Edn)*, Blackwell Scientific Publications, Oxford.

Motluk, A. (1997). 'Pollution may lead to a life of crime.' *New Scientist* 31 May. 4.

Mueller-Dombois, D. (1984). 'Classification and mapping of plant communities: a review with emphasis on tropical vegetation.' In Woodwell, G. M. (Ed), *The Role of Terrestrial Vegetation in the Global Carbon Cycle: Measuring by Remote Sensing*, John Wiley & Sons, New York. pp. 21–88.

Muller, P. (1980). *Biogeography*, Harper & Row, New York.

Murie, (1944). *The Wolves of Mount McKinley: Fauna of National Parks of U.S.*, Fauna Series **5**, Government Prints Office. Washington D.C.

Myers, N. (Ed) (1994) *The Gaia Atlas of Planet Management*, Gaia, London.

Myers, N. (1995). 'Population and biodiversity.' *Ambio* **24**. 56–7.

Naeem, S., Thompson, L. J., Lawlers, S. P., Lawton, J. H. & Woodfin, R. M. (1995). 'Empirical evidence that declining species diversity may alter the performance of terrestrial ecosystems.' *Philosophical Transactions of the Royal Society of London B* **347**. 249–62.

Nagy, K. A. (1987). 'Field metabolic rate and food requirement scaling in mammals and birds.' *Ecological Monographs* **57**. 111–28.

National Power (1996). *Environmental Performance Review*, National Power, Swindon.

National Water Council (1981). *River Quality – the 1980 Survey and Future Outlook*, National Water Council, London.

Neill, S. R. St J. & Cullen, J. M. (1974). 'Experiments on whether schooling by their prey affects the hunting behaviour of cephalopods and fish predators.' *Journal of Zoology, London* **172**. 549–69.

Nève, G., Barascud, B., Hughes, R., Aubert, J., Descimon, H., Lebrun, P. & Baguette, M. (1996). 'Dispersal, colonization power and metapopulation structure in the vulnerable butterfly *Proclossiana eunomia* (Lepidoptera: Nymphalisae).' *Journal of Applied Ecology* **33**. 14–22.

New, T. R. (1994). 'Butterfly ranching: sustainable use of insects and sustainable benefit to habitats.' *Oryx* **28**. 169–172.

Nichol, S. (1996). 'Life on the edge of catastrophe.' *Nature* **384**. 218–9.

Norman, M. J. T. (1979). *Annual Cropping Systems in the Tropics*, University of Florida Press, Gainesville, Florida.

Norton, T. A. (1985). 'The zonation of seaweeds on rocky shores.' In Moore, P. G. & Seed, R. (Eds), *The Ecology of Rocky Coasts*, Hodder & Stoughton, London. pp. 7–21.

O'Brien, S. J., Roelke, M. E., Marker, L., Newman, A., Winkler, C. A., Meltzer, D., Colly, L., Evermann, J. F., Bush, M. & Wildt, D. E. (1985). 'Genetic basis for species vulnerability in the cheetah.' *Science* **227**. 1428–34.

O'Brien, S. J., Wildt, D. E., Goldman, D., Merril, C. R. & Bush, M. (1983). 'The cheetah is depauperate in genetic variation.' *Science* **221**. 459–62.

Odum, E. P. (1971). *Fundamentals of Ecology (3rd Edn)*, Saunders, Philadelphia.

Odum. H. T. (1957). 'Trophic structure and productivity of Silver Springs, Florida.' *Ecological Monographs* **27**. 55–112.

Oeschger, H. & Stauffer, B. (1986). 'Review of the history of atmospheric CO_2 recorded in ice cores.' In Trabalka, J. R. & Reichle, D. E. (Eds), *The Changing Carbon Cycle: A Global Analysis*, Springer-Verlag, New York. pp. 89–108.

Ogden, J. (1968). *Studies on Reproductive Strategy with Particular Reference to Selected Composites*, Ph.D. thesis, University of Wales.

Oinonen, E. (1967a). 'Sporal regeneration of bracken (*Pteridium aquilinum* (L.) Kuhn.) in Finland in the light of the dimensions and the age of its clones.' *Acta Forestalia Fennica* **83** (1). 1–96.

Oinonen, E. (1967b). 'The correlation between the size of Finnish bracken (*Pteridium aquilinum* (L.) Kuhn.) clones and certain periods of site history.' *Acta Forestalia Fennica* **83** (2). 1–51.

Oksanen, L. (1987). 'Interspecific competition and the structure of bird guilds in boreal Europe: the importance of doing fieldwork in the right season.' *Trends in Ecology and Evolution* **2**. 376–9.

Olsson, L. (1993). 'On the causes of famine – drought, desertification and market failure in the Sudan.' *Ambio* **22**. 395–403.

Ono, Y. (1965). 'On the distribution of ocypoid crabs in the estuary.' *Mem. of the Faculty of Science, Kyushu University, Series E (Biology)* **4**. 1–60.

Oosting, H. J. (1956). *The Study of Plant Communities (2nd Edn)*, W. H. Freeman & Co., San Francisco.

Orpin, C. G. & Anderson, J. M. (1988). 'The animal environment.' In Lynch, J. M. & Hobbie, J. E. (Eds), *Micro-organisms in Action: Concepts and Applications in Microbial Ecology*, Blackwell Scientific Publications, Oxford. pp. 163–92.

Oster, G. F. & Wilson, E. O. (1978). *Caste and Ecology in the Social Insects*, Princeton University Press, Princeton.

Owen-Smith, R. N. (1988). *Megaherbivores: The influence of very large body size on ecology*, Cambridge University Press, Cambridge.

Packer, C. (1986). 'The ecology of sociality in felids.' In Rubenstein, D. I. & Wrangham, R. W. (Eds), *Ecological Aspects of Social Evolution: Birds and Mammals*, Princeton University Press, Princeton. pp. 429–51.

Packer, C. R. & Pusey, A. E. (1982). 'Cooperation and competition within coalitions of male lions: kin selection or games theory.' *Nature* **296**. 740–42.

Packer, C., Herbst, L., Pusey, A. E., Bygott, J. D., Hanby. J. P., Cairns, S. J. & Borgerhoff Mulder, M. (1988). 'Reproductive success of lions.' In Clutton-Brock, T. H. (Ed), *Reproductive Success: Studies of Individual Variation in Contrasting Breeding Systems*, University of Chicago Press, Chicago. pp. 363–83.

Pagel, M. D. & Greenough, J. A. (1987). 'Explaining species size ratios in nature.' *Trends in Ecology and Evolution* **2**. 114–15.

Pain, S. (1993). 'Valdez spill wasn't so bad, claims Exxon.' *New Scientist* 8 May. 4.

Pandian, T. J. (1967). 'Intake, digestion, absorption and conversion of food in the fishes *Megalops cyprinoides* and *Ophiocephalus striatus*.' *Marine Biology* **1**. 16–32.

Park, T. (1948). 'Experimental studies of interspecies competition. I. Competition between populations of the flour beetles *Tribolium confusum* Duval and *Trifolium castaneum* Herbst.' *Ecological Monographs* **18**. 265–307.

Park, T. (1954). 'Experimental studies of interspecies competition. II. Temperature, humidity, and competition in two species of *Tribolium*.' *Physiological Zoology* **27**. 177–238.

Payne. A. I. (1986). *The Ecology of Tropical Lakes & Rivers*, John Wiley & Sons, Chichester.

Peacocke, A. R. (1979). *Creation and the World of Science: The Bampton Lectures, 1978*, Clarendon Press, Oxford.

Pearce, F. (1993). 'The high cost of carbon dioxide.' *New Scientist* 17 July. 26–9.

Pearce, F. (1995). 'Global warming "jury" delivers guilty verdict.' *New Scientist* 9 December. 6.

Pearce, F. (1996a). 'Sit tight for 30 years, argues climate guru.' *New Scientist* 20 January. 7.

Pearce, F. (1996b). 'Will Kew collect after lottery win?' *New Scientist* 6 January. 9.

Pearce, F. (1997a). 'Lightning sparks pollution rethink.' *New Scientist* 25 January. 15.

Pearce, F. (1997b). 'The concrete jungle overheats.' *New Scientist* 19 July. 14.

Pearce, F. (1997c). 'Wispy trails could warm the Earth.' *New Scientist* 29 March. 5.

Pearce, F. (1997d). 'Mighty mite saves Africa's staple food.' *New Scientist* 24 May. 12.

Peet, R. K. (1984). 'Twenty-six years of change in a *Pinus strobus, Acer saccharum* forest, Lake Itasca, Minnesota.' *Bulletin of the Torrey Botanical Club* **111**. 61–8.

Pellew, R. (1997). 'Our plans to save India's tigers.' *WWF News* Summer. 3.

Pennington, W. (1974). *The History of the British Vegetation (2nd Edn)*, The English Universities Press.

Perrill, S. A., Gerhardt, H. C. & Daniel, R. E. (1982). 'Mating strategy in male green treefrogs (*Hyla cinerea*): an experimental study.' *Animal Behaviour* **30**. 43–8.

Perrin, R. M. S., Willis, E. H. & Hodge, C. A. H. (1964). 'Dating of humus podzols by residual radiocarbon activity.' *Nature* **202**. 165–6.

Perring, F. H. & Walters, S. M. (Eds) (1962). *Atlas of the British Flora*, Botanical Society of the British Isles, Nelson, London.

Peters. C. M., Gentry, A. H. & Mendelson. R. O. (1989). 'Valuation of an Amazonian Rainforest.' *Nature* **339**. 655.

Peters, R. H. (1983). *The Ecological Implications of Body Size*, Cambridge University Press, Cambridge.

Petridou, E. *et al.* (1996). 'Infant leukaemia after *in utero* exposure to radiation from Chernobyl.' *Nature* **382**. 352–3.

Phillips. D. H. (1982). *Living with Huntington's Disease*, Junction Books, London.

Phillipson, J. (1966). *Ecological Energetics*, Edward Arnold, London.

Pianka, E. R. (1970). 'On *r* and *K* selection.' *American Naturalist* **104**. 592–7.

Pianka, E. R. (1975). 'Niche relations of desert lizards.' In Cody, M. L. & Diamond, J. M. (Eds), *Ecology and Evolution of Communities*, Belknap Press of Harvard University Press, Cambridge, Massachusetts. pp. 292–314.

Pianka, E. R. (1976). 'Competition and niche theory.' In May, R. M. (Ed), *Theoretical Ecology: Principles and Applications*, Blackwell Scientific Publications, Oxford. pp. 114–41.

Pianka, E. R. (1988). *Evolutionary Ecology (4th Edn)*, Harper & Row, New York.

Pielou, E. C. (1979). *Biogeography*, J. Wiley & Sons, New York.

Pietsch, T. W. (1975). 'Precocious sexual parasitism in the deep sea ceratioid anglerfish, *Cryptopsarus couesi* Gill.' *Nature* **256**. 38–40.

Pigott, G. D. (1989). 'The growth of lime *Tilia cordata* in an experimental plantation and its influence on soil development and vegetation.' *Quarterly Journal of Forestry* **83** (1). 14–24.

Pimental, D., Levin, S. A. & Soans, A. B. (1975). 'On the evolution of energy balance in some exploiter–victim systems. *Ecology* **56**. 381–90.

Pimm, S. L. (1982). *Food Webs*, Chapman & Hall, London.

Pimm. S. L. & Lawton, J. H. (1977). 'The number of trophic levels in ecological communities.' *Nature* **268**. 329–31.

Pimm, S. (1995). 'Seeds of our own destruction.' *New Scientist* 8 April. 31–5.

Pinxten, R. & Eens, M. (1997). 'Copulation and mate-guarding patterns in polygynous European starlings.' *Animal Behaviour* **54**. 45–58.

Plowden, C. & Bowles, D. (1997). 'The illegal market in tiger parts in northern Sumatra, Indonesia.' *Oryx* **31**. 59–66.

Postgate, J. (1988). 'The ghost in the laboratory.' *New Scientist* 4 February. 49–52.

Potts, G. R. (1967). 'Urban starling roosts in the British Isles.' *Bird Study* **14**. 35–42.

Prance, G. (1996). *The Earth Under Threat: A Christian Perspective*, Wild Goose Publications, Glasgow.

Prance, G. T. & Schubart, H. O. R. (1978). 'A preliminary note on the origin of the open white sand campinas of the lower Rio Negro.' *Notes on the Vegetation of Amazonia* **30**. 60–63.

Prime, R. (1992). *Hinduism and Ecology: Seeds of Truth*, Cassell, London.

Putnam, C. (1994). 'Dead whales: stepping stones for creatures of the deep.' *New Scientist* 20 August. 16.

Rackham, O. (1980). *Ancient Woodland: its history, vegetation and uses in England*, Edward Arnold, London.

Rackham, O. (1986). *The History of the Countryside*, J. M. Dent, London.

Rahmstorf, S. (1997). 'Ice-cold in Paris.' *New Scientist* 8 February. 26–30.

Rankama, K. & Sahama. T. G. (1950). *Geochemistry*, University of Chicago Press, Chicago.

Rappoldt, C. & Hogeweg. P. (1980). 'Niche packing and number of species.' *American Naturalist* **116**. 480–92.

Ratcliffe, D. A. (1963). 'The status of the peregrine in Great Britain.' *Bird Study* **10**. 56–90.

Raven, R. H. & Axelrod, D. I. (1974). 'Angiosperm bio-geography and past continental movement.' *Annals of the Missouri Botanic Garden* **61**. 539–673.

Rayner, A. (1997). *Degrees of Freedom: Living in Dynamic Boundaries*, Imperial College Press, London.

Reiss, M. J. (1985). 'The allometry of reproduction: why larger species invest relatively less in their offspring.' *Journal of theoretical Biology* **113**. 529–44.

Reiss, M. J. (1989). *The Allometry of Growth and Reproduction*, Cambridge University Press, Cambridge.

Reiss, M. J. & Straughan, R. (1996). *Improving Nature? The Science and Ethics of Genetic Engineering*, Cambridge University Press, Cambridge.

Remmert, H. (Ed) (1994) *Minimum Animal Populations*, Springer-Verlag, Berlin.

Reynolds, J. C. (1985). 'Details of the geographic replace-ment of the red squirrel (*Sciurus vulgaris*) by the grey squirrel (*Sciurus carolinensis*) in Eastern England.' *Journal of Animal Ecology* **54**. 149–62.

Richards, O. W. & Waloff, N. (1954). 'Studies on the biology and population dynamics of British grasshoppers.' *Anti-locust Bulletin* **17**. 1–182.

Ricklefs, R. E. (1980). *Ecology (2nd Edn)*, Nelson. Sunbury-on-Thames, Middlesex.

Ricklefs, R. E., White, S.C. & Cullen, J. (1980). 'Energetics of postnatal growth in Leach's storm-petrel.' *Auk* **97**. 566–75.

Rohani, P. & Earn, D. J. D. (1997). 'Chaos in a cup of flour.' *Trends in Ecology and Evolution* **12**. 171.

Romer, A. S. (1966). *Vertebrate Palaeontology (3rd Edn)*, University of Chicago Press, Chicago.

Ronsheim, M. L. (1988). 'Determining the pattern of resource allocation in plants.' *Trends in Ecology and Evolution* **3**. 30–31.

Root, R. B. (1967). 'The niche exploitation pattern of the blue-gray gnatcatcher.' *Ecological Monographs* **37**. 317–50.

Rosenzweig, M. L. (1968). 'Net primary productivity of

terrestrial communities: prediction from climatological data.' *American Naturalist* **102**. 67–74.

Rosswall, T. (1983). 'The nitrogen cycle.' In Bolin, B. & Cook, R. B. (Eds), *The Major Biogeochemical Cycles and Their Interactions, Scope 21*, John Wiley, Chichester. pp. 46–50.

Rowell, T. A. (1987). 'History and experimentation in the management of Wicken Fen.' *Nature in Cambridgeshire* **29**. 14–19.

Roy, K. & Foote, M. (1997). 'Morphological approaches to measuring biodiversity.' *Trends in Ecology and Evolution* **12**. 277–81.

Rozema, J., van de Staaij, J., Björn, L. O. & Caldwell, M. (1997). 'UV-B as an environmental factor in plant life: stress and regulation.' *Trends in Ecology and Evolution* **12**. 22–8.

Ruether, R. R. (1997). 'Ecofeminism: first and third world women.' *Ecotheology* **2**. 72–83.

Russell, G. & Morris, P. (1970). 'Copper tolerance in the marine fouling alga *Ectocarpus siliculosus*.' *Nature* **228**. 288–9.

Russell, M. J. & Hall, A. J. (1997). 'The emergence of life from iron monosulphide bubbles at a submarine hydrothermal redox and pH front.' *Journal of the Geological Society, London* **154**. 377–402.

Rymer, L. (1976). 'The history and ethnobotany of bracken.' *Botanical Journal of the Linnean Society* **73**. 151–76.

Sahlins, M. D. (1976). *The Use and Abuse of Biology: An Anthropological Critique of Sociobiology*, University of Michigan Press. Ann Arbor.

Sakai, A. & Weiser, C. J. (1973). 'Freezing resistance of trees in North America with reference to tree regions.' *Ecology* **54**. 118–61.

Salisbury, F. B. & Ross, C. W. (1985). *Plant Physiology (3rd Edn)*, Wadsworth, Belmont, California.

Salt, D. T., Brooks, G. L. & Whittaker, J. B. (1995). 'Elevated carbon dioxide affects leaf-miner performance and plant growth in docks (*Rumex* spp.).' *Global Change Biology* **1**. 153–6.

Sarmiento, G. (1986). 'Ecological features of climate in high tropical mountains.' In Vuilleumier, F. & Monasterio, M. (Eds), *High Altitude Tropical Biogeography*, Oxford University Press, Oxford. pp. 11–46.

Sarrazin, F. & Barbault, R. (1996). 'Reintroductions: challenges and lessons for basic ecology.' *Trends in Ecology and Evolution* **11**. 474–8.

Savchenko, V. K. (1995) *The Ecology of the Chernobyl Catastrophe: Scientific Outlines of an International Programme of Collaborative Research*, Unesco, Paris.

Schaffer, W. M. & Kot, M. (1986). 'Chaos in ecological systems: the coals that Newcastle forgot.' *Trends in Ecology and Evolution* **1**. 58–63.

Schaffer, W. M. & Schaffer, M. V. (1977). 'The adaptive significance of variation in reproductive habit in the Agavaceae.' In Stonehouse, B. & Perrins, C. (Eds), *Evolutionary Ecology*, Macmillan, London. pp. 261–76.

Schaller, G. B. (1972). *The Serengeti Lion: A Study of Predator–Prey Relations*, University of Chicago Press, Chicago.

Schemske, D. W. (1981). 'Floral convergence and pollinator sharing in two bee-pollinated tropical herbs.' *Ecology* **62**. 946–54.

Schluter, D., Price, T. D. & Grant, P. R. (1985). 'Ecological character displacement in Darwin's finches.' *Science* **227**. 1056–9.

Schmidt-Nielsen, K. (1964). *Desert Animals: Physiological Problems of Heat and Water*, Oxford University Press, Oxford.

Schmidt-Nielsen, K. (1983). *Animal Physiology: Adaptation and Environment (3rd Edn)*, Cambridge University Press, Cambridge.

Schmidt-Nielsen, K. (1984). *Scaling: Why is Animal Size so Important?*, Cambridge University Press. Cambridge.

Schmidt-Nielsen, K., Schmidt-Nielsen, B., Jarnum, S. A. & Houpt, T. R. (1957). 'Body temperature of the camel and its relation to water economy.' *American Journal of Physiology* **188**. 102–12.

Schneirla, T. C. (1971). *Army Ants: A Study in Social Organization*, W. H. Freeman, San Francisco.

Schonbeck, M. & Norton, T. A. (1978). 'Factors controlling the upper limits of fucoid algae on the shore.' *Journal of Experimental Marine Biology and Ecology* **31**. 303–13.

Schonbeck, M. & Norton, T. A. (1980). 'Factors controlling the lower limits of fucoid algae on the shore.' *Journal of Experimental Marine Biology and Ecology* **43**. 131–50.

Schultz, J. C. (1988). 'Many factors influence the evolution of herbivore diets, but plant chemistry is central.' *Evolution* **69**. 896–7.

Shaffer, J. A. (1993). 'Closed ecological systems.' *Carolina Tips* **56** (4). 13–5.

Sharp, R. (1997). 'The African elephant: conservation and CITES.' *Oryx* **31**. 111–19.

Sheail, J. (1985). *Pesticides and Nature Conservation: The British Experience 1950–1975*, Clarendon Press, Oxford.

Shelford, V. E. (1943). 'The abundance of the collared lemming (*Dicrostonyx groenlandicus* (Tr) var. *richardsoni* Mer.) in the Churchill Area, 1929 to 1940.' *Ecology* **24**. 472–84.

Shelford, V. E. (1945). 'The relation of snowy owl migration to the abundance of the collared lemming.' *Auk* **62**. 592–6.

Shotyk, W., Cheburkin, A. K., Appleby, P. G., Fankhauser, A. & Kramers, J. D. (1996). 'Two thousand years of atmospheric arsenic, antimony, and lead deposition recorded in an ombrotrophic peat bog profile, Jura Mountains, Switzerland.' *Earth and Planetary Science Letters* **145**. E1–7.

Sibley, C. G. & Ahlquist. J. E. (1982). 'The relationships of the Hawaiian honeycreepers (Drepaninini) as indicated by DNA–DNA hybridisation.' *Auk* **99**. 130–40.

Sibly, R. M. & Calow, P. (1986). *Physiological Ecology of Animals: An Evolutionary Approach*, Blackwell Scientific Publications, Oxford.

Silvertown, J. (1985). 'History of a latitude diversity gradient: woody plants in Europe 13 000–1000 years BP.' *Journal of Biogeography* **12**. 519–25.

Silvertown, J. & Dodd, M. (1996). 'Comparing plants and connecting traits.' *Philosophical Transactions of the Royal Society of London B* **351**. 1233–9.

Silvertown, J. & Law, R. (1987). 'Do plants need niches? Some recent developments in plant community ecology.' *Trends in Ecology and Evolution* **2**. 24–6.

Silvertown, J. W. & Lovett Doust, J. (1993). *Introduction to Plant Population Biology*, Blackwell Scientific Publications, Oxford.

Silvertown, J., Wells, D. A., Gillman, M., Dodd, M. E.,

Robertson, H. & Lakhani, K. H. (1994). 'Short-term effects and long-term after-effects of fertilizer application on the flowering population of green-winged orchid *Orchis morio*.' *Biological Conservation* **69**. 191–7.

Simberloff, D. (1986). 'Design of nature reserves.' In Usher, M. B. (Ed), *Wildlife Conservation Evaluation*, Chapman & Hall Ltd., London. pp. 315–37.

Simberloff, D. & Boecklen, W. (1981). 'Santa Rosalia reconsidered: size ratios and competition.' *Evolution* **35**. 1206–28.

Simms, E. (1985). *British Warblers*, Collins, London.

Sinclair, A. R. E. (1979). 'Dynamics of the Serengeti ecosystem: process and pattern.' In Sinclair, A. R. E. & Norton-Griffiths, M. (Eds), *Serengeti: Dynamics of an Ecosystem*, University of Chicago Press, Chicago. pp. 1–30.

Sinclair, A. R. E. (1985). 'Does interspecific competition or predation shape the African ungulate community?' *Journal of Animal Ecology* **54**. 899–918.

Sinclair, A. R. E. & Norton-Griffiths, M. (Eds) (1979). *Serengeti: Dynamics of an Ecosystem*, University of Chicago Press, Chicago.

Slade, A. J. & Hutchings, M. J. (1987). 'The effects of nutrient availability on foraging in the clonal herb *Glechoma hederacea*.' *Journal of Ecology* **75**. 95–112.

Smart, N. (1989). *The World's Religions: Old Traditions and Modern Transformations*, Cambridge University Press, Cambridge.

Smith, R. L. (1980). *Ecology and Field Biology (3rd Edn)*, Harper & Row, New York.

Solbrig, O. R. & Simpson, B. B. (1974). 'Components of regulation of a population of dandelions in Michigan.' *Journal of Ecology* **62**. 473–86.

Soulé, M. E. (1980). 'Thresholds for survival: maintaining fitness and evolutionary potential.' In Soulé, M. E. & Wilcox, B. A. (Eds), *Conservation Biology*, Sinauer Associates Inc., Massachusetts. pp. 151–70.

Southward, A. J. (1989). 'Animal communities fuelled by chemosynthesis: life at hydrothermal vents, cold seeps and in reducing sediments.' *Journal of Zoology, London* **217**. 705–9.

Spicer, R. A. & Chapman. J. L. (1990). 'Climatic change and the evolution of high-latitude terrestrial vegetation and floras.' *Trends in Ecology and Evolution* **5**. 279–84.

Spicer, R. A. & Parrish, J. T. (1986). 'Paleobotanical evidence for cool north polar climates in Middle Cretaceous (Albian–Cenomanian) time.' *Geology* **14**. 703–6.

Spicer, R. A., Burnham, R. J., Grant, P. & Glicken, H. (1985). '*Pityrogramma calomelanos*, the primary post-eruption coloniser of Volcan Chichonal, Chipas, Mexico.' *American Fern Journal* **75** (1). 1–5.

Spring, G. M., Priestman, G. H. & Grime, J. P. (1996). 'A new field technique for elevating carbon dioxide levels in climate change experiments.' *Functional Ecology* **10**. 541–5.

Spurgeon, D. (1997). 'Canada's cod leaves science in hot water.' *Nature* **386**. 107.

Spurr, S. H. & Barnes, B. V. (1980). *Forest Ecology (3rd Edn)*, John Wiley, New York.

Stafford. J. (1971). 'Heron populations of England and Wales, 1928–70.' *Bird Study* **18**. 218–21.

Staheli, P. (1995). 'The new face of protest.' *Livewire* Aug/Sept. 22–6.

Stanier, M. (1993). 'The restoration of grassland on the Devil's Ditch, Cambridgeshire.' *Nature in Cambridgeshire* **35**. 13–17.

Stark, N. M. & Jordan, C. F. (1978). 'Nutrient retention by the root mat of an Amazonian rain forest.' *Ecology* **59**. 434–7.

Stewart, D. (1988). 'Ankarana damaged.' *Oryx* **22**. 240–41.

Stewart, J. (1993). 'Propagation and translocation of orchids.' *Botanical Society of the British Isles News* **64**. 51–3.

Stewart, J. W. B., Cole, C. V. & Maynard, D. G. (1983). 'Interactions of biogeochemical cycles in grassland ecosystems.' In Bolin, B. & Cook, R. B. (Eds), *The Major Biogeochemical Cycles and Their Interactions*, Scope 21, John Wiley, Chichester. pp. 247–69.

Stokes. G. M. & Barnard, J. C. (1986). 'Presentation of the 20th century atmospheric CO_2 record in Smithsonian spectrographic plates.' In Trabalka, J. R. & Reichle, D. E. (Eds), *The Changing Carbon Cycle: A Global Analysis*, Springer-Verlag, New York. pp. 50–65.

Strong, K. W. & Daborn. G. R. (1979). 'Growth and energy utilisation of the intertidal isopod *Idotea baltica* (Pallas) (Crustacea: Isopoda).' *Journal of Experimental Marine Biology and Ecology*. **41**. 101–23.

Stutt, A. D. & Willmer, P. (1998). 'Territorial defence in speckled wood butterflies: do the hottest males always win?' *Animal behaviour* **55**. 1341–47.

Sugihara, G., Grenfell, B. & May, R. M. (1990). 'Distinguishing error from chaos in ecological time series.' *Philosophical Transactions of the Royal Society of London B* **330**. 235–51.

Sumption, K. J. & Flowerdew, J. R. (1985). 'The ecological effects of the decline in rabbits (*Oryctolagus cuniculus* L.) due to myxomatosis.' *Mammal Review* **15**. 151–86.

Sunquist, M. (1985). 'Tiger.' In Macdonald, D. (Ed). *The Encyclopaedia of Mammals: 1*, Guild Publishing, London. pp. 36–8.

Sutherland, W. J. (1995). 'Introduction and principles of ecological management.' In Sutherland, W. J. & Hill, D. A. (Eds) *Managing Habitats for Conservation*, Cambridge University Press, Cambridge. pp. 1–21.

Swihart, R. K., Slade, N. A. & Bergstrom, B. J. (1988). 'Relating body size to the rate of home range use in mammals.' *Ecology* **69**. 393–9.

Swinbanks, D. (1 989a). 'China blamed for high pH.' *Nature* **340**. 671.

Swinbanks. D. (1989b). 'Ivory trade.' *Nature* **342**. 7.

Tanahashi, K. (Ed) (1988). *Dogon. Moon in a Dewdrop*, Element Books, Shaftesbury.

Tans, P. P. & Bakwin, P. S. (1995). 'Climate change and carbon dioxide.' *Ambio* **24**. 376–8.

Tansley. A. G. (1935). 'The use and abuse of vegetational concepts and terms.' *Ecology* **16**. 284–307.

Tansley, A. G. (1949). *The British Islands and their Vegetation*, Cambridge University Press, Cambridge.

Tansley. A. G. & Adamson, R. S. (1925). 'Studies of the vegetation of the English chalk. III. The chalk grasslands of the Hampshire-Sussex border.' *Journal of Ecology* **13**. 177–223.

Taylor, C. (1988). *Introduction and commentary to 'The Making of the English Landscape'*, Hoskins, W. G. (1955), Guild Publishing, London.

Taylor, K. C., Lamorey, G. W., Doyle, G. A., Alley, R. B., Grootes, P. M., Mayewski, P. A., White, J. W. C. & Barlow, L. K. (1993). 'The "flickering switch" of late

Pleistocene climate change.' *Nature* **361**. 432–6.

Taylor, P. W. (1986). *Respect for Nature: A Theory of Environmental Ethics*, Princeton University Press, Princeton.

Teal, J. M. (1962). 'Energy flow in the salt marsh ecosystem of Georgia.' *Ecology* **43**. 614–24.

Templeton. A. R. & Read. B. (1983). 'The elimination of inbreeding depression.' In Schonewald-Cox. C. M., Chambers, S. M., MacBryde, B. & Thomas, W. L. (Eds), *Genetics and Conservation*, Benjamin Cummings, London. pp. 241–62.

Thomas, J. A. (1980a). 'Why did the Large Blue become extinct in Britain?' *Oryx* **15**. 243–7.

Thomas, J. A. (1980b). 'The extinction of the Large Blue and the conservation of the Black Hairstreak butterflies (a contrast of failure and success).' *Annual Report, The Institute of Terrestrial Ecology*, 1979. 19–23.

Tilman, D. & Downing, J. A. (1994). 'Biodiversity and stability in grasslands.' *Nature* **367**. 363–5.

Tittensor, A. M. (1977a). 'Red squirrel *Sciurus vulgaris*.' In Corbet, G. B. & Southern, H. N. (Eds), *The Handbook of British Mammals (2nd Edn)*, Blackwell Scientific Publications, Oxford. pp. 153–64.

Tittensor, A. M. (1977b). 'Grey squirrel *Sciurus carolinensis*.' In Corbet, G. B. & Southern, H. N. (Eds), *The Handbook of British Mammals (2nd Edn)*, Blackwell Scientific Publications, Oxford. pp. 164–72.

Transeau, E. N. (1926). 'The accumulation of energy by plants.' *Ohio Journal of Science* **26**. 1–10.

Trenberth, K. E. (Ed) (1992). *Climate System Modeling*, Cambridge University Press, Cambridge.

Trivers, R. L. (1971). 'The evolution of reciprocal altruism.' *Quarterly Review of Biology* **46**. 35–57.

Tucker, C. J., Dregne, H. E. & Newcomb, W. W. (1991). 'Expansion and contraction of the Sahara Desert from 1980 to 1990.' *Science* **253**. 299–301.

Tucker, M. E. (1997). 'The Emerging Alliance of Religion and Ecology.' *Worldviews: Environment, Culture, Religion* **1**. 3–24.

Turesson, G. (1925). 'The plant species in relation to habitat and climate.' *Hereditas* **6**. 147–236.

Turner, I. M. & Corlett, R. T. (1996). 'The conservation value of small, isolated fragments of lowland tropical rain forest.' *Trends in Ecology and Evolution* **11**. 330–3.

Usher, M. B. (1973). *Biological Management and Conservation*, Chapman & Hall, London.

van der Leun, J. C., Tang, X. & Tevini, M. (1995). 'Environmental effects of ozone depletion: 1994 assessment.' *Ambio* **24**. 138–42.

Vanclay, J. K. (1993). 'Saving the tropical forest: needs and prognosis.' *Ambio* **22**. 225–31.

Van Valen, L. (1973). 'A new evolutionary law.' *Evolutionary Theory* **1**. 1–30.

Varley, G. C. (1970). 'The concept of energy flow applied to a woodland community.' In Watson, A. (Ed), *Animal Populations in Relation to the Food Resources: A Symposium of the British Ecological Society, Aberdeen 24–28 March 1969*, Blackwell Scientific Publications, Oxford. pp. 389–404.

Veggeberg, S. (1993). 'Escape from Biosphere 2.' *New Scientist* 25 September. 22–4.

Vines, G. (1997). 'The plight of the bumblebee.' *New Scientist* 21 June. 26–8.

Vitousek, P. M., Loope, L. L. & Adsersen, H. (Eds) (1995)

Islands: Biological Diversity and Ecosystem Function, Springer, Berlin.

Wagner, D. (1997). 'The influence of ant nests on *Acacia* seed production, herbivory and soil nutrients.' *Journal of Ecology* **85**. 83–93.

Wakelin, D. (1987). 'Parasite survival and variability in host immune responsiveness.' *Mammal Review* **17**. 135–41.

Walker, D. (1970). 'Direction and rate in some British post-glacial hydroseres.' In Walker. D. & West, R. G. (Eds) *Studies in the Vegetational History of the British Isles*, Cambridge University Press, Cambridge. pp. 117–39.

Walker, G. (1996). 'Secrets from another Earth.' *New Scientist* 18 May. 31–5.

Waller. D. M. (1986). 'The dynamics of growth and form.' In Crawley, M. J. (Ed), *Plant Ecology*, Blackwell Scientific Publications, Oxford. pp. 291–320.

Walter, H. (1979). *Vegetation of the Earth and Ecological Systems of the Geobiosphere (2nd Edn)*. Translated from the third, revised German Edition by Joy Wieser. Springer-Verlag, New York.

Wania, F. & Mackay, D. (1993). 'Global fractionation and cold condensation of low volatility organochlorine compounds in polar regions.' *Ambio* **22**. 10–8.

Ward, P. & Zahavi, A. (1973). 'The importance of certain assemblages of birds as "information-centres" for food-finding.' *Ibis* **115**. 517–34.

Wassink, E. D. (1959). 'Efficiency of light energy conversion in plant growth.' *Plant Physiology* **34**. 356–61.

Watt. A. S. (1957). 'The effect of excluding rabbits from grassland B (mesobrometum) in Breckland.' *Journal of Ecology* **45**. 861–78.

Weihs, D. (1973). 'Hydromechanics of fish schooling.' *Nature* **241**. 290–91.

Werdelin, L. (1996). 'Community-wide character displacement in Miocene hyaenas.' *Lethaia* **29**. 97–106.

West, G. B., Brown, J. H. & Enquist, B. J. (1997). 'A general model for the origin of allometric scaling laws in biology.' *Science* **276**. 122–6.

Wetselaar, R. & Farquhar, G. D. (1980). 'Nitrogen losses from tops of plants.' *Advances in Agronomy* **33**. 263–302.

White, J. (1980). 'Demographic factors in populations of plants.' In Solbrig, O. T. (Ed), *Demography and Evolution in Plant Populations*, Blackwell, Oxford.

White, L. (1967). 'The historical roots of our ecologic crisis.' *Science* **155**. 1203–7.

Whittaker, J. B. & Tribe, N. P. (1996). 'An altitudinal transect as an indicator of responses of a spittlebug (Auchenorrhyncha: Cercopidae) to climate change.' *European Journal of Entomology* **93**. 319–24.

Whittaker. R. H. (1967). 'Gradient analysis of vegetation.' *Biological Reviews of the Cambridge Philosophical Society* **42**. 207–64.

Whittaker. R. H. (1975). *Communities and Ecosystems (2nd Edn)*, Macmillan, New York.

Whittaker. R. H., Levin, S. A. & Root, R. B. (1973). 'Niche, habitat, and ecotope.' *American Naturalist* **107**. 321–38.

Wielgolaski, F. E., Bliss, L. C., Svoboda, J. & Doyle, G. (1981). 'Primary production of tundra.' In Bliss, L. C., Heal, O. W. & Moore, J. J. (Eds), *Tundra Ecosystems: A*

Comparative Analysis, Cambridge University Press, Cambridge. pp. 187–225.

Wigley, T. M. L., Richels, R. & Edmonds, J. A. (1996). 'Economic and environmental choices in the stabilization of atmospheric CO_2 concentrations.' *Nature* **379**. 240–3.

Wijesinghe, D. K. & Hutchings, M. J. (1997). 'The effects of spatial scale of environmental heterogeneity on the growth of a clonal plant: an experimental study with *Glechoma hederacea*.' *Journal of Ecology* **85**. 17–28.

Wilkinson, G. S. (1984). 'Reciprocal food sharing in the vampire bat.' *Nature* **308**. 181–4.

Williamson, M. (1981). *Island Populations*, Oxford University Press, Oxford.

Wilson, E. O. (1975). *Sociobiology: The New Synthesis*, Belknap Press of Harvard University Press, Cambridge, Massachusetts.

Wilson, E. O. (1978). *On Human Nature*, Harvard University Press, Cambridge, Massachusetts.

Wilson, E. O. (1992). *The Diversity of Life*, Belknap Press of Harvard University Press, Cambridge, Massachusetts.

Wilson, E. O. & Willis, E. O. (1975). 'Applied biogeography.' In Cody, M. L. & Diamond, J. M. (Eds), *Ecology and Evolution of Communities*, Harvard University Press, Cambridge, Massachusetts. pp. 522–34.

Wilson, J. B., Ullmann, I. & Bannister, P. (1996). 'Do species assemblages ever recur?' *Journal of Ecology* **84**. 471–4.

Wilson, J. T. (1965). 'A new class of faults and their bearing on continental drift.' *Nature* **207**. 343–7.

Woodward, F. I. (1987). *Climate and Plant Distribution*, Cambridge University Press, Cambridge.

World Council of Churches (1990). *Giver of Life – Sustain Your Creation. Report of the Pre-Assembly Consultation on Sub-theme 1*, Kuala Lumpur, Malaysia. Reprinted in *The Ecumenical Review* July–October 1990.

Wynne-Edwards, V. C. (1929). 'The behaviour of starlings in winter.' *British Birds* **23**. 138–53; 170–80.

Wynne-Edwards, V. C. (1962). *Animal Dispersion in Relation to Social Behaviour*, Oliver & Boyd, Edinburgh.

Yampolsky, E. & Yampolsky, H. (1922). 'Distribution of sex forms in phanerogamic flora.' *Bibl. Genet.* **3**. 1–62.

Yates, M. G. (1980). 'Biochemistry of nitrogen fixation.' In Miflin, B. J. (Ed), *The Biochemistry of Plants: Amino Acids and Derivatives Vol. 5*, Academic Press, New York. pp. 1–64.

Young, A., Boyle, T. & Brown, T. (1996). 'The population genetic consequences of habitat fragmentation for plants.' *Trends in Ecology and Evolution* **11**. 413–8.

Zach, R. (1979). 'Shell-dropping: decision-making and optimal foraging in northwestern crows.' *Behaviour* **68**. 106–17.

Zeleny, L. (1969). 'Starlings versus native cavity-nesting birds.' *Atlantic Naturalist* **24**. 158–61.

Zeyl, C. & Bell, G. (1997). 'The advantage of sex in evolving yeast populations.' *Nature* **388**. 465–8.

Zimmerman, E. C. (1970). 'Adaptive radiation in Hawaii with special reference to insects.' *Biotropica* **2** (1). 32–8.

Index